More Than Darwin

More Than Darwin

An Encyclopedia of the People and
Places of the Evolution-Creationism
Controversy

Randy Moore and Mark D. Decker

GREENWOOD PRESS
Westport, Connecticut • London

Library of Congress Cataloging-in-Publication Data

Moore, Randy.
   More than Darwin : an encyclopedia of the people and places of the evolution-creationism
controversy/Randy Moore and Mark D. Decker.
      p.   cm.
   Includes bibliographical references and index.
   ISBN: 978-0–313–34155–7 (alk. paper)
   1. Evolution (Biology)—Encyclopedias.   2. Creationism—Encyclopedias.
I. Decker, Mark D., 1960–   II. Title.
   QH360.2.M66   2008
   576.803—dc22   2007044406

British Library Cataloguing in Publication Data is available.

Library of Congress Catalog Card Number: 2007044406
ISBN: 978-0-313-34155-7

First published in 2008

Greenwood Press, 88 Post Road West, Westport, CT 06881
An imprint of Greenwood Publishing Group, Inc.
www.greenwood.com

Printed in the United States of America

The paper used in this book complies with the
Permanent Paper Standard issued by the National
Information Standards Organization (Z39.48–1984).

10 9 8 7 6 5 4 3 2 1

To Mom and Dad,
the heroes of my life.
—R.M.

To Pat and Zack,
for their patience and support.
—M.D.

Contents

List of Entries

Illustrations

Preface

Many of the transforming discoveries in science are intimately linked to the people who made those discoveries—for example, special relativity is linked to Einstein, heliocentrism to Copernicus and Galileo, genetics to Mendel, and the structure of DNA to Watson and Crick. Although most textbooks present only the conclusions and implications of scientists' work, these conclusions were influenced profoundly by the scientists' personal lives—that is, who they were and where they worked—as well as their social contexts. Understanding the history, people, and places associated with science helps us understand science because it shows us how science is done. As biologist Ernst Mayr noted, "Most scientific problems are far better understood by studying their history than their logic."

Although today's participants in the evolution-creationism controversy may seem different from those of the past, their ideas are remarkably similar. For example, many creationists claim that evolution is a destructive, atheistic "theory in crisis" that is being replaced by creationism as the story of life's history:

> The science of 20 or 30 years ago was in high glee at the thought of having almost proved the theory of biological evolution. Today, for every careful, candid inquirer, these hopes are crushed; and with weary, reluctant sadness does modern biology now confess that the church has probably been right all the time . . . science is in complete harmony with the opening chapters of the ancient Hebrew Scriptures: "In the beginning god created the heaven and the earth."
>
> – George McCready Price, 1922

> We have heroically and triumphantly delivered the knockout blow against evolution and evolutionists.
>
> – J. Frank Norris, 1926

> The world has had enough of evolution . . . in the future, evolution will be remembered only as the crowning deception which the arch-enemy of human souls foisted upon the race in his attempt to lead man away from the savior. The science of the future will be creationism.
>
> – Harold W. Clark, 1929

> The chain of evidence that purports to support the theory of evolution is a chain indeed, but its links are formed of sand and mist. Analyze the evidence and it melts away; turn the light of true

investigation upon its demonstrations and they fade like fog before the freshening breeze. The theory [of evolution] stands today positively disproved.

– Harry Rimmer, 1935

All the real facts of science support this biblical position . . . more and more scientists are becoming creationists all the time.

– Henry Morris, 1984

Scientists are leaving Darwinian evolution in droves.

– John Morris, 1994

Much of the history of the 20th century will be seen in retrospect as a failed experiment in scientific atheism.

– Phillip Johnson, 1995

Darwinian evolution is little more than a historical footnote in biology textbooks.

– Jonathan Wells, 2004

I see [Darwinian evolution] disintegrating very quickly.

– William Dembski, 2006

Scientists and others have waged a counterattack, noting that scientific evidence overwhelmingly supports evolution, and that creationism—along with its variants such as "creation science" and "intelligent design"—is merely superstition:

The modern theory of evolution has met with remarkably rapid acceptance by those best qualified to judge its merits.

– Edward Drinker Cope, 1872

Of course, like every other man of intelligence and education, I do believe in organic evolution. It surprises me that at this late date such questions should be raised.

– Woodrow Wilson, 1922

The antievolution movement is dull, but dead.

– Julian Huxley, 1931

It is hard to believe that the religious views of the Middle Ages can persist in this age of railroads, steamboats, the telephone, the radio, the airplane, all the great mechanistic discoveries.

– Arthur Garfield Hays, 1957

I believe that the Dayton trial marked the beginning of the decline of fundamentalism . . . Restrictive legislation on academic freedom is forever a thing of the past.

– John Scopes, 1965

This is the death knell for the entire creation-science movement across the country.

– The ACLU, in response to the *Edwards v. Aguillard* decision, 1987

These differences between creationists and scientists have produced the evolution-creationism controversy, one of the most enduring conflicts in American history.

This is a book about the people and places of the evolution-creationism controversy. Instead of describing evolution and creationism *per se*, we have described the controversy in terms of the people and places that produced and influenced the controversy. In doing so, we have tried to neither condemn nor praise either "side" of the controversy, nor have we attempted to reconcile the views of science and religion. We agree with Charles Darwin's comment in 1880 to his cousin Hensleigh Wedgwood that "there have been too many attempts to reconcile Genesis and science." Our only goal has been to present—as best we can—an objective, interesting, accurate, and accessible description of the people and places associated with the controversy.

Unlike the ill-fated laws passed in the 1980s in Arkansas and Louisiana requiring that evolution and creationism be given "equal time" in science classrooms, we have not given "equal time" to all of the people and places in the evolution-creationism controversy. Some people associated with the controversy—for example, Charles Darwin and William Jennings Bryan—achieved fame that far exceeded their roles in the evolution-creationism controversy (e.g., Bryan was a three-time candidate for president of the United States). Although these individuals are covered in significant detail, we have tried to situate them and their work in new contexts. Similarly, we have not included everyone who has commented—but had little impact otherwise—on the controversy. For example, country music legend Merle Haggard caused a minor fuss in 1990 by claiming that "evolution is a laughing matter for anybody that's got a rational mind," and former Senator Bill Frist of Tennessee—a state with a prominent role in the controversy—claimed that giving "equal time" to creationism "is the fairest way to go about educating and training people for the future." Although you'll not find entries for Haggard and Frist in this book, you will encounter several lesser-known people—legislators, theologians, judges, and scientists alike—who played important roles in the controversy. These people may be unfamiliar, but their ideas and roles in the controversy are not. For example,

Nicolas Steno, a scientist who, after helping found modern geology, became a priest, took an oath of poverty, and never wrote another scientific paper.

Billy Sunday, a baseball-player-turned-evangelist who, while denouncing evolution and other social ills, spoke directly to more people than anyone in history.

Frank Norris, an antievolution crusader who brought fundamentalism to the South, pastored the largest congregation in the nation, and was repeatedly arrested for felonies ranging from perjury to arson and murder.

Harry Sinclair, a dinosaur aficionado who triggered the public's fascination in prehistoric life and whose misbehavior helped establish the right of Congress to compel testimony. Today, people are reminded of Sinclair's influence when they see the familiar *Apatosaurus* on the logo at Sinclair gas stations.

In most instances, we have not tried to separate the science, theology, and politics from the friendships, feuds, tragedies, and other events that were associated with—and often motivated—the people you will read about. In doing so, we hope that you will be able to better appreciate how these people's lives and circumstances shaped their contributions to the controversy.

You'll also visit important places associated with controversy, including some geological sites in Scotland and Italy, a sprawling estate outside London where the reclusive Charles Darwin wrote *On the Origin of Species*, an old courthouse in a tiny town in Tennessee, and several churches involved in the controversy. You may not yet be familiar with these places, but we hope you will enjoy learning about how what happened at these places shaped the controversy. Because we could not have possibly described all of the connections between the individuals and places in each entry of this book, we have included an extensive index that will help interested readers understand how the people and places of the controversy are linked. We hope you will find the index useful.

We have provided three sources of additional information about the people and places discussed in this encyclopedia:

For More Information: At the end of many entries you'll find a section titled "For More Information" that includes references that directly address the topics you've read about. These references will provide additional sources of information, photographs, and quotations about the respective topics.

Bibliography: The second source of additional information is the Bibliography, which lists books that address more general topics (e.g., evolution, creationism).

Web site: Finally, visit our Web site to see additional information, references, and photographs regarding topics in this encyclopedia.

Given the breadth of this encyclopedia, these sources are not comprehensive. However, they will enable readers to find much additional information about the people and places associated with the evolution-creationism controversy.

Throughout this book, you'll learn more than just the information directly associated with evolution, creationism, and the controversy. You'll also see glimpses of people's personal lives—how some of the iconic figures attained wealth and fame, whereas others' involvement with the controversy portended poverty, prison, and suicide; how some people associated with the controversy were honored with tombs and monuments in sacred sites, while others were buried in nameless graves; and how some of the people and places associated with the controversy have been memorialized, whereas others have long been forgotten.

We hope you will enjoy learning about the people and places of the evolution-creationism controversy.

Randy Moore
Mark Decker
December 2007

Acknowledgments

We could not have written this book without the help of the many people who advised us, opened their homes to us, and granted our requests for interviews, documents, and photographs. We are especially grateful to Richard Cornelius of Bryan College, Vickie Bryant of Arlington Baptist College, Betsy Dunbar of the American Baptist Historical Society, The University of Texas Center for American History, Angus Miller of GeoWalks, The Library of Congress, The Billy Sunday Museum, Rick Duhrkopf, Bob Baldridge, Jerry Tompkins, Mark Borello, The University of Tennessee Sue Hicks Collection, Oregon Historical Society, David Cantrell, Randal Keynes, David Fankhauser, Down House, Robert Dennison, Susan and Jon Epperson, Clifford Ford, Smithsonian Institution, David Kring, Norman Butcher, Jeff Dixon, Jack Chick, Janice Moore, Grace College, George Beccaloni, Creation Evidence Museum, Layton Brueske of First Baptist Church in Minneapolis, Linda Rust and Jerry Beilby of Northwestern College, University of Albany, Lee Pierce and Lawrence Ford of the Institute for Creation Research, Michael Behe, InterVarsity Press, The Foursquare Church, The British Museum (Natural History), Westminster Abbey, Don Aguillard, Ken Hubert, Rod LeVake, Richard Dawkins, and Eugenie Scott, Glenn Branch, and the other helpful people at the National Center for Science Education. We especially thank Kevin Downing for his encouragement, advice, and help with our work. With one exception, everyone we contacted was exceedingly helpful and gracious.

Finally, we thank our families and friends for their patience with our preoccupation with this book. We look forward to having more time to spend with them.

Randy Moore
Mark Decker
December 2007

Introduction

On July 21, 1925, John Scopes was found guilty of teaching human evolution in violation of Tennessee's Butler Act. Although Scopes' now-legendary trial accomplished nothing from a legal perspective (his conviction was later set aside on a technicality), it became a watershed event in the relationship between science and religion. Both "sides"—religion (especially Christianity based on a literal interpretation of the Bible) and science—claimed victory. The legendary confrontation at the Scopes Trial between William Jennings Bryan and Clarence Darrow became nearly synonymous with the evolution-creationism controversy.

The Scopes Trial is the best-known event in a long and ongoing series of clashes among alternative approaches for understanding the universe. Socrates' execution for impiety, Galileo's arrest for heresy, and John Scope's conviction all resulted from the dominance of religion. But just as some religious dictums have contradicted scientific findings, so too has scientific progress continued to challenge the authority of religious dogma. Nowhere in American society is this struggle manifested so clearly as in the evolution-creationism conflict.

The proposals by Charles Darwin and Alfred Russel Wallace that natural processes are sufficient for explaining the history of life were turning points in the development of science and its relationship to religion. Suddenly, a theory that explained many otherwise confusing observations and that made testable predictions was available that conflicted with some versions of religious belief. Relevant scientific information continued to accumulate, including demonstration of the antiquity of the Earth, which further challenged a literal interpretation of the Bible and ultimately manifested itself at John Scopes' trial in tiny Dayton, Tennessee. Today, just as in 1925, scientists, theologians, politicians, and others continue to argue passionately about evolution and creationism. The controversy rages.

As feminist author Kate Millet has noted, revolutions—intellectual or otherwise—are best viewed as "another ring in the tree of history," and the Darwinian revolution is no exception. Neither Darwin nor his detractors worked in a historical vacuum, and the roots of the current evolution-creationism controversy extend centuries into the past. Consequently, it is important to not only examine the current "ring" as it is laid down, but to also take a cross-sectional view across the entire trunk. This book is an attempt to do just that by introducing the people, places, and events that have shaped the conflict.

A

ADAM AND EVE

Adam was the first human created by God "from the dust of the Earth," and the story of Adam's creation is described in Genesis 2–5 of the Bible. When God noticed that Adam was lonely, God created Eve from Adam's rib as a helpmate (Flemish anatomist Vesalius' report in 1543 that men and women have the same number of ribs created a public controversy). According to the Bible, Adam was 930 years old when he died. The story of Adam and Eve has had a profound cultural impact; they are depicted at the Sistine Chapel and Notre Dame Cathedral, are described in C.S. Lewis' *The Chronicles of Narnia* and Milton's *Paradise Lost*, and are the basis for the Adam's apple, so-named because the forbidden fruit became lodged in Adam's throat. Linnaeus described Adam as "the first naturalist."

LOUIS AGASSIZ (1807–1873)

> *Natural History must in good time become the analysis of the thoughts of the Creator of the Universe.*

Jean Louis Rodolphe Agassiz was born on May 28, 1807, in Môtier, Switzerland, the first of five children to survive infancy (Figure 1). Agassiz, who had been called Louis from birth, spent much of his childhood in small villages, first in Môtier where his father, Rodolphe, was an assistant pastor, and then in Orbe, when his father became pastor of his own congregation. Louis began his formal education at the age of eleven, studying in Bienne and Lusanne, before moving to the Medical School of Zurich in 1824. Agassiz's parents hoped he would enter business or the ministry, but Louis wanted to study natural history, and thought an education in medicine was a possible route.

After Zurich, Agassiz continued his studies at the University of Heidelberg in 1826, and from 1827 to 1830 was at the University of Munich, ostensibly continuing his medical studies, but spending much more time on zoology. He produced his first publication (which he dedicated to Georges Cuvier) during this period, a description of fish collected by a just-returned expedition to Brazil. His parents correctly interpreted this event as functionally ending their son's career as a physician. In 1829, Agassiz earned his PhD from the University of Erlangen, followed by his medical degree from Munich in 1830, although he never practiced medicine.

During the late 1820s, Agassiz started what would become a long-term analysis of fossil fish. In 1831, he went to Paris, where he met naturalist and explorer Alexander von Humboldt. (Humboldt remained Agassiz's mentor, advocate, and patron for many years.) Humboldt, in turn, introduced Agassiz to Cuvier, and Agassiz worked

1. Harvard's Louis Agassiz was one of the last great biologists to reject evolution. (*Library of Congress*)

closely with Cuvier until Cuvier's death the following year.

In 1832, after declining a job at the Jardin des Plantes, Agassiz accepted a professorship at a newly created college in Neuchâtel, Switzerland. After selling his collections to the college, he used the proceeds to complete his study of fossil fish, efforts which produced one of Agassiz's most influential and acclaimed works, the five-volume *Recherches sur les Poissons Fossiles* (1833–1834). Agassiz viewed the fossil record as the orderly genesis of species by the Creator and concluded that there is a direct correspondence between sequences in the fossil record and the embryonic development of individual organisms. Agassiz married in 1832, although his wife, Cécile (whom he had met as a student in Heidelberg), later left him because of neglect and financial problems.

In 1836, Agassiz collaborated with his friend Jean de Charpentier on Agassiz's "hobby," the study of glaciers (although de Charpentier and others would later criticize Agassiz for not acknowledging their involvement). After examining the topography and geology of Europe,

Agassiz concluded that much of central Europe was once covered with a thick sheet of ice. This work led to *Études sur les Glaciers* (1840) and *Système Glaciaire* (1847), within which Agassiz outlined the proposal of a universal ice age at the end of the Tertiary. Invoking the catastrophism of his mentor Cuvier, Agassiz concluded that the Creator had repeatedly destroyed and recreated life. Agassiz replaced the biblical flood with glaciers, although Agassiz was by no means a biblical literalist.

Agassiz's lectures in the United States in 1846 were received warmly by both the public and scientific community. When civil war and reorganization efforts began in Switzerland in the late 1840s, Agassiz moved to Harvard University, where he would remain for the rest of his life. (When he left Europe, he brought his collection of 5,000 glacial boulders or "erratics" with him, which are now housed in the Natural History Museum at Princeton University.) He married Elizabeth Cary in 1849, and when his first wife died of tuberculosis, his children moved to the United States to live with him.

Agassiz encouraged Charles Darwin to study barnacles (and sent him several boxes of specimens), and Darwin sent Agassiz a copy of the first edition of *On the Origin of Species*. However, Agassiz rejected evolution, and he characterized Darwin's proposal as "a scientific mistake, untrue in its facts, unscientific in its method, and mischievous in its tendency." Late in life, Agassiz visited the Galápagos Islands, and argued that the islands were too young to have allowed the degree of speciation Darwin suggested. Agassiz was one of the last great scientists to reject evolution, although most of his students (including his son, Alexander) accepted an evolutionary perspective.

When first in the United States, Agassiz noted that blacks "are not of the same blood as us." In 1850, Agassiz—who vehemently opposed interracial marriage—proposed a separate origin for all human races (a common perspective in the United States in the mid-1800s), thereby providing racial supremacists with their primary arguments. When Agassiz's statements were criticized in proabolitionist New England, Agassiz retracted some of his comments. But Agassiz still had doubts, and he studied blacks in South Carolina. These studies, which produced fifteen daguerreotypes of naked slaves, led Agassiz to claim separate racial origins of humans which correlated with climatic zones. (Agassiz's daguerreotypes were discovered at Harvard's Peabody Museum in 1975, and in 1992 became part of a major exhibition of nineteenth-century photography.) Agassiz's claims about race helped convince geologist Charles Lyell of the importance of Charles Darwin's ideas; as Lyell noted in 1860, Agassiz "drove me far over into Darwin's camp…for when he attributed the original of every race of man to an independent starting point, or act of creation, and not satisfied with that, created whole 'nations' at a time, every individual out of 'earth, air, and water' as Hooker styles it…I could not help thinking Lamarck must be right, for the rejection of his system led to such license." Oliver Wendell Holmes praised Agassiz's views of natural history: "I look with ever increasing admiration on the work you are performing for our civilization."

Agassiz worked tirelessly for a natural history museum at Harvard, which led to the establishment in 1859 (the year Darwin's *Origin* was published) of the Museum of Comparative Zoology. He organized the museum along three themes—zoology, geography, and embryology—which in his view were the main relationships around which life is organized. Agassiz served as the first director until his death. Alfred Wallace later noted that "it is surely an anomaly that the naturalist most opposed to the theory of evolution should be the first to arrange the museum in such a way as best to illustrate that theory."

In 1871, while on a lengthy sea voyage (which included the Galápagos), Agassiz suffered a cerebral hemorrhage which left him temporarily paralyzed. He did not survive the next hemorrhage two years later, and Agassiz died on December 14, 1873. Agassiz's funeral was attended by various dignitaries, including Henry Wilson, the Vice President of the United States. Agassiz is buried at Mount Auburn Cemetery in Cambridge; a 2,500-pound glacial boulder, brought from the Alps, marks his grave.

Following the retreat of the North American ice sheet 12,000 years ago, a huge glacial lake arose that extended through the upper midwest of the United States and into Canada. In honor of Agassiz's elucidation of past ice ages (although our contemporary understanding of them differs markedly from Agassiz's perspective), this lake is called Lake Agassiz. Today, statues of Agassiz and von Humboldt adorn the entry to Stanford University's Jordan Hall.

For More Information: Lurie, E. 1988. *Louis Agassiz: A Life in Science.* Baltimore, MD: Johns Hopkins University.

DON AGUILLARD (b. 1954)

I knew that this was important. I was not ready to pretend that "creation science" was legitimate or that evolution didn't matter.

Don Aguillard was born in Ville Platte, Louisiana, on June 20, 1954. After graduating in 1975 from the University of Southwest Louisiana (now the University of Louisiana at Lafayette), Don began teaching biology at Acadiana High School in Lafayette, Louisiana. There, in 1980, Aguillard saw an advertisement in *The American Biology Teacher* asking teachers to call the American Civil Liberties Union (ACLU) if they wanted to challenge the Louisiana law requiring "balanced treatment" for evolution and creationism in public schools. Aguillard, who knew that he "couldn't stand by and do nothing," became the lead plaintiff in a lawsuit that later became known as *Edwards v. Aguillard*. Although many of Louisiana's biology teachers were prepared to avoid teaching evolution rather than teach creationism, fewer than ten biology teachers agreed to participate in the lawsuit.

On January 10, 1985, U.S. District Judge Adrian Duplantier ruled that Louisiana's "balanced treatment" law was unconstitutional. Duplantier's ruling, which blocked Louisiana from implementing the law, noted that (1) creationism is a religious belief "tailored to the principles of a particular religious sect," (2) there can be no valid secular reason for banning the teaching of evolution, and (3) the statute's avowed purpose of protecting academic freedom was inconsistent with the statute's legislative history and its requirement for teaching creation science whenever evolution is taught. Duplantier's decision noted that no conceivable trial testimony could save the law.

When Louisiana appealed the decision, the U.S. Court of Appeals for the Fifth Circuit—noting the ongoing influence of William Jennings Bryan and the Scopes Trial—affirmed by a vote of 8-7 Duplantier's ruling. As the Court's appellate panel noted, "Through the years, religious fundamentalists have publicly scorned the theory of evolution and worked to discredit it ... [Louisiana intended] to discredit evolution by counterbalancing its teaching at every turn with the teaching of creationism, a religious belief." Although the dissenting opinion by Judge Thomas Gee (1925–1994) that "there are two bona fide views" of origins was the first published judicial support for creationist claims since the Scopes Trial, the case wasn't over, for when a federal court rules that a state law is unconstitutional, the U.S. Supreme Court must consider hearing the case.

Edwards v. Aguillard was heard before the U.S. Supreme Court on December 10, 1987 (at the time, Edwin Edwards [b. 1927] was governor of Louisiana). Although Wendell Bird (representing the state of Louisiana) admitted that "some legislators had a desire to teach religious doctrine in the classroom," he claimed that the law had a primary secular purpose based on "fairness" and "academic freedom."

On June 19, 1987, in a 7-2 decision, the Supreme Court—noting that evolution has been "historically opposed by some religious denominations"—affirmed that (1) Louisiana's "balanced treatment" law was unconstitutional because it impermissibly endorsed religion by advancing the religious belief that a supernatural being created humankind, (2) it is unconstitutional to mandate or advocate creationism in public schools, for creationism is a religious idea, (3) the contention that "a basic concept of fairness" for requiring the teaching of creation

science is "without merit," (4) banning the teaching of evolution when creation science is not also taught undermines a comprehensive science education, and (5) the law is "facially invalid as violative of the Establishment Clause of the First Amendment, because it lacks a clear secular purpose." The Court also noted that it was not happenstance that the legislature required the teaching of a theory that coincided with a particular religious view, adding that the "purpose of the Creationism Act was to restructure the science curriculum to conform with a particular religious viewpoint ... Requiring schools to teach creation science with evolution does not advance academic freedom." Invoking the Scopes legacy, *Time* magazine described the *Edwards* decision as "a major setback for fundamentalist Christians." When he heard of the decision, Aguillard said, "I hope my case made a difference. Biology teachers have to speak up. In-service programs are not the answer, nor are better textbooks; neither of those things will make any difference if teachers refuse to teach evolution thoroughly and unapologetically."

Although *Edwards* diminished subsequent attempts to pass "balanced treatment" laws, many creationists were encouraged by a part of Justice William Brennan's (1906–1997) majority opinion: "Teaching a variety of scientific theories about the origins of humankind to school children might be validly done with the clear secular interest of enhancing the effectiveness of science instruction." They were also encouraged by the dissenting opinions of Justices Antonin Scalia (b. 1936) and William Rehnquist (1924–2005) (both appointed to the Supreme Court by Ronald Reagan), who argued that "the people of Louisiana, including those who are Christian fundamentalists, are quite entitled, as a secular matter, to have whatever scientific evidence there may be against evolution presented in their schools, just as Mr. Scopes was entitled to present whatever scientific evidence there was for it.... The body of scientific evidence supporting creation science is as strong as that supporting evolution. In fact, it may be stronger. Evolution is merely a scientific theory or 'guess.' ... Although creation science is educationally valuable and strictly scientific, it is now being censored from or misrepresented in the public schools.... Teachers have been brainwashed by an entrenched scientific establishment composed almost exclusively of scientists to whom evolution is like a 'religion.' These scientists discriminate against creation scientists so as to prevent evolution's weaknesses from being exposed."

After his trial, Aguillard continued to stress evolution in his classes, and in 1998 he earned a PhD from Louisiana State University. Aguillard, an award-winning school administrator, is currently the Superintendent of St. Mary Parish schools in Louisiana. Meanwhile, creationism remains popular in Louisiana. Late in 2007, Senator David Vitter (R-LA) tried to earmark $100,000 for the pro-creationism Louisiana Family Forum to "develop a plan to promote better science education." Louisiana Family Forum has been advised by creationists such as Kent Hovind.

CARL AKELEY (1864–1926)

Most of my worshipping has been done in the cathedral forests of the African jungles, with the voices of birds and animals as music.

Carl Ethan Akeley was born on May 19, 1864, in Clarendon, New York (Figure 2). He attended school for only three years, but became famous for inventing a new type of taxidermy to preserve and display animals. Whereas taxidermists before Akeley stuffed skins with cotton and straw, Akeley built carefully measured models, over which he stretched the animal's skin. The so-called "Akeley Method" revolutionized exhibits at museums, and brought distant, unimagined animals to life for the viewing public.

Akeley began his career by stuffing his neighbors' pets. However, in 1867 Akeley began working at Wards Natural Science Establishment (a biological supply company) in Rochester, New York, where he helped mount P.T. Barnum's (1810–1891) famous elephant Jumbo, who had died in a macabre accident in St. Thomas, Canada. (Akeley's model of Jumbo went to Tufts University, where it was the school's mascot until it burned in 1975; Jumbo's skeleton is

2. Carl Akeley revolutionized taxidermy and, in the process, brought distant, unimagined animals to life for the public. The Akeley Hall of African Mammals at the American Museum of Natural History is among the world's greatest museum displays. (*Library of Congress*)

stored in the American Museum of Natural History.) Akeley then worked at the Milwaukee Public Museum, where in 1890 he helped create the museum's first habitat diorama.

Akeley further developed his new taxidermy techniques while working at Chicago's Field Museum. Akeley's method soon became the standard at most major museums. Akeley later turned his attention to saving gorillas, and he was the first to film wild gorillas. Today, Akeley's work is displayed in the Akeley Hall of African Mammals at the American Museum of Natural History, an exhibit considered to be among the world's greatest museum displays.

In 1924, Akeley sculpted *The Chrysalis*, a bronze sculpture that depicted a human emerging from the cracked-open skin of a gorilla. The sculpture depicted Akeley's feelings of kinship with animals; he knew that humans had not literally sprung from gorillas, but that "they undoubtedly had a common ancestor." *The Chrysalis*, in which humans' ascent replaced a primal fall, was commissioned for New York's West Side Unitarian Church, and its display there outraged many creationists.

Akeley died of dysentery in the Congo while collecting for the American Museum of Natural History on November 17, 1926, which was ten years before the completion of the Akeley Hall of African Mammals. He was buried in a place depicted in the Hall's famous Gorilla Diorama. During the hostilities between gorilla poachers and Dian Fossey in 1979, Akeley's grave was vandalized and his bones removed. Some were later recovered, and his memorial was later repaired by Penelope Bodry-Sanders, Akeley's photographer and biographer. The Carl Akeley Award is given annually to the world's top taxidermists.

For More Information: Bodry-Sanders, P. 1998. *The Life and Legacy of Carl Akeley*. Jacksonville, FL: Batax Museum.

ALABAMA BOARD OF EDUCATION

The Alabama Board of Education voted 6-1 in 1995 to require that all 40,000 biology textbooks used in its public schools include a "Message from the Alabama State Board of Education" that claimed, among other things, that evolution is "a controversial theory." In Georgia, the use of a similar disclaimer led to a lawsuit (*Selman v. Cobb County School District*) that in 2006 declared these disclaimers unconstitutional.

FREDERICK LEWIS ALLEN (1890–1954)

Legislators might go on passing antievolution laws, and in the hinterlands the pious might still keep their religion locked in a science-proof compartment of their minds; but civilized opinion everywhere had regarded the Dayton trial with amazement and amusement, and the slow drift away from Fundamentalist certainty continued.

Frederick Lewis Allen was born in Boston, Massachusetts, on July 5, 1890, the son of Alberta and Rev. Frederick Allen. Rev. Allen helped found the Watch and Ward Society, an organization active in banning books and public performances that it deemed inappropriate. Frederick Allen graduated from Harvard in 1912 and earned a master's degree in 1913. After graduation, Allen taught English at Harvard, and in 1914, he became assistant editor of the newly founded *The Atlantic Monthly* magazine.

Except for stints with Harvard and the Council of National Defense, Allen was a magazine journalist for the rest of his career. After working for *The Atlantic*, he moved to *Harper's* in 1923, becoming editor-in-chief in 1941. Allen recognized the stock market crash of 1929 was a turning point in the history of the United States, and decided to record the popular history of the 1920s. These efforts produced the influential *Only Yesterday: An Informal History of the 1920s* in 1931.

In *Only Yesterday*, Allen tried to convey the feel of the decade and emphasized the "trivial happenings with which [the public] was preoccupied." Allen's book sold more than 500,000 copies during Allen's lifetime. It also became assigned reading in many college history courses, and is still in print. Although Allen claimed that "religion had become a debatable subject instead of being accepted without question," the "prestige of science was colossal."

No event better captured the tension Allen perceived between religion and science than the Scopes Trial of 1925. *Only Yesterday* portrayed the trial solely as a clash between William Jennings Bryan and Clarence Darrow. Bryan was depicted as the "perfect embodiment of old-fashioned American idealism," while Darrow was "a radical, a friend of the underdog." By describing Bryan as "the Fundamentalist champion," Allen established the trial as the inevitable and conclusive showdown between Fundamentalism and Modernism. Bryan ended up "covered in humiliation" and was only spared further shame when Judge Raulston "mercifully refused to let the ordeal of Bryan continue." Allen judged the Scopes Trial to be the beginning of the end for Fundamentalism.

Allen's portrayal of Scopes' trial as Bryan vs. Darrow, and blind fundamentalism vs. enlightened skepticism, became the standard interpretation of events in Dayton, later reinforced by

Inherit the Wind. Critics have noted that Allen's treatment oversimplified the actual events ("cartoonlike simplicity" in the words of historian Edward Larson) and included several inaccuracies. Although subsequent authors used Allen's account to perpetuate the claim that Fundamentalism died at Dayton, membership in Fundamentalist churches continued to grow after the 1920s.

Following the success of *Only Yesterday*, Allen published *Since Yesterday* in 1940, which similarly chronicled the 1930s. Working with his second wife Agnes (his first wife, Dorothy, to whom *Only Yesterday* is dedicated, died before that book was finished), Allen published several books that examined life in the United States from the Civil War to the Second World War. Allen died in 1954 of a cerebral hemorrhage.

LUIS (1911–1988) AND WALTER ALVAREZ (b. 1940)

Luis W. Alvarez (1911–1988) was a physicist who, with his son Walter (b. 1940), suggested in 1980 that the Cretaceous-Tertiary extinction was caused by a giant asteroid that hit Earth 65 million years ago. The Alvarez' claim—which newspapers described as "the worst weekend in the history of the world"—was initially ridiculed, but there was much supporting evidence, including the "iridium anomaly" at the Cretaceous–Tertiary boundary (Figure 3). Iridium is rare on Earth (most of the iridium that was part of the molten Earth 4.5 billion years ago sank into its core with other metals), but is abundant in interplanetary particles. The Alvarezs, using limestone collected from Gubbio in Northern Italy, discovered that clay deposited at the end of the Cretaceous had thirty-times more iridium than did sediments above or below the clay. Other evidence for the Alvarezs' theory included the presence of shocked quartz grains characteristic of a powerful impact, sediments suggesting that a large tsunami was triggered by a massive impact, and—most importantly—the impact crater itself (Chicxulub) on the Yucután Peninsula. Walter Alvarez—who in 2005 won the Penrose Medal, The Geological Society of America's highest award—described their

discovery in his 1998 bestseller *T. Rex and the Crater of Doom*. Luis Alvarez, who helped investigate the Kennedy assassination, won a Nobel Prize in physics in 1968 for his studies of subatomic particles, and was presented with the Collier Trophy (the highest honor in aviation) by President Harry Truman in 1945 for developing a system for landing aircraft in bad weather. Luis Alvarez died of cancer in Berkeley, California in 1988.

For More Information: Alvarez, L.W. 1987. *Alvarez: Adventures of a Physicist*. New York: Basic Books; Alvarez, W. 1997. *T. Rex and the Crater of Doom*. Princeton, NJ: Princeton University.

AMERICAN ANTI-EVOLUTION ASSOCIATION

The American Anti-Evolution Association was formed by Arkansas preacher Ben Bogard as an organization open to everyone except "Negros [sic], and persons of African descent, Atheists, Infidels, Agnostics, such persons as hold to the theory of Evolution, habitual drunkards, gamblers, profane swearers, despoilers of the domestic life of others, desecrators of the Lord's Day, and those who would depreciate feminine virtue by vulgarly discussing relationships." During the 1920s, the Association gathered more than 19,000 signatures, thereby ensuring that the Arkansas legislation banning the teaching of human evolution would be decided by a public referendum. However, after a popular vote approved the law, the Association had little impact, and soon dissolved.

AMERICAN ASSOCIATION FOR THE ADVANCEMENT OF SCIENCE (AAAS) (Est. 1848)

The American Association for the Advancement of Science (AAAS) was founded in Philadelphia in 1848, and today is best known for its publication of *Science*, the most widely distributed general science journal in the world (circulation ~150,000). *Science* was founded in July 1880 by a group that included Thomas Edison, and the second issue of *Science* reported a

3. The chalk-colored K-T (Cretaceous-Tertiary) Boundary was deposited 65 million years ago after a meteor struck Earth. The iridium-rich layer separates the Age of Reptiles from the Age of Mammals. The "K" symbol for the Cretaceous is derived from the equivalent German word *Kreide*, meaning *chalk*. (*David Kring*)

lecture by Thomas Huxley entitled "The Coming of Age of the *Origin of Species*." Near the end of that lecture, Huxley noted that "evolution is no longer a speculation, but a statement of historical fact." The first technical presentation about evolution in *Science* was a description of protective coloration in 1886 by Alfred Wallace, who acknowledged Charles Darwin as the first to propose a convincing mechanism for evolution. The affiliation of *Science* with AAAS began in 1900.

In 1922, AAAS adopted the first of several resolutions supporting the teaching of evolution and opposing the teaching of creationism. That resolution, titled "Present Scientific Status of the Theory of Evolution," noted that "no scientific generalization is more strongly supported by thoroughly tested evidences than is that of organic evolution." AAAS also passed evolution-related resolutions in 1972, 1982, 1999, 2002, and 2006.

Science, whose members' contributions helped pay the defense's bills at the Scopes Trial, reported the Scopes Trial and published Henry Fairfield Osborn's prepared testimony for the defense. Two months after Scopes' trial, the sister publication of *Science—The Scientific Monthly* (1915–1957), with which *Science* merged in 1958—published several statements prepared for Scopes' defense, including one entitled "The Truth of Evolution" in which biologist Maynard Metcalf noted that it would be malpractice to teach biology without evolution. In 1952, Scopes' defender Kirtley Mather claimed in *Science* that the conflict between evolution and religion in the nineteenth century had produced an antiscience bias, and urged scientists to confront antiscience attitudes.

In 1972, AAAS denounced creationism as "neither scientifically grounded nor capable of performing the roles required of scientific theories." *Science* also reported *Edwards v. Aguillard*, and—in an unprecedented move—published in its February 19, 1982, issue the entire opinion of Judge William Overton in *McLean v. Arkansas*. After Kansas removed evolution from its science education standards in 1999, AAAS (with the

National Science Teachers Association and the National Research Council) publicly disassociated itself from the Kansas Board of Education.

In 1995, AAAS established the Program of Dialogue Between Science and Religion to facilitate communication between scientific and religious communities. In 1999, this program was renamed The Dialogue on Science, Ethics, and Religion, and in 2006 it released *The Evolution Dialogues: Science, Christianity, and the Quest for Understanding*. AAAS presidents have included Louis Agassiz, Asa Gray, David Jordan, and Stephen Gould.

AMERICAN CIVIL LIBERTIES UNION (ACLU) (Est. 1920)

The origin of the American Civil Liberties Union, better known as the ACLU, can be traced to 1917 when the National Civil Liberties Bureau (NCLB) was established by Crystal Eastman and Roger Baldwin. The NCLB was an offshoot of the American Union Against Militarism, which had opposed America's entry into World War I. The ACLU was chartered in 1920 in New York as a national nonprofit organization that works through litigation, legislation, and community education "to defend and preserve the individual rights and liberties guaranteed to every person in this country by the Constitution and the laws of the United States."

On March 22, 1925, Lucille Milner of the ACLU read a newspaper article entitled "Tennessee Bans the Teaching of Evolution" describing Tennessee's Butler Law, which made it unlawful to teach human evolution in Tennessee's public schools. Milner gave the article to Roger Baldwin, the ACLU's Executive Director, who began raising money to challenge the law. At the time of the Scopes Trial in 1925, the ACLU was still looking for its first court-victory.

On May 4, 1925, the ACLU placed an ad in the *Chattanooga Daily Times* and other Tennessee newspapers: "Looking for a Tennessee Teacher Who Is Willing to Accept Our Services on Testing the Law in the Courts." Dr. George Rappleyea, manager of Dayton's ailing Cumberland Coal and Iron Company, saw the ad and was

intrigued. With newspaper in hand, Rappleyea went to Robinson's Drug Store, where he and other community leaders devised a test case to boost the area's struggling economy. When first-year coach and teacher John Scopes agreed to be the law-breaking villain, Rappleyea sent a telegram to the ACLU: "Professor J.T. Scopes, teacher of science at Rhea County high school, Dayton, Tennessee, will be arrested and charged with teaching evolution. Consent of superintendent of education for test case to be defended by you. Wire me collect if you wish to co-operate and arrest will follow."

The ACLU agreed to finance the case, and Rappleyea swore out a warrant for Scopes' arrest. From July 10 to 21, 1925, Dayton hosted the Scopes Trial (i.e., *State of Tennessee v. John Thomas Scopes*), the most famous event in the history of the evolution-creationism controversy.

By the time Scopes' verdict was announced, the ACLU was in debt. It had spent $8,993.01 on the case, most of which went to expert witnesses. Scopes' bill for court costs totaled $343.87, all of which was paid by the ACLU. Darrow and Scopes' other attorneys paid all of their expenses; Darrow, who spent about $2,000 of his own money for the trial, told the ACLU that "I don't want you to think about my expenses. I could afford it and I never got more for my money." Dudley Field Malone raised $1,350 for the trial, but the ACLU didn't pay off its Scopes-related debts until 1926, thanks to a mass-mailing to AAAS members which turned the deficit into a surplus of more than $2,000. For several years after Scopes' conviction, the ACLU searched for another challenger to the Butler Law, but could find no volunteers. Laws banning the teaching of human evolution in Tennessee, Arkansas, and Mississippi remained in place for more than forty years.

In the 1960s, the ACLU offered to help Susan Epperson challenge Arkansas' law banning the teaching of evolution (their offer was declined), and defended Gary Scott, a high school biology teacher in Tennessee who was fired for teaching evolution. In the 1980s, the ACLU defended Don Aguillard's challenge to a Louisiana law allowing the teaching of creation science. The

ACLU has fought other attempts by creationists to undermine the teaching of evolution, including those associated with *McLean v. Arkansas Board of Education* and *Kitzmiller et al. v. Dover Area School District*, in which federal judge John Jones ruled that "intelligent design" is religion, not science.

Throughout its history, the ACLU has been involved in many important civil rights cases, including *Brown v. Board of Education* (which in 1954 banned racial segregation in public schools), *Engel v. Vitale* (which in 1962 banned state-sponsored prayer in public schools), and *Abington School District v. Schempp* (which in 1963 banned state-sponsored reading of the Bible in public schools).

In 1990, evangelist Pat Robertson founded the American Center for Law and Justice to counter the ACLU, which Robertson characterized as "liberal" and "hostile to traditional American values." After the September 11, 2001, attacks on the World Trade Center, Jerry Falwell remarked that the ACLU's attempts to "secularize America" had provoked the wrath of God, and thereby had caused the terrorist attacks. Falwell later apologized for the comment.

For More Information: Walker, S. 1999. *In Defense of American Liberties: A History of the ACLU.* Carbondale, IL: Southern Illinois University.

AMERICAN SCIENTIFIC AFFILIATION (ASA) (Est. 1941)

The American Scientific Affiliation (ASA) was formed by Irwin Moon (1907–1986) and Will Houghton (1887–1947), then the president of the Moody Bible Institute. Houghton believed that evolution misleads and damages society and Christianity, and Moon—who in 1945 founded the Moody Institute of Science—influenced the thinking of famed creationist Henry Morris. ASA's goal was to correlate "the facts of science and the Holy Scriptures." ASA's *Modern Science and Christian Faith*, a handbook for students, appeared in 1948, at a time when many of ASA's members questioned George Price's flood geology. At an ASA meeting in 1953, Morris met John Whitcomb, Jr., and this partnership later

produced *The Genesis Flood*, which became a foundation of the modern "creation science" movement. Morris and other flood geologists left the ASA in 1963 to form the Creation Research Society because ASA was "soft on evolution" and had "capitulated to theistic evolution." When the National Academy of Sciences distributed more than 40,000 copies of its *Science and Creationism: A View from the National Academy of Sciences* (1984), the ASA responded with its *Teaching Science in a Climate of Controversy: A View from the American Scientific Affiliation*. Among the several thousand members of ASA is Christian apologist Francis Collins, the Director of the NIH National Human Genome Research Institute.

ANAXIMANDER (c. 611–547 BC)

Anaximander was a Greek philosopher who provided one of the earliest ideas about the origin of humans. Anaximander, the first Western philosopher to claim that life originated from nonliving matter, proposed that humans were "like another animal, namely, a fish, in the beginning."

MARY ANNING (1799–1847)

Mary Anning was born on May 21, 1799, in the southern England coastal town of Lyme Regis. When her father died of tuberculosis in 1810, the family was strapped for money, and Mary and her brother Joseph began selling fossils collected from nearby cliffs. In 1811, Mary uncovered the first complete *Ichthysaurus*, which she sold for $23. When wealthy fossil-collector Thomas Birch heard of Mary's financial problems, he sold his own collection of fossils and gave the receipts—about $400—to Mary.

In 1824, Mary discovered a near-intact skeleton of *Plesiosaurus*, which she sold to the Duke of Buckingham; this skeleton is the type specimen of the genus. Four years later, Anning found the first *Pterodactylus*, which was later renamed *Dimorphodon marconyx* by Richard Owen; this was the first pterodactyl found outside of Germany. In 1832, Mary found the large *Ichthyosauus* that is displayed today at the British Museum of

Natural History in London. The scientists who described Anning's discoveries seldom named her in their publications.

Although Anning was self-taught, famous geologists sought her help. Anning's discoveries helped convince people of extinction, and in 1846 she was made an honorary member of the Geological Society of London. Anning, the source of the tongue-twister "She sells sea shells by the sea shore," remains one of the most influential and picturesque people in the history of paleontology. In the last decade of her life, Anning—by then a national celebrity—received an annuity from the British Association for the Advancement of Science. Anning's *Plesiosaurus*, which she discovered in 1823, was the basis for one of the "reconstructions" of the Loch Ness Monster. That specimen remains on display in the British Museum of Natural History.

Anning died of breast cancer in Lyme Regis on March 9, 1847, and was buried at St. Michael the Archangel Church in Lyme Regis. Her brother Joseph is buried beside her. Her obituary was published in the *Quarterly Journal of the Geological Society*, an organization that barred women from its membership until 1904. After Anning's death, The Royal Society donated funds for a stained-glass window to Anning's memory in the Parish Church at Lyme Regis. That window, which shows Mary helping poor and sick people, is "in commemoration of [Anning's] usefulness in furthering the science of geology, also of her benevolence of heart and integrity of life." Every year, Lyme Regis celebrates Mary Anning Day.

For More Information: Brown, D. 2003. *Rare Treasure: Mary Anning and Her Remarkable Discoveries*. New York: Houghton Mifflin.

ANSWERS IN GENESIS (AIG) (Est. 1994)

The Scriptural accounts of the Creation, Noah's flood, and other major events of biblical history can be trusted.

Answers in Genesis (AIG) is an antievolution organization whose defense of Christianity focuses primarily on questions related to the book of Genesis, which it claims is the most-attacked book of the Bible. AIG, which began operating in a small shopping center in Florence, Kentucky, proclaims the Bible to be the inerrant "history book of the universe," claims that "in six 24-hour days God made a perfect creation," believes that evolution undermines the Christian foundations of America, and seeks "to expose the bankruptcy of evolutionary ideas, and its bedfellow, a 'millions of years old' Earth." AIG also claims that evolution is "the anti-God religion of death," and a presentation of AIG claims that predation "is what sin looks like." Answers in Genesis is the most dominant of the many antievolution organizations; it controls almost 60 percent of the more than $20 million in donations to (and purchases from) the ten largest creationist organizations.

In May 2007, just three weeks after three Republican candidates for president publicly rejected evolution, AIG opened near Petersburg, Kentucky, a 60,000 square foot, $27 million Creation Museum filled with life-sized dinosaur models, live exhibits, and other collections that "proclaim the authority of the Bible from its very first verse" (Figure 4). In the museum, Methuselah—Noah's grandfather who died just before the flood at age 969—warns visitors of God's upcoming judgment. Exhibits compare "human reason" with "God's word" and show children cavorting with dinosaurs, which were aboard Noah's Ark after being created on the sixth day of the biblical creation week. The museum was visited by more than 160,000 people during its first three months of operation. Just south of the AIG Creation Museum is Big Bone Lick State Park, which is touted as the birthplace of American vertebrate paleontology. In the early 1800s, several skeletons were collected there by Lewis and Clark, who were dispatched by President Thomas Jefferson.

The Creation Museum proclaims "that there is a Creator, and that this Creator is Jesus Christ, who is our Savior." War, death, and the pains of childbirth are portrayed as the wages of primal sin. The museum endorses James Ussher's claim that creation occurred in 4004 BC, and includes

11

4. In 2007, Answers in Genesis—an evangelical organization that promotes Young-Earth creationism— opened a $27 million Creation Museum in Kentucky. The museum's Main Hall features humans cavorting with 40'-long dinosaurs. More than 100,000 people visited the museum during its first seven weeks of operation. Answers in Genesis was founded by Ken Ham (Figure 42). (*Randy Moore*)

31 rooms, 200 exhibits, 52 video presentations, a 2,600 square foot bookstore, a 150-seat Noah's Café (with dinosaur footprints embedded in the floor), an 84-seat planetarium, a 60-seat theater, and a refreshment area. Visitors are also told that most fossils were created by the Flood (God's worldwide judgment), that all animals were vegetarian before Adam's sin, and that Young-Earth creationism is the only view of creationism that does not destroy churches. Soon after the Creation Museum opened, the Big Valley Creation Museum—the first permanent creation museum in Canada—opened midway between Calgary and Edmonton.

AIG has a mailing list of 500,000 names and a newsletter that reaches more than 50,000 readers. As part of its mission to return the "increasingly anti-Christian" United States to a belief in the Bible, AIG sponsors a "Research Paper Challenge." The winner of this contest gets a $50,000 scholarship to Jerry Falwell's Liberty University.

The CEO of the U.S. branch of AIG is antievolutionist Ken Ham, a former biology teacher from Australia. AIG also has offices in Canada, New Zealand, South Africa, Australia, and the United Kingdom. In 2007, the Australian branch of AIG (now known as Creation Ministries International) sued the U.S. branch of AIG, claiming—among other things—that AIG stole subscribers for its magazine, *Answers*. Ham responded to the lawsuit by claiming that the Australian group was not following the Bible. In 1987, the Australian group's split with co-founder John Mackay included accusations of necrophilia and demonic possession.

ANTI-EVOLUTION LEAGUE OF AMERICA (Est. 1924)

The Anti-Evolution League of America was formed in 1924, one year after William Bell Riley founded the Anti-Evolution League of Minnesota. The League's first president was Kentucky theologian J.W. Porter, who helped get an anti-evolution bill before the Kentucky legislature. T.T. Martin was the League's field secretary and editor of *Conflict*, the organization's official publication. The League was endorsed and supported by William Jennings Bryan, and soon after its formation began a national "Bible-and-Christ-and-Constitution Campaign against Evolution in Tax-Supported Schools" campaign. When Bryan died five days after the Scopes Trial, Bryan's son—William Jennings Bryan, Jr.—became president of the League.

ST. THOMAS AQUINAS
(c. 1225–1274)

Some intelligent being exists by which all natural things are directed to their end; and this being we call God.

Thomas Aquinas (Thomas of Aquin) studied at the Benedictine monastery of Monte Cassino, where his uncle was an abbot, and at the University of Naples, where he learned of Aristotle. By 1243, Thomas was determined to enter (at that time) the somewhat radical Dominican order, a decision that upset his family. His parents had Thomas abducted while he was traveling to Rome, and held him in seclusion to change his mind (even, supposedly, tempting him with women). When his parents released him a year later, he went to Cologne to study with the renowned Scholastic Albertus Magnus (1193/1206–1280), who had fused Aristotle's philosophy with Christian thought and sought accommodation between religion and science.

By 1252, Thomas was in Paris studying theology, and was teaching five years later. He returned to Naples in 1272 to found a university. During Mass in 1272, he underwent an undetermined experience, which prompted him to stop writing, stating that "All that I have written seems to me like straw compared to what has now been revealed to me." Early in 1274, Pope Gregory X (1210–1276) appointed Thomas to the Council of Lyon, which was to meet to discuss Church doctrine. En route, Thomas became ill and died several weeks later, on March 7, 1274. Thomas was canonized in 1323, and was buried in Milan's Sant' Eustorgio, not far from Peter the Martyr, the patron saint of inquisitors.

Thomas is the patron saint of academics, philosophers, universities, and book sellers. The thumb of Thomas' right hand is displayed at his grave.

The philosophy and theology of Thomas have influenced Christianity, especially Roman Catholicism, for over 700 years. Thomas, in his most influential works, *Summa contra Gentiles* and *Summa Theologica*, proposed that there is no conflict between Church doctrine and natural science because there exists both a primary cause (the act of the Creator) as well as secondary causes. Knowledge of the primary cause is available only via revelation, while secondary causes are open to scientific examination. Thomas also reformulated the medieval view that the primary purpose of sex and marriage is to produce children.

Thomas' proposals lay at the heart of Scholasticism that reigned in Europe from the twelfth century to the Renaissance. His influence is evident in the development of the "Great Chain of Being" that, drawing upon the perspectives of ancient Greeks, viewed the universe as composed of objects that fit together into a perfect series, from minerals, up to humans, on to angels, and ending with God. The fixity of each level in the hierarchy was absolute, which meant that evolution of one living form into another was impossible. This concept of the "fixity" of species remains influential, especially among biblical literalists who interpret Genesis as meaning each species was divinely created and incapable of change. In 1879, Pope Leo XIII issued an encyclical, the *Aeterni Patris* (On the Restoration of Christian Philosophy), which acknowledged the influence of "the chief and master ... Thomas Aquinas," and urged the Church "to restore the golden wisdom of St. Thomas, and to spread it far and wide for the defense and beauty of the Catholic faith, for the good of society, and for the advantage of all the sciences."

Proponents of intelligent design often invoke the writings of Thomas Aquinas to support their perspective. However, contemporary Thomist philosophers reject the claim that the lack of a naturalistic explanation for a feature of an organism is somehow itself evidence for a designer.

13

Thomists do not concede that a lack of understanding of secondary causes can inform one about the action of the primary cause, but they likewise disagree with those (e.g., Richard Dawkins, Daniel Dennett) who argue that an understanding of secondary causes precludes the existence of a Creator.

ROBERT ARDREY (1908–1980)

Robert Ardrey was an anthropologist and screenwriter who wrote *African Genesis*, a 1961 bestseller depicting early humans as "killer apes" who used lethal weapons to kill each other. Ardrey argued that aggression distinguishes human ancestors from other primates, thereby explaining the violent nature of modern humans. As Ardrey told his readers, "We are Cain's children … whose natural instinct is to kill with a weapon." Ardrey, who believed that he was describing "a contemporary revolution in the natural sciences," won several awards for his work in anthropology. However, many of his speculations about the similarities of baboon behavior (as it was understood at that time) and human behavior were inaccurate. Famed anthropologist Richard Leakey noted that "nobody would have heard of the killer-ape concept if it hadn't been for Robert Ardrey, but … it should have been abandoned by everybody long ago." Ardrey's work, along with that of Raymond Dart, helped inspire the man-apes of *2001: A Space Odyssey*. Ardrey died of natural causes in Kalk Bay, Africa, in 1980.

ARISTOTLE (384–322 BC)

Nature proceeds little by little from inanimate things to living creatures, in such a way that we are unable, in the continuous sequence to determine the boundary line between them or to say which side an intermediate falls.

Aristotle was born in 384 BC in Chalcidice (now the Greek region of Macedonia). Aristotle went to Athens when he was 17 to study at Plato's Academy. He remained there, first as student and then as teacher, until 347 when Plato

died. Aristotle then moved to Assus, where he remained for three years and married Pythias, with whom he had a daughter. In 345, Aristotle moved to the island of Lesbos. There he met Theophrastus, with whom he studied the flora and fauna of the surrounding islands.

In 340 BC, Aristotle began tutoring the son of Phillip II, Alexander, later known as Alexander the Great, who replaced his father as king in 336. By 335, Aristotle had returned to Athens to found his own school in the Lyceum, a sacred grove dedicated to Apollo. Over the next thirteen years, Aristotle spent his mornings lecturing to students (usually while walking the grounds of the school, and hence Aristotle's followers became known as "peripatetics"), and in the afternoon would give public lectures. While in Athens, Pythias died, and Aristotle eventually developed a relationship with Herpyllis, with whom he had a son.

When Alexander was killed in battle in 323, anti-Macedonian forces came to power in Athens. Aristotle, due to his birthplace and association with Alexander, was charged with "impiety" (the same charge for which Socrates was executed in 399 BC). Referring to Socrates' fate, Aristotle stated that he did not want Athens to "sin twice against philosophy," and he was allowed to return to Chalcidice. He died there the following year.

A major disagreement between Plato and Aristotle concerned the basis of knowledge. Plato proposed that reality was unknowable and that our imperfect perceptions never allow us to completely understand reality. Aristotle rejected this perspective, instead proposing that knowledge is based on empirical information gathered from the universe. (The difference between the two men is famously depicted in Raphael's fresco *The School of Athens*, in which Plato points heavenward as Aristotle gestures toward Earth.) Basing conclusions on careful observation—a hallmark of science—revolutionized philosophy, and made Aristotle the "father of science."

Aristotle's view of knowledge ultimately led him to develop his "four causes," introduced in *Physica*. Whereas Plato had invoked an illusion of a changing world due to our inability to discern reality adequately and consistently, Aristotle believed that the universe (which is without beginning or end) was constantly changing. The four causes for why a thing exists as it does—material, formal, efficient, and final—range from the constituent substance (material cause) to the purpose for the thing (final cause or *telos*). Hence, objects exist as we see them, but they also exist as the objects they could become. Aristotle's final cause, which was critical for Aristotle's view of the universe, was fundamentally teleological: the constant change we observe in the universe is real and directed toward activating the potential within all objects.

In applying his four causes to the biological world, Aristotle interpreted life as an orderly and perfect series of forms, with one type grading into another. He ordered life from the lowest form (e.g., worms) to highest (humans), with each step in this "Great Chain of Being" representing a closer approximation of perfection (e.g., apes are an incomplete, imperfect realization of the human form). Christians later extended the Great Chain to include angels and God.

Aristotle categorized life into "lower" and "higher" forms. It would take many centuries before biology would start to shed this bias by recognizing that "success" in an evolutionary sense is merely a measure of reproductive fitness, meaning "lower" forms (i.e., bacteria) are equally as advanced as "higher" forms. Also, because Aristotle accepted that this ordering of life was perfect, there was no possibility for change and no room for additional forms; the "fixity" of the species in the Great Chain of Being precludes evolutionary change within a form or creation of new forms.

The impact of Aristotle's thinking on Western thought in areas as diverse as politics, logic, ethics, and science has been monumental and enduring. Aristotle's influence on evolutionary biology was especially significant because his view of the unchanging nature of species was held for centuries afterward. Sir Richard Owen, an opponent of Charles Darwin's proposals, affirmed the Aristotelian perspective of biology when he proclaimed that "zoological science sprang from [Aristotle's] labours ... like Minerva from the head of Jove, in a state of noble and splendid maturity."

HERBERT ARMSTRONG (1892–1986)

Herbert W. Armstrong was a famous evangelist who in 1933 founded the Worldwide Church of God, which adamantly opposed evolution. Armstrong, who decided in 1927 that the rejection of evolution was one of seven "conclusions" that would guide his life, claimed that evolution was "a false theory" because it contradicted the Bible. Armstrong endorsed gap creationism (i.e., that there was a long period of time between Genesis 1:1 and Genesis 1:2) and believed that Earth was debris from a battle between God and Satan. Following a funeral service attended by 4,000 mourners, Armstrong was buried in Mountain View Cemetery in Altadena, California.

RUSSELL ARTIST (b. 1911)

Russell Artist was a biology professor at Tennessee's David Lipscomb College and a member of the Creation Research Society who in 1970 coauthored the creationist textbook *Biology: A Search for Order in Complexity*. When his book was rejected by the Tennessee State Textbook Commission for use in the state's public schools, Artist helped initiate Tennessee's "Genesis Bill," which in 1973 required textbooks presenting a theory of human origins to include evolution as "a theory ... not represented as a scientific fact" and allocate "an equal amount of emphasis on" other theories "including, but not limited to, the Genesis account in the Bible." One of the bill's sponsors was House Speaker Ned McWherter (b. 1930), a Democrat who was governor of Tennessee from 1987 to 1995. The "Genesis Bill" was overturned in 1975 by *Daniel v. Waters*.

AUGUSTINE OF HIPPO (354–430)

Reckless and incompetent expounders of Holy Scripture bring untold trouble and sorrow on their wiser brethren when they are caught in one of their mischievous false opinions and are taken to task by those who are not bound by the authority of our sacred books. For then, to defend their utterly foolish and obviously untrue statements, they will try to call upon Holy Scripture for proof and even recite from memory many passages which they think support their position, although they understand neither what they say nor the things about which they make assertion.

Aurelius Augustinus (Augustine) was born on November 13, 354, in Tagaste, at the southern extent of the Roman Empire (now Algeria). Augustine did not take Christianity seriously as a young man, and for many years he questioned much of the Bible. Augustine's parents provided an excellent education for their son, including study in Carthage, then the leading educational center in northern Africa. In 383, Augustine went to Rome to teach, and later taught in Milan. The death of his son and an overall lack of direction in his life led Augustine to an increased dependence on religion. In Milan, he met Bishop Ambrose (339–397), under whose influence Augustine converted to Catholicism.

In 391, Augustine became a priest in the city of Hippo, and then a bishop, a position he retained the rest of his life. Augustine, a forceful public speaker and a prolific writer (more than 5 million of his words have survived), became one of the most influential Christian figures. His influence was felt for centuries after his death in 430, particularly in the thinking of Martin Luther and John Calvin.

Two of Augustine's most influential works are *Confessions* and *City of God*. *Confessions* is part autobiographical, although Augustine's primary intent was to discuss faith and the difficulties of contending with one's imperfections. *City of God* considers the central tenets of Christianity and was written in response to the overthrow of Rome, an event Augustine used to urge readers to pay less attention to politics (as in Rome, the "city of man") and more to spiritual matters (as in Jerusalem, the "city of God").

Augustine famously stated that "Nothing is to be accepted save on the authority of Scripture, since greater is that authority than all the powers of the human mind." Many subsequent creationists, including Henry Morris and Kurt Wise, based their views of nature on Augustine's claim. However, Augustine also warned that there is real danger in reading the Bible literally. For Augustine, scripture and scientific understanding were not independent, but when science provides incontrovertible evidence for some aspect of nature, the Bible necessarily will need to be reinterpreted.

Augustine did not view Genesis as a factual account of how God created the universe. Instead, he believed that the universe was created all at once, if for no other reason than before there was anything, time did not exist, and so any creation necessarily had to be instantaneous. (He also rejected the eternal nature of the universe espoused by the ancient Greeks.) However, Augustine proposed that only the *rationes seminales*, or the latent capacity to become, was formed at the instant of creation: "In the seed, then, there was invisibly present all that would develop in time into a tree. And in this same way we must picture the world when God made all things together." This perspective extends to the appearance of humans too, in that the Earth was peopled "according to the productive potency scattered like a seed in the world by the word of God." For Augustine, the divine word of God was a given, and accommodation between science and faith was not only possible but necessary as both reflected the action of God.

Although Augustine believed that species have gradually appeared since the time of Creation, it is unlikely that he accepted the transformation of one species into another. Even though all species were not immediately present, their essence was present, and his views are one of species stability rather than change.

Augustine is buried near the altar of Saint Pietro in Ciel D'Oro Church in Pavia, Italy.

For More Information: O'Donnell. J.J. 2005. *Augustine: A New Biography*. New York: HarperCollins.

B

CHARLES BABBAGE (1791–1871)

We take the highest and best of human faculties, and, exalting them in our imagination to an unlimited extent, endeavour to attain an imperfect conception of that Infinite Power which created every thing around us.

Charles Babbage was born on December 26, 1791, in London, to wealthy parents Benjamin and Elizabeth Babbage. By the time Babbage entered Cambridge in 1810, he was already an accomplished mathematician, and his years at Cambridge were filled with academic success as well as the diversions afforded a student with a sizable allowance from a wealthy father. Babbage befriended John Herschel, son of astronomer William Herschel, and the two remained companions for many years. (Herschel, in his book *Physical Geography*, would later derogate Darwin's theory as the "law of higgledy-piggledy.") Babbage and Herschel formed the Analytical Society at Cambridge in 1811, through which they published papers about mathematics in the Society's *Memoirs* and encouraged Cambridge to embrace the mathematical advances then taking place in continental Europe. Babbage graduated from Cambridge in 1814, and in 1817 earned an MA.

In 1814, Babbage married Georgiana Whitmore. Around this time, he also began publishing papers on applied mathematics and gave a series of lectures on astronomy at the Royal Institution. These activities garnered Babbage enough attention that he became a fellow of the Royal Society in 1816. (Babbage would later ridicule the Royal Society as "a collection of men who elect each other to office, and then dine together at the expense of this society to praise each other over wine and give each other medals.") In 1824, Babbage directed a new life insurance company and immersed himself in the study of actuarial science. Although the business never opened, the experience provided Babbage with enough background to publish *A Comparative View of the Various Institutions for the Assurance of Lives*, which provided the public with an insider's view of the industry. He also published *On the Economy of Machinery and Manufacture*, which introduced the scientific study of manufacturing as a means to increase efficiency, and was later cited by Karl Marx.

In 1827, Babbage's father, wife, and two children died. Babbage left his surviving children with his mother, and spent a year traveling through Europe. When he returned in 1828, he accepted the Lucasian chair of mathematics—once held by Isaac Newton (1642–1727)—at Cambridge. He held this position until 1839, although he never gave a lecture. He also launched two unsuccessful bids for election to Parliament.

Babbage invented several useful devices, including the cowcatcher and the opthalamoscope. He published articles on topics ranging from meteorology and astronomy to chess, and he appreciated the value of experimental data (e.g., he was once lowered into a volcano to make observations, and had himself baked in an oven at 265° F to study the effects of heat). Babbage was also interested in geology, and during his trek across Europe, after the loss of his family members, he visited the Temple of Serapis near Naples, Italy. In 1831, when Charles Lyell became professor of geology at King's College, London, he and Babbage discussed the Temple frequently. With encouragement from Lyell, Babbage read a paper to the Geological Society in 1834 that discussed how gradual geological processes could cause the subsidence and elevation evident at the Temple. Babbage subsequently studied the effects of heating and cooling of rocks to further test his proposals, and these efforts, combined with his analysis from the Temple of Serapis, were published in the *Journal of the Geological Society of London* in 1847. Babbage is credited with demonstrating that geological processes could be studied experimentally, and later editions of Lyell's *Principles* were modified to accommodate Babbage's conclusions.

Charles Babbage is best known for building machines that could perform analytical functions, and has become known as "the father of the computer." By 1823, Babbage had developed a prototype of his "difference engine," so named for its ability to calculate finite differences. Babbage secured governmental funding to produce a larger version, and by 1832 the engine was one-seventh finished. However, by 1838, Babbage had spent all of the grant money and had estranged his project manager, and construction of the difference engine was halted.

Instead of abandoning the project, Babbage envisioned building a more complicated version, the analytical engine, which would be programmable with punched cards. Furthermore, the analytical engine could perform iterations ("loops") based on the outcome of previous calculations. Efforts to secure governmental funding were unsuccessful and an embittered Babbage eventually abandoned the project.

Babbage, who wanted to unify his deism with science, rejected William Whewell's claim in his *Bridgewater Treatise* that "we may thus, with the greatest propriety, deny to the mechanical philosophers and mathematicians of recent times any authority with regard to their views of the administration of the universe." The Earl of Bridgewater had commissioned *Bridgewater Treatises* (eight were produced) to unite natural theology with current scientific understanding. Babbage produced an unofficial *Ninth Bridgewater Treatise* (1837) to counter not only Whewell, but also the claim throughout *Treatises* that miraculous intervention through the suspension of natural laws is apparent. In contrast, Babbage argued that unusual events could result from natural laws, and to demonstrate this, he used his inventions. Lyell and Charles Darwin observed Babbage use the prototype of his difference engine to calculate a series of digits that established a predictable pattern. However, at any time, a seemingly anomalous digit might appear: Is this "anomaly" due to divine intervention? Babbage answered that it is not, because it was simply a function of the mathematical principles specified by the user but unknown to the observer. Babbage claimed that nature could work this way, in that God has programmed a pattern for the universe that follows the natural laws he has created, but the propensity of the human intellect to attribute unpredicted events to the action of miracles makes it appear that God is intervening. Babbage's claim that nature is governed by predictable and intelligible laws influenced Robert Chambers' *Vestiges of the Natural History of Creation* (1844) and Darwin as he considered a mechanism for "descent with modification."

Charles Babbage died on October 18, 1871, of "suppuration of the kidney." He is buried in Kensal Green Cemetery in London.

For More Information: Collier, B., & J. MacLachlan 1998. *Charles Babbage and the Engines of Perfection*. New York: Oxford University.

ROGER BALDWIN (1884–1981)

Roger Nash Baldwin was the Harvard-educated director of the American Civil Liberties Union who, in 1925, instigated the ACLU's involvement in the Scopes Trial. Baldwin proclaimed that "we shall take the Scopes case to the United State Supreme Court if necessary to establish that a teacher may tell the truth without being thrown in jail." Baldwin, who opposed Clarence Darrow's participation in the trial, later described the Scopes Trial as "the most widely reported trial on a public issue ever to have taken place in the United States." Baldwin did not participate in the Scopes Trial, and in 1981 received the Presidential Medal of Freedom in recognition of his lifetime of service to the nation.

JAMES BATEMAN (1811–1897)

James Bateman was a botanist who specialized in growing orchids. Darwin's studies of some of these orchids resulted in his 1862 book titled, *On the Various Contrivances by Which British and Foreign Orchids are Fertilized by Insects, and on the Good Effects of Intercrossing*. In this book, Darwin describes how various orchids are elaborately adapted for sexual reproduction via interactions with insects. Darwin was so confident of this coevolution that he made a bold prediction about one of the orchids that he had received from Bateman. That orchid was *Anagraecum sesquipedale*, a Madagascar orchid having an 11″ tube, at the bottom of which is nectar. Because the pollen of these orchids "would not be withdrawn until some huge moth, with a wonderfully long probiscus, tried to drain the last drop" of nectar, Darwin predicted that such a long-snouted moth must exist. Alfred Wallace shared Darwin's confidence: "Naturalists who visit [Madagascar] should search for the giant moth with as much confidence as astronomers searched for the planet Neptune, and I venture to predict they will be equally successful." (Wallace was referring to German astronomer Johann Galle [1812–1910], who had searched for and found Neptune after French mathematician

Urbain Le Verrier predicted its existence and position from calculations involving the orbit of Uranus.) In 1903—forty-one years after Darwin's prediction and twenty-one years after his death—entomologists in Madagascar discovered just such a moth: *Xanthopan morgani praedicta*. The *praedicta* was added to honor Darwin's prediction. Bateman died in Worthing, Sussex in 1897.

HENRY WALTER BATES (1825–1892)

The contemplation of nature alone is not sufficient to fill the human heart and mind.

Henry Walter Bates was born on February 8, 1825, in Leicester, England. When he was young, he collected insects, and in 1843 he published a short article about beetles in *The Zoologist*. In 1848, Bates accompanied his friend Alfred Wallace to the Amazon. When Bates returned to England eleven years later, he brought with him almost 15,000 specimens, 8,000 of which were unknown to science.

In 1863, Bates wrote *The Naturalist on the River Amazons*, which Darwin claimed was "the best work on natural travels ever published in England." Bates, who encouraged Wallace to develop his theories of organic evolution, discovered that closely related species often were separated geographically by rivers; Bates later realized that this was evidence of geographical speciation and evolution by natural selection. Bates' classic *Naturalist*, which is still in print, has become a standard against which biologists measure the ecological impact of the past century.

Bates' 1862 study of coloration in butterflies established Batesian mimicry, in which some good-tasting nonpoisonous animals mimic the warning colors of bad-tasting poisonous animals. Bates argued that this mimicry is "a most beautiful proof of natural selection," and Darwin used Bates' evidence to support his theory of evolution. In 1863, Darwin—who praised Bates' mimicry paper as "one of the most remarkable and admirable papers I ever read in my life"—anonymously reviewed one of Bates' papers about insects in the Amazon valley; that review was one of only two reviews that Darwin ever published. Bates was elected as a Fellow of the Royal Society in 1881.

Bates died on February 16, 1892, of complications resulting from bronchitis. Many of his collections are in The British Museum.

For More Information: Beddall, B.G. 1969. *Wallace and Bates in the Tropics*. London: Macmillan.

WILLIAM BATESON (1861–1926)

Modern research lends not the smallest encouragement or sanction to the view that gradual evolution occurs by transformation of masses of individuals, though that fancy has fixed itself upon the popular imagination.

William Bateson was born into a wealthy family on August 8, 1861, in Whitby, England. Although Bateson struggled academically before entering Cambridge in 1879, as an undergraduate he excelled, especially in science, and graduated in 1882. Bateson spent two summers studying the development of acorn worms (*Balanoglossus*) in the United States, and based on his research, proposed that the worms are primitive chordates and are important for understanding evolution of the chordate lineage. W.K. Brooks, Bateson's advisor, had recently published *The Law of Heredity: A Study of the Cause of Variation and the Origin of Living Organisms*. In this book, Brooks proposed a model that, following ideas by T.H. Huxley and Francis Galton, described evolution proceeding by "saltations" or jumps, and not by slow, gradual change.

Bateson found Brooks' ideas influential, and he decided to test this model by examining the nature of variation: Is it continuous or is much of the variation within a species in the form of large discrete differences? After several years of research, Bateson published his findings in 1894 as *Materials for the Study of Variation*, which was intended to support the saltational model. Bateson argued that because physical environments exist along a smooth continuum while species exist as discontinuous entities, speciation must not be a product of the environment. Instead, it must "be in the living thing itself." Although Galton and Huxley praised Bateson's book, W.F.R. Weldon (a friend of Bateson's who had become a professor at University College, London) wrote a negative review in *Nature*. Weldon dismissed Bateson's conclusions, while reiterating Darwin's original model of gradual change within continuous variation. Alfred Russel Wallace also wrote a negative review of Bateson's book.

In 1895, Bateson (still without a permanent academic position) began hybridizing plants to demonstrate that discontinuous variation is retained across generations even when "sports" crossbreed with individuals lacking these traits. Soon thereafter, Bateson read Hugo de Vries papers that referenced work done thirty years earlier by Gregor Mendel. After reading Mendel's original work, Bateson realized that here was an explanation for how discontinuous variation could arise and be maintained. Bateson had Mendel's paper translated and published by the Royal Society, which finally provided Mendel's work to the English-speaking scientific community. Because Bateson associated Mendel's results with his views on speciation, the label "Mendelism" became attached to the saltational model of evolution.

Meanwhile, Weldon and his University College colleague Karl Pearson had appropriated a model (developed by Francis Galton) known as the Law of Ancestral Inheritance. This model accounted for inheritance from all of an individual's ancestors, devalued by the number of generations between the individual and the ancestor (i.e., parents collectively contribute 50 percent, grandparents contribute 25 percent, etc., with the complete backward-running series summing to one). For biometricians, intent on finding continuous variation, this model predicted exactly what they wanted. With two discrete and competing models, a bitter conflict between the "Mendelists" and the "biometricians" ensued. Ultimately, the conflict was resolved by Thomas Morgan's demonstration that the hereditary units, genes, are located linearly on chromosomes. (This also explained the linkage of traits that Bateson, working with R.C. Punnett, had discovered independently of Morgan.) Mendel's results finally had a cytological explanation. Bateson, however, could not accept that differences in something as seemingly minor as the form of a gene could explain speciation. Instead he initially proposed that there is a "residue—a basis—upon which the unit characters are imposed" that determines the concept of "species." By 1921, however, Bateson had abandoned this proposal.

In 1908, Bateson became professor of biology at Cambridge, but he resigned in 1910 to direct the newly created John Innes Horticultural Institute, where he remained for the rest of his career. Bateson is now remembered for two aspects of his career: being wrong about the process of evolution, and coining the term *genetics*. (He also introduced many other genetics terms, including *homozygote*, *heterozygote*, F_1, F_2, and *allelomorph*, now shortened to *allele*.) Subsequent biologists have blamed Bateson for impeding advances in genetics and evolution (e.g., Ronald Fisher referred to Bateson's influence being "chiefly retrogressive," and G. Ledyard Stebbins criticized him for "delaying the [modern] synthesis"). However, Bateson was an excellent experimentalist, and he recognized the significance of Mendel's results. He was awarded the Royal Society's Darwin Award in 1904, and the Royal Medal in 1920, but he declined knighthood. Bateson, who noted that "the campaign against the teaching of evolution is a terrible example of the way in which truth can be perverted by the ignorant," died of heart failure on February 8, 1926.

H.M.S. *BEAGLE* (1820–1870)

The H.M.S. (His Majesty's Ship) *Beagle* was a 90' long, double-masted, 235-ton brig that carried ten guns and up to 120 men as a ship-of-war in the British Royal Navy. The ship was a "Cherokee Class" warship, a class of ship referred to by British sailors as "coffin brigs" because 26 of the 107 ships were lost at sea. The *Beagle*—built at a cost of £7,803—was launched on May 11, 1820, from the Royal Naval Dockyard in Woolwich. In July of that year, the *Beagle* led a naval review to celebrate King George IV's coronation, and was the first ship to sail under the new London Bridge.

In 1825, after being moored for five years, the *Beagle* was converted to an exploration ship. On May 22, 1826, under the leadership of Commander Pringle Stokes, the *Beagle* began a surveying voyage to Tierra del Fuego. In 1828, after Stokes killed himself, Robert FitzRoy assumed command of the *Beagle*. The *Beagle* returned to England in 1830, and on December 27, 1831, it left again for South America under the guidance of Captain FitzRoy. A young Charles Darwin was aboard (Figure 5).

In September of 1835, the *Beagle* docked for a month-long stay in the Galápagos Islands, which straddle the equator about 620 miles west of Ecuador. The *Beagle* returned to England in October 1836. Darwin's book about his *Beagle* voyage, today titled *The Voyage of the Beagle* (1839), is one of the world's great travel books, and remains a steady seller.

In 1837, John Wickham captained the *Beagle* on its third surveying voyage, this one to Australia. In 1845, the *Beagle* was sold to the Coast Guard Service, and it was moored in the River Roach in Essex. By 1851 the *Beagle* had been renamed *Southend Work Vessel Number 7*, and in 1870 it was sold for £525. It continued to be used as a patrol boat along the Essex coast and in River Roach, and several years later disappeared from public records.

On June 2, 2003, the European Space Agency launched *Beagle 2* to search for life on Mars. The mission was prefaced with this announcement: "H.M.S. *Beagle* was the ship that took Darwin on his voyage around the world in the 1830s and led to our knowledge about life on Earth making a real quantum leap. We hope *Beagle 2* will do the same thing for life on Mars." Although *Beagle 2* was scheduled to land on Mars on December 25, 2003, no communications were received from *Beagle 2*, and on February 6, 2004, the spacecraft was declared lost. The European Space Agency plans to launch an improved spacecraft, provisionally named *Beagle 2: Evolution*, in 2009.

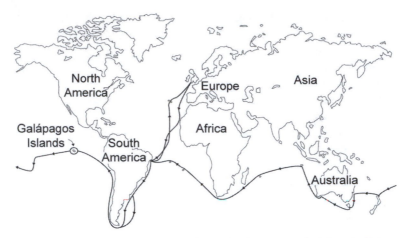

5. From 1831 to 1836, the H.M.S. *Beagle*—under the command of Captain Robert FitzRoy—carried Charles Darwin around the world and, in the process, profoundly affected Darwin's views of nature. During the *Beagle*'s 4,000-mile voyage, Darwin spent eighteen months at sea; the rest of the time was spent on land investigating geology and, to a lesser extent, wildlife. Darwin's book about that voyage—*The Voyage of the Beagle*—remains one of the world's greatest travel books. (*Jeff Dixon*)

No ship, except for Noah's Ark, figures more prominently in the evolution-creationism controversy than the *Beagle*. A replica of the *Beagle* is set to launch in 2009, the 200th anniversary of Charles Darwin's birth.

FRANCIS BEAUFORT (1774–1857)

Francis Beaufort was a Captain in the British Royal Navy who was in charge of the H.M.S. *Beagle*. After being made Captain of the *Beagle*, Robert FitzRoy asked Beaufort to recommend a naturalist for the voyage. Beaufort asked Rev. George Peacock of Cambridge for a recommendation, who then asked his friend Rev. John Stevens Helsow, who recommended Charles Darwin for the job. Because Darwin paid his own expenses while aboard the *Beagle*, Beaufort allowed Darwin to keep all of the specimens that he collected. Beaufort died in Hove, Sussex in 1857, and was buried in the church garden of St. John at Hackney, London.

ANTOINE BECQUEREL (1852–1908)

Antoine Henri Becquerel was a French physicist whose discovery of radioactivity in 1896 ultimately led to recalculations of the age of the Earth. Rock-dating techniques based on Becquerel's discovery showed that the Earth is approximately 4.5 billion years old, which was enough time for Darwin's theory of evolution by natural selection to produce life's diversity. Becquerel's discovery also enabled British geologist Arthur Holmes (1890–1965) in 1913 to propose a geologic time scale. Becquerel, the recipient of the 1903 Nobel Prize in Physics, died in 1908. The SI unit for radioactivity, the Becquerel (Bq), is named after him.

HENRY WARD BEECHER (1813–1887)

Evolution is the method of God in the creation of the world . . . I am a cordial evolutionist.

Henry Ward Beecher was born on June 24, 1813, in Litchfield, Connecticut (Figure 6). Beecher—whose father, Lyman, was one of the last great Puritan preachers in America—graduated from Amherst College in 1834 and earned a degree from Lane Theological Seminar in Ohio in 1837. After working as a pastor in Indiana, he became pastor of the Plymouth

6. Henry Ward Beecher, the most famous preacher of his era, was one of the first prominent theologians to embrace evolution. Beecher's sermons were printed in newspapers throughout America, and his church was often attended by famous politicians and celebrities. Beecher's affair with Elizabeth Tilton produced one of the most famous scandals and trials of the nineteenth century. (*Library of Congress*)

Congregational Church in Brooklyn, New York, where he became one of the most famous preachers in America. Beecher's congregations included celebrities such as Mark Twain, Walt Whitman, and Abraham Lincoln, and Beecher was described as "the most admired man in America after Abraham Lincoln." Even Lincoln expressed his "profound admiration for the talents of the famous pastor of Plymouth Church." Beecher's sermons attracted enormous crowds and were printed in newspapers throughout America.

Beecher advocated rational thought and progressive causes, including women's suffrage and the abolition of slavery. He raised money for antislavery activists, and guns bought with the money he raised became known as "Beecher's Bibles." During the Civil War, Beecher's church equipped an infantry regiment, and was a vital philosophical and geographical link in the Underground Railroad. Beecher was the main speaker when the U.S. flag was raised at Fort Sumter near the end of the war.

In the 1870s, when many preachers were denouncing evolution as atheistic, Beecher fused religion and evolution into a new form of spiritual evolution. Beecher, who taught that Genesis "is a poem, not a treatise on cosmogony," claimed that it was inefficient for God to design each species separately, so He designed laws that generated everything. Beecher often preached about evolution, noting that "while Evolution is certain to oblige theology to reconstruct its system, it will take nothing away from the grounds of true religion," and a popular collection of his sermons titled *Evolution and Religion* was published in 1885. Not surprisingly, Beecher was condemned by fundamentalists. For example, Jonathan Blanchard (1811–1892)—who wanted the United States to pass a constitutional amendment making the nation "a Christian Nation"—branded Beecher a "crafty leader of degeneracy and corruption." In 1919, Blanchard's son Charles drafted the doctrinal statement of the World's Christian Fundamentals Association, and in 1925 William Jennings Bryan delivered the eulogy at Charles' funeral.

At the peak of his popularity, Beecher—a married man—had an affair with Elizabeth Tilton, a member of his congregation and his best friend's wife. Although Beecher described the affair as a sacred "highly religious love," it produced one of the most famous scandals and trials of the nineteenth century. Beecher's trial, which began on January 11, 1875, lasted six months. There were fifty-two jury ballots, but jurors could not reach a verdict. Beecher's church exonerated Beecher, but excommunicated Elizabeth Tilton and her husband. Throughout his life, Beecher remained close to his older sister, Harriet Beecher Stowe, who wrote *Uncle Tom's Cabin*.

Beecher often proclaimed that "I have more health than I know what to do with." But on March 8, 1887, after whispering, "now comes the mystery," Beecher died of a brain hemorrhage. Brooklyn declared a day of mourning, and condolences poured in from throughout the world, including from President Cleveland. Beecher was buried on March 11, 1887, in Green-Wood Cemetery in Brooklyn beneath the epitaph, "He thinketh no evil" (the same sentence used by Herman Melville to introduce his *Confidence Man*). Nearby is the grave of Elizabeth Tilton, who died ten years after Beecher (and one month after Beecher's wife). Beecher is memorialized by a statue in Brooklyn's Columbus Park, as well as by a statue by Gutzon Borglum (who created the Mount Rushmore memorial) in the garden of Beecher's church, which is now called Plymouth Church of the Pilgrims and is a National Historic Landmark.

For More Information: Applegate, D. 2006. *The Most Famous Man in America: The Biography of Henry Ward Beecher*. New York: Doubleday.

MICHAEL BEHE (b. 1952)

If the evidence was there, people would be quieter, but the evidence isn't there, and people know it.

Michael J. Behe was born on January 18, 1952, in Altoona, Pennsylvania (Figure 7). Behe was taught evolution in the Catholic schools he attended, and as a young man accepted a universe that was created by a god working through observable natural laws, including natural selection: "Here was Darwin's theory, and it looks like God set up the world to begin producing life. I remember thinking 'That's cool.'"

Behe earned a BS from Drexel University in chemistry in 1974 and a PhD in biochemistry from the University of Pennsylvania in 1978. Following postdoctoral work at the National Institutes of Health, Behe became an assistant professor in the chemistry department at Queens College (part of the City University of New York) in 1982. There he met his future wife, Celeste, with whom he eventually had nine children. In 1985, Behe became an associate professor of chemistry at Lehigh University in Bethlehem, Pennsylvania. In 1995, he moved to the biology department at Lehigh, and became full professor in 1997.

By the 1990s, Behe had published several dozen articles in peer-reviewed journals. During this phase of his career, Behe accepted the central role of evolution in producing the processes he studied. However, in 1987 he read Michael Denton's *Evolution: A Theory in Crisis* (1985). Denton, a biochemist, concluded that there is no mechanism that would cause speciation, and rejected Darwin's proposal of common descent. Behe's response to Denton's book was that it "startled me...there was a good chance [Darwin's theory] was incorrect; it could not really describe how life came to be. I got mad." Behe searched for research that demonstrated how complex biochemical systems could have evolved. He concluded that such support was lacking, which led him to believe that complex biological systems must have been designed by an external agent.

In the early 1990s, Behe read Phillip Johnson's *Darwin on Trial* (1991), finding in it much that agreed with his new perspective on evolution. After a letter by Behe in support of Johnson's book was published in *Science*, Johnson invited Behe to join others in advocating intelligent design (ID). Behe accepted, and contributed material on biochemistry to the 1993 version of the pro-ID textbook *Of Pandas and People: The Central Question of Biological Origins*. In 1996, Behe also became a senior fellow at the pro-ID Discovery Institute.

In 1996, Behe described his opposition to standard evolutionary biology in his iconoclastic *Darwin's Black Box: The Biochemical Challenge to Evolution*. This book sold more than 200,000 copies, was named Book of the Year for 1996 by *Christianity Today*, and was hailed by the *National Review* as one of the top 100 nonfiction books of the twentieth century. The "black box" of the title is the basic unit of all life, the cell. Behe claims that when Darwin proposed his thesis, the cell was relatively unknown and was considered a simple building block upon which the complexity of life was constructed. However, contemporary research has shown that cells are complicated, meaning that life is not "complexity at the top and simplicity beneath, but...complexity at the top and more complexity underneath." Behe proposed that it is impossible to imagine how the intricate "molecular machines" of the cell could have evolved by natural selection because

7. Michael Behe is a biology professor whose award-winning book *Darwin's Black Box* provided the "scientific" foundation for the modern "intelligent design" movement. Behe, who acknowledges that Earth is old and that humans and apes share a common ancestor, is one of the most recognizable creationists of the early twenty-first century. (*Michael Behe*)

only the final version with all the necessary parts in place produces a working system: cellular components (like Behe's favorite example, the flagella) are therefore "irreducibly complex" and are scientific evidence for the action of an intelligent designer. *Darwin's Black Box* soon became a foundation for ID's "scientific" opposition to evolution.

Response to *Darwin's Black Box* from the scientific community was swift and severe. Behe was criticized for not understanding basic evolutionary biology (e.g., he has described the action of natural selection as "random Darwinian processes" and has noted that "cells are simply too complex to have evolved randomly"), as well as for mixing science and religion by invoking a supernatural cause that he claims can be studied scientifically. His concept of irreducible complexity has been equated with both the discredited "argument from design" proposal (i.e., traits appear to be designed because they *are*

designed) and the weak "God in the gaps" argument for the existence of God (i.e., the inability of science to explain all aspects of the natural world is proof of God). Behe dismissed his critics' evolutionary explanations as "materialism in the gaps." Although Behe claims that ID is "one of the greatest achievements in science," most scientists regard it as religion, not science.

Behe is unusual among ID proponents in that he accepts an evolutionary explanation for many traits, agrees that the Earth is ancient, and recognizes common descent among organisms, including between humans and apes. But critics charge that he inexplicably treats evolution at the cellular level differently (e.g., "Darwin's theory encounters its greatest difficulties when it comes to explaining the development of the cell") and promotes the need for entirely different explanations at these levels. Although Behe claims otherwise, the scientific community has deemed ID and irreducible complexity as not scientific because the concepts do not generate testable hypotheses. The lack of publication of original, peer-reviewed research in this area is also cited as evidence for the nonscientific nature of ID.

Behe counters that he has tried to publish ID-based research and to obtain grant money for study of ID, but his efforts are consistently rejected. In 2004, Behe and David Snoke, a professor in the department of Physics and Astronomy at the University of Pittsburgh, did publish a paper in the peer-reviewed journal *Protein Science* titled "Simulating Evolution by Gene Duplication of Protein Features That Require Multiple Amino Acid Residues." In their introduction, Behe and Snoke make clear the purpose of their article: "Although many scientists assume that Darwinian processes account for the evolution of complex biochemical systems, we are skeptical." The authors do not mention "intelligent design" or "irreducible complexity," but do report that for mutation to produce novel protein functions, large populations would be required. They conclude that "such numbers seem prohibitive," but urge caution in drawing conclusions from this single study. A critical response in the same journal by evolutionary biologist Michael Lynch of Indiana University deemed the model used by Behe and Snoke as non-Darwinian because it assumed intermediate steps between the original and new protein to be nonfunctional. Lynch created a "Darwinian version" of the model and concluded that "the origins of new protein functions are easily explained in terms of well-understood population-genetic mechanisms."

Behe was queried extensively about the *Protein Science* paper during his multiday testimony in the *Kitzmiller v. Dover Area School District* trial in 2005. Behe testified that he considered the article to be an "intelligent design article" that "seems to present...problems for Darwinian evolution," and that the original version did in fact include the term "irreducible complexity," but a reviewer required the term be removed. When cross-examined by the plaintiffs' attorney, Behe confirmed that the model was limited to only one particular type of mutation and that the model did indeed predict that advantageous mutations could become fixed in the population

in a reasonable number of generations within typically large microbial populations. Other aspects of Behe's testimony figured prominently in Judge John Jones' final decision: "Consider, to illustrate, that Professor Behe remarkably and unmistakably claims that the plausibility of the argument for ID depends upon the extent to which one believes in the existence of God. As no evidence in the record indicates that any other scientific proposition's validity rests on belief in God, nor is the Court aware of any such scientific propositions, Professor Behe's assertion constitutes substantial evidence that in his view, as is commensurate with other prominent ID leaders, ID is a religious and not a scientific proposition."

Behe's viewpoints remain controversial; even his home department's Web site includes a disclaimer stating that "intelligent design has no basis in science, has not been tested experimentally, and should not be regarded as scientific." Behe, along with fellow creationists William Dembski and Stephen Meyer, edited *Science and Evidence for Design in the Universe* (2001), a collection of essays addressing the role of design in nature. In 2007, Behe published *The Edge of Evolution: The Search for the Limits of Darwinism*, a follow-up to *Darwin's Black Box*, which propounds Behe's views on the limited creative power of mutation and natural selection. Like *Darwin's Black Box*, *The Edge of Evolution* was harshly criticized by scientists. Behe continues to write about ID for the popular press and has produced several videos about ID that are distributed through the Access Research Network.

For More Information: Behe, M.J. 1996. *Darwin's Black Box*. New York: Free Press; Behe, M.J. 2007. *The Edge of Evolution*. New York: Free Press.

BRUCE BENNETT (1917–1979)

This is the Bible, buddy. I intend to defend it.

Bruce Bennett was born on October 31, 1917, to Oakley and Anita Bennett in Helena, Arkansas. After serving in the Army during World War II, Bennett graduated from the Vanderbilt University Law School in 1949. Bennett was the prosecuting attorney in Arkansas' 13th Judicial Circuit from 1952 to 1956, after which he served as Arkansas Attorney General from 1957 to 1960 and from 1963 to 1966.

In 1958, Bennett—an ardent segregationist—joined Governor Orval Faubus to write legislation "designed to harass" civil-rights protesters he considered "the enemies of America." The legislation bypassed federal desegregation orders, opened the NAACP (National Association for the Advancement of Colored People) records to state scrutiny, and prevented NAACP members from getting state jobs. Bennett, who blamed Communists for racial unrest in Arkansas, ran segregationist campaigns for governor in 1960 and 1968, but lost each election. In the 1960 campaign, Bennett accused Faubus of being a secret ally of state NAACP leader Daisy Bates.

Bennett, who was equally known for flaunting the law as for upholding it, defended Arkansas' antievolution law when it was challenged by Susan Epperson in 1965. Bennett claimed that the law was a simple exercise of administrative control of the curriculum, and that Epperson could not freely "substitute her judgment for that of her employer." Bennett—who had promised to present "scientific witnesses to dispel the offbeat theory of evolution"—attacked Epperson, claiming that she believed that humans "evolved from monkeys, apes, sharks, porpoises, seaweed, or any other form of animal or vegetable." While questioning Epperson, Bennett referred to "crackpots" who would teach the "God is Dead" theory, the "Man Came from a Gorilla" theory, and the "Ham and Eggs Theory of California." Bennett claimed Epperson wanted "to advance an atheistic doctrine," despite the fact that Epperson was a devout Christian. Bennett repeatedly sought an opening that would allow him to introduce witnesses questioning evolution, to which the prosecution objected. Murray Reed, the presiding judge, sustained all of Epperson defense lawyer Eugene Warren's sixty-three objections.

After Epperson's trial, Bennett lost his bid for reelection. In 1969, Bennett was charged with twenty-eight counts of securities violations, postal fraud, and wire fraud for his involvement with Arkansas Loan and Thrift, a company he had helped found and which went bankrupt in 1967 because of its questionable investments. Bennett, who was nearing the end of a ten-year battle with throat cancer, never stood trial. He died on August 26, 1979, and is buried in Arlington Cemetery in El Dorado, Arkansas.

GERALD "JERRY" BERGMAN (b. 1946)

Gerald "Jerry" Bergman was a biology instructor and Young-Earth Creationist at Bowling Green State University who, when he was denied tenure in 1978, sued the university, alleging that his firing was due to his religious views. Bergman's case was dismissed in 1985, and his appeal was rejected in June 1987, when the court ruled that Bergman's denial of tenure was not due to his religious views. Bergman's *"Vestigial Organs" Are Fully Functional* (1990) claimed that vestigial organs are the Creator's "handiwork and design," and his *The Criterion: Religious Discrimination in America* (1984) chronicled cases of alleged discrimination similar to his own.

GEORGE THOMAS BETTANY (1850–1892)

George Thomas Bettany was a botanist who in 1887 published *Life of Charles Darwin*. Aside from obituaries and the published lecture delivered by Louis Miall to the Leeds Philosophical and Literary Society, this was the earliest biography of Darwin. Darwin's life later became the basis of a large industry.

BIBLE CRUSADERS OF AMERICA (Est. 1925)

Bible Crusaders of America continued William Jennings Bryan's crusade against the teaching of human evolution in public schools. The Crusaders were funded by real estate tycoon George Washburn, a friend of Bryan. Washburn enlisted virtually all of the leading antievolutionists for his cause, including John Straton and William Riley. The Crusaders' campaign director was T.T. Martin, who helped convince the Mississippi legislature to pass a law banning the teaching of human evolution. The Crusaders promoted the work of creationists such as Arthur Brown (1875–1947), a surgeon who claimed that evolution—"the greatest hoax ever foisted on a credulous world"—is a tool used by Satan to attack the Bible.

BIG BONE LICK

Big Bone Lick is a swampy area in Kentucky that was described in 1744 as "the place where they found the elephant bones." In 1795, President William Harrison collected thirteen barrels of bones at Big Bone Lick, all of which were subsequently lost in the Ohio River. In 1803, Meriwether Lewis (of Lewis and Clark fame) visited Big Bone Lick on his way to join William Clark. In 1807, Thomas Jefferson—an avid paleontologist—sent William Clark (of Lewis and Clark fame) to Big Bone Lick, from where he shipped 300 boxes of bones of mastodons, mammoths, and other organisms to the White House; bones from Big Bone Lick remain displayed at Monticello (Jefferson's home). These bones were primarily from mammals from the Pleistocene Epoch (more than 15,000 years ago; often called the Ice Age). Clark's three-week excavation—the first organized vertebrate paleontology expedition in the United States—established American vertebrate paleontology.

By 1836, Big Bone Lick was a famous health resort, and in 1841 it was visited by Charles Lyell. In 1960, Big Bone Lick State Park was established, in 1971 the area was designated a National Historic Site, and in 2002 the National Park Service designated Big Bone Lick a Lewis & Clark Heritage Trail Site.

BIOLOGICAL SCIENCES CURRICULUM STUDY (BSCS) (Est. 1958)

The history of the Biological Sciences Curriculum Study (BSCS) can be traced to the October 4, 1957 launch of the small (23″ diameter, with one watt of power) Soviet satellite *Sputnik*, which awakened the American political and educational establishment to the importance of improving science education. The following year, Congress passed the National Defense Education Act, which encouraged the National Science Foundation (NSF) to develop state-of-the-art science textbooks. In the same year, NSF allocated $143,000 to establish the BSCS to educate "Americans in general to the acquisition of a scientific point of view." By 1959, BSCS had established its headquarters at the University of Colorado (Figure 8).

In the early 1960s, BSCS created new biology textbooks that, unlike other textbooks, stressed concepts rather than facts, and investigations rather than lectures. The three BSCS books published in 1963 became known by the color of their covers: *Blue* emphasized molecular biology, *Green* emphasized ecology, and *Yellow* emphasized cellular and developmental biology. Approximately 70 percent of the content of each book was identical, but the material was presented using different themes. Although BSCS wanted to avoid the criticism that it was trying to establish a national curriculum, their books—for all practical purposes—did exactly that, for in the 1960s, most schools in the United States used BSCS textbooks.

When John Scopes was convicted of teaching human evolution in 1925, publishers feared that discussing evolution in biology textbooks would hurt sales. As a result, biology textbooks published after Scopes' conviction did not include the word *evolution*. However, BSCS books were different. Instead of relying on professional writers to prepare their textbooks, BSCS recruited the best scientists and teachers in the United States as authors. Not surprisingly, all of the BSCS books stressed evolution. Today, BSCS is credited with "putting evolution back into the biology curriculum." BSCS books were an agent in the U.S. Supreme Court's ruling that laws banning the teaching of human evolution are unconstitutional (i.e., *Epperson v. Arkansas*), as well as in cases involving issues such as instruction about human reproduction and the use of live animals in biology classrooms. Some states, such as Texas (in 1970) and Kentucky (in 1965), banned

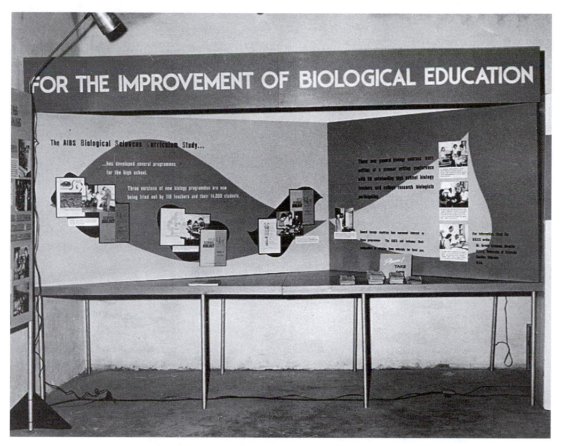

FOR THE IMPROVEMENT OF BIOLOGICAL EDUCATION

8. The Biological Sciences Curriculum Study (BSCS) had humble beginnings, but by the 1970s its books were used in most U.S. high school biology classrooms. This photo, taken in 1960, shows a BSCS exhibit that announced "the first chapters" of the new BSCS textbooks. Those books, which "put evolution back into the biology curriculum," transformed biology books and science education. (*BSCS*)

the BSCS books. Evangelist Reuel Lemmons of Austin condemned the textbooks as the "most vicious attack we have ever seen on the Christian religion."

Today, BSCS is a nonprofit corporation headquartered in Colorado Springs, Colorado, which continues to publish high-quality biology textbooks; domestic sales of BSCS *Green* have exceeded 2.6 million copies, and sales of *Blue* have exceeded 1.6 million copies. BSCS materials have been printed in more than twenty-five languages for use in more than sixty countries. Since its inception, more than 20 million students have used BSCS materials.

For More Information: Engelman, L. 2001. *The BSCS Story: A History of the Biological Sciences Curriculum*

Study. Colorado Springs, CO: Biological Sciences Curriculum Study.

ROLAND BIRD (1899–1978)

Roland Thaxter "R.T." Bird was a junior-high dropout and drifter who became a famous dinosaur-hunter. When Bird sent paleontologist Barnum Brown an amphibian jaw that he had discovered, Brown hired Bird as an assistant. Throughout the 1930s, Bird worked for the American Museum of Natural History, and his fossil-hunting expeditions were often funded by oil magnate Harry Sinclair. Bird, who's *Bones for Barnum Brown: Adventures of a Dinosaur Hunter* was published in 1985, is famous for discovering and preserving the dinosaur track-ways in the

Paluxy Riverbed. Bird was buried in 1978 in Grahamsville, New York, with dinosaurs' gizzard bones in his pocket. On his tombstone is carved the likeness of a brontosaur and the inscription, "R.T. Bird, Discoverer of Sauropod Dinosaur Footprints."

For More Information: Bird, R.T. 1985. *Bones for Barnum Brown: Adventures of a Dinosaur Hunter.* Fort Worth, TX: Texas Christian University.

WENDELL BIRD (b. 1954)

Public secondary and elementary schools must give balanced treatment to the theory of scientific creationism and the theory of evolution.

Wendell R. Bird was born in 1954 in Atlanta, Georgia, and in high school was named one of the top forty science students in the nation by the Westinghouse Science Talent Search (Figure 9). Bird, the first three-time winner of the NASA-NSTA Science Competition, was the first student in Vanderbilt University history whose scores on entrance exams allowed him to exempt the freshman year. Bird received a BA (*summa cum laude*, and in the top 1 percent of his class) from Vanderbilt University in 1975, and in 1978 received a JD from Yale Law School.

In 1978, Bird published in *Yale Law Journal* a strategy for incorporating the teaching of creationism into public schools. The article won the Egger Prize (awarded by Yale Law School for the best student legal article), and claimed that teaching only evolution denied "academic freedom" and violated the free exercise of religion by forcing students to learn heretical views. Bird, citing the Creation Science Research Society, Duane Gish, Henry Morris, *Biology: A Search for Order in Complexity*, and the Institute for Creation Research (ICR), argued that "treatment of either the theory of evolution or the theory of scientific creationism must be limited to scientific evidences and must not include religious doctrine." Bird also justified his resolution by noting that evolution "is contrary to the religious convictions and moral convictions of many students and parents."

9. Wendell Bird's ideas formed the basis for legislation requiring that evolution and creationism be given "equal time" and "balanced treatment" in science classrooms of public schools in Arkansas and Louisiana. The laws in both states were declared unconstitutional in the 1980s. (*AP Images*)

After graduating from law school, Bird worked as a legal adviser (and later as a staff attorney) at ICR. In 1979, ICR distributed thousands of Bird's four-page resolution in the May issue of *Impact*, noting that it was "a suggested resolution, to be adopted by boards of education, not legislation proposed for enactment as law." Paul Ellwanger, a respiratory therapist from South Carolina, molded Bird's resolution into legislation, and founded Citizens for Fairness of Education for people "who favor academic freedom and are opposed to suppression of information about evolution and creation." In Arkansas, A.A. Blount, a preacher in Little Rock, gave Ellwanger's draft to his state senator, James L. Holsted. Holsted, a self-described "born-again fundamentalist" who noted that Bird's resolution "represents my beliefs and the beliefs of the majority of my constituents," introduced the legislation, despite acknowledging that it favored biblical literalists. The legislation passed, as one legislator noted, "This is a terrible bill, but it's worded so cleverly that none of us can vote against it if we want to come back up here." The entire legislative process, from its introduction to its signing on March 19, 1981, by Governor Frank White (despite not having read it), took less than a week. A modified version of Bird's resolution was also introduced in 1980 by state Senator Bill Keith in Louisiana, where it also became law.

Arkansas' "balanced treatment" law was first tested in *McLean v. Arkansas Board of Education*, in which Judge William Overton referred to Bird's paper as "a student note." When Bird—general counsel to ICR—was excluded from the Arkansas lawsuit, he convinced several witnesses to leave town before testifying. Overton later ruled that the Arkansas "balanced treatment" law was unconstitutional and that "creation science" is religion, not science. The "balanced treatment" law was also tested in Louisiana in *Edwards v. Aguillard*. This law, which Bird claimed was not as restrictive as the Arkansas law, was also ruled unconstitutional (in this instance, by the U.S. Supreme Court). Bird's two-volume *The Origin of Species Revisited*, which in 1989 introduced the "abrupt appearance" theory (for which some creationists would later demand "equal time"), was an outgrowth of Bird's experiences in *Edwards v. Aguillard*.

In 2005, Bird—on behalf of the Association of Christian Schools International and the Calvary Chapel Christian School of Marietta, California—alleged in a lawsuit (*Association of Christian Schools International v. Roman Stearns*) that the University of California violated the constitutional rights of applicants from creationism-based Christian schools whose coursework was deemed inadequate for college. (The University of California rejected curricula based on creationist textbooks published by Bob Jones University as "inconsistent with viewpoints and knowledge generally accepted in the scientific community.") Bird published books and articles about taxes and constitutional law, and won courtroom judgments unrelated to

the teaching of creationism totaling hundreds of millions of dollars.

For More Information: Bird, W. 1991. *The Origin of Species Revisited*. Nashville, TN: Thomas Nelson.

EDWARD BIRGE (1851–1950)

Edward Birge was a scientist and president of the University of Wisconsin who, after William Jennings Bryan gave his "Brute or Brother?" speech there in 1921, claimed that Bryan would destroy students' faith by identifying religion as incompatible with science.

BISHOP v. ARONOV

Bishop v. Aronov (1991) challenged biology professor Phillip Bishop's claim that he could use class meetings to promote "intelligent design" and his religious faith ("evidence of God in Human Physiology"). The court ruled that (1) academic freedom is not an independent First Amendment right; (2) a university can control the style and content of speech in school-sponsored events; and (3) a university can control its curriculum, provided the university's "actions are reasonably related to legitimate pedagogical concerns."

HUGO BLACK (1886–1971)

Hugo Lafayette Black was a U.S. Supreme Court Justice who, like the other Justices, supported Susan Epperson's lawsuit when it reached the Court (*Epperson v. Arkansas*). However, Black was the only Justice who expressed any reluctance with the decision, noting that "there is no case or controversy here." The 82-year-old Black—who earlier in his career had been supported by the Ku Klux Klan—disagreed with other Justices' views of evolution, and was the only Justice to question the validity of evolution, noting that "certainly the Darwinian theory precisely like the Genesis story of creation of man is not above challenge." Black's majority opinion in *McCollum v. Board of Education* (1948) held that government could not provide religious instruction in public schools. Black also

31

wrote the majority opinion in *Engel v. Vitale* (1962), which banned state-sponsored prayer in public schools. Black is buried in Arlington National Cemetery.

JOSEPH BOEHM (1834–1890)

Sir Joseph Edgar Boehm sculpted the statue of Charles Darwin that was unveiled by Thomas Huxley on June 9, 1895, at The British Museum. When Darwin supporter Ray Lankester replaced Darwin foe Richard Owen as director of the museum, he moved the statue to the top of the stairs of the Great Hall, where it looked down on every visitor. (Today, a statue of Richard Owen occupies that position, and the statue of Darwin—as well as a similar one of Thomas Huxley—has been relegated to a downstairs cafeteria.) Boehm was paid £2,100 for sculpting the statue of Darwin. Boehm, whose life-size relief bust of Darwin in the North Choir aisle of Westminster Abbey was unveiled in 1888, is best known for sculpting the head of Queen Victoria (1819–1901, sovereign 1837–1901) on British coinage and the Duke of Wellington at Hyde Park Corner.

BENJAMIN BOGARD (1868–1951)

Benjamin Marcus Bogard was a Baptist preacher who proclaimed that "people of the right sort want to live in a state where the faith of their children will not be attacked in the free schools." Bogard believed that evolution had "been brought about by the John D. Rockefeller Foundation, which has been controlled by skeptics and infidels and atheists." Bogard, who participated in several highly publicized debates with fellow antievolutionist Aimee McPherson about miracles, helped place antievolution legislation on the November 1928 ballot in Arkansas. In that election, voters endorsed this legislation by a vote of 108,991 to 63,406, thereby making Initiative Act No. 1 (banning the teaching of human evolution in public schools) the only antievolution law ever passed by a popular vote. Bogard's American Anti-Evolution Association threatened to "black list" legislators who did not want to ban the teaching of evolution.

BERTRAM BOLTWOOD (1870–1927)

Bertram Borden Boltwood discovered that radioactive uranium decays to stable lead, and in 1907 used this information to determine that some rocks were up to 2.2 billion years old. Although critics such as John Joly argued that Boltwood's conclusions were unreliable because radioactive decay had not been consistent over geological time (a claim that continues to be promoted by many creationists), Boltwood's method became a standard for radiometric dating. While recuperating from a nervous breakdown, the 57-year-old Boltwood committed suicide.

GEORGE BRADLEY (1821–1903)

George Granville Bradley, Dean of Westminster Abbey from 1881–1902, approved Charles Darwin's burial in Westminster Abbey. Bradley's name appeared on admission cards for Darwin's funeral, but Bradley was abroad when Darwin was buried.

BRIGHTS

Brights are members of a movement based on a naturalistic worldview that is "free of supernatural and mystical elements." The Brights Movement, which was created to avoid stereotypes and conflicts associated with terms such as "atheist," aims to (1) promote the public's understanding of a naturalistic worldview; (2) gain public recognition that people who hold such a worldview can contribute to society; and (3) promote the full and equitable civic participation of all such individuals. Richard Dawkins and Daniel Dennett have aligned themselves with the Brights.

BRITISH MUSEUM OF NATURAL HISTORY (LONDON)

The Natural History Museum originated in The British Museum, which was established in the will of Sir Hans Sloane (1660–1753), a naturalist, a collector, and a physician. During

10. The Natural History Museum, also known as the British Museum of Natural History, is one of the most renowned museums in the world. The museum's Great Hall is dominated by an 80'-long skeleton of *Diplodocus* that was donated by American businessman and eugenicist Andrew Carnegie. The statue atop the stairs in the background commemorates Robert Owen, the museum's first director and opponent of Darwinian evolution (also see Figure 65). (*Randy Moore*)

his life, Sloane collected more than 70,000 objects as well as a library and herbarium. Sloane bequeathed his collection (which Linnaeus described as being in "complete disorder") to King George II for Britain in return for a payment of £20,000 to Sloane's heirs. Although King George II wasn't overly interested in Sloane's collection, Arthur Onslow—the Speaker of Parliament—helped pass an Act establishing The British Museum on June 7, 1753, to house Sloane's donation. Funded by more than £90,000 raised in a scandal-plagued lottery, The British Museum opened to the public on January 15, 1759. Except for parts of World Wars I and II, the Museum has been open ever since. In 2003, The British Museum marked its 250th anniversary as the first national, public, and secular museum in the world.

In 1856, famed anatomist Richard Owen became Superintendent of the Natural History Departments of The British Museum. Owen, a foe of Charles Darwin, was well-known for coining the word *dinosaur* and his reconstructions of extinct animals (Figure 10). In 1881, the natural history collection of the British Museum was moved to a new building designed by architect Alfred Waterhouse (1830–1905), and was named the British Museum of Natural History. In 1896, the Museum absorbed the adjacent Geological Museum of the British Geological Survey, and in 1963 the British Museum of Natural History became independent of The British Museum. In 1992, the title of the British Museum of Natural History was formally changed to the Natural History Museum. Its newly developed Darwin Centre contains millions of specimens, including many collected by Charles Darwin, Alfred Wallace, and John Gould. In 1912, the Museum accepted the Piltdown skull from Charles Dawson, who described it as the most important fossil ever. However, in 1953 the skull was proven a fake.

The British Museum of Natural History memorializes several famous British biologists. On June 9, 1885, Thomas Huxley unveiled the Museum's first statue, a marble monument to Charles Darwin. In 1897, Owen—who had opposed Darwin's statue—was also memorialized with a statue, and three years later a statue of Huxley was seated in the Museum. Since then, the Museum has incorporated memorials to Wallace and others. Antievolutionist Ken Ham considers the museum "a church of atheism."

Today, the Natural History Museum includes more than 70 million items that span botany, entomology, mineralogy, paleontology, and zoology. The Museum has a wildlife garden and a staff of more than 900 people, of which almost 300 are scientists and researchers. It is famed for its skeletons of dinosaurs, especially the 84'-long *Diplodocus carnegii* at the Museum's entrance. That skeleton can be traced to King Edward VII, who visited the United States in 1903 as a guest of eugenicist and steel magnate Andrew Carnegie. After King Edward VII saw the *Diplodocus* skeleton at the Carnegie Museum of Natural History in Pittsburgh, Carnegie shipped a replica to Britain.

For More Information: Thackray, J. & B. Press 2004. *The Natural History Museum*. London: Natural History Museum.

SHARON BROOME (b. 1956)

Sharon Weston Broome was a Louisiana state legislator who in 2001 claimed that evolution had "provided the main rationale for racism." Broome—a Democrat—attracted national attention by introducing legislation linking evolution and Darwin with Hitler, racism, and eugenics. (Antievolution organizations such as Answers in Genesis have made similar claims.) References to evolution and Darwin were later removed from Broome's legislation, leaving only a condemnation of racism.

BARNUM BROWN (1873–1963)

Barnum Brown, who was named after the illustrious showman P.T. Barnum, was a curator and the chief bone-hunter at the American Museum of Natural History in the early 1900s. Brown, who often worked with Roland Bird, gathered most of the museum's many tons of dinosaur fossils. While working in Hell Creek, Montana in 1902, Brown discovered the first *Tyrannosaurus rex*; today that specimen—which has a 4'-skull and 1'-wide neck vertebrae—glares menacingly at visitors to the American Museum of Natural History. Many of Brown's expeditions were funded by Harry Sinclair, who incorporated an image of *Diplodocus* into his company's logo to emphasize corporate gigantism; in return, Brown—the "Father of the Dinosaurs"—wrote the dinosaur booklets that were distributed at Sinclair gas stations in the 1930s and 1940s.

In 1934, Brown's discovery of a giant "dinosaur graveyard" in the Bighorn Mountains led him to speculate that the dinosaurs had died while seeking water in a drying lakebed; six years later, this scene was popularized in Walt Disney's *Fantasia*. Brown's work with Sinclair also produced the dinosaur exhibits at the Chicago World's Fair in the 1930s and the New York World's Fair in the 1960s, and Brown's *Apatosaurus* ("deceptive lizard") continues to be displayed in the Jurassic Hall of the American Museum of Natural History. The eccentric Brown, who was awarded an honorary doctorate in science from Lehigh University in 1934, often visited fossil collection sites in a fancy hat and a full-length fur coat. Brown also funneled geological information gleaned from his fossil-hunting expeditions to governmental agencies such as the Office of Strategic Services, the forerunner of today's CIA. Among paleontologists, Brown is almost universally recognized as the greatest fossil-hunter of all time. Dozens of crates of his fossils remain unopened at the American Museum of Natural History.

For More Information: Bird, R.T. 1985. *Bones for Barnum Brown: Adventures of a Dinosaur Hunter*. Fort Worth, TX: Texas Christian University.

ROBERT BROWN (1773–1858)

Robert Brown was a botanist whose death was announced at the Linnean Society meeting at which the Darwin-Wallace paper was read on July 1, 1858. This announcement may have overshadowed the importance of the Darwin-Wallace paper.

GEORGE BROWNE (1850–1945)

George Buckston Browne was a surgeon who bought Down House from the Darwin heirs in 1927 for £4,250. After spending £10,000 on repairs and providing an endowment of £20,000, Browne gave Down House to the British Association for the Advancement of Science (est. 1831) in 1929.

W.A. BROWNE

W.A. Browne was a friend of Charles Darwin while Darwin was an undergraduate at the University of Edinburgh. In 1927, Browne read a paper with a materialistic perspective to the Plinian Society. The paper upset many people, and all references to the paper (including the minutes from the previous meeting announcing Browne's intention to deliver it) were removed from the minutes. Darwin learned from Browne's experience, noting that "to avoid stating how far, I believe, in Materialism, say only that emotions, instincts, degrees of talent, which are hereditary, are so because the brain of a child resembles parent stock."

WILLIAM JENNINGS BRYAN (1860–1925)

The contest between evolution and Christianity is a duel to the death. If evolution wins, Christianity goes.

William Jennings Bryan was born on March 19, 1860, in Salem, Illinois, the same town in which young John Scopes would later attend high school and learn biology (Figure 11). Bryan's wealthy parents were devout Christians.

11. William Jennings Bryan, a three-time candidate for president of the United States, represented the World's Christian Fundamentals Association at the Scopes Trial in 1925. Bryan's death five days after the trial, and his subsequent portrayal in *Inherit the Wind* (Figure 47), made him an icon of the evolution-creationism controversy. Bryan is also shown in the upper right corner of the cover of this book. (*Library of Congress*)

After graduating with honors from Illinois College in 1881 and studying law at Union College of Law in Chicago, Bryan was admitted to the bar in 1883. The following year, he married Mary Baird, who was admitted to the bar in 1888, but never practiced. When John Scopes graduated from high school in Salem, Bryan delivered the commencement address.

Bryan practiced law in Illinois, and then in Nebraska, before being elected to the U.S. Congress in 1890. In Congress, Bryan advocated a variety of progressive causes, including a graduated income tax, women's suffrage, and the free coinage of silver. After losing a bid for the Senate in 1894, Bryan edited the *Omaha World-Herald* and was a popular public speaker. Many of Bryan's speeches stressed his belief that the dollar should be backed by silver rather than gold. Bryan, who greatly admired Thomas Jefferson, often urged Christians to solve the problems created by "the arrogance of wealth."

When the 36-year-old Bryan went to the Democratic National Convention in Chicago in 1896, he was not widely known. The Convention deadlocked on a presidential nominee for four ballots, at which time Bryan rose to speak. There, late in the evening of July 9, Bryan defied his party's conservative leader (Grover Cleveland) with his magnificent "Cross of Gold" speech: "You shall not press down upon the brow of labor this crown of thorns, you shall not crucify mankind upon a cross of gold." The speech electrified the convention, and Bryan was nominated for president. At the time, "Boy Bryan" was the youngest person ever nominated for the presidency. During the campaign, Bryan—who was outspent 20-to-1 by his Republican opponent William McKinley—traveled almost 20,000 miles, and was the first presidential candidate to take his message directly to voters, often from the back of railroad cars (this was a new tactic, since presidential candidates had traditionally stayed home and let others speak on their behalf). Although Bryan gave 600 speeches in twenty-seven states during the campaign, McKinley defeated Bryan by an electoral vote of 271–176 and by a popular vote of 51–47 percent. (Many other Democratic candidates for national office lost during that election, including Clarence Darrow.) In 1900, Bryan—nicknamed "The Great Commoner" because of his faith in the goodness of common people—lost again to McKinley, and in 1908 he lost to William Taft. Throughout his life, Bryan's belief in the majority remained strong; as he often asked, "By what logic can the minority demand privileges that are denied to the majority?"

In 1898, Bryan served with a Nebraska regiment in the Spanish-American War. When Woodrow Wilson became president in 1912 (after a convention battle that blocked Bryan's fourth nomination), Wilson appointed Bryan as Secretary of State. Bryan had promised that "there will be no war while I am Secretary of State," and resigned on June 9, 1915, over "war

preparedness" that led the United States into World War I.

In 1901, Bryan founded *Commoner*, a weekly newspaper that he published for twelve years. At the height of its popularity, *Commoner* had more than 140,000 subscribers. Bryan continued to advocate social reforms such as prohibition, all of which were based on Bryan's deep religious faith. Bryan did not separate politics and religion, and his policies were often described as "applied Christianity." At Bryan's Sunday school classes, which attracted thousands of worshippers, Bryan often attacked evolution, claiming that "more of those who take evolution die spiritually than die physically from smallpox."

As a young man, Bryan had investigated evolution and decided "to have nothing to do with it." However, in 1916, Bryan was alarmed by *The Belief in God and Immortality*, in which Bryn Mawr psychology professor James Leuba showed that most scientists were nonbelievers and that college eroded students' religious faith. Bryan was especially troubled by human evolution, dismissing the rest because it "does not affect the philosophy upon which one's life is built." Interestingly, all laws banning the teaching of evolution banned only the teaching of *human* evolution.

By 1920, Bryan had labeled evolution "the most paralyzing influence with which civilization has had to contend during the last century." Bryan, who had little use for theistic evolution ("it deadens the pain while the Christian's religion is being removed"), began promoting the evils of evolution, and his pamphlet *The Menace of Darwinism* was distributed nationwide. Bryan claimed that evolution was merely "a guess," that "not one syllable in the Bible" supports evolution, and that "neither Darwin nor his supporters have been able to find a fact in the universe to support their hypothesis." Bryan also claimed that science must bow before religion; as he often noted, "If the Bible and the microscope do not agree, the microscope is wrong." Bryan toured the country proclaiming his message, often appearing with other antievolution crusaders such as Billy Sunday, Frank

HEAR
HONORABLE WILLIAM
JENNINGS BRYAN
At the State Fair
HIPPODROME
(TAKE COMO AVE. STREET CAR)
Sunday Oct. 22nd At 2:30 P. M.
Subject
"EVOLUTION"
A Menace to Christianity Education and Civilization
MASSED CHOIR
650 VOICES
Under Direction of F. V. Steel Will Sing
Doors Open At 2 P. M.   ADMISSION FREE
8000 Seats. 3000 Reserved for Students
AUSPICES OF NORTHWESTERN BIBLE AND MISSIONARY TRAINING SCHOOL, MINNEAPOLIS

12. In the early 1920s, Bryan's speeches condemning evolution attracted huge crowds. This speech in Minneapolis attracted more than 10,000 people. (*First Baptist Church, Minneapolis*)

Norris, and William Riley. In response, Reverend Harry Emerson Fosdick told reporters that "the real enemies of the Christian faith are not the evolutionary biologists, but folks like Mr. Bryan who insist on setting up artificial adhesions between Christianity and outgrown scientific opinions."

In 1924, Bryan—seeking to strengthen his scientific credentials—joined the American Association for the Advancement of Science, and the following year he delivered his famed "Is the Bible True?" speech in Nashville. Soon thereafter, John Butler introduced legislation to ban the teaching of human evolution in Tennessee's public schools. When the Butler Law was passed, Bryan telegrammed Governor Austin Peay that "The Christian parents of the State owe you a debt of gratitude for saving their children from the poisonous influence of an unproven hypothesis." Bryan's speeches drew huge crowds (Figure 12).

On May 12, 1925—just five days after John Scopes was "arrested" in Dayton for teaching human evolution—Riley asked Bryan to represent the World's Christian Fundamentals Association in Scopes' upcoming trial. When Bryan reached his next stop in Pittsburgh, Pennsylvania, Bryan wired Riley that he would "be pleased to act for your great religious organizations and without compensation." Sue Hicks, a local attorney in Dayton on the prosecution team, told Bryan, "we will consider it a great honor to have you with us in this prosecution." Although Bryan was not as theologically rigid as many fundamentalists (he did not have *The Fundamentals* in his home library), he looked forward to the trial, and thanked Riley for "the opportunity the Fundamentalists have given me to defend the faith."

Bryan asked several other fundamentalist leaders to come to Dayton to assist him. However, flood geologist George Price was in

London, and J. Gresham Machen politely declined Bryan's request, as did Sunday. John Stratton promised to attend, but never showed up. Frank Norris also promised to be there, but instead sent a stenographer, and William Riley went to Seattle to fight modernists at the Northern Baptist Convention. The absence of these fundamentalists may have explained Bryan's opposition to expert witnesses for the defense (since he had none to offer for the prosecution). Although Bryan's allies did not come to Dayton, Bryan's entry into the trial made it a world-class event.

In 1923, when editorials in the *Chicago Tribune* condemned Bryan's antievolution campaign, Bryan wrote a letter to the newspaper that prompted a response from famed defense lawyer, Clarence Darrow. Darrow's letter was published on the front-page of the July 4, 1923 issue of the newspaper, and included more than fifty questions to which Bryan never responded. Darrow, realizing that Dayton might be his chance to confront Bryan, volunteered his services for Scopes' defense. Throughout the pretrial publicity, Bryan stood firm about his beliefs: "We cannot afford to have a system of education that destroys the religious faith of our children." Bryan deplored Social Darwinism (be it by German militarists or American tycoons such as Andrew Carnegie), believing that it would lead to exploitation of workers, corruption of government, and moral collapse of the country. Bryan believed that natural selection was "the law of hate," and told Hicks that the Scopes Trial would "end all controversy."

Bryan spent the first four days of the trial listening and fanning himself with a large palm leaf fan, which he claimed was evidence of "the great eternal plan of adapting all nature to man's use." Bryan then attacked Darwin's theory and ridiculed biology textbook author George Hunter for grouping humans into a category with "3,499 other mammals—including elephants!" Bryan then argued that evolution threatened morality. When the defense called their expert witnesses, Bryan proclaimed that "the Bible is not going to be driven out of this court by experts." Defense attorney Dudley

Malone responded with his famed "We Are Not Afraid of the Truth" speech, which Bryan later said was "the greatest speech I ever heard."

Throughout the trial, Mary Bryan sent "bulletins" to her absent daughters. To Mary, Dayton's citizens were "mountain people" who were "pathetic," "do not shave every day," and who "marry and intermarry until the stock is very much weakened." Mary described John Scopes was "a long-jawed mountain product," Darrow as having "a weary, hopeless expression," and Hays as being "as self-asserting as the New York Jews can be."

On July 20, Bryan was confronted by Darrow in the most memorable event of the Scopes Trial. Darrow's relentless questioning wore down Bryan, and Bryan's admission that "days" of Genesis might have been "periods" caused many of Bryan's followers to question Bryan's convictions. The next day, Bryan's testimony was expunged from the record because, Judge John Raulston ruled, it could not help determine Scopes' guilt or innocence. Decades later, evangelist Jerry Falwell claimed that Bryan "lost the respect of fundamentalists when he subscribed to the idea of periods of time for creation rather than twenty-four-hour days."

After the trial, Bryan remained in Dayton, during which time he pledged $50,000 to create a college in Dayton that would promote fundamentalists' ideals, and scouted possible sites for the college. Bryan then fulfilled a promise to fellow prosecutor A.T. Stewart by speaking at the county fair in Winchester, Tennessee, the home of Stewart and Judge Raulston. After having lunch with Stewart and Raulston in Winchester, Bryan told reporters, "If I should die tomorrow, I believe that on the basis of the accomplishments of the last few weeks I could truthfully say, well done." He then returned to Chattanooga, where he was told by a physician to rest (Bryan had diabetes and a heart condition). Bryan had made arrangements with George Milton of the *Chattanooga News* to publish his much-anticipated, 15,000-word closing speech (when the defense waived its closing argument, the prosecution was also barred from offering a closing argument). Bryan's speech stressed majoritarianism and that evolution contradicts the Bible, destroys faith in God, and "is not truth." Bryan replaced some of the text with the last words he would ever write, "With hearts full of gratitude to God." Bryan also wrote a letter to Frank Norris, thanking him for getting Bryan involved with the case and noting that, "Well, we won the case. It woke up the country ... Sorry you were not there."

On Sunday morning, Bryan drove to Dayton. He appeared in good health and spirits, noting that he would "stay [involved in Scopes' appeal] and see it through." Bryan then made his last public appearance at morning worship services at First Southern Methodist Church. At lunch, he told his wife that a doctor had examined him the previous day and found him in excellent condition, noting repeatedly that "I never felt better in my life" and that there was "nothing to worry about." After making a few telephone calls and arranging a vacation in the Smoky Mountains for the following week, the 65-year-old Bryan went upstairs at 3:00 PM for a nap. He never woke up. Mary Bryan sent William "Jimmy" McCartney (the family chauffeur) to awaken her husband at 4:30 PM so Bryan could prepare for his sermon that evening at the First Southern Methodist Church. McCartney was the first to see Bryan's corpse.

Bryan died in his sleep in the home of Richard Rogers on South Market Street, just a few blocks from the Rhea County Courthouse. Walter F. Thomison was the attending physician who signed Bryan's death certificate. Prosecutors Sue Hicks, Herbert Hicks, Wallace Haggard, and Gordon McKenzie watched Bryan's body through the night, after which an honor guard of servicemen ringed the house. The Ku Klux Klan burned crosses in Bryan's memory, and Mary Bryan received thousands of messages of condolences, including those from President Calvin Coolidge, U.S. Supreme Court Chief Justice William Taft, and several presidents of foreign countries. Coolidge, who dismissed the Scopes Trial as a regional event, praised Bryan's many years of public service and ordered flags at all national buildings to be flown at half-mast. Famed evangelist Billy Sunday declared Bryan

to be "God's Napoleon ... one of the world's immortals ... who fell with his face to the enemy." Bryan's much-anticipated *Last Message*, which was originally titled *Fighting to Death for the Bible* and touted by Bryan as "the mountain peak of my life's efforts," was distributed on July 28 as his body lay in state in Dayton. The final words of Bryan's last speech came from one of his favorite hymns: "Faith of our fathers—holy faith, we will be true to thee till death."

Many people wrote songs immortalizing Bryan, and others made Bryan a martyr; Governor Austin Peay, "The Maker of Modern Tennessee" who had signed John Butler's legislation into law, proclaimed that Bryan had died "a martyr to the faith of our fathers" and announced a state holiday to commemorate Bryan's funeral. Fundamentalists compared Bryan to Jesus Christ, and Scopes' defenders to Pontius Pilate and other biblical villains. William Riley, who had been instrumental in getting Bryan into the Scopes trial, described Bryan as "the great outstanding man of our movement" and the Scopes trial as "Bryan's best and last battle." Others, however, continued to attack Bryan. For example, socialist Eugene Debs, one of Darrow's former clients, announced that "the cause of human progress sustains no loss in the death of Mr. Bryan." Darrow expressed sorrow about Bryan's death and commended Bryan's ability, courage, and strong convictions, and Ben McKenzie hailed Bryan as "the noblest hero of these times." Later, however, Darrow expressed pity that Bryan had been "obliged to show his gross ignorance" and that Bryan's vanity and failures had driven him to "a state of hallucination that would impel [Bryan] to commit any cruelty that he believed would help his cause." Reporter H.L. Mencken began Bryan's obituary in the *Baltimore Sun* by asking, "Has it been duly marked by historians that William Jennings Bryan's last secular act on this globe of sin was to catch flies?"

Bryan's memorial service in Dayton was held on the lawn of the Rogers' home where he died, and was officiated by Reverend C.R. Jones, the pastor of Dayton's First Southern Methodist Church. Dayton's mayor proclaimed a day or mourning, and Dayton's flags were flown at half-mast. Crowds of mourners lined railroad tracks to see the special Pullman car that carried Bryan's body to Virginia for burial. The train's conductor had taken Bryan along a similar route during Bryan's presidential campaign in 1896.

Bryan's funeral in Washington, DC, on July 31, was held at New York Avenue Presbyterian Church, "The Church of the Presidents," and was broadcast nationwide on radio. Bryan was buried atop a tree-covered hill in the south end of Arlington National Cemetery, where he rests with his wife Mary (1861–1930) beneath the tiny inscription, "He Kept the Faith."

Knowing that "my family history does not promise a long life," Bryan had prepared his will on July 5, 1925, just before coming to Dayton. The will, which opened with "In the name of God, farewell," was settled in early August 1925. Bryan's estate was valued at $860,500. Bryan's wife received their home in Florida (Marymount in Cocoanut Grove) and one-third of the remaining estate. Each of Bryan's children received $100,000. Bryan noted that he had "hoped to aid in the establishment of an academy that would embody my idea [of a Christian school]" at which "boy students [would] wear a uniform of blue and gray, to symbolize the union of north and south." Bryan left all publication rights of his writings to his widow and children, who were authorized to publish Bryan's official biography. Bryan insisted that the biography describe his wife's life, "setting forth particularly the aid which Mrs. Bryan has rendered to her husband in all his work during his life." Bryan's daughter, Ruth Bryan Owen (1885–1954), later became the first woman from Florida to serve in the U.S. House of Representatives.

Soon after Bryan's death, F.E. Robinson, Judge John Raulston, A.T. Stewart, Walter White, Ben McKenzie, and others formed a memorial association to help create the college that Bryan proposed in the final days of his life. Although almost half of the $1,000,000 gathered for William Jennings Bryan University was erased by the Great Depression, ground was broken on November 5, 1926, by Governor Peay for

the conservative Christian school (whose name was changed to William Jennings Bryan College in 1958, and then shortened to Bryan College in 1993). The ceremony was attended by more than 10,000 onlookers, and Bryan College opened in 1930 in the old high school building where John Scopes allegedly taught evolution. The first class included thirty-one students, and by 1939–1940, the college enrolled ninety students. Today, Bryan College is a coeducational liberal arts college located atop a hill overlooking Dayton. The college enrolls about 600 students and is based on "Christ Above All" and the "unequivocal acceptance of the inerrancy and authority of the Scriptures." Scopes Trial instigator George Rappleyea wanted to create a liberal college to offset Bryan College, but that college was never built.

After Bryan's death, several people (including New York fundamentalist John Straton) claimed Bryan's place in the antievolution crusade, but none ever matched Bryan's stature or credibility. Although religious conservatives such as George Price, Henry Morris, and Jerry Falwell have criticized Bryan's performance at Dayton, most of Bryan's life served the public. Indeed, few statesmen have been more vindicated by history than Bryan, for many of Bryan's political causes were subsequently enacted into law, including the sixteenth (graduated income tax), seventeenth (direct election of senators), eighteenth (prohibition of liquor), and nineteenth (woman suffrage) Amendments to the Constitution, various labor laws (e.g., eight-hour workday, minimum wage), the Federal Reserve Act, and tariff reform.

In 1986, Bryan was commemorated on a $2 U.S. postage stamp, and today he is memorialized with a variety of museums (e.g., the William Jennings Bryan Birthplace Museum in Salem, Illinois; Fairview, the Bryan Museum in Lincoln, Nebraska), statues (e.g., including one on the Rhea County Courthouse lawn in Dayton, and another that was designed by famed sculptor Gutzon Borglum, who sculpted the presidents on Mount Rushmore), libraries, schools, and parks throughout the United States.

Borglum's 12′-high statue of Bryan was dedicated in 1934 by President Theodore Roosevelt in Washington, DC, and today stands in Salem, Illinois.

For More Information: Kazin, M. 2007. *A Godly Hero: The Life of William Jennings Bryan*. New York: Anchor Books; Olson, L.R. 1995. *Legacy of Faith: The Story of Bryan College*. Dayton, TN: Bryan College.

WILLIAM JENNINGS BRYAN, JR. (1889–1978)

To permit an expert to testify [about evolution] would be to substitute trial by experts for trial by jury, and to announce to the world your honor's belief that this jury is too stupid to determine a simple question of fact.

William Jennings Bryan, Jr. was born on June 24, 1889, in Lincoln, Nebraska, and was the only son of William Jennings Bryan. Bryan attended the University of Nebraska, after which he graduated from the University of Arizona in 1912. After studying law at Georgetown University, Bryan served as a Regent of the University of Arizona and as an assistant U.S. Attorney for Arizona from 1915 to 1920. In 1921, Bryan moved to Los Angeles and practiced law.

During the Scopes Trial, the younger Bryan assisted the prosecution. Bryan, who on several days went swimming with John Scopes to combat Dayton's sweltering heat, had a minor role in the trial, and on the last weekend of the trial he returned to his law practice in California. Bryan was not in Dayton to hear Darrow question his famous father.

After the trial, the soft-spoken Bryan was involved with the Anti-Evolution League of America for a few years, and in 1939 he served as federal commissioner for the San Francisco Exposition. In 1938, President Roosevelt appointed him Collector of Customs for the Port of Los Angeles. After serving four terms, Bryan retired in 1953.

In 1975, Bryan—who John Scopes described as "a pleasant fellow"—moved to Santa Fe, New Mexico, to be near his two daughters. He died there on March 27, 1978.

BRYAN BIBLE LEAGUE (Est. 1925)

Bryan Bible League was organized to perpetuate William Jennings Bryan's religious views. The league, which was endorsed by Bryan's widow, was one of the few antievolution organizations to form on the West Coast. Its founder, Paul Rood, claimed that God told him to form the organization honoring Bryan.

PATRICK BUCHANAN (b. 1938)

Patrick Joseph Buchanan was a famous political commentator and candidate for president of the United States in 1996, who reignited the evolution-creationism debate by claiming that "I think [parents] have a right to insist that Godless evolution not be taught to their children or their children not be indoctrinated in it."

CARRIE BUCK (1906–1983)

Carrie Buck was a 17-year-old woman from Charlottesville, Virginia, who was forcibly sterilized after being declared "feebleminded" by the state of Virginia (Figure 13). State officials had claimed that Buck was a mentally retarded "incorrigible" genetic threat to society. Buck resisted the sterilization, and sued to stop the process. During her trial, Buck's attorney called no witnesses, and Buck was denounced as "shiftless, ignorant and worthless." Eugenics advocate Harry Laughlin declared Buck to be morally "delinquent," despite the fact that he had never met her.

Buck's lawsuit produced *Buck v. Bell*, in which the U.S. Supreme Court on May 2, 1927, ruled (by a vote of 8-1) that Virginia could forcibly sterilize Buck. Justice Oliver Wendell Holmes Jr. (1841–1935) wrote the decision upholding the statute instituting compulsory sterilization "for the protection and health of the state," noting that "it is better for all the world if, instead of waiting to execute degenerate offspring for crime, or to let them starve for their imbecility, society can prevent those who are manifestly

13. Carrie Buck, shown here in 1924 with her mother Emma at the Virginia Colony for Epileptics and Feebleminded, was forcibly sterilized by the state of Virginia. (*Arthur Estabrook Papers, M.E. Grenander Department of Special Collections and Archives, University of Albany, SUNY*)

unfit from continuing their kind ... Three generations of imbeciles are enough."

On October 19, 1927, Buck was sterilized in Virginia's State Colony for Epileptics and Feebleminded, which was headed by Dr. James H. Bell. Buck was later paroled from the colony and worked in Bland, Virginia. Buck, an active reader, later married William Eagle. Buck's daughter, Vivian, who was also sterilized because she was "below average" and "not quite normal," earned high grades in school, but died when she was 8 years old.

Indiana had enacted the first law allowing forced sterilization in 1907, but forced sterilization did not gain widespread popular approval until the 1920s. The leading advocate of forced sterilization was Laughlin, who was superintendent of Charles Davenport's Eugenics Record Office in Cold Spring Harbor, New York. Laughlin's "Model Eugenical Sterilization Law" encompassed the "feebleminded, insane, criminalistic, epileptic, inebriate, diseased, blind, deaf; deformed; and dependent" and included "orphans, ne'er-do-wells, tramps, the homeless, and paupers." Virginia's law, which had been passed

in 1924, was based on Laughlin's model. By 1941, more than 35,000 people had been forcibly sterilized.

The last forced sterilization occurred in the early 1970s, and Virginia repealed its law in 1974. Although in 1942 the Supreme Court struck down a law allowing the forced sterilization of criminals, it never reversed the general concept of eugenic sterilization established by *Buck v. Bell*. In 1933, Germany used Laughlin's model to rationalize its more than 350,000 sterilizations, and Laughlin was awarded an honorary degree in 1936 from the University of Heidelberg for his work in "the science of racial cleansing."

Carrie Buck died in 1983, alone in a nursing home. She is buried in Charlottesville near Vivian, her only daughter. In 1994, Buck's story was made into a movie, *Against Her Will: The Carrie Buck Story*. On May 2, 2002, Virginia unveiled a historic marker commemorating *Buck v. Bell* in Charlottesville, at which time Governor Mark Warner apologized for Virginia's participation in eugenics.

WILLIAM BUCKLAND (1784–1856)

Organic beings must have had a beginning, and where is that beginning to be found, but in the will and fiat of an intelligent and all-wise Creator?

William Buckland was born on March 12, 1784, in Devon, England. He attended schools in Tiverton and Winchester, and earned a scholarship in 1801 to study for the ministry at Oxford University. After earning his BA in 1804 and his MA in 1808, Buckland became an ordained Anglican priest.

After becoming the first geology professor at Oxford University, Buckland often brought his pet bear and hyena to college functions (Charles Darwin described Buckland as "a vulgar man ... craving for notoriety"). Buckland—who referred to geology as "undergroundology"— corresponded with French anatomist Georges Cuvier about fossils, and in 1818 Cuvier visited Buckland at Oxford. During that visit, Cuvier saw the giant bones that Buckland had collected that would later be recognized as *Megalosaurus*.

Buckland collected fossils in a large blue bag, which he carried with him almost everywhere. Several portraits of Buckland show him holding that bag.

In 1818, Buckland was elected a Fellow of the Royal Society, and two years later he published *Vindiciae Geologiae; or the Connexion of Geology with Religion Explained*, which used gap creationism to reconcile geology with Noah's flood and biblical creation. Buckland claimed that the word *beginning* in Genesis referred to an unspecified period between the formation of the Earth and the creation of today's organisms, during which there were many extinctions and successive creations. Buckland's so-called "diluvium theory," which was based on Cuvier's "successive catastrophies theory," was embraced by theologians searching for evidence of a worldwide flood. Although Buckland's *Reliquiae Diluvianae* (1823) argued that a flood had covered the entire Earth, Buckland later concluded that it is "impossible to ascribe the formation of these strata to...the single year occupied by the Mosaic deluge...The strata...must be referred...to periods of much greater antiquity." Buckland later used the sixth *Bridgewater Treatises* to advocate a flood that was short, gradual, and left few visible effects. This so-called "tranquil flood theory" had been first suggested by Carolus Linnaeus, and introduced to the public in 1826 by Scottish minister John Fleming. Buckland later distanced himself from flood-based geology.

In 1824, Buckland—then president of the Geological Society of London and England's foremost geologist—became the first person to publish a description of a dinosaur ("Notice of the Megalosaurus or great Fossil Lizard of Stonesfield," published in *Transactions of the Geological Society of London*). Buckland named the fossil *Megalosaurus*, or Great Lizard, a large meat-eating theropod of the Middle Jurassic. (Robert Plot, a chemistry professor at Oxford University, had described a similar bone in 1677, but did not recognize it as belonging to a dinosaur.) In 1826, Cuvier gave the dinosaur a complete binomial: *Megalosaurus bucklandi*, or "Buckland's Big Lizard" (Richard Owen did not coin the term *dinosaur* until 1842 in an article published in the

Report of the British Association for the Advancement of Science). Replicas of *Megalosaurus* footprints adorn the lawn of the Oxford University Museum of Natural History.

In 1832, Buckland presided over the second meeting of the British Association for the Advancement of Science (held at Oxford), and in 1836 he published the two-volume *Geology and Mineralogy, Considered with Reference to Natural Theology*, which was part of *Bridgewater Treatises on the Power, Wisdom and Goodness of God, as Manifested in the Creation*. In this book, Buckland described his ideas about day-age and theistic evolution. *Bridgewater Treatises*, of which Buckland's was the only one about geology, were funded by the Earl of Bridgewater who, when he died in 1828, left £8,000 to further the work of Anglican archdeacon William Paley in natural theology. Unlike many scriptural geologists of his day, Buckland—a Gap creationist who rejected evolution—was not a biblical literalist, nor did he assume that Earth exists only for humans.

In the 1840s, Buckland—an Anglican clergyman—was appointed Dean of Westminster; while serving in that position, Buckland helped maintain Westminster Abbey. He was later a trustee in The British Museum, won the Copley Medal in 1822, and was awarded the Wollaston Medal in 1848 by the Geological Society of London.

Buckland died on August 24, 1856, and is buried in a spot he selected in Islip churchyard. Buckland's *Megalosaurus* remains on display at The British Museum.

GEORGES-LOUIS LECLERC BUFFON (COMTE DE BUFFON) (1707–1788)

We will no longer consider a species as a collection or a continuation of similar individuals, but as one living whole, independent of number, independent of time.

Georges-Louis Leclerc was born on September 7, 1707, in Montbard, in the Burgundy region of France. Leclerc's family moved to Dijon when Georges-Louis was young, and it was here that he began his formal education, at the

College des Godrans. In 1728, Georges-Louis moved to Angers to study mathematics, but was forced to depart Angers after dueling a fellow student.

As was customary for a young man from a wealthy family, Georges-Louis toured Europe to interact with the great intellectuals of the day, especially scientists and mathematicians. While he was traveling, his mother died. Leclerc's father quickly remarried and tried to divert Georges-Louis' inheritance of his mother's estate. Georges-Louis eventually regained the property, and by 1740 was a wealthy young man, as was then often necessary to be a scientist. Upon receipt of his mother's estate, Georges-Louis adopted the additional last name of "Buffon," and in 1773 the government officially conferred upon him the title of Comte. He married Françoise de Saint-Belin-Malain in 1752, with whom he had a son in 1764. Buffon later hired Jean-Baptiste Lamarck to tutor his son.

Buffon was eager to join the Académie Royale des Sciences, and to gain entrance he published an analysis of a gambling game which involved participants betting on the probability of a playing piece landing within or across floor tiles. The problem and his analysis became known as "Buffon's needle," a classic case still discussed in the discipline of mathematics known as geometrical probability. Buffon was admitted to the Académie in 1734. He was also elected to the Royal Society of London, and became director of the Jardin du Roi in 1739.

While at the Jadin du Roi, Buffon became increasingly interested in natural history. In the 1740s, he began examining the Earth's history, from its formation through the rise of humans. Buffon's *l'Histoire naturelle, générale et particulière avec la description du cabinet du Roi*—one of the most widely read scientific books of the eighteenth century—eventually included forty-three volumes, thirty-five written by Buffon and eight by a Jardin du Roi colleague. Buffon, who rejected the proposition that God intervened in the functioning of nature, marginalized divine action without being openly atheistic. This was a bold position (and one that would require the issuance of several apologies to French theologians), but established a philosophical perspective of science based on explaining nature strictly through observable events.

In the 1760s, Buffon investigated a proposal by Isaac Newton that the primordial Earth was originally a "red hot" body that had subsequently cooled. Buffon, who owned an iron factory, heated iron spheres in a forge and recorded the time it took for these spheres to cool to ambient temperature. Extrapolating these data to a body the size of the Earth, he initially calculated the Earth to be approximately 75,000 years old. Buffon later extended the age to 7 million years to account for the fossil record, which to Buffon was too extensive to have been produced in less than 100,000 years. He reported these findings in *Epoques de la nature* in 1788, and described seven *epoques* (curiously the same number of days as found in Genesis) through which the Earth had progressed. Each *epoque* was a "milestone on the eternal road of time."

Buffon ascribed the origin of life to spontaneous generation (at the North Pole), rather than divine creation. He viewed species as real, fixed entities ("Species are the only beings in Nature; they are perpetual beings, as old, as permanent as Nature itself"). Buffon, who suggested that closely related species might share a common ancestor, accepted change within different populations of a species, as was apparent from widely separated geographic areas having similar yet divergent animals (i.e., the variety of cat species scattered across different continents). But to Buffon, this change from the original form represented degeneration: less than ideal environmental conditions caused populations to lose their original identity and to become weakened as their *moules intérieurs* was affected. Hence, such change did not create new species, and Buffon's conception of organic change fit nicely with the Great Chain of Being that disallowed the addition of intermediate forms to an already perfect progression. But Buffon refused to accept that God originated this process because a deity would not concern himself with, for example, "the particular fold in a beetle's wing." Buffon eventually concluded that there were thirty-eight basic types ("families") in nature that have been modified by local environmental conditions, and that the original form can be made to reappear by placing "degenerate" types back into the ideal environment. This was true for humans, who were treated as just another type; Buffon claimed that if Africans were removed from their extreme (hot) environment, they would lose their "degenerate" characteristics and would, presumably, become more like Europeans.

Buffon created a minor international incident when he remarked that the poor environment of North America had caused that continent's organisms (including humans) to be smaller (and inferior) than those found in the Old World. This proposition so infuriated Thomas Jefferson (then serving in France) that he arranged for the skeleton of a huge moose to be delivered to Buffon. Unfortunately, the moose arrived in such bad shape (and was clearly cobbled together from more than one individual) that Buffon was unconvinced. Buffon also suggested that because higher temperatures caused organisms to grow larger (i.e., larger mammals tend to be found in equatorial regions), early life forms that lived on a still-cooling Earth must have been gigantic.

For Buffon, the concept of "species" was the underlying ideal form, which may or may not be expressed in any particular individual. Consequently, individuals are unimportant, as is variation within a population: "an individual, no matter the species to which it belongs, is nothing in the Universe." However, Buffon—who speculated that humans are related to apes—recognized that individuals can be grouped according to type and he proposed that the ability to interbreed is the primary feature for such classification. Such a proposal anticipated the biological species concept developed by Ernst Mayr early in the twentieth century.

Buffon's proposed classification system conflicted with the system developed by Linnaeus, a contemporary of Buffon's. The Linnaean system was based on morphology (e.g., in plants, on reproductive structures), which Linnaeus acknowledged created some artificial groupings. Buffon lobbied against the already popular Linnaean system by demonstrating that the entire

system, especially taxonomic units above the species level, was arbitrary and therefore useless. ("This manner of thinking has made us imagine an infinity of false relationships between natural beings.") Buffon—a theist but not a practicing Christian—further argued that the only natural basis on which to order life is in relation to humans: "One judges the objects of Natural History in terms of the relations they have with him. Those which are most necessary and useful to him will take the first rank." Linnaeus ignored Buffon's comments, perhaps contenting himself with the knowledge that he had already classified a foul-smelling plant into the genus *Buffonia*.

Buffon—the first scientist to advocate day-age creationism—died in his Paris apartment on April 16, 1788. His funeral in Paris attracted thousands of spectators, and his body was interred in the family cemetery in Montbard. (Supposedly, his heart was removed and buried separately, in Paris.) Buffon's reliance on a scientific approach that invoked only natural causes was revolutionary and influenced later generations of scientists, including Charles Darwin, who commented on Buffon's contributions in *On the Origin of Species*. Buffon is commemorated by *rue Buffon*, so-named in 1790 to recognize Buffon's accomplishments and his purchase of the house at Number Eight for the director of the French Museum of Natural History.

For More Information: Roger, J., & S.L. Bonnefoi, 1997. *Buffon: A Life in Natural History*. Ithaca, NY: Cornell University.

LUTHER BURBANK (1849–1926)

Luther Burbank was a famous plant-breeder and an ardent supporter of Darwin's theory of evolution by natural selection. Burbank introduced more than 800 new varieties of plants (McDonald's French fries are "Russet Burbank" potatoes). Burbank, who was well known to Tennessee farmers at the time of the Scopes Trial, was an agnostic who claimed that the Scopes Trial was "a great joke, but one which will educate the public and thus reduce the number of bigots." Two weeks before the Scopes Trial,

Clarence Darrow published a letter from Burbank about science and academic freedom.

After telling his young wife that "I feel better now," Burbank died at the age of 77, and is buried in the lawn of his Santa Rosa, California, home (now the Luther Burbank Home and Gardens), which is a registered National, State, and City Historic Landmark. In California, Burbank's birthday (March 7) is celebrated as Arbor Day, and trees are planted in his memory.

CLIFFORD BURDICK (1894–1992)

Clifford L. Burdick was an energetic researcher in the Creation Research Society whose controversial claims continue to be promoted by many creationists. In the 1950s, Burdick qualified to take comprehensive exams for a doctorate in geology at the University of Arizona (despite the fact that he had never been formally admitted to the program), but during the exam Burdick claimed that he "browned out." The committee's refusal to grant Burdick another chance to pass the exam was subsequently used by creationists as evidence of academic discrimination against their ideas.

Burdick's claim that he had discovered human and dinosaur tracks alongside each other in the Paluxy Riverbed—the famed "Burdick Track" (Figure 14)—was included in *The Genesis Flood*, but was later discredited by scientists and creationists alike. In 1966, Burdick described his alleged discovery of pollen from conifers in Precambrian rocks as "science-shaking original-pioneer work," but this work was also subsequently discredited. Burdick claimed a variety of false education degrees (e.g., an MS from the University of Wisconsin). As Burdick's controversial claims mounted, he was marginalized by creationists; for example, Walter Lammerts criticized Burdick as someone who was "weak," "slow," and who "has not kept up with his reading."

BURGESS SHALE OUTCROP

The Burgess Shale Outcrop in the Canadian Rockies of British Columbia (near the Burgess

14. Clifford Burdick's claim that he discovered human tracks along dinosaur tracks in the Paluxy riverbed became a common argument used by creationists to "disprove" evolution. Shown here is the 14"-long "Burdick Track," which was featured in *The Genesis Flood*, the book that founded the modern "creation science" movement. Burdick's claims were subsequently discredited by scientists and creationists alike. (*Creation Evidence Museum*)

Pass) was the site of the discovery of thousands of fossils that greatly expanded the understanding of life's history. The exquisitely preserved fossils, which were discovered in September 1909 by geologist Charles Doolittle Walcott, are about 505 million years old, and include thousands of bizarre, soft-bodied organisms that were trapped in mud deposited by underwater landslides. Walcott, with the help of his wife and children, collected 65,000 specimens. In 1910, Walcott helped open the Smithsonian's new building housing the National Museum of Natural History. Walcott coined the term *geochrone*, from which the term *geochronology* evolved.

The 170 species of fossils in the shale were unlike anything previously known on Earth. One of the organisms—this one having seven tentacles on its back and seven pairs of stilt-like legs—was named *Hallucigenia* by its discoverer, who could not believe the bizarreness of the organism he was examining. These species evolved from unknown ancestors during the Precambrian, and all later became extinct. Fossils of the Burgess Shale, along with similar discoveries in China and Greenland, have given biologists insights into the earliest inhabitants of Cambrian oceans. Today the Burgess Shale locality is a World Heritage Site.

After their discovery by Walcott, the Burgess Shale specimens were stored in the National Museum of Natural History at the Smithsonian Institution. They were later popularized by American biologist Stephen Gould's best-selling book, *Wonderful Life: The Burgess Shale and the Nature of History*. Gould's book, which attacked the notion of preordained "progress" in evolution, was named after Frank Capra's classic 1946 movie, *It's a Wonderful Life*, in which Jimmy Stewart's character sees what the world would have been like if he had never been born.

For More Information: Gould, S.J. 1989. *Wonderful Life: The Burgess Shale and the Nature of History*. New York: W.W. Norton; Whittington, H. 1985. *The Burgess Shale*. New Haven, CT: Yale University.

THOMAS BURNET (1636–1715)

Thomas Burnet was an Anglican clergyman whose popular series of four books titled, *Sacred Theory of the Earth: Containing an Account of Its Original Creation, and of All the General Changes which It Hath Undergone, Or Is to Undergo, until the Consummation of All Things* (1680–1690) founded scriptural geology by reconciling geology with the Bible. Burnet, who argued that all events in the Earth's history could be explained by natural processes, advocated natural theology—namely, that God's plan for Earth and humans is understandable by studying nature. Burnet often argued that no "truth concerning the Natural World can be an enemy to religion." Despite his emphasis on natural explanations, Burnet relied primarily on reasoning rather than observation, and Noah's Flood was the foundation of his ideas. Since oceans did not contain enough water to cover the mountains, Burnet concluded that there must have been underground oceans that flooded the Earth when Earth's original crust cracked open (this was consistent with the Bible's claim in Genesis 8:2 that "fountains of the great deep" erupted). Unlike Isaac Newton, who was Burnet's contemporary at Cambridge, Burnet refused to invoke miracles to explain nature. According to Burnet, who reworked French Catholic natural philosopher René Descartes' (1596–1650) speculation about Earth's age to fit the Bible, the antediluvian Earth was a smooth ovoid, but its drying surface cracked and released the waters of Noah's flood. Although *Sacred Theory of the Earth* was criticized for describing Noah's flood as a natural event rather than a punishment for sin, Burnet's type of scriptural geology remained popular in England until the nineteenth century. In the late 1700s, James Hutton read Burnet's work, but noted that "it surely cannot be considered in any other light than as a dream, formed upon a poetic fiction of a golden age." Burnet later became the private chaplain of King William III.

GEORGE BUSH (b. 1946)

George Walker Bush was the forty-third president of the United States who, during a campaign in 2000, claimed that "on the issue of evolution, the verdict is still out on how God created the Earth." During his reelection campaign four years later, Bush's advocacy of intelligent design was supported by Secretary of Education Margaret Spelling (b. 1957), an architect of the "No Child Left Behind" legislation.

GEORGE BUSK (1807–1886)

George Busk was a surgeon and friend of Thomas Huxley who, in November 1864, with paleontologist Hugh Falconer, nominated Charles Darwin for the Copley Medal of the

Royal Society. Although Lyell and Hooker wrote the introduction and prepared the text of the famous Darwin–Wallace paper, it was Busk who read the paper to the Linnean Society of London on the evening of July 1, 1858. At that time, Busk—a charter member of the X-Club—was the Hunterian Professor of the Royal College of Surgeons, a position held at other times by Richard Owen and Huxley. In 1863, Busk diagnosed Darwin's sickness as "waterbrash," a "diseased secretion of the stomach." Busk and Huxley were friends, and in 1853 they coauthored *Manual of Human Histology*. Busk, a recipient of the Geological Society's Wollaston and Lyell Medals, is buried in London's Kensal Green Cemetery.

JOHN W. BUTLER (1875–1952)

The evolutionist...cannot be a Christian... evolution [is] the greatest menace to civilization in the world today. The Dayton trial is the beginning of a great battle between infidelity and Christianity.

John Washington Butler was born in 1875 in Macon County, Tennessee. He grew up on his family's farm, which he began operating in his early twenties. Butler taught school half of the year and spent the remainder of the year working at his farm. In 1921, after hearing a preacher describe a local woman who became an atheist after being taught evolution, the prosperous Butler began to worry that the teaching of evolution would corrupt his five children. The following year, he ran for state legislator, and his campaign pamphlets urged voters to ban the teaching of evolution in its public schools. He won the election, but didn't introduce any evolution-related legislation during his first term. However, when he ran for reelection, he told voters that the teaching of evolution "robs the Christian of his hope and undermines the foundation of our government." Late in 1924, Butler promised his constituents that he would introduce antievolution legislation if reelected. Butler was reelected and placed on a committee that oversaw public schools.

On the morning of his forty-ninth birthday, the amicable Butler—who claimed to have read Darwin's *On the Origin of Species*—drafted what came to be known as the Butler Law (House Bill No. 185). This law made the teaching of evolution (i.e., ANY THEORY THAT DENIES THE STORY OF THE DIVINE CREATION OF MAN AS TAUGHT IN THE BIBLE, AND TO TEACH INSTEAD THAT MAN HAS DESCENDED FROM A LOWER ORDER OF ANIMALS) in any of Tennessee's public schools a misdemeanor punishable by a fine of $100–500. Although editors of the *Chattanooga Times* urged that Butler's legislation be ignored, the proposed ban on teaching human evolution was otherwise unopposed. On March 21, 1925, Butler's legislation was signed into law by Tennessee Governor Austin Peay, thereby making Tennessee the first state to ban the teaching of human evolution.

During Scopes' Trial, Butler—who believed that his newly enacted law would "protect our children from infidelity"—was a guest-reporter for a press syndicate, and filed several reports of the trial's activities. After the trial, Butler—who disagreed with Judge Raulston's decision to exclude testimony of Darrow's expert witnesses—finished his term as state representative, and in 1927 returned to life on his farm. Butler died on September 24, 1952, and was buried in Butler Cemetery in Lafayette, Tennessee.

C

LARRY CALDWELL v. ROSEVILLE JOINT UNION HIGH SCHOOL DISTRICT

Larry Caldwell was a lawyer and parent who in 2003 proposed that Roseville (a suburb of Sacramento, California) High School adopt his Quality Education Policy and related instructional materials that presented the scientific weaknesses of evolution. The following year, Caldwell "expressed praise to God" when the district agreed to discuss his policy, and asked his supporters to pray for his policy's success. When the school district rejected Caldwell's proposal (by a 3-2 vote), Caldwell charged the district with misconduct and religious discrimination and filed a lawsuit which became known as *Larry Caldwell v. Roseville Joint Union High School District*. Caldwell's lawsuit was publicized by The Discovery Institute. On September 7, 2007, Judge Frank C. Damrell, Jr., ruled against Caldwell, noting that Caldwell had "failed to proffer evidence sufficient to demonstrate a triable issue of fact as to any of his constitutional claims based upon this alleged discrimination."

CALIFORNIA STATE BOARD OF EDUCATION

The California State Board of Education mandated in 1969 in its *Science Framework for California Public Schools* the equal scientific standing of evolution and "creation theory." Six years earlier, the California Superintendent of Public Instruction had ordered that textbooks identify evolution as "a theory."

CAMBRIDGE PHILOSOPHICAL SOCIETY

The Cambridge Philosophical Society was founded in 1819 by, among others, John Henslow and Adam Sedgwick. The Society published Charles Darwin's *Letters on Geology* in 1835, and reprinted them in 1960. In 1879, the Society paid Sir William Richmond (1842–1921) £400 to paint a portrait of Charles Darwin. Although Darwin considered the portrait to be "horrid," it still hangs in the Society's library. Darwin was never a member of the Society.

CARDIFF GIANT (b. 1868)

Cardiff Giant was a 10'-high, 2,990-pound statue of a human conceived of, and purchased for $2,600 by, cigar manufacturer George Hull in 1868. Hull buried the statue behind the barn of his cousin William "Stub" Newell in Cardiff, New York, and on October 16, 1869, Newell "discovered" the giant while digging a well there. Some people speculated that the statue was the petrified remains of a giant, extinct species of humans that once roamed upper New York, and biblical literalists claimed that it confirmed the claim in Genesis 6:4 that "there were giants in the Earth in those days."

A Syracuse newspaper hailed the giant as "taller than Goliath whom David slew," and Hull and Newell set up a small museum that charged visitors 50 cents to see the giant. Although paleontologists such as Othniel Marsh denounced the giant as "remarkable—a remarkable fake," thousands of people came to see the giant. A group of businessmen paid Hull $37,500 to move the statue to Syracuse, where it was displayed even more prominently. Special trains brought visitors to see the giant, and showman P.T. Barnum offered $60,000 to lease the giant for ninety days. When Hull refused Barnum's offer, Barnum created a giant of his own that he displayed in Brooklyn. Hull then sued Barnum, and reporters began to investigate Hull's giant. In Iowa, they discovered Hull's purchase of a 5-ton piece of gypsum, and in Chicago they found the giant's sculptor, Edward Salle. Hull confessed that his giant was fake, which cleared Barnum of forgery (i.e., he could not be prosecuted for showing a fake of a fake). Hull and Newell made more than $30,000 from their fraud, and Barnum more than $150,000. It was not clear whether Hull intended to use the giant to cheat people out of money or, as he later claimed, to expose theologians who insisted on the literal truth of the Bible.

The giant eventually ended up in Des Moines, Iowa, after which it was purchased by the New York Historical Association for $30,000. Today, the Cardiff Giant is displayed as "America's Greatest Hoax" at Farmer's Museum in Cooperstown, New York, not far from the Baseball Hall of Fame. Barnum's replica of Hull's statue is displayed—with hundreds of curious coin-operated machines—at Marvin's Marvelous Mechanical Museum just outside of Detroit.

ANDREW CARNEGIE (1835–1919)

I remember that light came as in a flood and all was clear. Not only had I got rid of theology and the supernatural, but I had found the truth of evolution.

Andrew Carnegie was born November 25, 1835, in Dunfremline, Scotland. His father, William, was a weaver who used handlooms to make cloth. The family's poverty was exacerbated as factory-based mechanical looms came into widespread use in the textile industry during the first half of the nineteenth century. In 1848, Carnegie's family immigrated to Allegany, Pennsylvania, near where Carnegie's mother had relatives.

Family poverty forced Carnegie to seek employment at the age of 13, and by the age of 20 he was a clerk and telegraph operator for Thomas Scott, assistant superintendent of the Pennsylvania Railroad Company. When Scott was promoted, he handpicked Carnegie to succeed him, during which time Carnegie distinguished himself as an effective manager. By 1863, Carnegie's shrewd investments were so successful that he quit his job to concentrate on managing his portfolio. Carnegie's marriage in 1887 to Louise Whitfield produced one child, Margaret.

During the American Civil War, Carnegie's financial interests spread to include not only railroads, but also holdings in oil and construction companies. After the war, Carnegie became convinced of the potential of the iron and steel industry in a rapidly growing America. By 1872, Carnegie had invested in the Thomson Steel Company near Pittsburgh that used Sir Henry Bessemer's process for manufacturing steel. In 1884, Carnegie partnered with Henry Frick, who had recently become sole owner of vast coal reserves, to form the Carnegie Steel Corporation. Frick served as chairman, while Carnegie, who would hold no official title, maintained a majority stake in the company. Carnegie and (especially) Frick's antiunion position led to the infamous strike at their factory in Homestead, Pennsylvania, in 1892.

Like many titans of business during the Gilded Age, Carnegie's views on the role of the wealthy industrialist in society were influenced by philosopher Herbert Spencer, the father of the political theory of Social Darwinism. Carnegie believed that capitalism was "not only benefi-cial, but essential for the future progress of the race" and that "the best means of benefiting the community is to place within its reach the lad-ders upon which the aspiring can rise." In turn, the wealthy should not be faulted for the tactics they use to produce this societal good.

However, Carnegie rejected Spencer's phi-losophy as to how to treat the poor. Instead of ignoring them as Spencer proposed, Carnegie advocated philanthropy. He believed his own rags-to-riches story demonstrated that the "epoch-makers" could arise among the wealthy and poor alike, and that philanthropists should "help those who will help themselves." In his essay "Wealth," published in the *North American Review* in 1889, Carnegie chastised a self-avowed follower of Spencer who admitted to giving money to a beggar, stating that "he only gratified his own feelings, saved himself from annoyance—and this was probably one of the most selfish and very worst actions of his life."

Carnegie believed that "the amassing of wealth is one of the worst species of idolatry." His "scientific philanthropy" listed the "wisest" targets of charity, with universities and libraries at the top of the list and churches at the bottom (even "swimming baths" were ranked higher than churches). Popular recipients of Carnegie's money were local libraries, and over the course of his life he established more than 2,800 public libraries in the United States and Britain. He also established the Carnegie Institution of Washing-ton, a research institution to aid the nation's uni-versities.

In 1895, in his adopted home of Pitts-burgh, Carnegie established the Carnegie Insti-tute. Carnegie was keen on procuring dinosaur fossils for the new museum, but was unable to obtain specimens from existing collections. In 1898, when newspapers reported the unearthing in Wyoming of a complete skeleton of a giant dinosaur, Carnegie sent recently hired Institute director William Jacob Holland to buy the fos-sil. Holland soon discovered that the newspa-per reports were a hoax, but he did learn of a source of fossils of large dinosaurs in the Freeze-out Mountains of Wyoming. After much political

wrangling (the University of Wyoming initially claimed ownership), the bones of a large sauro-pod were on their way to Pittsburgh in 1899. With the aid of prominent paleontologists hired away from the American Museum of Natu-ral History, the 84'-long dinosaur skeleton was reconstructed and officially named *Diplodocus carnegii* (but called "Dippy" by museum staff). Complete casts of the giant skeleton were pre-pared and given by Carnegie to European roy-alty, and one of these casts dominates the British Museum of Natural History. The Carnegie Insti-tute funded other expeditions to search for di-nosaur remains, and today the museum houses some of the most important dinosaur fossils, in-cluding the first *T. rex* ever found.

By the time he died on August 11, 1919, An-drew Carnegie had given away more than $350 million, trying to live up to his motto of "the man who dies rich dies disgraced" (although he was still worth $30 million). The outbreak of World War I left Carnegie despondent, and he stopped work on his autobiography the day the war started (although he lived for another five years). Many of the organizations and foun-dations Carnegie started are still active today. Carnegie is buried in Sleepy Hollow Cemetery in Tarrytown, New York.

For More Information: Rea, T. 2001. *Bone Wars: The Excavation and Celebrity of Andrew Carnegie's Dinosaur.* Pittsburgh, PA: University of Pittsburgh.

JIMMY CARTER (b. 1924)

James Earl Carter, Jr., the thirty-ninth Presi-dent of the United States, entered the evolution-creationism controversy in January 2004, when he responded to Georgia State School Superin-tendent Kathy Cox's removal of references to evolution from the state's science curriculum. Carter, a theistic evolutionist, who predicted that Cox's move would bring ridicule to the state, noted that "as a Christian, a trained engineer and scientist, and a professor at Emory Univer-sity, I am embarrassed by Superintendent Kathy Cox's decision to censor and distort the educa-tion of Georgia's students…There is no need to teach that stars can fall out of the sky and land on a flat Earth in order to defend our religious faith." When Ronald Reagan—Carter's opponent for the presidency in 1980—claimed that evolution "is a scientific theory only" and that teachers should teach "the biblical theory of creation," Carter announced that he found evolution to be "convincing...I don't see anything wrong with God having created the evolutionary process." Carter lost the election.

JAMES CATTELL (1860–1944)

James McKeen Cattell was the owner-editor of *Science* who met with John Scopes in New York in 1925 to reaffirm the support of the AAAS for Scopes' upcoming trial. Cattell had a long, acrimonious fight with flood geologist George Price, whose views Cattell considered "not of interest to scientific men" (Price responded that evolution had become "a violent, anti-Christian religion" to scientists). Cattell worked with Maynard Shipley's Science League of America and Watson Davis' Science Service to support the scientists who were to testify for Scopes. Cattell, who worked briefly with Francis Galton, was the world's first professor of psychology. In 1904, he founded *Popular Science Monthly*, which later became *Popular Science*. Cattell died in Lancaster, Pennsylvania in 1944.

ROBERT CHAMBERS (1802–1871)

That God created animated beings, as well as the terraqueous theatre of their being, is a fact so powerfully evidenced, and so universally received, that I at once take it for granted. But in the particulars of this so highly supported idea, we surely here see cause for some re-consideration.

Robert Chambers was born on July 10, 1802, in Peebles, Scotland, 20 miles south of Edinburgh (Figure 15). Robert and his older brother William shared the unusual trait of having six digits on each hand and foot. Although their parents had the extra fingers and toes removed from both boys, Robert was left lame and in periodic pain, and was relatively inactive throughout his

15. Robert Chambers' sensational *Vestiges of the Natural History of Creation* started the public controversy about evolution.

childhood, and he became a voracious reader (i.e., he read the entire *Encyclopedia Britannica* as a child). Robert's parents supported his academic pursuits and planned for him to enter the ministry. By 1813, however, Robert's father had declared bankruptcy and moved his family to Edinburgh. Robert had planned to enter the University of Edinburgh, but due to his family's ongoing financial difficulties, he was forced to end his academic training in 1816.

As a young man, Robert worked at a number of jobs and struggled to support himself. With encouragement from William, Robert began selling books from his own collection and that of his parents. This enterprise quickly proved profitable, and the brothers joined forces to not only sell books, but also to publish. Robert enjoyed writing, and in 1821 the brothers starting publishing a weekly literary newspaper, *The Kaleidoscope*, which lasted for only eight editions. Undeterred, the brothers next wrote and published books about Scottish lore and history. These books sold well enough that, by the late 1820s, Robert was able to concentrate on writing. In 1830, Robert became editor of Edinburgh's *Advertisor* newspaper while continuing to write popular books (twenty-three by 1832) about Scotland. Robert's success attracted the attention of Scottish novelist Walter Scott, and the two men remained friends until Scott's death a few years later. Robert married Anne Kirkwood in 1829, with whom he had nine children.

In 1832, William Chambers introduced *Chambers' Edinburgh Journal,* a low-cost weekly newspaper written for the working class. When the first edition of the newspaper sold an astounding 30,000 copies, Robert saw not only a financial opportunity, but also a way to provide high-quality journalism to the masses, and he joined his brother as partner in the newspaper. Robert interpreted the success of the *Journal* as evidence that the working class wanted reliable information and educational materials. In response, the brothers produced an encyclopedia (*Information for the People*) and a series of educational texts that covered history, science, and literature that grew to include 100 volumes. By the 1860s, the brothers had produced *Chambers' Encyclopedia of Universal Knowledge for the People*, which would be published continuously for the next 100 years.

Robert Chambers' publications spanned a wide range of topics, and science (in particular geology) was among these interests. By 1840, he had not only educated himself about geology and paleontology, but had also befriended many prominent scientists in Edinburgh. As his background in science expanded, Chambers felt that an understanding of the geologic history of Earth necessarily led to questions about the origins of life and the progression life had taken ever since. Eventually, he considered the audacious proposal of uniting scientific knowledge to explain the history of life on the planet. He became convinced that such a synthesis was possible, and in 1841 he moved to the smaller and more secluded town of St. Andrews to concentrate on this effort.

By 1844, Chambers (acknowledging Lamarck, via Lyell, as a source of his inspiration) had developed his ideas into the hugely influential *Vestiges of the Natural History of Creation*, a 390-page book that was "the first attempt to connect the natural sciences into a history of creation." Chambers proposed that organisms arise by spontaneous generation and are transformed into more complex forms. Changes in species through time—clearly recorded in the fossil record—were based on extensions of individual development: a species could transmute into a new, more complex form through extended development of the individual. Chambers' theory extended von Baer's laws of embryonic development, in which linear progress follows the embryonic sequence of the highest organism, and the higher-level cause of progress pushes organisms up the sequence. An individual organism, therefore, replays its species' history as it develops. Chambers' concept of speciation thus did not produce the branching evolutionary trees of common descent proposed later by Darwin, but rather to parallel lines of development.

Chambers published the book anonymously (his wife transcribed the entire manuscript sent to the publisher so as to disguise the handwriting) to protect his business interests from the backlash likely to follow his controversial ideas. Chambers, who was known as "Mr. Vestiges," dismissed Lamarck's theory as an "inadequate...folly of the wise." Joseph Hooker's article tracing the authorship of *Vestiges* to Chambers was published in *Chambers' Edinburgh Journal*, and Lyell publicly rebuked "Mr. Vestiges" to show him that his "reasonings...would not be tolerated among scientific men."

Vestiges, the most controversial book of its time, began public controversies about evolution. The first edition sold out in a few days, and by 1860 had sold almost 24,000 copies. Novelist and politician Benjamin Disraeli (1804–1881) noted that *Vestiges* "is convulsing the world," and Abraham Lincoln became "a warm advocate of the doctrine." However, the first edition was also rife with scientific errors, and Chambers invoked concepts (e.g., spontaneous generation, transmutation of plowed-under oats into rye) already rejected by science. Thomas Huxley berated Chambers for his presentation of complex scientific concepts for a mass audience and cited the book's popularity as evidence of the "ignorance of the public mind as to the methods of science and the criterion of truth," and later called *Vestiges* "pretentious nonsense" and "charlantanerie." (Ironically, it was Chambers who, in 1860, convinced Huxley to attend the meeting of the British Association for the Advancement of Science where Huxley engaged Samuel Wilberforce in their famous debate.) Equally opposed to *Vestiges* was geologist Adam Sedgwick, who attempted to "pulverize" it in a biting eighty-page review of the book in the *Edinburgh Review*. Sedgwick dismissed the book for various reasons, but especially for failing to interpret the fossil record correctly in that the paleontological evidence did not then support the gradual biological change proposed by Chambers. Chambers responded to his critics by publishing *Explanations: A Sequel to "Vestiges of the Natural History of Creation" by the Author of that Work* the following year. Chambers considered the objections raised by critics in the ten revisions of *Vestiges*, the last appearing in 1860, a year after the publication of *On the Origin of Species*.

Vestiges—one of the greatest sensations of the Victorian era—also engendered negative reactions because of the materialism implicit in the author's theory: humans are simply one more, albeit very complex, species created by natural processes. Chambers had foreseen this response and noted the possibility of congruence between Scripture and his proposal. But the damage was done, and Chambers had downgraded the role of God in creation. Huxley, however, criticized even this perspective for its leaving the possibility open for design.

Charles Darwin read *Vestiges* at the British Museum in November 1844, at which time he noted that "I must allude to this." Darwin's *On the Origin of Species* presented *Vestiges* as a failed attempt to establish a mechanism for speciation. Annual sales of *Origin* did not consistently overtake those of *Vestiges* until the twentieth century.

Chambers knew his proposals would be incendiary, but he had hoped for an objective hearing from the scientific community. By 1860, when he and wife Anne returned to Edinburgh, he continued to write, but never again on scientific topics. In 1863, Anne died, followed a few weeks later by one of their daughters. In 1864, Chambers published his last major work, *Book of Days*, a collection of "popular antiquities" associated with each day of the year. He returned to live in St. Andrews and married again in 1867, a period marked by an increased interest in spirituality, ranging from séances to Christianity. His second wife died in 1870, and a year later, on March 17, Robert Chambers died. He is buried in the old Church of St. Regulus.

For More Information: Secord, J.A. 2000. *Victorian Sensation: The Extraordinary Publication, Reception, and Secret Authorship of* Vestiges of the Natural History of Creation. Chicago, IL: University of Chicago.

SERGEI S. CHETVERIKOV (1880–1959)

No matter how small the newly arisen improvement be . . . it will emerge and gradually become a property of the individuals of the species.

Sergei Sergeevich Chetverikov was born in Russia on May 6, 1880. He graduated from Moscow University in 1906, immediately entered graduate school there, and graduated with an advanced degree in 1909. Chetverikov was also jailed for two months in 1905 for participating in student protests.

Chetverikov's early interest was natural history, and he collected butterflies even before starting college (shortly before his death, Chetverikov donated his personal collection of more than 300,000 butterfly specimens to the Academy of Science's Zoological Museum in Leningrad). His first scientific paper (published in 1906) proposed that periodic "waves" of population decline could enable chance events to overcome selection, clearly anticipating the work of Sewall Wright on genetic drift.

From 1909 to 1919, Chetverikov taught entomology and systematics at Moscow University. Chetverikov joined the Institute of Experimental Biology in 1921, and, although not a geneticist, he began directing the Department of Genetics in 1924. He convened a group of colleagues to discuss current research in genetics, and this group ultimately predicted that because mutations are random events that tend to produce recessive alleles, mutations might persist in populations (populations should "soak up mutations like a sponge"), providing significant genetic variation for selection to use. These ideas formed the basis of Chetverikov's most influential paper, "On several aspects of the evolutionary process from the viewpoint of modern genetics," published in 1926. These predictions were confirmed when Chetverikov's group found that wild-caught *Drosophila* had dozens of hidden mutations that were revealed when the flies were inbred.

Chetverikov's research, which bridged laboratory-based experimental genetics and field research on evolution in natural populations, verified that natural populations house significant genetic variability, thereby linking genetics with Darwin's ideas on adaptive evolutionary change. However, because his relatively few publications (only twenty-six during his career) appeared only in Russian journals, Chetverikov's work was unknown by most of the biologists who founded the modern synthesis. J.B.S. Haldane, however, met Chetverikov in 1927 and cited his work in his own early publications.

Theodosius Dobzhansky, who left Russia in 1927 to work with T.H. Morgan, interacted with Chetverikov frequently before leaving for the United States. After working with Morgan and becoming an experimental geneticist interested in natural populations, Dobzhansky integrated Chetverikov's research, which influenced Dobzhansky's 1937 book *Genetics & the Origin of Species*. Although colleagues and students of Chetverikov disseminated their work to a wider scientific audience, Chetverikov remained a relatively unknown but important architect of the modern synthesis.

In 1929, when Russia was controlled by Joseph Stalin, Chetverikov was arrested, spent months in prison, and was exiled for several years. The reason for his arrest is unclear. Some, including Dobzhansky, claim that Chetverikov was arrested because the group he assembled at the Institute met in secrecy, and that the group's discussions frequently strayed into politics. Another possibility is that because Chetverikov opposed the Lamarckian model of inheritance (a model the Marxist leaders supported), the government needed to remove him. During his exile, Chetverikov served as a zoo consultant and taught mathematics. Afterward, he reentered academia in 1935 as head of the Department of Genetics at Gorky University. Although Chetverikov abandoned his research in genetics, in 1948 he still became a victim of Lysenko's official declaration of genetics as "bourgeois pseudoscience," which purged universities of geneticists. Around this time, Chetverikov's health declined due to a series of heart attacks, and in 1959 he died, shortly after it was announced that he was to be one of eighteen recipients of the Darwin Badge from the German Academy of Sciences Leopoldina in honor of his leading contributions to evolution on the hundreth anniversary of the publication of *On the Origin of Species*.

JACK CHICK (b. 1924)

We didn't evolve! The system has been feeding us THE BIG LIE!

Jack Chick was born on April 13, 1924, in Los Angeles, California, as the first of two children to artist Thomas Chick and his wife Pauline. In 1943, Chick enlisted in the Army and served in the Pacific theater during World War II. Chick, who failed the first grade, became a Christian fundamentalist after hearing an antievolution sermon that was broadcast on a radio program called *Charles Fuller's Old Fashioned Revival Hour*.

After working as an illustrator in the aerospace industry, Chick began a kitchen-based business built on publishing religious tracts. One of Chick's most popular publications is *Big Daddy?*, a small, twenty-four-page comic book written with the help of Young-Earth creationist Kent "Dr. Dino" Hovind (Figure 16), *Big Daddy?* is the most widely distributed antievolution publication in history.

Big Daddy? begins with an arrogant biology professor humiliating a student who dares to

16. Jack Chick, a reclusive artist and publisher, produced *Big Daddy?*, the most widely distributed antievolution booklet in history. (*Copyright 2000 by Jack T. Chick. Reproduced by permission of Chick Publications, Web site www.chick.com*)

question evolution. In a classic David-versus-Goliath confrontation, the enraged professor screams that the student is a fanatic and threatens him with jail, but the tide turns when the student tells the professor about "amazing findings which are rarely made public." These "amazing findings" expose evolution as a "big lie" and force the professor to admit that the student is "destroying me." The humiliated and repentant professor then pleads for the student to answer the questions that science can't answer. The story ends when everyone becomes a creationist and the professor—admitting that he can no longer teach evolution—is fired by heathen administrators (Figure 17).

Big Daddy? was originally published in 1972, and today appears in many different languages; there are millions of copies in print, and thousands of new copies are distributed each year. More people have read *Big Daddy?* than all other antievolution publications combined.

Chick's opposition to evolution also appears in *In the Beginning*, which uses the Paluxy dinosaur footprints and a talking serpent in the Garden of Eden to attack evolution and promote Young-Earth creationism. Biologists are described as "brainwashed" and evolution as "the religion of scientists who laugh at God." Some of Chick's other publications (e.g., *The Collapse of Evolution*) provide "undeniable scientific evidence" proving that "evolution is false," and videos that "force even the most devout evolutionists to reconsider their beliefs." These videos—which are promoted with Lenin's quote, "Give me your four-year-olds and in a generation I will build a socialist state"— describe how God cures cancer, how Noah packed dinosaurs and other animals on the Ark, and how readers can get evolution's "lies" out of textbooks.

Chick has more than 500 million books and booklets in circulation, some of which appear in the Smithsonian Institution. His publications, which have been translated into more than 150 languages, link evolution with social ills such as pornography (*Wounded Children*), gay rights (*Doom Town, The Story of Sodom*), rock 'n roll (*Angels*), Catholicism (*The Attack*), Islam (*Allah Had No Son*), and Halloween (*Satan's Big Night*). Chick's *The Ark* describes Russia's alleged attempts to block Christians from finding Noah's Ark atop Mount Ararat, and his *Primal Man?* depicts brainwashed scientists who are making

I could have you jailed for that!!* How dare you even mention the word Bible in this school. You know it's unscientific?? - - If you talk to me, it will be **ONLY** in scientific terms! *Do you understand?*

*It has never been against the law to teach the Bible or creation in public schools. See, Public School Presentation video from Creation Science Evangelism. www.drdino.com.

17. In *Big Daddy?*, a fanatical biology teacher denigrates a student who asks about life's origins. In the end, the teacher is humiliated and becomes a creationist. (*Copyright 2000 by Jack T. Chick. Reproduced by permission of Chick Publications, Web site www.chick.com*)

a film about evolution despite the fact that many viewers "will lose their souls."

Chick, a recluse who avoids publicity, lives in California.

CHICXULUB CRATER

Chicxulub Crater in the Yucután Peninsula is a 180-km-wide crater formed 65 million years ago by a 10-km-wide asteroid that struck Earth while moving at a speed of approximately 30 km sec^{-1} (a speed 150 times faster than that of a jetliner). This asteroid, if laid on its side, would have been higher than Mount Everest. The impact of the asteroid, which is estimated to have been 10 billion times more powerful than the atomic bomb dropped on Hiroshima, ejected rocks from several kilometers beneath Earth's surface and produced debris hundreds of meters thick near the crater (the debris is a centimeter thick 2,000 miles away in Colorado). The impact produced climatic changes that eliminated more than 60 percent of all species. All over the world, the noble metal iridium—an element rare on Earth but abundant in meteors—is found at the boundary of the Cretaceous and the Tertiary (the K-T boundary). The so-called K-T extinction remains the most famous mass-extinction, not because of its magnitude (the Permian extinction 250 million years ago wiped out many more species), but instead because of its most famous victims—the dinosaurs.

CHRISTIAN IDENTITY MOVEMENT

The Christian Identity Movement is an antievolution organization that denounced Henry Morris and "creation science" because each allowed change among the different "kinds" of organisms. Like the Ku Klux Klan, Christian Identity uses the Bible to justify its claim that blacks and other "inferior" races were originally created as "beasts."

CICERO (106–43 BC)

Cicero, Marcus Tullius, was a Roman philosopher who anticipated the "watchmaker" argument of William Paley when he noted that "when you see a sundial or a water-clock, you see that it tells the time by design and not by chance. How then can you imagine that the universe as a whole is devoid of purpose and intelligence?" Cicero was later declared an enemy of the state and, when he was caught, his last words were said to have been, "There is nothing proper about what you are doing, soldier, but do try to kill me properly." Cicero's head and hands were displayed in the Forum.

HAROLD CLARK (1891–1986)

Harold Willard Clark was a student of George McCready Price who believed that the fossil record was a fabrication. However, after visiting oilfields in Texas and Oklahoma in 1938, Clark was shocked to learn that the predictable arrangement of fossils in the geological column was the basis for the success of the petroleum exploration. In *Back to Creationism* (1929), which Clark dedicated to Price ("teacher, friend, fellow-warrior, and prophet of the new catastrophism"), Clark described Price's ideas as "creationism." Before Clark, antievolutionists had described themselves as fundamentalists or antievolutionists.

Clark's geology textbook *The New Diluvialism* (1946) claimed that rising floodwaters had successively destroyed ecological zones, which resulted in the predictable arrangement of fossils visible today. Clark also discussed the age of the Earth, biological change, the fossil history of humans, and other topics in a framework that Clark labeled "neo-creationism." Clark's so-called "ecological zonation theory," which remains popular among some creationists, was mocked by Price, and enabled creationists to accept the validity of the geological column while rejecting an ancient Earth. However, Clark's idea was rejected by scientists for many reasons, including the fact that some species appear in virtually all strata (e.g., corals, clams) and that organisms living in the same ecological zone (e.g., fish and whales, birds and flying reptiles) appear in different strata. When Clark substituted a non-Adventist text for Price's *The New Geology* in his geology course at Pacific Union College, Price denounced Clark as being influenced by Satan and filed heresy charges against Clark. Nevertheless, in 1966, Clark published *Crusader for Creation*, a sympathetic view of Price's life and work.

For More Information: Clark, H.W. 1966. *Crusader for Creation*. Omaha, NE: Pacific Press.

STEVE CLARK (b. 1947)

The theory that the universe was created is at least as scientific as the theory of evolution.

Steve Clark, the Attorney General of Arkansas from 1979 to 1990, headed Arkansas' defense of its law requiring "balanced treatment" of "creation science" in public schools. In 1982, the state lost the case (*McLean v. Arkansas Board of Education*), and Clark, who described the law's religious overtones as an "insurmountable problem," did not appeal the ruling. Clark's handling of the case was denounced by Wendell Bird, Pat Robertson, Jerry Falwell, and other creationists. Robertson, who claimed that Clark had contributed to the ACLU, encouraged listeners to call Arkansas governor Frank White to express their disgust with Clark's work. Nearly 500 telephone calls followed Robertson's plea.

In 1990, Clark announced that he would challenge then-governor Bill Clinton for reelection, but quit the campaign a few weeks later when the *Arkansas Gazette* reported that his office had spent more than $150,000 on travel and meals (including $80 drinks). Clark was convicted of felony theft in 1990, but did not go to prison, and was pardoned by Governor Mike Huckabee in 2004. Thereafter, Clark practiced law in Texas, declared bankruptcy, and moved to Miami Gardens in Florida, where he taught law at the St. Thomas School of Law.

EDWARD CLARKE

Edward Young Clarke was a membership director of the Ku Klux Klan who in January 1926, formed the Supreme Kingdom in Atlanta to continue William Jennings Bryan's crusade against the teaching of evolution. Clarke was indicted in 1922 for fraud, and was arrested the following year for violating the Mann Act (an antiprostitution measure enacted in 1910). Clarke organized the Supreme Kingdom like the Klan (i.e., with sovereigns, princes, and prime ministers), hired Fred Rapp—Billy Sunday's former business manager—to promote the organization, and offered to pay militant antievolution evangelist John Straton $30,000 to give sixty antievolution lectures. The Supreme Kingdom dissolved soon after the *Macon Telegraph* reported that Clarke was pocketing two-thirds of every $12.50 membership fee.

BAINBRIDGE COLBY (1869–1950)

Bainbridge Colby, who succeeded William Jennings Bryan as Secretary of State under Woodrow Wilson, was asked by George Rappleyea and the ACLU to help defend John Scopes. Colby accepted the invitation, and urged John Neal to try to transfer the trial to a federal court, where he could argue the constitutionality of the newly passed Butler Law. Colby resigned from the defense team on July 8, 1925. Colby, who left the Republican Party with Theodore Roosevelt to found the National Progressive Party, practiced law with Woodrow Wilson after Wilson left office and was Mark Twain's personal attorney. Colby is buried in Ellery, New York, beneath the inscription, "Faithful Public Servant."

FAY COOPER COLE (1881–1961)

Fay Cooper Cole was an anthropologist from the University of Chicago who attended the Scopes Trial as an expert witness. Cooper's affidavit noted that human development would be difficult to explain without referring to the many similarities that humans share with other animals. Cole was prepared to testify that evolution is fundamental to understanding anthropology, and noted several vestigial structures whose presence support the theory of evolution by natural selection. Cole's testimony was read into the court record by Arthur Hays. Two years after the trial, Cole noted that "Mr. Darwin once wrote: 'It is my fondest hope to make people think.' Mr. Darrow has achieved Mr. Darwin's aim." Cole cofounded the Society for American Archaeology.

JOHN "JACK" COLLIER (1850–1934)

John "Jack" Collier was a famous painter who was commissioned by the Linnean Society to paint a three-quarter length oil-based portrait of Charles Darwin in 1881. Darwin liked the painting, which hangs in the headquarters of the Linnean Society in London. Collier married Marian Huxley (a daughter of Thomas Huxley) in 1879, and Ethel Gladys Huxley (another daughter of Thomas Huxley) in 1889. Collier, who also painted portraits of Rudyard Kipling and his father-in-law Huxley (both of which are displayed in London's National Portrait Gallery), was on the "Personal Friends invited" list of guests for Charles Darwin's funeral.

FRANCIS COLLINS (b. 1940)

Francis Collins is Director of the National Human Genome Research Institute of the U.S. National Institutes of Health. Although he considered himself an atheist when he was in graduate school, he later became an evangelical Christian after witnessing the faith of seriously ill patients, and is now one of the most visible Christians in science. In 2006, Collin's book, *The Language of God: A Scientist Presents Evidence for Belief*, discussed his personal melding of religion and science (i.e., DNA sequences are "God's instruction book"). He considers the sequencing of the human genome an "occasion of worship" and believes that God and science should not be viewed as threats to each other. Collins is, however, an outspoken critic of creationism, including intelligent design.

JOHN CONLAN (b. 1930)

John Bertrand Conlan was a Republican Congressman from Arizona who in 1976 added an amendment to an education appropriations bill specifying that no funds could be used to support "secular humanism," including the teaching of evolution. The amendment passed the House of Representatives, but died in the House-Senate conference committee.

EDWARD DRINKER COPE
(1840–1897)

Edward Drinker Cope was a prominent American paleontologist and neo-Lamarckian who believed that his discoveries of dinosaurs would vindicate his belief in progress and perfectability. Cope, who excavated hundreds of fossilized dinosaurs, began his career as a zoology professor at Haverford College, but soon quit to devote all of his time to searching for dinosaur bones. Cope, who had a fierce twenty-year competition with fellow paleontologist Othniel Marsh, was a deeply religious Quaker, but he often used questionable methods to obtain his discoveries; he stole a whale skeleton from Louis Agassiz, settled disputes by fighting, and dynamited his sites so that others would not be able to make additional discoveries. Cope bought controlling interest in the journal *American Naturalist* so that he would have a fast way to publish his discoveries.

Within days of Custer's Last Stand in 1876, Cope began excavating fossils just a few miles away. Cope's *The Origin of the Fittest* (1877) argued that natural selection might cull poorly adapted organisms, but that Lamarckism produced variation.

After losing more than $250,000 in a mining fraud, Cope sold most of his dinosaur collection to Henry Osborn at the American Museum of Natural History for $32,000. In 1897, Cope died of syphilis in Philadelphia in a house packed floor-to-ceiling with fossils. He willed his body to a museum, hoping that it would be used as a "type specimen" to represent the human species, but his syphilitic lesions made his body unsuitable for such use.

The dinosaur "Drinker" was named in 1990 by U.S. paleontologist Robert Bakker (who first drew dinosaurs as upright and not dragging their tails) and others in tribute to Cope. Charles Knight, who worked with Cope and Marsh, produced paintings of dinosaurs that produced the public's stereotypes of dinosaurs for several decades; his paintings and reproductions circulated to schools and museums, and were featured on covers of magazines such as *Scientific American*. Knight's most famous painting—a confrontation between *Tyrannosaurus rex* and *Triceratops*—hangs in the Field Museum of Chicago.

For More Information: Ottaviani, J., Z. Cannon, S. Petosky, K. Cannon, and M. Schultz. 2005. *Bone Sharps, Cowboys, and Thunder Lizards: A Tale of Edward Drinker Cope, Othniel Charles Marsh, and the Gilded Age of Paleontology.* Ann Arbor, MI: G.T. Labs.

ARCHIE COTTON

Archie Cotton was a coal mine operator and member of Tennessee's Campbell County Board of Education who in April 1967 convinced his colleagues to fire biology teacher Gary Scott for teaching evolution. Cotton's action ultimately led to the repeal of the state's ban on teaching human evolution.

SYMS COVINGTON (1816–1861)

Syms Covington was Charles Darwin's servant and near-constant companion from 1832 to 1839, during the voyage of the *Beagle* and for more than two years afterward. Darwin, whose father paid Covington £30 per year, taught Covington how to shoot and stuff birds. After the *Beagle* returned to London, Covington continued to work as Darwin's servant until February 1839, and in 1842 organized Darwin's notes about volcanic islands. Covington later joined the British migration to Australia, where Darwin corresponded with him until 1859. Covington died "of a paralysis" on February 19, 1861.

KATHY COX

Kathy Cox was Georgia's State Schools Superintendent who in 2004 proposed removing the word evolution from the state science curriculum "to ensure that our kids are getting a quality science education." Cox, who advocated the teaching of intelligent design because it "is a scientific theory," believed that evolution is "a buzzword that causes a lot of negative reactions" that brings up "that monkeys-to-man sort of thing." Cox was rebuked by scientists and others, including former president Jimmy Carter.

CREATION EVIDENCE MUSEUM (Est. 1984)

People driving along Highway 205 on their way to Dinosaur Valley State Park just outside of Glen Rose, Texas, are often surprised to encounter the Creation Evidence Museum, located just a few hundred yards from the park's entrance (Figure 18). The popular museum consists of a small group of trailers and a larger building that bills itself as a "scientifically chartered museum." The museum's founder is Carl Baugh, a Baptist preacher, archaeologist, and Trinity Broadcasting Network personality who uses the museum to discredit evolution by proving that people lived contemporaneously with dinosaurs. Baugh began his excavations along the Paluxy River on March 15, 1982, and two days later announced discoveries of human and dinosaur tracks having "unparalleled historic significance."

In the museum, visitors watch a forty-minute *Creation in Symphony* video in which Baugh describes his story of creation that includes water being sprayed 70 miles into the air and God stretching the "space fabric" to a point where faraway stars exploded. Baugh, a Young-Earth creationist, claims that evolution offers no explanation for our existence. Baugh's creationism, on the other hand, provides hope and a happier ending. The museum has a variety of research programs, including expeditions that claim to have found living pterodactyls in New Guinea.

Museum officials claim that the fossilized human footprints displayed in the Museum were made in what people have been "educated" to believe are 113-million-year-old deposits of limestone in nearby Dinosaur Valley State Park. Baugh claims to have excavated almost 100 footprints and 475 dinosaur footprints. (Researchers in Dinosaur Valley State Park have found thousands of dinosaur tracks, but no contemporaneous human footprints.) One of the largest footprints on display—the 14"-long "Burdick Track"—was found by Clifford Burdick, a

founder of the Deluge Society, one of America's first creationist groups. The footprints from the Paluxy site are featured in *The Genesis Flood*, a book by Henry Morris and John Whitcomb that in 1961 launched the modern "creation science" movement in the United States.

The Creation Evidence Museum also includes a large magenta-windowed "hyperbaric biosphere" in which Baugh claims to have recreated "Earth's original pre-flood environment." According to Baugh, the biosphere increases organisms' lifespans by 300 percent; it also detoxifies copperheads' venom. Near the biosphere is an aquarium in which Baugh grows "vegetarian piranhas." Baugh believes these findings support the long lifespan of Biblical patriarchs such as Noah, and the harmonious environment (e.g., no carnivores or death) that existed before Eve introduced sin into the world. Baugh hopes to grow dinosaurs in the biosphere. On the walls, visitors see paintings in which pre-flood children play with a baby *Apatosaurus* in the nearby Paluxy River. Visitors can purchase these replicas, as well as books, posters, and other materials such as certificates honoring recipients as "visionaries" for "supporting truth in education."

CREATION RESEARCH SOCIETY (Est. 1963)

The Bible is the written Word of God, and because it is inspired throughout, all its assertions are historically and scientifically true in the original autographs.

The Creation Research Society (CRS) was founded in Michigan in June 1963 as a professional organization of scientists and interested laypersons who were dissatisfied with the American Scientific Affiliation (ASA) and who advocated "creation science." Five of the ten founders of CRS held earned doctorates in biology, and two others had doctorates in science or engineering. CRS is not officially affiliated with any religious group, and its members must have a postgraduate degree in science and endorse the literal truth of the Bible. Among the founders of

18. The Creation Evidence Museum is an evangelical organization that claims to have uncovered more than 100 human footprints alongside dinosaur footprints in the Paluxy Riverbed in Glen Rose, Texas. The museum, which is toured by thousands of visitors each year, sponsors archeological expeditions and tries to recreate Earth's early atmosphere. The museum also exhibits the famous "Burdick track." (*Randy Moore*)

the CRS were Henry Morris and Duane Gish. Whereas the ASA was concerned with the relationship of Christianity and science, the CRS concerned itself with the creationism-evolution controversy.

In July 1964, CRS began publishing its journal—*Creation Research Society Quarterly*—because its ten founders had been unable to publish their creationism-based ideas in established scientific journals. In 1970, CRS published John Moore's creationism-based biology textbook, *Biology: A Search for Order in Complexity*, which claimed that "the most reasonable explanation for the actual facts of biology as they are known scientifically is that of Biblical creationism." Although creationists such as Morris hoped that the book "would be of interest to public school systems desiring to develop a genuine scientific attitude in their students," the book's use was hindered by legal challenges for more than a decade. In 1996, CRS began publishing the journal *Creation Matters*, which contains "somewhat lighter" articles. All members of CRS must endorse the CRS Statement of Belief, which states that:

1. The Bible is the written Word of God, and because it is inspired throughout, all its assertions are historically and scientifically true in the original autographs. To the student of nature this means that the account of origins in Genesis is a factual presentation of simple historical truths.

2. All basic types of living things, including man, were made by direct creative acts of God during the Creation Week described in Genesis. Whatever biological changes have occurred since Creation Week have accomplished only changes within the original created kinds.

3. The great flood described in Genesis, commonly referred to as the Noachian Flood, was an historic event worldwide in its extent and effect.

4. We are an organization of Christian men and women of science who accept Jesus Christ as our Lord and Savior. The account of the special creation of Adam and Eve as one man and one woman and their subsequent fall into sin is the basis for our belief in the necessity of a Savior for all mankind. Therefore, salvation can come only through accepting Jesus Christ as our Savior.

Critics claim that such creeds, which are part of Answers in Genesis, Institute for Creation Research, and several other antievolution organizations, prove that creationists accept only what they already assume.

CRS distributes books such as Robert Kofahl's (b. 1924) *Handy Dandy Evolution Refuter*, Kelly Segraves' *The Creation Explanation: A Scientific Alternative to Evolution*, and various creationism-based guides to natural history. CRS sponsors speakers who lecture on topics such as "Is Evolution a Fact?" and "Theistic Evolution! No way! No way!" CRS has also established the Van Andel Creation Research Center (named after Jay Van Andel, founder of the Amway Company) in north-central Arizona to help CRS members and others with their research. In 2007, CRS had approximately 1,700 members.

CREATION SCIENCE RESEARCH CENTER (Est. 1970)

The Creation Science Research Center (CSRC) was established by Henry Morris (Director), Nell Segraves (Research Librarian), and son Kelly Segraves (Assistant Director) to produce teaching materials and "extension ministries" (e.g., seminars and radio programs) for schools wanting to include flood-based creationism in their science curricula. The CSRC was originally part of the Christian Heritage College, a religious school near San Diego founded by fundamentalist and antievolutionist Tim LaHaye. When the CSRC split from the college in 1972 (primarily because of conflicts between Morris and Kelly Segraves), the Segraves kept the CSRC name and Morris founded the Institute for Creation Research to continue his creationism-based

research. Although CSRC went into debt, it continued to fight evolution, along with abortion, women's rights, and gay rights. In 1977, Kelly Segraves reported that "evolutionary interpretations of science data result in widespread breakdown in law and order" leading to "divorce, abortion, and rampant venereal disease."

The most dramatic example of CSRC's impact was *Segraves v. state of California*, which started in 1979 when Kelly Segraves—then a parent with children of his own—sued to eliminate dogmatism about evolution in California's *Science Framework*. The lawsuit attracted much media attention, and was quickly named "Scopes II." Although Segraves demanded equal time for creationism, he later backtracked by asking only that the state "stop posing the theory that man and all life on Earth developed from a common ancestor, as a fact." Segraves claimed victory, noting that the settlement "will stop the dogmatic teaching of evolution and protect the rights of the Christian child." The judge—claiming that "everybody won"—ordered California to distribute its 1972 "anti-dogmatism" policy to school officials, textbook publishers, and science teachers, and to include the policy in all future versions of *Science Framework*. Many creationists and others questioned CSRC's publication of Kelly Segraves' *Sons of God Return* (1975), which linked demons, flood geology, and UFOs.

FRANCIS CRICK (1916–2004)

The age of the Earth is now established beyond any reasonable doubt as very great, yet in the United States millions of Fundamentalists still stoutly defend the naive view that it is relatively short, an opinion deduced from reading the Christian Bible too literally. They also usually deny that animals and plants have evolved and changed radically over such long periods, although this is equally well established. This gives one little confidence that what they have to say about the process of natural selection is likely to be unbiased, since their views are predetermined by a slavish adherence to religious dogmas.

Francis Harry Compton Crick was born on June 8, 1916, to Harry Compton Crick and

Annie Elizabeth Crick. Francis' grandfather was an avid amateur naturalist, and in early 1882 discovered a water beetle with a small mollusk attached to its leg. He was so enthralled by this discovery that he wrote to the greatest naturalist of the day, Charles Darwin. Darwin was also impressed, having earlier proposed that mollusks may be dispersed in this way, and had the specimen sent to him. Darwin published the findings in *Nature* in April of 1882, only thirteen days before he died.

By the time Francis was born, World War I had pushed England into an economic slump that caused the family business to fail, and the Cricks were forced to move to London. In 1930, Francis won a scholarship to the Mill Hill boarding school in London, where he exhibited an affinity for science. Even so, he was denied entry to Cambridge and Oxford, and instead entered University College, London, in 1934, graduating three years later with a BS in physics. He immediately started work toward a PhD in physics at University College, but the outbreak of World War II forced him to suspend his graduate education. Crick joined the British Admiralty Research Laboratory and remained there until 1947, developing mines for the military. During this period, he married his first wife, Ruth, with whom he had a son, Michael. The Cricks divorced in 1947.

A reading of physicist Edwin Schrödinger's book, *What Is Life? The Physical Aspects of the Living Cell* (1944), led Crick away from particle physics and into biology. He transferred to the Strangeways Research Laboratory at Cambridge, where he learned X-ray diffraction, a tool that would be crucial for identifying the structure of DNA. In 1949, Crick moved again, this time to the Medical Research Council Unit for Molecular Biology at the Cavendish Laboratory at Cambridge University. The Cavendish Laboratory was famous for research in physics (James Clerk Maxwell was its first director), and in the late 1930s had added molecular biology as an area of study. It was here that Crick's interest in genetics began. Crick married his second wife, Odile, in 1949, with whom he had two daughters.

American biologist James Watson joined the Cavendish Laboratory in 1951. Watson was keen on using X-ray crystallography to investigate the structure of DNA, and Crick and Watson immediately started working together. Combining recent research by others (including data that may have been improperly obtained from ex-colleague Rosalind Franklin) with their own work, Watson and Crick built models (using both mathematics and cardboard) showing the possible configuration of the DNA molecule. Ultimately, they determined the structure of the now well-known double helix, and published their results in *Nature* in 1953. The following year, Crick earned his PhD for research on helical protein structures. Crick, Watson, and colleague Maurice Wilkins (a physicist who worked with Rosalind Franklin on X-ray diffraction studies of DNA) shared the Nobel Prize for Physiology or Medicine in 1962 for their work on DNA.

In 1957, Crick began studying mutations in viruses with biologist Sydney Brenner (who would win a Nobel Prize in 2002). These experiments eventually linked nucleotide sequences in DNA with amino acid sequences in proteins. Work by others, showing the involvement of RNA in protein synthesis, allowed Crick to elucidate the triplet nature of the genetic code and to propose the "central dogma" of molecular biology: DNA $\rightarrow$ RNA $\rightarrow$ protein. In 1962, Crick became codirector of Cambridge University's Molecular Biology Laboratory with Brenner.

Crick's understanding of molecular biology—what he originally referred to as "the borderline between living and the nonliving"—encouraged him to consider how life may have arisen. Crick concluded that the production of life from nonliving molecules was an extremely rare event and unlikely to have happened on Earth. This idea was expanded in Crick's 1981 book *Life Itself: Its Origin and Nature*, which invoked the concept of "directed panspermia"—that life had arisen elsewhere in the universe and had been dispersed to Earth by intelligent alien life forms. By the 1990s, however, discovery of enzymatic RNA had fostered development of the "RNA world" hypothesis for the origin of life on Earth, and Crick acknowledged that he had underestimated the probability of life arising on Earth without outside influence. Regardless, Crick's statements (e.g., "The probability of life originating at random is so utterly minuscule as to make it absurd") have been cited repeatedly by antievolutionists as evidence that science has proven the need for a supernatural explanation for life. The rise of intelligent design (ID) provided yet another opportunity for Crick's ideas, particularly panspermia, to be misused. For example, during the *Kitzmiller et al. v. Dover Area School District* trial, ID advocate Michael Behe used Crick's writings to suggest that scientists naturally believe the universe is designed.

When he was 12 years old, Crick stopped attending church, and he remained an atheist throughout his life. In his 1967 book, *Of Molecules and Men*, Crick confronted the concept of "vitalism"—that is, that life cannot be explained entirely by natural phenomena open to scientific inquiry. He opposed mixing religion and science, and resigned as fellow of Churchill College, Cambridge, when the college decided to build a chapel on campus. His refusal to allow religion to dictate how science is taught also spurred him to join over seventy other Nobel laureates in signing an amicus brief stating " 'creation-science' simply has no place in the public-school science classroom" for the *Edwards v. Aguillard* case in 1987.

Crick relied on what he called the "gossip test"—what you talk about in everyday conversation likely reflects your primary interests—to determine his main scientific interests: molecular biology and the workings of the brain. Having successfully studied the first area for several decades, he decided in 1977 to investigate the second by accepting the position of Kieckhefer Distinguished Research Professor at the Salk Institute of Biology in La Jolla, California, where he would remain for the remainder of his career. In 1994, he published *The Astonishing Hypothesis: The Scientific Search for the Soul*, which concluded that self-awareness results from interactions of the multitude of neurons in the brain,

and that the human "soul" is not an entity separate from the understandable workings of the brain.

Francis Crick died on July 28, 2004, of colon cancer.

For More Information: Ridley, M. 2006. *Francis Crick: Discoverer of the Genetic Code*. New York: Atlas Books.

CROWLEY v. SMITHSONIAN INSTITUTION

Crowley v. Smithsonian Institution (1980) established that the federal government (1) can fund public exhibits that promote evolution, and (2) is not required to provide money to promote creationism. Dale Crowley, Jr., was a retired missionary and Executive Director of the National Foundation for Fairness in Education who filed the lawsuit in early 1978 after learning that Congress had funded an exhibit entitled "Dynamics of Evolution" at the National Museum of Natural History, a division of the Smithsonian Institution. Crowley claimed that the Smithsonian was using tax money to establish the religion of "secular humanism" and that the exhibit forced fundamentalists to choose between violating their religious views by entering the museum or abandoning the right to access public property. In December 1978, U.S. District Judge Barrington Parker rejected Crowley's claim that the Smithsonian restricted the exercise of religion, noting that "the plaintiffs can carry their beliefs into the Museum with them, though they risk seeing science exhibits that are contrary to that faith." Crowley's appeal ended in 1980 when the U.S. Supreme Court refused to hear Crowley's final appeal. Crowley, who was influenced by creationist Harry Rimmer, hosted a radio show called *The King's Business*, in which he often discussed the "lies and wild guesses" of "evolutionism." Like many other creationists, Crowley claimed that evolution is an evil lie, a pagan religion, and a "Satanic plot."

MARIE CURIE (1867–1934)

In 1898 Marie Curie discovered (and coined the term) radioactivity, and her husband Pierre discovered that radium releases heat as it decays. This discovery ultimately helped show that the Earth is much older than many people believe. That same year, John Joly accurately predicted that the discovery of radioactivity would invalidate Lord Kelvin's estimate of the Earth's age. The Curies shared the 1903 Nobel Prize with Henri Becquerel, and in 1911 Marie won a second Nobel Prize for her work with radium. Marie Curie was the first person to win or share two Nobel Prizes, is one of only two people who have been awarded a Nobel Prize in two different fields (the other is Linus Pauling), and is the only woman to have won two Nobel Prizes. Marie died of leukemia in 1934; her original lab notebooks are stored in a lead-lined safe. In 1995, Marie became the first person to be interred under the dome of the Pantheon on her own merits.

WINTERTON C. CURTIS (1875–1966)

Winterton C. Curtis was an invertebrate zoologist from the University of Missouri who attended the Scopes Trial as an expert witness. Curtis argued that evolution is a fundamental tool for understanding biological, geological, and cosmological questions. Curtis' testimony—which included letters from Woodrow Wilson and William Bateson—was read into the court record by Arthur Hays. Wilson's letter, written in 1922, noted that "of course, like every other man of intelligence and education, I do believe in organic evolution. It surprises me that at this late date such questions should be raised." After the trial, Curtis noted that "Dayton was more like a town prepared for a Billy Sunday revival than for a court trial." Curtis is buried in Union Cemetery in Fairfield County, Connecticut.

GEORGES CUVIER (1769–1832)

All of these facts, consistent among themselves, and not opposed by any report, seem to me to prove the existence of a world previous to ours, destroyed by some kind of catastrophe.

Jéan-Léopold-Nicholas-Frédéric Cuvier was born on August 23, 1796, in Montbéliard

19. Georges Cuvier, who attributed animals' similarities to common functions rather than shared ancestry, established the discipline of vertebrate paleontology and documented extinctions. These extinctions, which Cuvier believed were caused by God-driven catastrophes ("revolutions"), "broke" the Great Chain of Being. Although Cuvier's discoveries supported Charles Darwin's theory of evolution by natural selection, Cuvier rejected evolution because it was "contrary to moral law, to the Bible, and to the progress of natural science itself." (*Library of Congress*)

(Figure 19). The name Georges, by which Cuvier preferred to be called, was added after an older brother with that name died in childhood. Georges attended the prestigious Académie Caroline in Stuttgart, and after graduating in 1788, began tutoring the children of noblemen in Normandy, where Cuvier became interested in geology and marine organisms, especially mollusks. Cuvier also served in the local government of his community, a foreshadowing of his later involvement in high levels of administration and government.

Following the Reign of Terror of the French Revolution from 1793 to 1794 (from which Cuvier was safely insulated in Normandy), the patronage system that once controlled science was dismantled, which afforded an unknown outsider a greater chance of joining the scientific community. Hence, when Cuvier left for Paris in 1795 to seek a career in the study of natural history, he was taking a significant but calculated risk. His work on mollusks attracted attention in Paris, including from Étienne Geoffroy Saint-Hilaire, professor of zoology at the Museum National d'Histoire Naturelle, and Cuvier secured a position as an assistant at the Museum. He was soon appointed professor of comparative anatomy.

The Museum's vast collections enabled Cuvier to focus on comparative anatomy. His earliest work examined the anatomy of elephants, and he concluded that not only was there more than one species of elephant, but that elephant-like bones found in northern regions represented extinct species (now recognized as mammoths and mastodons). Cuvier's subsequent studies of the extinct megafauna of South America (e.g., *Megatherium*, a giant ground sloth) allowed Cuvier to conclude that fossils generally represent extinct species, and that the fossil record indicates "the existence of a world previous to ours, destroyed by some kind of catastrophe."

Although fossils had been studied for some time, natural historians had generally assumed species were immutable and everlasting. The Great Chain of Being endorsed by the Church demonstrated a continuous sequence of increasing perfection, which not only disallowed transmutation, but also extinction. Cuvier's endorsement of extinction—including mass extinction—was nearly heretical because it "broke" the Great Chain. But as fossils that had no living counterpart continued to be found (e.g., Irish elk, mastodons, giant sloths), Cuvier's catastrophism became more widely accepted. He further proposed that the perceived gaps in the fossil record indicated that catastrophes (he preferred the less theologically loaded term "revolutions") had occurred repeatedly, which also implied a vast age for the Earth. As Cuvier noted in 1836, "The surface of our globe has suffered a great and sudden revolution, the period of which cannot be dated further back than 5 or 6,000 years." When Cuvier's work was translated into English, Robert Jameson of Edinburgh University presented Cuvier's theory as though the most recent catastrophe was the biblical flood.

Cuvier originally believed extinctions were not global in scale, and that areas became recolonized from relict populations. Later, as gaps in the fossil record seemed to document worldwide events, Cuvier proposed the separate creations of new species after each "revolution." Cuvier suggested that floods were the most likely cause of mass extinctions (Louis Agassiz would later invoke global ice ages as the cause), but even as a devout Christian, Cuvier did not use science to understand events described in the Bible. Later discoveries of mosasaurs and pterodactyls led Cuvier to propose that mammals may not have always been the dominant life form on Earth, but he was also certain that human fossils would never be found. Cuvier believed that humans originated after the biblical flood.

Cuvier expanded Linnaeus' taxonomy by grouping living and extinct species into classes based on functional similarities. Cuvier believed that organisms are integrated units: no one body part could be modified without negatively affecting the rest of the organism. Each individual body-part therefore reflected the organism as a whole, which meant that an organism could be identified based only on a small fragment of its anatomy, a feat Cuvier was famous for achieving. Cuvier identified four basic "integrated" body plans (*embranchments*) into which all life is grouped: Vertebrata, Articulata, Mollusca, and Radiata. (This idea of separate, unrelated body plans influenced paleontologist Richard Owen, who studied with Cuvier, and who became a chief nemesis to Charles Darwin.) Because each body plan was an integrated unit, and because change disrupted the entire organism, neither intermediary nor rudimentary forms were possible, and evolution was effectively excluded from consideration. Cuvier's classification of animals into four discrete types was controversial, just as his proposal of extinction had been, because it likewise conflicted with the relationship among species encapsulated in the Great Chain of Being.

Cuvier, who established the discipline of vertebrate paleontology, criticized Buffon's earlier suggestions that species could degenerate because of environmental influences. Buffon's ideas clearly contradicted Cuvier's typological concept of species and, to Cuvier, reduced species to "variable forms and fleeting types." Cuvier also disagreed with his Museum colleague, Jean-Baptiste de Lamarck, about evolution through inherited characteristics. Cuvier argued that bits and pieces of organisms cannot change, and that the fossil record (Cuvier's trump card, as he was the best paleontologist of the time) contained no transitional forms.

Most famously, Cuvier's ideas conflicted with those of Étienne Geoffroy Saint-Hilaire. In contrast to Cuvier's discrete categories of body plans, Geoffroy proposed that life was unified, that the environment can produce change, and that transmutation of species was possible. Geoffroy's proposed mechanism of transmutation was alteration in the developmental pattern of organisms in response to changes in the environment. He believed that Cuvier's four embranchments could be unified, even suggesting that the exoskeleton of invertebrates was the same as the internal vertebral column of the vertebrates.

The famous Geoffroy-Cuvier debates took place in 1830 at meetings of the Paris Académie des Sciences. By most accounts, Cuvier—with his vast knowledge of comparative anatomy and paleontology—won, although his proposals were eventually overturned by more sophisticated evolutionary perspectives. The history of these debates is recorded by the street names surrounding the French Museum of Natural History: *rue Cuvier* is on the east of the museum, while *rue Geoffroy-St. Hilaire* is on the west.

During his final years, Cuvier became involved with government service. His appointments to the posts of Inspector-General of public education (1808), State Councilor (1813; an appointment made by Napoleon), and vice president of the Ministry of Interior (1817), led him to spend increasing amounts of time away from the Museum, although he maintained his position there. He was also elected to the Legion of Honour and raised to the rank of peer of France.

On May 8, 1832, Cuvier gave a lecture denouncing Saint-Hilaire's "pantheism" and "useless scientific theories." That night, Cuvier had a stroke, and he died six days later. Cuvier rests in a family plot in Paris' Cimetière du Père Lachaise, not far from composer Frédéric Chopin, who taught Emma Darwin how to play piano.

Although Cuvier is often remembered for his staunch adherence to the immutability of species, he made tremendous contributions to comparative anatomy and is considered the "father of paleontology." Furthermore, Cuvier's influence forced acceptance of extinction as a natural occurrence, and his work figured prominently in the development of Darwin's ideas on evolution. Even though Cuvier was devoutly religious, he felt that religion is a personal matter that should be kept separate from science.

For More Information: Rudwick, M.J.S. 1997. *Georges Cuvier, Fossil Bones, and Geological Catastrophes.* Chicago, IL: University of Chicago.

D

VERNON DALHART (1883–1948)

Vernon Dalhart (born Marion Try Slaughter) was a popular country singer who helped popularize the Scopes Trial by recording Carson Robison's *The John T. Scopes Trial* for Columbia Phonograph Company (#15037-D). Dalhart, who earlier had auditioned for Thomas Edison, recorded the song on July 10, 1925 (the first day of Scopes' trial), and it was released two weeks later (it sold almost 80,000 copies). Dalhart recorded the same song for three other companies. On August 10, 1925, Dalhart also recorded *Bryan's Last Fight* (Columbia Records #15039), which proclaimed that Bryan "stood for his own convictions, and for them he'd always fight." Dalhart, who ended up working as a night watchman, was posthumously inducted into the Country Music Hall of Fame in 1981. Dalhart is buried in Mountain Grove Cemetery in Bridgeport, Connecticut.

DANIEL v. WATERS

Daniel v. Waters (1975) overturned the Tennessee law requiring equal emphasis on evolution and the Genesis version of creation. The lawsuit originated when Joseph Daniel and others challenged Tennessee's "Genesis Law" requiring biology textbooks presenting human origins to include the concept as a theory "not represented as a scientific fact" and that they include "equal space" to other theories "including, but not limited to, the Genesis account in the Bible." The law, which excluded "all occult or Satanical beliefs of human origin," defined the Bible not as a textbook, but instead as "a reference work," and therefore not subject to the law. The popular legislation—the first statute requiring "equal time" for evolution and creationism—became law in 1973 when Governor Winfield Dunn (b. 1927) refused to sign or veto the bill. After the Tennessee Attorney General refused to appeal the *Daniel* decision, the Tennessee Supreme Court issued an opinion agreeing that the law was unconstitutional. In 1975, the *Daniel* decision was cited in *Steele v. Tennessee*, another lawsuit involving the constitutionality of the "equal time" law passed in Tennessee in 1973. Hugh Waters was named in the lawsuit in his role as Chairman of the Textbook Commission of the State of Tennessee. When the Genesis Law was ruled unconstitutional in 1975, Tennessee Governor Ray Blanton (1930–1996) praised the decision because it "saves me from administering a law I did not believe in anyway."

CLARENCE DARROW (1857–1938)

Nero tried to kill Christianity with persecution and law. Bryan would block enlightenment with law.

20. Clarence Darrow was the most famous attorney of the twentieth century. He became famous for his participation in sensational trials such as those of thrill-killers Leopold and Loeb, labor organizer William "Big Bill" Haywood, and biology teacher John Scopes. (*Library of Congress*)

Had we Mr. Bryan's ideas of what a man may do towards free thinking, we would still be hanging and burning witches and punishing persons who thought the Earth was round ... America is founded on liberty and not on narrow, mean, intolerable, and brainless prejudice of soulless religio-maniacs.

Clarence Seward Darrow was born on April 18, 1857, in Kinsman, Ohio, and was raised by agnostic, politically involved parents (Figure 20). Darrow's father was an undertaker and furniture maker. After attending Allegheny College and the University of Michigan Law School, Darrow was admitted to the Ohio Bar in 1878, after which he began practicing law in Youngstown, Ohio. In 1887, Darrow moved to Chicago, and in 1890 became an attorney for the Chicago and North Western Railway. During the Pullman Strike, Darrow sympathized with the trade union workers and offered them his services. In 1894, after resigning his position with the railway, Darrow defended socialist Eugene V. Debs (1855–1925; President of American Railway Union) when Debs was arrested for contempt of court. Although Debs and his fellow unionists were convicted, Darrow established himself as an effective labor attorney. In subsequent years, Darrow continued to defend union leaders. For example, in 1905 Darrow helped obtain an acquittal for William "Big Bill" Haywood (leader of the Industrial Workers of the World and the Western Federation of Miners) when he was charged with murdering Frank Steunenberg, a former governor of Idaho. In 1902, President Theodore Roosevelt appointed Darrow to arbitrate a Pennsylvania coal strike. Although Darrow supported the Allies involvement in World War I, he defended several antiwar activists charged with violating state sedition laws.

In 1896, Darrow attended the Democratic National Convention, where he heard William Jennings Bryan give his famous "Cross of Gold" speech. That same year, Darrow lost his race for Congress by about 100 votes. In 1908, Bryan asked Darrow to help his third campaign for the presidency, but Darrow declined.

In 1911, Darrow defended unionist John McNamara and his brother Jim McNamara, who were charged with bombing the *Los Angeles Times* building during a labor strike. Although the bombing killed twenty people, Darrow obtained prison sentences (rather than death sentences) for the McNamaras. When Darrow was charged with attempting to bribe jurors, he told jurors that "I have committed one crime: I have stood for the weak and poor," and he obtained a hung jury. Darrow agreed that if he would not be retried, he would never again practice law in California. The 56-year-old Darrow returned to Chicago, tired and broke.

When corporations and labor unions dropped Darrow from their list of favored attorneys, Darrow shifted to criminal law. In 1924, Darrow participated in what would become one of his most famous trials—the defense of teenaged thrill-killers Nathan Leopold, Jr., and Richard

Loeb, the youngest graduate in the history of the University of Michigan. In May of 1924, Leopold and Loeb—whose parents were millionaires—kidnapped 13-year-old Bobby Franks on his way home from school, killed him with a hammer, mutilated his body with acid, and stuffed his body into a culvert, merely to see if they could get away with it. They tried to collect a $10,000 ransom, but were caught and confessed. Darrow's arguments, which included psychological determinism, helped save Leopold and Loeb from the death penalty (but not prison) by convincing the judge that they were not entirely responsible for their behavior.

When Darrow arrived in Dayton, Tennessee, in July 1925 to defend John Scopes, he was at the height of his powers and fame; he was America's greatest criminal lawyer and its most famous champion of anticlericalism. After giving the commencement address at John Neal's law school ("some time we may have a movement to let people alone"), Darrow went to Dayton and denounced Tennessee's ban on the teaching of human evolution "as brazen and bold an attempt to destroy learning as was made in the Middle Ages."

Darrow was anxious to confront Bryan in Dayton. In 1923, Darrow had published a letter on the front page of the *Chicago Tribune* questioning Bryan about his beliefs. For example, did Bryan believe in the literal truth of the Bible? What about the flood? The origin of man? Did Jonah live inside a whale for three days? Bryan did not respond, and Darrow hoped to confront Bryan about these beliefs in Dayton. Darrow's objective in Dayton was to prevent "bigots and ignoramuses from controlling the education of the United States, and that is all." Darrow later stated that "My object, and my only object, was to focus the attention of the country on [Bryan] and the other fundamentalists." When Darrow heard that Bryan would be at Dayton, "at once I wanted to go . . . I realized that there was no limit to the mischief that might be accomplished unless the country was aroused to the evil at hand."

On July 20, 1925, after defense attorney Arthur Hays finished reading the statements of expert witnesses, Darrow objected to the "Read Your Bible" banner on the courthouse and had it removed. Then almost 2,000 spectators watched the trial reach its climax on the front lawn of the Courthouse. After earlier baiting Bryan by saying that "Bryan has not dared test his views in open court under oath," Hays announced to the court that "the defense desires to call Mr. Bryan as a witness." Bryan did not have to testify, and Judge Raulston left the decision to Bryan. Bryan—falling for Darrow's trap—took the witness chair (Figure 21).

During his ninety-minute examination, Darrow referred to Bryan's "fool religion" and questioned Bryan about his "fool ideas" (e.g., Jonah being swallowed by a whale, Joshua's commanding the sun to stand still to lengthen the day, the worldwide flood). Bryan eventually admitted that he didn't believe in a literal interpretation of the Bible. Prosecutor A.T. Stewart tried to stop the questioning more than a dozen times, but Bryan refused to step down: "I am simply trying to protect the word of God against the greatest atheist or agnostic in the United States. I want the papers to know I am not afraid to get on the stand in front of him and let him do his worst." When Bryan claimed that "the only purpose Mr. Darrow has is to slur the Bible," Darrow shouted back, "I am examining your fool ideas that no intelligent Christian on Earth believes." Raulston then adjourned court for the day. In 1930, writer William Sweet's popular book *The Story of Religion in America* inaccurately depicted Darrow's questioning of Bryan as "fundamentalism's last stand."

On the final day of the trial, Darrow asked Judge Raulston "to bring in the jury and instruct [them] to find [Scopes] guilty." When the jury did just that, Darrow thanked the jurors by telling them: "We wanted you folks to do just what you did." Darrow, who believed the Butler Law was "silly and senseless . . . even in the state of Tennessee," had promised that Scopes would appeal his verdict, and Scopes' lawyers were confident that Scopes' conviction would be overturned and the law declared unconstitutional.

After the trial, newspapers printed a list of questions that Bryan had given to Darrow, as

21. Darrow helped defend John Scopes at the Scopes Trial. Here, in the most famous event of the trial, Darrow (standing and holding his suspenders) questions William Jennings Bryan (seated) on an outdoor dais. (*Smithsonian Institution Archives Image #2005-26202*)

well as Darrow's responses. In those answers, Darrow said that he did not believe in Bryan's god, did not believe in miracles, and could find no evidence of the immortality of the soul. During the trial, Darrow had described his agnosticism this way: "I do not pretend to know where many ignorant men are sure; that is all agnosticism means." Darrow later elaborated, "I feel that the Earth is the home and the only home of man, and I am convinced that whatever he is to get out of his existence he must get while he is here . . . I am an agnostic because I am not afraid to think. I am not afraid of any god in the universe who would send me or any other man or woman to hell. If there were such a being, he would not be a god; he would be a devil."

Throughout the trial, Darrow held Scopes in high regard. Darrow, who believed that Scopes had been indicted "for the crime of teaching the truth," noted that "Scopes was trying to do for Dayton, Tennessee, what Socrates did

for Athens." Darrow also described Scopes as "a man of courage," a phrase which became Scopes' epitaph.

Many people opposed Darrow's involvement in Scopes' appeal. ACLU officials pleaded with Scopes to let someone besides Darrow handle the appeal. Despite the protests, Scopes stood by Darrow, and Darrow remained on the appeal team. During the appeal, Darrow linked science with progress and stressed that religion should be a personal issue. Darrow argued that science teachers should teach science, not religion, and concluded—to applause—with a defense of individualism: "We are once more fighting the old question, which after all is nothing but a question of intellectual freedom of man." True to form, Darrow chided the fundamentalists with his now famous line: "With flying banners and beating drums, we march back to the glorious ages of medievalism." Darrow's villainy at the Scopes Trial became fodder for countless fundamentalist sermons.

When the Tennessee Supreme Court set aside Scopes' conviction, Darrow scoffed at the decision. Darrow—whom Stewart said was "the greatest menace present-day civilization has to deal with" and had to be suppressed to prevent "more evil"—told Scopes that "I am pretty well satisfied that the law is dead but we want to make sure, if possible." When motions for a new hearing were rejected, Darrow admitted that "it will probably take another case to clear up the matter." That case would not come along until 1965, when Arkansas biology teacher Susan Epperson challenged the Arkansas law banning the teaching of human evolution.

Darrow and other attorneys worked without pay and paid all of their expenses during Scopes' trial and appeal. Darrow, who spent about $2,000 of his own money for the trial, noted simply that "I really wanted to take part in it...I never got more for my money."

Soon after the trial, Darrow and his wife Ruby left Dayton. However, Darrow was growing tired and participated in only a few more cases. In Detroit, Darrow was assisted by Arthur Hays while defending Ossian Sweet, an African-American charged with murdering a member of an angry mob that had assembled outside his house. Sweet was acquitted by an all-white jury after Darrow gave an eight-hour closing argument. Darrow then, again with Hays, helped obtain acquittals for two Italians charged with murder in New York, after which he spent a year in Europe and "learned to loaf." After the Depression wiped out most of his retirement savings (from $300,000 to $10,000), Darrow went to Honolulu for his last case—the "Massie Case," in which Darrow defended four men accused of killing a young Hawaiian man. Darrow's closing argument was broadcast live throughout the United States.

Darrow continued to champion anticlericalism laws and questioned "testing every fact in science by a religious dictum." Darrow, who Bryan described as "an able man, and, I think, an honest man," predicted that Tennessee's Butler Law would soon be repealed, but he was wrong; the law remained on the books for more than forty years. Darrow was challenged by New York fundamentalist John Straton to a series of debates about evolution, but Darrow declined.

The Scopes Trial generated more mail for Darrow than any of Darrow's previous cases, including his defense of Leopold and Loeb. Until his death, Darrow continued to get offers of salvation, advice, and get-rich-quick schemes. When John Scopes moved to Chicago in September 1925, to start graduate school at the University of Chicago, Scopes stayed with (and later often visited) Darrow.

Darrow, who believed that Bryan's religion made Bryan the "idol of all Morondom," continued to write and speak about evolution, prohibition, and personal liberty, and in 1931 he narrated *The Mystery of Life*, a film released by Universal Pictures that covered much of the same material as the scientific testimony in Dayton. When the film was shown in Dayton, local ministers called for a boycott of the "anti-Biblical and anti-Christian" movie, while also noting that Darrow was an agnostic who had defied prohibition laws. In 1932, Darrow published *The Story of My Life*. When Darrow returned to Dayton and saw a church being built across the street from Robinson's Drug Store (i.e., where the Scopes Trial originated), he admitted, "I guess I didn't do much good here after all."

When Scopes last saw Darrow in Houston, Texas, in the mid-1930s, Scopes "knew [Darrow] wasn't going to last much longer." On his seventy-ninth birthday Darrow made a sentimental return to his childhood home in Ohio, after which he became increasingly ill and confused. Darrow's last days were painful and ugly.

Darrow died of heart disease at age 80 on March 13, 1938, in his home at 1537 East 60th Street, overlooking Jackson Park in Chicago. Darrow's wife Ruby told a friend that her husband "didn't care whether he went to heaven or hell because he has so many good friends in either place." As tributes poured in from around the world, Hays declared Darrow one of "the greatest of them all." Darrow's funeral was held at the Bond Chapel at the University of Chicago, after which his ashes were scattered from a bridge into Jackson Park Lagoon, which is just south of Chicago's Museum of Science and

Industry. At a ceremony held on May 1, 1957, the bridge was named the Clarence Darrow Memorial Bridge. On March 13 of every year, a wreath is placed at the bridge to commemorate Darrow's life. In the decades after his death, Darrow's fame did not decline, and in the 1970s his life was the subject of a one-man play starring Henry Fonda. Today, Darrow stands as the most celebrated lawyer of the twentieth century.

For More Information: Darrow, C. 1932. *The Story of My Life.* New York: Charles Scribner's Sons; Stone, I. 1941. *Clarence Darrow for the Defense.* New York: Doubleday, Doran & Company.

RAYMOND DART (1893–1988)

It is mentally difficult, if not revolting, for civilized people to look back upon their primate past. The general attitude of many . . . Americans to the remote primate past of mankind seems to have reminded [us] of the snobbish behavior of some socially successful folk. Their lowly origin offends their newfangled ideas of what is fitting to their present estate.

Raymond Arthur Dart was born on February 4, 1893, in Brisbane, Australia. After attending the University of Queensland and the University of Sydney, Dart served in a medical corps during World War I. He then enrolled at the University of Manchester, where he was a student of Arthur Keith. Dart then became a professor of anatomy at the University of Witwatersrand in Johannesburg, South Africa.

In 1924, Raymond Dart chipped the first known skull of an australopithecine child from a piece of limestone from the Buxton Limeworks quarry in Harts Valley near Taung, South Africa. Dart, who had been given the stone by a student (Josephine Salmons), suspected that it represented an important find, and asked for more samples from the quarry. He got two more crates of samples just a few months later, and when he opened the first crate, "I knew at a glance that what lay in my hand was no ordinary anthropoidal brain . . . It is logically regarded as a man-like ape." Dart's discovery had a cranium larger than an immature ape, but smaller than a child, and its milk canines were smaller than

those of a chimp, suggesting that it ate a different diet than chimps. Moreover, its foramen magnum—the opening at the base of the skull through which the spinal cord connects to the brain—was more forward than in apes.

In the February 7, 1925, issue of *Nature*, Dart described the discovery as "an extinct link between man and his simian ancestor." Dart named his discovery *Australopithecus africanus* ("australis" meaning "south," and "pithecus" meaning "ape"). Although Dart's "Taung Child" skull had human-like teeth, it was neither confined to southern Africa nor was it an ape. Nevertheless, because of the rules governing the naming of taxa, Dart's name cannot be changed.

In Dart's era, many paleontologists believed that humans originated in Asia, and others believed that Piltdown Man indicated that humans originated in Europe. These paleontologists rejected Dart's claim. Indeed, the next issue of *Nature* included four letters dismissing Dart's discovery as being merely an unusual chimpanzee or juvenile gorilla. However, Dart was supported by renowned Scottish paleontologist Robert Broom (1866–1951), who noted that, "In *Australopithecus* we have a connecting link between the higher apes and one of the lowest human types." Broom, who considered himself the "scientific son" of Richard Owen, agreed with Dart that the discovery vindicated "the Darwinian claim that Africa would prove to be the cradle of mankind." Soon after Dart's discovery, Broom visited Dart's lab, where Broom knelt before Taung Child "in adoration of our ancestor."

Dart went to London in 1930 to promote his discovery, but his trip was overshadowed by the recent discovery of Peking Man (*Homo erectus*; sometimes called Beijing Man). In 1936, Broom found more specimens of *Australopithecus* in a cave at Sterkfontein, South Africa. Some of Broom's fossils looked like Dart's discovery, but others looked more robust, and were therefore named *Australopithecus robustus*. In 1959, paleontologists Louis and Mary Leakey discovered an even more robust *Australopithecus* that they named *Australopithecus boisei*. In 1947, Keith—originally one of Dart's critics—admitted "Dart was right, and I was wrong." In the 1950s, the

Leakeys confirmed Dart's claim that modern humans originated in Africa, and in the 1970s Donald Johanson discovered older and more complete remains of australopithecines in Ethiopia, the most famous of which was "Lucy."

Dart, with Robert Ardrey, also advocated the "killer ape" theory regarding early humans (e.g., "the persistently bloodstained progress of man"). In 1953, Dart's paper titled "The Predatory Transition from Ape to Man" argued that some apes had branched from their fruit-eating kin and became carnivores. After making weapons, the human creatures turned the weapons on their own kind. These ideas were consistent with lingering thoughts about *King Kong* (1933) and the killer ape in Edgar Allen Poe's "Murders in the Rue Morgue." Dart's ideas about killer apes, which were extended by ethologist Konrad Lorenz, inspired the opening scenes of man-apes in the movie *2001: A Space Odyssey*.

Dart taught at the University of Witwatersrand until 1958. He died on November 22, 1988, at the age of 95. Dart's autobiography, *Adventures with the Missing Link*, was published the following year.

For More Information: Dart, R. 1959. *Adventures with the Missing Link*. New York: Viking.

ANNE DARWIN (1841–1851)

Anne Elizabeth "Annie" Darwin was born on March 2, 1841, and was Charles and Emma Darwin's second child and eldest daughter. Like the other Darwin children, Annie was raised a Christian; she read the Bible and attended church regularly. Charles drew strength from his family, and he confessed to his cousin William Fox that Annie was his favorite child. She often accompanied Charles as he walked along the Sandwalk behind Down House.

In 1849, Annie contracted tuberculosis, and her health began to deteriorate. Hoping to find a cure, Charles took Annie to the spa town Great Malvern, where she was treated with Dr. James Gully's "water cure." Although Gully predicted her recovery, Annie's health continued

22. Annie Darwin's death in 1851 removed the last traces of theism from her famous father, Charles Darwin. Charles, who did not attend Annie's funeral, never saw her grave. (*Randal Keynes*)

to deteriorate. Watching Annie fade away was excruciating for Charles, who wrote to Emma (who remained in Downe because she was pregnant) that "It is a relief to me to tell you [about Annie], for whilst writing to you, I can cry tranquilly."

Ten-year-old Annie died in Malvern at noon on April 23, 1851 (Figure 22). Charles left the following day to be with Emma; neither of Annie's parents was present for her burial in the Priory Churchyard in Great Malvern. On April 25 (the day Annie was buried), Charles wrote to a friend that, "Sometime I should wish to know on which side and part of the church-yard, as far as you can describe it, the body of our once joyous child rests."

Annie rests beneath a cedar of Lebanon tree facing the north side of the chancel. Annie's tombstone has a simple inscription, "A dear and good child." Annie's death removed the last vestiges of Charles' religious faith.

Charles remembered Annie by writing a memorial on April 30, 1851, a week after she died. The memorial ends poignantly: "We have lost the joy of the household, and the solace of our old age: she must have known how we loved her; oh that she could now know how deeply, how tenderly we do still and shall ever love her dear joyous face. Blessings on her." Emma remembered Annie by gathering some of Annie's treasures in Annie's writing box. That box contained Annie's letters, Charles' daily notes on Annie's illness and death, Annie's sewing supplies, a lock of Annie's hair, and a note about Annie's gravestone at Malvern. In 2000, Randal Keynes (Charles Darwin's great-great-grandson; b. 1919) discovered this box and used it as a basis for writing *Annie's Box*, the most personal of the many books about Darwin. Annie's box and its contents are displayed at Down House. A movie based on *Annie's Box* is scheduled for 2009.

For More Information: Keynes, R. 2001. *Annie's Box*. London: Fourth Estate.

CHARLES DARWIN (1809–1882)

We must, however, acknowledge, as it seems to me, that man with all his noble qualities ... still bears in his bodily frame the indelible stamp of his lowly origin.

Charles Robert Darwin was born in Shrewsbury, Shropshire, England (about 160 miles northwest of London), on February 12, 1809, into an affluent, well-known family. Charles' father, Robert Waring Darwin (1766–1848), was a physician. His mother, Susannah Wedgwood Darwin (1765–1817)—the heir to £25,000—was the daughter of Josiah Wedgwood, one of the founders of the Wedgwood pottery company and noted opponent to slavery. Charles—the

fifth of six children—was baptized in Shrewsbury's Anglican church. Charles had one brother (Erasmus, named after his eccentric grandfather), three older sisters (Marianne, Caroline, and Susanne), and one younger sister (Emily Catherine). Charles attended Shrewsbury Grammar School, but preferred hunting and collecting insects to learning Greek and Latin. In the autumn of 1825, Charles enrolled in Edinburgh University to study medicine. There, he learned taxidermy from a freed Guyanan slave, John Edmonstone, which would be useful for him later. However, Darwin was sickened by watching surgery, and dropped out of college.

In 1827, after being scolded by his father that "you care for nothing but shooting, dogs, and rat-catching, and you will be a disgrace to yourself and your family," Charles enrolled at Cambridge University and began studying to join the clergy. However, he remained interested in collecting insects, and it was at Cambridge that Charles began to appreciate the vast diversity of species, an appreciation that is inescapable if insects are the focus. While at Cambridge, Charles was mentored by Robert Grant, an admirer of Charles' grandfather Erasmus and a supporter of Lamarck's theory of inheritance of acquired characteristics. Darwin also learned about natural theology, which sought to understand God by studying God's creation (i.e., nature). To advocates of natural theology, God's goodness was visible in the way He had created a progression of life from "lower" to "higher" forms, culminating in the special creation of humans. At Cambridge, Darwin read William Paley's *Natural Theology*, which argued that nature's complexity was evidence of design, and that design required a designer. Although Darwin studied and enjoyed Paley's book, Darwin later rejected Paley's arguments, "now that the law of natural selection has been discovered." While at Cambridge, Darwin lived in the same dormitory rooms that had housed William Paley seventy years earlier. Today, those rooms—which are now offices—are marked by a statue of Darwin in the adjacent garden.

At Cambridge, Darwin had traditional religious beliefs, but a noticeable lack of religious

23. Charles Darwin, the discoverer of evolution by natural selection, is one of the most famous and influential scientists in history. Although Darwin was not the first person to propose evolution, his *On the Origin of Species* documented a mechanism for evolution and, in the process, changed the course of biology. This photo shows Darwin in his old age. (*Image courtesy History of Science Collections, University of Oklahoma Libraries; copyright the Board of Regents of the University of Oklahoma*)

zeal. Although he later claimed that some of his time at Cambridge "was sadly wasted," Darwin made his first scientific presentation there (about *Flustra*, a bryozoan), and in 1831 he earned a bachelor's degree in theology, ranking tenth among 178 nonhonors students. However, it was a great era of exploration, and Darwin was drawn to the prospect of travel; he hoped to delay his religious career until he had visited exotic locales.

In the 1830s, the British government commissioned the H.M.S. *Beagle* to conduct expeditions "devoted to the noblest purpose,

the acquisition of knowledge." The *Beagle*'s captain was the temperamental 26-year-old Robert FitzRoy, and FitzRoy needed a naturalist for the voyage. FitzRoy wanted the naturalist to be Cambridge cleric-botanist John Henslow, a friend of Darwin's. Henslow declined FitzRoy's offer, but recommended Darwin for the job. Charles' father did not want him to go on the voyage, but Josiah Wedgwood (Charles' uncle) convinced Charles' father to let his son go on the cruise. Charles later noted that he "was resolved to go at all hazards."

The *Beagle* sailed from Plymouth, England, on December 27, 1831. Darwin's experiences were overwhelming—he hiked through a jungle in Brazil, dug up fossils in Argentina, witnessed an exploding volcano, withstood an earthquake that raised shellfish beds 3' above the shoreline, and studied coral reefs. By the end of his voyage, Darwin had written 1,383 pages of notes about geology, 368 pages about zoology, entered 770 pages in his diary, preserved 1,529 specimens, and labeled 3,907 skins, bones, and other specimens. He had also seen numerous interesting geological formations and new habitats (e.g., the Andes and tropical islands). His voyage lasted fifty-eight months, forty-three of which were spent in South America.

While at sea, Darwin read Volume 1 of Charles Lyell's *Principles of Geology*, which had been given to him by Captain FitzRoy. Lyell documented that ancient Earth had been molded by the same slow, directionless forces that shape Earth today. When the *Beagle* docked at St. Jago, Lyell's claim that land slowly rises and falls was confirmed when Darwin found a band of coral fragments and shells high among the volcanic cliffs. Later, when exploring the Andes, Darwin found fossil trees that had grown on a sandy, shell-littered beach. Lyell showed that Earth is very old, and that Earth's history had been characterized by the extinction and appearance of innumerable species. This was the world that Darwin sought to explain (Figure 23).

In September of 1835, the *Beagle* docked for a month-long stay at a group of thirteen volcanic islands called the Galápagos Islands. These islands, which straddle the equator, are about 620 miles west of Ecuador. Darwin later noted that the islands' songbirds—called finches—appeared as if "one species had been taken and modified for different ends." Darwin was told by Nicholas Lawson that local inhabitants could tell the home island of a tortoise just by examining its shell. As Darwin noted later, "By far the most remarkable feature in the natural history of this archipelago [is] that the different islands to a considerable extent are inhabited by a different set of beings." This intrigued Darwin, for it suggested that each island had its own group of organisms. Did each island really have a unique species of tortoise? Did each island have a unique species of finch? Had there been a separate creation event at each island? Darwin was initially skeptical: "I never dreamed that islands 50 or 60 miles apart, and most of them in sight of each other, formed of precisely the same rocks, placed under a quite similar climate, rising to nearly equal height, would have been differently tenanted."

After leaving the Galápagos, the *Beagle* sailed to Tahiti, New Zealand, and Australia, during which time Darwin pondered the questions he had been developing about life's diversity. Soon after Darwin returned to England in October 1836, ornithologist John Gould of the London Zoological Society examined the birds that Darwin and others had collected on the Galápagos Islands and told Darwin that each island *did*, in fact, house a separate species of finch. Back in England, Darwin lived on Fitzwilliam Street in Cambridge from 1836 to 1837. Today, his residence is marked by a small plaque.

In September of 1838, Darwin read "for amusement" Thomas Malthus' *An Essay on the Principles of Population*. Malthus argued that populations could grow exponentially (e.g., 2, 4, 8, 16, 32, 64, 128, 256, etc.), but that resources such as food can increase only linearly (e.g., 1, 2, 3, 4, 5, 6, 7, etc.). Malthus' book had a tremendous influence on Darwin: What would be the consequences of a constant struggle for existence (as Malthus proposed) that persisted for millions of years (as Lyell proposed)?

As Darwin pondered this question, he realized that Malthus' struggle throughout the history of Lyell's ancient Earth might explain the great diversity of plants and animals that he had encountered on his travels. As he later wrote in his autobiography, "Being well prepared to appreciate the struggle for existence...it at once struck me that under those circumstances, favourable variations would tend to be preserved and unfavourable ones destroyed. The result of this would be the formation of a new species. Here, then, I had at last got a theory by which to work." Combining Malthus' idea with what he had seen at the Galápagos (which he called the primary source of all his views) gave Darwin a new insight. He now knew how species evolved. "Did [the] Creator make all new [species on oceanic islands], yet [with] forms like [on] neighboring continent? This fact speaks volumes. My theory explains this but no other will." Despite his confidence, however, Darwin was not ready to announce his discovery.

As experts continued to sort through Darwin's enormous collection of specimens from his *Beagle* voyage, Darwin began to think more and more about his radical idea. The finches collected from the Galápagos suggested to Darwin that new species could evolve from a common ancestor. Darwin began to think of humans not as an ultimate and special creation, but instead as merely one more species, albeit one with unusual mental powers.

In 1837, Darwin moved to a residence on Great Marlborough Street in London, where he began writing about his "dangerous" idea in a secret notebook labeled "Transmutation of Species." This notebook contains Darwin's first "tree of life," in which he depicts life not as a hierarchical ranking of "higher" and "lower" forms (as Aristotle and other naturalists had claimed), but instead as a branching tree showing shared origins. The branches of the tree did not necessarily lead anywhere; they just spread. Instead of marching up a chain or ladder as Lamarck and other naturalists had suggested, Darwin's tree showed that species evolved; in some cases, one species could give rise to many species (as had occurred on the Galápagos Islands). Although

Darwin would not publish his theory for twenty-two years, his "tree of life" would become a metaphor for Darwin's view of how species evolve.

While at the Galápagos, Darwin failed to record all the information he later needed to make full use of his data. For instance, Darwin understood—once his finches were properly identified after his return to England—that they were all closely related. However, while at the Galápagos, Darwin had paid them little mind, and because he had neglected to record the precise island from which each was taken, he could not reconstruct their probable relationships. Returning to the Pacific was out of the question, so he asked FitzRoy and other shipmates if he could borrow the Galápagos birds they had donated to the British Museum. Darwin received six sets of bird skins, and Gould's conclusion was strengthened: Each island housed a different species of finch. Meanwhile, Thomas Bell, who had been identifying Darwin's reptiles, provided a parallel conclusion: Each island of the Galápagos chain had produced its own distinct species of iguana.

Although Darwin's theory consumed much of his time, Darwin began considering marriage. He listed the advantages of marrying ("constant companion and a friend in old age . . . better than a dog anyhow"), as well as the disadvantages ("less money for books," and "terrible loss of time"). On January 29, 1839—just five days after he was elected Fellow of the Royal Society—Darwin married his first cousin Emma Wedgwood at a ceremony officiated by Rev. John Allen Wedgwood (Emma's cousin) at the Church of St. Peter near the Wedgwood mansion. For the rest of his life, Charles referred to Emma as "my greatest blessing." The newlyweds moved into a house near Charles' brother Erasmus in London to start their family; they eventually had ten children, but only seven reached adulthood. Darwin was wealthy; after returning from his *Beagle* voyage, Darwin's father gave him stocks and a yearly allowance of £400, which was raised to £500 when Darwin married. Emma's dowry brought in another £5,000, and the Darwins inherited £45,000 when Charles' father died. Later in his life, Darwin—who invested primarily in railways and government bonds—had his son William Erasmus handle his finances. By 1881, Darwin's income had risen to £17,299, of which he spent £4,880 (he invested £10,218 and gave £3,000 to his children). Late that year, William informed his father that his estate was worth £282,000. The Darwins had no financial worries, and never had to work.

While living in London, Darwin used a journal he originally wrote for his family as the basis for a book documenting his voyage aboard the *Beagle*. The book had a ponderous title: *Narrative of the Surveying Voyages of His Majesty's Ships* Adventure *and* Beagle, *between the Years 1826 and 1836 Describing their Examination of the Southern Shores of South America and The* Beagle's *Circumnavigation of the Globe in Three Volumes*. Darwin's book was Volume 3 of the set (another volume was written by FitzRoy). The volumes could be purchased separately, and Darwin's became a bestseller. When Henry Colburn, the publisher, reprinted Darwin's book, he gave it a grander title: *Journal of Researches into the Geology and Natural History of the Various Countries Visited by HMS* Beagle *Under the Command of Captain FitzRoy, R.N. from 1832 to 1836*. Darwin's book, retitled *The Voyage of the Beagle* at its third printing, was reprinted many times. This book, one of the world's great travel books, remains a steady seller.

Although Darwin, like Lyell, was praised as a scientist and writer, FitzRoy's volume—which included comments about geology and biblical history—was ridiculed, and Darwin's accomplishments soon relegated FitzRoy to a historical footnote. FitzRoy had taken Darwin around the world and made Darwin's discoveries possible, but he would regret his role in Darwin's achievements for the rest of his life.

Meanwhile, the Darwins soon tired of life in "dirty" London. In 1842, they paid Rev. J. Drummond £2,020 for a large house near the village of Downe; Charles and Emma Darwin lived contentedly in Down House for the rest of their lives. Despite his subsequent fame, Darwin never again left England.

In 1842, Darwin developed the ideas in his "Transmutation" notebook into a thirty-five-page outline of "descent with modification" (as evolution was called in Darwin's day). Darwin discussed his idea with a few of his close friends, most notably Joseph Hooker, the founder of plant geography, and Charles Lyell, who along with James Hutton founded modern geology. After confiding to Hooker that he had discovered "the simple way which species become exquisitely adapted to various ends," Darwin likened his idea to "confessing a murder."

Darwin had witnessed the 1844 firestorm caused by Robert Chambers' *Vestiges of the Natural History of Creation*, and was reluctant to announce his theory. Instead, he continued to do research that produced eight more books on topics that included insectivorous plants, earthworms, pigeons, barnacles, climbing plants, orchids, and plants' movements. Everywhere he looked, and regardless of what he studied, Darwin found evidence that supported his theory. In 1844, he expanded his thirty-five-page outline into a 231-page (about 50,000 words) manuscript, which he stored in a hallway closet in Down House.

As an adult, Darwin was chronically ill, prompting his friend Thomas Huxley to note that Darwin "might be anything if he had good health." Although Darwin continued to refuse to announce his idea, he set aside £400 for Emma to publish his expanded manuscript if he died unexpectedly.

On June 18, 1858, Darwin received a twenty-page letter from British naturalist Alfred Russel Wallace, who was halfway through eight years of collecting specimens across the Malay Archipelago. (Wallace had written the letter in February, but it had taken four months to reach Darwin.) In that letter, Wallace described the same idea for evolution that Darwin had been secretly writing about. As Darwin noted after reading Wallace's letter, "I never saw a more striking coincidence. All my originality, whatever it may amount to, will be smashed."

After consulting with friends, Darwin outlined his ideas, and Darwin and Wallace's letters, along with part of Darwin's 1844 essay and an earlier letter from Darwin to Asa Gray, were read on July 1, 1858, at a meeting of the Linnean Society, a leading society of professional scientists in England. Darwin did not attend the meeting (he was mourning the death of his son Charles Waring Darwin, who had died two days earlier of scarlet fever), and Wallace did not know about the meeting. The presentation generated little interest among those who attended the meeting. The paper was then published under the impressive title, *"On the Tendency of Species to Form Varieties; And on the Perpetuation of Varieties and Species By Natural Means of Selection* by Charles Darwin Esq., FRS, FLS, & FGS and Alfred Wallace Esq., communicated by Sir Charles Lyell, FRS, FLS, and J. D. Hooker Esq., MD, VPRS, FLS, &c." Darwin began writing a book describing his idea, and he finished the final chapter on March 19, 1859. Darwin understood the importance and potential impact of his idea—"It is no doubt the chief work of my life." Darwin's handwriting was poor, and publication of his book was delayed slightly because of questions raised by Ebenezer Norman, a schoolmaster at Downe who was the copyist of Darwin's manuscripts. On November 24, 1859—twenty-two years after Darwin had opened his secret "Transmutation of Species" notebook—John Murray Publishing of London (which had published all of Lyell's books) released Darwin's 502-page book, *On the Origin of Species by Means of Natural Selection, Or The Preservation of Favored Races in the Struggle for Life.*

Murray printed 1,250 copies of *On the Origin of Species*, 139 of which were distributed as promotional copies. Publishers bought all of the remaining copies, each of which was sold for 15 shillings apiece. On the day of its publication, Murray asked Darwin if he wanted to make any changes before the second printing (Darwin changed about 7 percent of the text, which appeared as the second edition in 1860). Darwin's subsequent revisions were published in several languages that took Darwin's idea throughout the world. Darwin updated his theory and addressed critics' concerns by revising the book five times, and the final (i.e., sixth) edition was published on February 19, 1872. Darwin's idea became known as the "survival of the fittest," a phrase coined in 1863 by British philosopher and economist Herbert Spencer. Although neither this phrase nor the word "evolution" were in the first edition of *On the Origin of Species*, Darwin liked Spencer's phrase and believed that it was "more accurate" than his own explanation of natural selection. Darwin first used the phrase "survival of the fittest" in the title of Chapter 4 ("Natural Selection, or the Survival of the Fittest") of the sixth edition of *Origin*. During Darwin's lifetime, Murray sold approximately 25,000 copies of the English version of *Origin*. From 1859 to 1881, Darwin's books earned him an average of £465 per year; in 1871, these royalties constituted only about 6 percent of Darwin's total income.

Darwin's *On the Origin of Species* was not the first book about evolution, but it was—and still is—the most influential. Unlike Lamarck's *Philosophie Zoologique*, which was purely a theoretical book, Darwin's *Origin* was an overwhelming compendium of facts. Darwin stressed that his book was not a denial of God's existence, but it did challenge biblical literalism and removed humans from their pinnacle as the ultimate purpose of God's creation. Not surprisingly, *Origin* was condemned by many religious leaders, and William Whewell, Master of Trinity College at Cambridge, refused to allow it into the college library. However, many others praised Darwin's book. Darwin himself saw his idea as enlightening, as he noted in the book's famous final paragraph:

> There is grandeur in this view of life, with its several powers, having been originally breathed into a few forms or into one; and that, whilst this planet has gone cycling on according to the fixed law of gravity, from so simple a beginning endless forms most beautiful and most wonderful have been, and are being, evolved.

Unlike earlier explanations of evolution, Darwin's theory was supported by a huge amount of evidence and included a workable, coherent, and testable mechanism that did not require a deity, miracles, or arbitrary purpose. Just as Newton had done in *Principia*, Darwin included an enormous number of detailed observations to create "one long argument" for his theory.

The cornerstone of Darwin's theory is natural selection, the differential survival and reproduction of organisms. Natural selection produces adaptations, which are traits that enable organisms to survive the "struggle for existence." Darwin was convinced that just as domestic animals evolve through selective breeding (i.e., artificial selection), species in the wild evolve "by means of natural selection." Darwin explained natural selection in Chapter 4 of *On the Origin of Species* ("I have called this principle, by which each slight variation, if useful, is preserved, by the term Natural Selection"). For Darwin, natural selection was the force that constantly adjusts the traits of future generations by sorting hereditary variations. Darwin did not discuss the origin of life in *On the Origin of Species*, and referred to human evolution in one sentence that could be the understatement of the nineteenth century: "Light will be thrown on the origin of Man and his history." The implications of Darwin's theory were clear to contemporary readers:

- Darwin replaced the notion of a perfectly designed and benign world with one based on an unending, amoral struggle for existence.
- Darwin challenged prevailing Victorian ideas about progress and perfectibility with the notion that evolution causes change and adaptation, but not necessarily progress and never perfection.
- Darwin's theory was theologically divisive, not because of what it implied about animal ancestry, but because it offered no purpose for humanity other than the production of fertile offspring.
- Darwin challenged the Providentially supervised creation of each species with the notion that all life—humans included—descended from a common ancestor. Humans are not special products of creation, but of evolution acting according to principles that act on other species.

Although Darwin knew that Wallace would have written an *Origin*-like book "if [he] had had my leisure," Wallace credited Darwin as the originator of the theory of evolution by natural selection. Darwin knew that his book would disturb many people; when he sent a copy of *On the Origin of Species* to Wallace late in 1859, he enclosed a note: "God knows what the public will think." However, Darwin had many defenders, most notably Harvard scientist (and evangelical Christian) Asa Gray in the United States, and Thomas Huxley in England. Throughout the uproar that followed publication of his book, Darwin stayed at Down House; he was interested in what was happening, but stayed out of the fray.

After Darwin published *Origin*, many people continued to claim that humans are exempt from evolution. These claims inspired Darwin to write the two-volume, *The Descent of Man, and Selection in Relation to Sex*, which Murray published in 1870–1871 to address what Darwin called "the highest and most interesting problem for the naturalist." Part I of the book described evidence for human evolution (e.g., homologies and vestigial structures shared with apes). Part II described sexual selection, which Darwin used in Part III to explain human diversity and the origin of unique human traits. In this book, the first comprehensive theory of human evolution, Darwin emphasized that humans, like all other species, are subject to evolution by natural selection. Darwin made an accurate prediction about humans' origin: "In each great region of the world the living mammals are closely related to the extinct species of the same region. It is, therefore, probable that Africa was formerly inhabited by extinct apes closely allied to the gorilla and chimpanzee; and as these two species are now man's nearest allies, it is somewhat more probable that our early progenitors lived on the African continent than elsewhere." As Darwin noted in the book's last sentence, "Man with all his noble qualities . . . still bears in his bodily frame the indelible stamp of his lowly origin." Darwin knew that his conclusions would "be highly distasteful to many" but that "there can hardly be a doubt that we are descended from barbarians."

Although Darwin is best known for *Origin* and *The Descent of Man*, he also published several other important books, all of which supported his theory of evolution by natural selection. Darwin's two-volume *Variation of Plants and Animals Under Domestication* was meant to be the first part of Darwin's planned "big book" expanding on the "abstract" he'd published as *Origin*. Despite its huge size, *Variation* sold well. Darwin's other books about orchids, insectivorous plants, climbing plants, and barnacles described numerous evolutionary adaptations and showed that slightly modified body parts could serve different functions in new environments (i.e., were adapted to "diversified places in the economy of nature"). Similarly, his *Expression of the Emotions in Man and Animals* discussed evolutionary aspects of emotions and behavior. In late 1881, Darwin's last major effort produced *Formation of Vegetable Mould, through the Action of Worms, with Observations on their Habits*, a small, quirky book that discusses how small events over long periods of time can produce major results. While writing this book, Darwin visited Stonehenge to see how far worms might have buried the "Druidical stones." (Although Darwin also set up a "worm-stone" in his backyard to study worms' movement of soil, the stone now at Down House was reconstructed by his son Horace's Cambridge Instrument Company in 1929.) In all, Darwin published seventeen books in twenty-one volumes consisting of more than 9,000 pages of text and an additional 170 pages of preliminary matter.

Late in his life, Darwin noted that, "I can indeed hardly see how anyone ought to wish Christianity to be true; for if so the plain language of the text seems to show that men who do not believe, and this would include my Father, Brother, and almost all my friends, will be everlastingly punished. And this is a damnable doctrine." Although Darwin argued against special creation and described himself as agnostic (a word coined by his friend Huxley), Darwin did not publicly argue against religion; he denied that species have separate origins, but did not deny the existence of God.

Late in 1877, in what was one of the proudest moments of his life, Darwin received an honorary doctorate from Cambridge University.

But his health continued to decline, and Darwin suspected that his death was imminent. After joining with Huxley to convince Queen Victoria to grant the financially strapped Wallace a lifelong government pension (£200 per year), Darwin made out a will in 1881 leaving £1,000 for his friends Hooker and Huxley "as a slight memorial of my lifelong affection and respect." When in his seventies, Darwin wrote a short autobiography for his family. Darwin did not intend the autobiography for publication, but a censored version of the manuscript was published in 1887 by Francis Darwin as *The Autobiography of Charles Darwin*. Years later, Darwin's granddaughter Nora published the entire original manuscript.

By early March 1882, Darwin's health began to fail. He took his last stroll along his beloved Sandwalk on March 7, after which he became increasingly ill. On April 19, 1882, Darwin told Emma to "remember what a good wife you have been" and that he "was not the least afraid to die." Later that afternoon, at age 73, Charles Darwin died in his upstairs bedroom at Down House.

The world noted the passage of Darwin and his towering intellect. A newspaper in Vienna noted that "Humanity has suffered a great loss...our century is Darwin's century," and in Paris, editors of *France* proclaimed that Darwin's work was "an epic—the great poem of the genesis of the universe, one of the grandest that ever proceeded from a human brain." In London, editors of the *Times* wrote that, "One must seek back to Newton, or even Copernicus, to find a man whose influence on human thought...has been as radical as that of the naturalist who has just died...Mr. Darwin will in all the future stand out as one of the giants in scientific thought and scientific investigation."

On the afternoon of April 25, Darwin's body was carried from Down House in a horse-drawn hearse. At noon the following day at a standing-room-only funeral service attended by Britain's leading politicians, clergy, and scientists (Hooker, Wallace, Huxley and Darwin's neighbor John Lubbock were among the pallbearers),

the choir sang an anthem specially composed by the abbey's organist Sir John Bridge (from Proverbs 3:13-17) to exalt Darwin's life of thought: "Happy is the man that findeth wisdom and getteth understanding..." Darwin was buried in a white-oak coffin (made by Downe undertaker and builder John Lewis) in the northeast corner of the nave of London's Westminster Abbey beside astronomer John Herschel, and near his friend Sir Charles Lyell. As Darwin's body was lowered into the abbey's floor, the choir sang, *His Body Is Buried in Peace, But His Name Liveth Evermore.* Darwin was the first and only naturalist to be buried in Westminster Abbey.

Darwin has been memorialized in many ways. More than 200 places, plants, animals, and awards are named for Darwin. A *darwin* is a SI unit of evolutionary change (proposed by J.B.S. Haldane) equal to a rate of change of 0.1 percent per thousand years (this unit is seldom used because there is little agreement about what should be measured to compute the rate of change). Darwin's time in Edinburgh is commemorated by a plaque at the Royal Museum of Scotland. Darwin College of Cambridge (est. 1964) honors Darwin, as do the Darwin Awards, which were originated in 1993 by Wendy Northcutt (b. 1963) and given to people whose "spectacularly stupid" behaviors either kill them or prevent them from reproducing. In 2000, Darwin's image replaced fellow Victorian Charles Dickens on the British £10 note.

For More Information: Browne, J. 1995. *Charles Darwin: Voyaging*. Princeton, NJ: Princeton University; Browne, J. 2002. *Charles Darwin: The Power of Place*. New York: Alfred A. Knopf; Darwin, C. 1859. *On the Origin of Species by Means of Natural Selection; Or the Preservation of Favoured Races in the Struggle for Life*. London: John Murray; Darwin, C. 1871. *The Descent of Man and Selection in Relation to Sex*. London: John Murray; Darwin, C. 1993. *Autobiography of Charles Darwin 1809–1882*. Edited by N. Barlow. New York: Norton; Darwin, C. 2002 [1835]. *The Voyage of the Beagle*. New York: Barnes & Noble edition; Desmond, A., and J. Moore. 1991. *Darwin: The Life of a Tormented Evolutionist*. New York: W.W. Norton.

EMMA DARWIN (1808–1896)

Emma Wedgwood was born on May 2, 1808, at the family estate in Staffordshire, England, and was the youngest of the eight children of Josiah Wedgwood II and his wife Elizabeth. The Wedgwoods were wealthy; Emma's grandfather Josiah Wedgwood had become rich selling pottery. Emma, who spoke English, French, German, and Italian, traveled extensively when she was young; for example, she had seen the Temple of Serapis that Charles Lyell made an icon of uniformitarianism in 1830 with his *Principles of Geology*. For a while, Emma studied piano in Paris with composer Frédéric Chopin. Emma's idyllic childhood gave her a tranquility and stability that characterized her entire life.

Charles Darwin was Emma's first-cousin; their shared grandparents were Josiah Wedgwood and his wife Sarah. As a child, Charles often visited the Wedgwood estate during the summers to hunt. On September 1, 1831, Emma was with her family when they helped Charles overcome his father's objections to Charles going on the voyage of the *Beagle*.

When he returned from his voyage aboard the *Beagle*, Charles—after making a list of the pros and cons of marriage—asked Emma to be his wife. The 30-year-old Emma—who described Charles as "particularly affectionate . . . perfectly sweet tempered . . . [and] the most open, transparent man I ever saw"—accepted his proposal on November 11, 1838. Charles and Emma were married in the Church of St. Peter in Maer on January 29, 1839, just five days after Charles was elected a Fellow of the Royal Society. The Reverend John Allen Wedgwood, their cousin, officiated the wedding, after which Charles and Emma moved into a house on London's Great Marlborough Street, near Charles' brother Erasmus. They next moved to nearby Upper Gower Street. (In 1941, the house was bombed, and today the site is part of the Biological Sciences Building of the University College, London. A blue plaque notes where the Darwins' home once stood.) While living in London, the Darwins began a family; William was born on

December 27, 1839, and Anne was born on April 2, 1841 (Emma would be pregnant for most of the next sixteen years). After living in London for three years, the Darwins moved to Down House in Downe, Kent. Nine days after moving into their new home, Emma gave birth to Mary Eleanor, who died three weeks later. Less than a year later, Emma gave birth to Henrietta, the daughter who later edited *Emma Darwin: A Century of Family Letters*. Charles and Emma eventually had ten children (the last when Emma was forty-eight), but only seven reached adulthood. After their son Francis' wife died, the Darwins also raised Francis' son Bernard.

The Darwins enjoyed living in Down House; as Emma noted, "I am as spoilt as much as a heart can wish." Charles especially liked the seclusion and routine of life there, noting that, "My life goes on like clockwork, and I am fixed on the spot where I shall end it." Emma translated and responded to letters that Charles received from France, Italy, and Germany. On most evenings, Emma played piano and two games of backgammon with Charles (Emma won 2,490 games, and Charles won 2,795). Emma protected Charles, and seldom left his side.

Emma was raised in the Unitarian church, a faith that Charles' grandfather Erasmus Darwin dismissed as "a featherbed to catch a falling Christian." Emma read the Bible and attended church, where she took sacrament. The Darwins' children were baptized and confirmed in the Church of England, and Emma took them to church at Downe's St. Mary the Virgin Church. Emma was upset by Charles' views of human evolution and Christianity. Although Charles was agnostic, he tried to express his differences to Emma gently and respectfully. Charles instructed (and left £400 for) Emma to publish his expanded manuscript about evolution if he died unexpectedly.

Emma was concerned about Charles' views of religion and the afterlife, noting in a letter to Charles near the time of their marriage that she would be "most unhappy if I thought we did not belong to each other forever." Charles could not affirm Emma's belief in an afterlife together. When Charles died, Emma found her

24. Charles Darwin is memorialized in the British Museum of Natural History by this statue, which sits beside a similar statue of Thomas Huxley in the museum's cafeteria (Figure 46). (*Randy Moore*)

earlier letter to Charles, at the bottom of which he had written, "When I am dead, know that many times, I have kissed and cried over this. C.D." Charles, who was exceedingly respectful of Emma's beliefs, was often sick, and Emma cared for him with patience and strength.

On April 19, 1882, Charles died in Emma's arms in their upstairs bedroom. Emma did not attend her husband's funeral. After Charles' death, Emma moved to Cambridge, where she often enjoyed visits by Charles' old friends. In June of 1885, Emma made a rare visit to London to see the statue of Charles in The British Museum of Natural History (Figure 24). Thomas Huxley had helped raise the money to pay for the statue. Emma returned to Down House only during the summers.

Emma Darwin died on October 2, 1896, sixty years to the day when Charles stepped ashore at

Falmouth after his epic voyage aboard the *Beagle*. After a quiet service, her five surviving sons and members of the household followed the hearse to nearby St. Mary the Virgin Church, where Emma was buried along High Elms Road in the same tomb as Charles' brother Erasmus (1804–1881).

For More Information: Healey, E. 2002. *Emma Darwin*. London: Headline Book Publishers.

ERASMUS DARWIN (1731–1802)

Would it be too bold to imagine, that all warm-blooded animals have arisen from one-living filament?

Charles Darwin's eclectic grandfather, Erasmus Darwin, was born on December 12, 1731, in Elston, England. He was the seventh child of lawyer Robert Darwin (1682–1754) and Elizabeth Hill Darwin (1702–1797). Darwin's baptism was a feast that included a special beer bottled in his honor. Two unopened bottles, and the feast's menu, survive.

Darwin, aided by a scholarship of £16 per year, attended Cambridge from 1750–1754, and obtained a medical degree from Edinburgh Medical School in 1756. He began practicing medicine at Lichfield, and became rich by attending to wealthy clients. Darwin promoted education for women and despised slavery. Darwin, an avid fossil collector who admired geologist James Hutton, declined King George III's invitation to be Royal Physician.

In the mid-1860s, Erasmus helped found the Lunar Society of Birmingham, an informal group of industrialists and philosophers who "were united by a common love of science, which we thought sufficient to bring together persons of all distinctions [including] Christians...and Heathens." The Lunar Society, which met at each full moon, was an important organization for England's intellectuals in the second half of the eighteenth century. Darwin's friends and associates in the Lunar Society included Joseph Priestley (1733–1804; the preacher who discovered oxygen), flamboyant industrialist Matthew Boulton (1728–1809), Boulton's partner James

Watt (1736–1819; inventor of the steam engine), and ambitious potter Josiah Wedgwood (1730–1795; Charles Darwin's other grandfather whose company—Josiah Wedgwood & Sons, Limited—remains a thriving company today). The Lunar Men blended science, commerce, and art, and, in the process, helped start the Industrial Revolution. In 1783, Erasmus also founded the influential Derby Philosophical Society. Some of Erasmus' work inspired Mary Shelley to write *Frankenstein* (Erasmus is mentioned in the preface to the 1818 edition, and in the introduction to the 1831 edition, of the novel).

Darwin, who translated parts of Linnaeus' books, invented, or helped invent, a horizontal windmill, a canal lift for barges, a copying machine, and a variety of weather-related gadgets. Erasmus never patented any of his inventions (he believed that doing so would damage his reputation as a physician), and he encouraged others to modify and use his inventions.

Darwin was also an accomplished poet, and his poems were admired by Shelley, Coleridge, and Wordsworth. For example, Coleridge—who coined the term *darwinizing* to refer to Erasmus' speculations—claimed that Erasmus possessed "perhaps, a greater range of knowledge than any other man in Europe, and is the most inventive of philosophical men." One of Darwin's most famous books of poetry, *The Botanic Garden* (1789), speculated about cosmological theories, and noted that "the ingenious theory of Dr. Hutton" implied an eternal nature of Earth. Darwin's practical *Phytologia, or the Philosophy of Agriculture and Gardening* (1800) claimed that plants have senses and volition. Darwin speculated that electricity is the basis for nerve impulses, and that there is a "resemblance between the action of the human soul and that of electricity."

Erasmus Darwin married twice and had fourteen children. His marriage in 1757 to Mary "Polly" Howard (1740–1770) produced a daughter and four sons, the third of which—Robert Waring Darwin (1766–1848)—was Charles Darwin's father. When Mary died in 1770, Erasmus fathered two illegitimate daughters with his

17-year-old mistress Mary Parker, after which he married widow Elizabeth Pole in 1781 and moved to her home near Derby. Erasmus and Elizabeth had four sons and three daughters. Violetta, the eldest of these daughters, was the mother of eugenics crusader Francis Galton.

In 1794, Erasmus published *Zoonomia, Or, the Laws of Organic Life*, his most important book. *Zoonomia* includes a system of pathology and a section that anticipated the views of Jean-Baptiste Lamarck. In *The Temple of Nature* (published posthumously in 1803), Erasmus used rhymed couplets to describe a gradual progress of life toward higher levels of complexity and greater mental powers. Erasmus never thought of natural selection, but he did suggest that all species have a common ancestor. Darwin also suggested that species' survival was governed by "laws of nature" rather than divine authority, and that new species arise because of competition and sexual selection. However, like others before him, Darwin had little evidence to support his claims; his arguments were not convincing, and he could not explain adaptations. Nevertheless, many people recognized his contributions. For example, the preface to George Bernard Shaw's *Back to Methuselah* describes Erasmus as one of the originators of evolutionary theory.

Charles Darwin never met his eccentric grandfather (Erasmus died seven years before Charles was born) and was not overly impressed by his scientific claims. Nevertheless, Charles admired Erasmus, and when Charles was 70 years old he wrote Erasmus' biography, *The Life of Erasmus Darwin*. Charles' daughter Henrietta edited the book, and removed parts of the text (about 16 percent of the total) that she considered too salacious for the Victorian audience. In his later years, Erasmus became a hedonist and was enormously fat; he had to carve a semicircle in his dining-room table to accommodate his expanding girth.

On March 25, 1802, Darwin moved to Breadsall Priory, just five miles north of Derby. He had inherited a house there from his son Erasmus, Jr. (1759–1799), who at age 40 had drowned—probably by suicide—in the Derwent River on December 29, 1799. Two weeks after moving to Breadsall Priory, Erasmus became ill, and a week later—on April 18, 1802—he died suddenly and painlessly. He was buried beside his son, Erasmus, Jr., on April 24 at All Saints Church in Breadsall; his wife was also buried there in February 1832. When the church was renovated in 1877, Darwin's coffin was opened, and his granddaughter Elizabeth Whelen described his remains as being "in perfect preservation. He was dressed in a purple velvet dressing-gown and his features unchanged." Today, Erasmus rests in the center of the nave of All Saints Church in Breadsall, and a monument erected by his widow honoring Erasmus' "zealous benevolence" adorns the church's south wall. Darwin is also commemorated with a medallion on the Exeter Bridge.

You can visit the Erasmus Darwin House on Beacon Street in Lichfield, Staffordshire. The impressive house, which was opened to the public in 1999, is a tourist attraction and research center, and includes a herb garden and a reconstruction of Darwin's medical office.

For More Information: Uglow, J. 2002. *The Lunar Men: Five Friends Whose Curiosity Changed the World*. New York: Farrar, Straus and Giroux.

DARWIN COLLEGE (Est. 1964)

Darwin College is a graduate and postgraduate college at Cambridge University whose alumni include Dian Fossey, who studied African mountain gorillas. A fictional version of "Darwin College" appeared in the 1932 movie *Horse Feathers*, starring the Marx Brothers. The film's story centers on a football game between two rival colleges, "Darwin College" and "Huxley College." The land occupied by Darwin College was donated to Cambridge by Charles Darwin's grandson (and Lord Kelvin's godson), Charles Galton Darwin (1887–1962).

DARWIN-WALLACE MEDAL

The Darwin-Wallace Medal, which is awarded by the Linnean Society, was designed

by Frank Bowcher and first struck in 1908. Medals were awarded that year to Wallace, Hooker, Haeckel, Weismann, and Galton.

JOHN DAVIS (1873–1955)

John W. Davis, the Democratic Presidential nominee in 1924, was the ACLU's first choice to defend John Scopes' test of the Butler Law. Davis declined the ACLU's request.

PERCIVAL DAVIS

Percival William Davis was an American writer, Young-Earth creationist, and community-college biology teacher who in August 1989—with fellow creationist Dean Kenyon—published the 166-page *Of Pandas and People: The Central Question of Biological Origins*, the first textbook to use intelligent design (ID) in its current sense. ("ID means that various forms of life began abruptly through an intelligent agency, with their distinctive features already intact.") In their book, which was meant to be used as a supplemental textbook, Davis and Kenyon used six case-studies to contrast evolution with intelligent design, and concluded that intelligent design was superior. Davis, who admitted that "of course my motives were religious," also wrote *Case for Creation* (1983), which was published by Chicago's Moody Bible Institute. After *Edwards v. Aguillard* made it unconstitutional to teach "creation science" in public schools, the publishers of *Pandas* changed "creation" and "creationism" to "intelligent design." Although *Pandas* sold only 23,000 copies in its first five years, Davis and Kenyon published a second edition of *Pandas* in 1993; this edition's chapter about biochemistry was written by Michael Behe. Both editions of *Pandas* were widely criticized by scientists and educators as being inaccurate, misleading, and poorly written. In 2004, the Dover, Pennsylvania, school board's attempt to use *Pandas* as an auxiliary textbook in biology classes led to *Kitzmiller v. Dover Area School District*, which established that ID is creationism, not science.

WATSON DAVIS (1896–1967)

Watson Davis was the managing editor and reporter for *Science Service* at the Scopes Trial. Davis helped the defense choose its expert witnesses, and during the trial, Davis shared a courtroom table with journalist H.L. Mencken.

RICHARD DAWKINS (b. 1941)

We are all atheists about most of the gods that societies have ever believed in. Some of us just go one god further.

Clinton Richard Dawkins was born in Nairobi, Kenya, on March 26, 1941, to Clinton and Jean Dawkins. When World War II began, Dawkins' parents moved to Kenya, where Richard was born, and remained there until 1949. The family moved to England after the war when they inherited a dairy farm in the countryside of Oxfordshire. Dawkins credits his parents' interest in natural history and ready access to nature as an early positive influence, although he claims he never particularly enjoyed rural living. While at boarding school at the age of 16, he was first exposed to Darwin's ideas, an event that caused him to question and ultimately lose his religious faith, except for perhaps a temporary "lingering feeling about the argument from design."

Dawkins earned an undergraduate degree in zoology from Oxford in 1962, and a doctoral degree from Oxford in 1966. He studied with ethologist Nikolaas (Niko) Tinbergen who, along with Karl von Frisch and Konrad Lorenz, won the Nobel Prize in Physiology or Medicine in 1973. In 1967, Dawkins moved to the University of California, Berkeley, to become assistant professor, the same year he married his first wife Marian Stamp (now a professor of biology at Oxford). They divorced in 1984. In 1969, Dawkins returned to Oxford as lecturer in zoology, and became reader in zoology in 1990. In 1984 he married his second wife, Eve Barham, with whom he had a daughter, Juliet. This second marriage also ended in divorce. In 1992 he married his current wife, Lalla Ward (an artist and actress

best known for her role as Romana in the BBC television series *Dr. Who*), after being introduced by Douglas Adams, author of the *Hitchhiker's Guide to the Galaxy* novels. Richard Dawkins is currently Charles Simonyi Professor of the Public Understanding of Science at Oxford, a position that frees him from most teaching and research to focus on dissemination of information about science to the general public through books and personal appearances. Dawkins' position is named after Hungarian-born Charles Simonyi, who became wealthy as head of International Programming at Microsoft. In 1995, Simonyi donated £1,500,000 to Oxford after meeting Dawkins, who he has referred to as "Darwin's Rottweiler," a reference to Thomas Huxley, a contemporary of Charles Darwin who was called "Darwin's bulldog."

Soon after returning to Oxford, Dawkins studied decision-making in animals using a device that recorded data as chicks pecked at different keys. Because the device relied on electricity, Dawkins was forced to suspend his research when Britain suffered electrical outages due to industrial unrest. This prolonged interruption in his work spurred a shift in attention away from basic research and toward writing a biology book for the general public. This book, *The Selfish Gene*, was published in 1976.

The Selfish Gene identified Dawkins as an excellent writer and maverick in evolutionary biology. Drawing upon work by others, particularly evolutionary biologists W.D. Hamilton and George Williams, Dawkins proposed that the unit of selection is the gene rather than the individual. This remains a controversial idea, even within biology. Starting with Darwin, it had become well accepted that natural selection causes adaptive evolutionary change through differential reproductive success of *individuals*. Dawkins proposed, however, that individuals are simply "vehicles," for the true "replicators" are the genes. In other words, individuals are merely a gene's way of copying themselves. Dawkins also proposed that "replicators" can exist in a variety of forms, and introduced the concept of the "meme" to describe ideas, concepts, and practices that can be replicated culturally. Genes and memes are therefore specific categories of "replicators."

The gene-centered view of evolution was elaborated in Dawkins' second book, *The Extended Phenotype* (1982), in which he argued that an individual's phenotype is not restricted to the corporal entity typically identified as the "individual." Beaver dams, termite mounds, and human-built structures, for example, favor replication of the genes that lead to the creation of such entities. The true phenotype, therefore, is anything the "vehicle" has, does, or creates that promotes replication of the genes it houses.

Not only were these proposals somewhat contradictory to the standard dogma in biology, but they were also viewed as potentially dangerous. Some biologists were uncomfortable with gene-centered selection because it could be interpreted as an extreme form of genetic determinism (i.e., we are at the mercy of our genes). Harvard paleontologist Stephen Gould was especially critical, labeling Dawkins a "Darwinian fundamentalist." These responses and counterresponses became known as the "Darwin Wars" that pitched the "Dawkinsians" (e.g., Dawkins, Daniel Dennett, John Maynard Smith) against the "Gouldians" (e.g., Gould, Richard Lewontin, Mary Midgely). The conflict centered on the role of the environment (especially "culture" in relation to humans) in shaping individuals as well the level at which selection occurs. Gould in particular supported the idea that selection can occur at multiple biological levels, from gene to ecosystem. Antievolutionists claimed the conflict was evidence that evolution was an idea in trouble.

In 1986, Dawkins published *The Blind Watchmaker* to counter claims about the implausibility of evolution alone producing complex organisms. The title refers to the proposition by nineteenth-century theologian William Paley that if one were to encounter a complex entity with an obvious function (like a watch), it is logical to conclude that there is a "watchmaker" responsible for its creation. Similarly, life must have a designer. Dawkins confronts Paley's "argument from design" by discussing how the nonrandom and cumulative features of

natural selection can produce organisms with complex traits. Dawkins has continued to speak out against all forms of creationism (including intelligent design, the latest version of Paley's argument), and has expressed amazement at the rate of acceptance of biblical literalism among the U.S. population. Dawkins has claimed this as evidence that the trend toward enlightenment has proceeded in Europe, but not in the United States.

Religion as a "virus of the mind" has been a common and recurring theme in Dawkins' writings, culminating in the 2006 bestseller, *The God Delusion*. Dawkins views religion as a particularly virulent meme that is easily passed from one individual to the next, especially to young children (he equates religious indoctrination of the young with child abuse). Unlike some other scientists and philosophers, Dawkins does not propose an adaptive advantage of religion for the believer (e.g., religion could provide believers stability because it makes sense of a hostile world): religion, not the individual person, is the replicator, so its effects only have to favor the survival of the meme itself, not that of the vehicle holding the beliefs. Dawkins notes that religion has caused many serious conflicts, from the Crusades through the 9/11 terrorist attacks. ("Bush and bin Laden are really on the same side: the side of faith and violence against the side of reason and discussion. This world would be a much better place without either of them.") His proposed remedy, then, is to eliminate the destructive "religion" meme with science and reason.

Dawkins has written several other books about evolution, including *River Out of Eden* (1995), *Climbing Mount Improbable* (1996), and *The Ancestor's Tale* (2004). In *Unweaving the Rainbow* (1997), he examined the relationship between science and society (especially religion) and confronted claims by the religious community that a naturalistic worldview is one necessarily devoid of meaning and morality. To mark the thirtieth anniversary of the publication of *The Selfish Gene*, biologists and philosophers contributed essays to *How a Scientist Changed the Way We Think* (2006), which discusses the contribution of Dawkins' work to evolutionary biology and science. Dawkins has received the Royal Society of Literature Award (for *The Blind Watchmaker*), the Bicentennial Kelvin Medal of The Royal Philosophical Society of Glasgow (2002), and the Shakespeare Prize for Contribution to British Culture (2005).

For More Information: Dawkins, R. 1976. *The Selfish Gene*. Oxford, UK: Oxford University; Dawkins, R. 1986. *The Blind Watchmaker: Why the Evidence of Evolution Reveals a Universe without Design*. New York: W.W. Norton; Grafen, A., & M. Ridley, 2007. *Richard Dawkins*. Oxford: Oxford University.

DAYTON, TENNESSEE

Dayton (Tennessee) was settled around 1820. The settlement's name was originally Smith's Crossroads, but in 1877 it was renamed Dayton after Dayton, Ohio.

In 1925, Dayton hosted the Scopes Trial. Although the trial became a landmark in American history, it was instigated by Dayton businessmen as a publicity stunt to attract investors to the area. As Congressman Foster Brown of Chattanooga noted, the trial was "not a fight for evolution or against evolution, but a fight against obscurity." Scopes' trial brought hundreds of visitors to Dayton, but within a week after the trial, virtually all of the spectators, street preachers, circus performers, and hucksters had left town, and Dayton returned to normal. Some people profited from the trial, but the long-term economic stimulus that the trial's instigators had sought never materialized. Several years after the trial, Bryan College opened in Dayton to honor the ideals of William Jennings Bryan, one of Scopes' prosecutors.

John Scopes' trial was held in the Rhea County Courthouse (Figure 25; also see Appendix). The courthouse, an Italian villa-style building built in 1891, was designated a National Historic Landmark in 1976, and was restored with the completion of the Scopes Trial Museum in the courthouse basement in 1978. At the front of the famous courtroom is posted a page from the *Congressional Record* listing the Ten Commandments. In 2005, a Cessna Decosimo statue of William Jennings Bryan was unveiled outside the courthouse; this statue depicts Bryan in 1891, when he began his Congressional career, and when the courthouse was built. Every July, the Dayton Chamber of Commerce hosts the Scopes Trial Play and Festival, which includes a reenactment of the trial in the courtroom in which Scopes' famous trial took place.

ROGER DEHART

Roger DeHart was a biology teacher at Burlington-Edison High School in Burlington, Washington, who in 1997 was reassigned to teaching Earth science after it was discovered he was using the pro-intelligent design (ID) textbook *Of Pandas and People*. Inquiry later indicated that DeHart had been discussing ID in his classroom for nearly ten years, with tacit approval from the administration. Legal action threatened by the ACLU finally caused the school to relieve him of teaching biology. The 2002 documentary film *Icons of Evolution* (based on Jonathan Wells' book of the same name) portrayed DeHart as "a modern-day John Scopes in reverse," and used DeHart's story to promulgate the Discovery Institute's "teach the controversy" campaign for inclusion of ID in public school science curricula.

TOM DELAY (b. 1947)

I don't believe there is a separation of church and state. I think the Constitution is very clear. The only separation is that there will not be a government church.

Tom DeLay was a Republican member of the U.S. House of Representatives from 1984 to 2006, serving as Majority Leader from 2002 to 2005 after serving as House Majority Whip for eight years. While in the House, DeLay advocated a conservative agenda—especially that of the religious right—including the outlawing of abortion, limiting the rights of homosexuals, protecting gun ownership, and instituting prayer in public schools. His domineering style earned him the nickname "The Hammer," which reflects his legendary ability to work the political system to further his vision for the United States.

25. In 1925, Dayton, Tennessee, hosted the Scopes Trial, the most famous event in the history of the evolution-creationism controversy. The trial occurred in the second-floor courtroom of the Rhea County Courthouse. Throughout Dayton are various trial-related sites (Appendix). (*Randy Moore*)

DeLay was born on April 8, 1947, in Laredo, Texas. DeLay was not close to his parents (claiming he "pretty much raised himself"), especially his father, an alcoholic who set exceedingly high expectations for his four children. Over time, DeLay became estranged from his parents and siblings. He earned a BS in biology from the University of Houston in 1970. In 1978, DeLay was elected to the Texas House of Representatives and remained there until 1985 when he was elected to represent Texas' twenty-second Congressional District in the U.S. House of Representatives.

While a state representative, DeLay developed a reputation as a playboy and carouser. However, after becoming a born-again Christian, DeLay began promoting "a biblical worldview" for America. According to DeLay, "Only Christianity offers a comprehensive worldview that covers all areas of life and thought, every aspect of creation." In June 1999, in the wake of the shootings at Columbine High School in Colorado, DeLay blamed the tragedy on "school systems [that] teach the children that they are nothing but glorified apes who evolutionized out of some primordial soup of mud." Although this quote has been attributed to DeLay, he read it in an article by Addison Dawson that had recently appeared in the *San Angelo Standard-Times*. (Nevertheless, his reading of it into the Congressional Record clearly indicates his approval of its content.) DeLay endorses intelligent design ("you must have a designer...God") and rejects evolution ("Give me one example that proves evolution. One example! You can't.").

In 2002, DeLay again attracted media attention (especially in his home state of Texas) when, after a speaking engagement, he responded to an audience member's request for guidance on where to send his child to college. The audience member bemoaned the fact that universities (even "here in Texas") generally do not include creationism in science curricula. DeLay responded with, "Don't send your kids to Baylor. And don't send your kids to A&M," referring to Baylor University and Texas A&M University, respectively, two major universities in Texas. Ironically, DeLay himself attended Baylor University (a private Baptist-affiliated university) from 1965 to 1967, but was expelled, and his daughter graduated from Texas A&M. Furthermore, up until 2000, Baylor University housed the Michael Polanyi Center for Complexity, Information, and Design, which studied intelligent design. Baylor is the only research university ever to create such a center.

In 2005, DeLay was indicted in Austin, Texas, on criminal charges of conspiracy to violate election laws and money laundering. Soon thereafter, he resigned his position as House Majority Leader and did not seek reelection. The charge of conspiracy was eventually dropped, but in 2008 he remained under indictment for money laundering. DeLay has denied any wrongdoing.

DELUGE SOCIETY

The Deluge Society, formally known as The Society for the Study of Deluge Geology and Related Sciences, was formed in 1938, and from 1941 to 1944 published twenty issues of the *Bulletin of Deluge Geology and Related Sciences*. Most members of the society, including founder George Price, were Seventh-day Adventists. By 1945, the Society included more than 600 members, all of whom were required to believe that creation lasted no more than "six literal days."

WILLIAM DEMBSKI (b. 1960)

Convinced Darwinists...need to block the design inference whenever it threatens to implicate God.

Once this line of defense is breached, Darwinism is dead.

William Albert Dembski was born July 18, 1960, in Chicago, Illinois. Dembski earned a BA in psychology in 1981 and an MS in statistics in 1983 from the University of Illinois at Chicago (UIC), a PhD in mathematics from the University of Chicago in 1988, MA and PhD degrees in philosophy from UIC in 1993 and 1996, respectively, and a Masters of Divinity from Princeton Theological Seminary in 1996. He also completed postdoctoral work at MIT (mathematics), the University of Chicago (physics), and Princeton University (computer science). He was awarded the Templeton Foundation's Book Prize ($100,000) for 2000–2001 for his writings on information theory.

Dembski did not question evolution as he was growing up, but became increasingly skeptical of the power of natural selection by the time he began college. In 1988, he attended a conference about randomness that noted that patterns are inevitably discovered within what initially appears random. Dembski concluded that randomness was "always a provisional designation until we found the pattern or design in it," a conclusion that prompted him to begin studying design in nature.

In 1992, Dembski met philosopher Stephen Meyer. Dembski, then in graduate school at the University of Chicago, was "doing the design work on the side," and had attracted Meyer's attention. Meyer and Dembski joined with other intelligent design (ID) proponents, including Phillip Johnson, Michael Behe, and Jonathan Wells, all of whom would soon become fellows (Meyer as director) of the Center for the Renewal of Science and Culture (later renamed the Center for Science and Culture) at the Discovery Institute. The duo of Behe and Dembski provided ID proponents with what they considered the necessary one-two punch: Behe's *Darwin's Black Box* (1996) purported to demonstrate the "irreducible complexity" of the basic unit of all life, the cell, while Dembski's work claimed evidence for the impossibility of life arising without the aid of a designer.

In 1998, Dembski's *The Design Inference: Eliminating Chance through Small Probabilities* introduced the use of an "explanatory filter" to identify design in nature. Dembski proposed that all patterns in nature have only three possible causes: regularity (the action of natural laws, e.g., gravity), chance, and design. The filter works as a simple flowchart: if the pattern can be explained by natural laws, the pattern is explained, otherwise move onto chance as an explanation, and then, finally, design.

Distinguishing patterns that are simply unlikely from those possible only through design—the heart of the matter for Dembski—requires identification of patterns that are not only unlikely (i.e., complex) but also demonstrate "specified complexity." For example, any particular sequence of heads and tails when flipping a coin 100 times is equally unlikely, and, therefore, any sequence so obtained would not require an explanation invoking anything beyond chance. Improbability alone, therefore, is not sufficient for identifying design. However, a specific sequence—say 100 consecutive heads—is extraordinarily unlikely, and, if encountered, would require an explanation that involves more than just chance. The *irreducible* complexity Behe claimed to have demonstrated was exactly the *specified* complexity Dembski required as evidence for design. Using a complicated formula that includes the number of particles in the universe and a physical constant known as Planck time, Dembski proposed that patterns less likely than one chance in 0.5×10^{150} are evidence for design.

Dembski's *Intelligent Design: The Bridge Between Science & Theology* (1999) continued his exploration of specified complexity, casting the search for design within the framework of information theory. In 2002, Dembski's *No Free Lunch: Why Specified Complexity Cannot Be Purchased Without Intelligence* extended his search for design through application of "No Free Lunch" (NFL) theorems. NFL theorems, in general, compare the overall performance of alternative "search algorithms" for accomplishing particular outcomes (e.g., the best route to take through a landscape to reach a desired objective). Dembski concluded that natural selection cannot perform better than random chance in generating specified complexity. Dembski's efforts, consequently, moved quickly from merely suggesting how design can be identified to a refutation of evolution.

Critics of Dembski's work noted that even though *Design Inference* passed inspection by the Cambridge University Press editorial board, none of Dembski's work had received standard peer evaluation. Dembski has claimed that he prefers to disseminate his work in book form because it is a faster process (Dembski has authored or coauthored an average of one book per year). Despite his training in mathematics, Dembski has been criticized harshly for his "mathematism," which relies on "pseudo-mathematical jargon" to cloak nonscientific ideas in scientific wrappings. Accusations of the misuse of probability, and, most recently, misapplication of the NFL theorems, have generated negative reviews (e.g., David Wolpert, who in part developed the NFL theorems, has described Dembski's work as "written in jello"). Dembski's proposed "filter" for identifying design has also been criticized as being inappropriate for ruling out the action of adaptive evolution (i.e., natural selection is not goal-directed) and for equating the action of natural selection with chance. ID proponents have, however, portrayed Dembski's work as a rigorous refutation of evolution (e.g., University of Texas philosopher and Discovery Institute fellow Robert Koons has called Dembski "the Isaac Newton of information theory").

Dembski's professional career started as a fellow with the Discovery Institute in 1996, where he is now a senior fellow. In 1999, Dembski joined Baylor University as Associate Research Professor, and soon thereafter was appointed by Baylor's president Robert Sloan to head the newly created Michael Polanyi Center (MPC) for Complexity, Information, and Design. The secretive nature of the creation of the MPC caused dissent among Baylor's faculty, especially when Dembski's agenda for the study of ID became apparent. In response, Sloan convened a committee of outside experts that ultimately recommended folding the MPC into an existing university entity. Dembski publicly declared the committee's findings a victory for ID over its opponents ("[they] have met their Waterloo"). When Sloan asked Dembski to retract the comments, Dembski refused, and Sloan removed Dembski as director of the MPC. Dembski, who had a multiyear contract with Baylor, remained at the school for five more years, during which time he worked off-campus on various books. Dembski later remarked that, "In a sense, Baylor did me a favor. I had a five-year sabbatical."

In 2005, Dembski moved briefly to the Southern Baptist Theological Seminary in Louisville, Kentucky, where he established the seminary's Center for Science and Theology. Less than a year later, he moved to the Southwestern Baptist Theological Seminary in Fort Worth, Texas, accepting the position of Research Professor in Philosophy. (Kurt Wise, formerly of Bryan College, replaced Dembski in Louisville.) In 2001, Dembski established the International Society for Complexity, Information, and Design, "a cross-disciplinary professional society that investigates complex systems apart from external programmatic constraints like materialism, naturalism, or reductionism."

Dembski also advocated the "wedge" strategy developed in the 1990s at the Discovery Institute. He later proposed that the strategy be morphed into a "vise" strategy (subtitled "Squeezing the Truth out of Darwinists") that would "make clear to those reading or listening to the Darwinists' testimonies that their defense of evolution and opposition to ID are prejudicial, self-contradictory, ideologically driven, and above all insupportable on the basis of the underlying science." Dembski is remarkably anti-science, claiming that "the scientific picture of the world championed since the Enlightenment is not just wrong, but massively wrong."

In the wake of the anti-ID decision in the *Kitzmiller v. Dover Area School District* case, Dembski posted on his blog an animation (that he helped create) showing Dover presiding judge John Jones as a flatulent puppet of the ACLU and evolutionary biologists. (Dembski originally agreed to testify in the Dover trial, but withdrew before the trial started.) Dembski was also

in "constant correspondence" with conservative commentator Ann Coulter during development of her confrontational book, *Godless: The Church of Liberalism* (2006), which claims that evolution is not science (e.g., it is no better supported than Coulter's mocking "Flatulent Raccoon Theory" for the origin of life) and represents part of the "religion" of liberalism.

For More Information: Dembski, W.A. 1998. *The Design Inference*. Cambridge: Cambridge University.

DANIEL DENNETT (b. 1942)

The kindly God who lovingly fashioned each and every one of us and sprinkled the sky with shining stars for our delight—that God is, like Santa Claus, a myth of childhood, not anything a sane, undeluded adult could literally believe in.

Daniel Clement Dennett was born in Boston, Massachusetts, in 1942. As an undergraduate at Wesleyan University, Dennett developed an interest in philosophy. A crucial event in his young academic career was reading *From A Logical Point of View* by Harvard philosopher and mathematician, Willard van Orman Quine (1908–2000). Dennett recalled that "I thought he [Quine] was wrong...but...I have to get where the action is" and he immediately transferred to Harvard. Quine's verificationist perspective (i.e., a statement can only be accepted if it can be empirically or analytically verified) influenced Dennett's work, especially his views of science and religion.

After receiving his undergraduate degree in philosophy from Harvard in 1963, Dennett and wife Susan, whom he married in 1962, moved to England for graduate study with Oxford philosopher Gilbert Ryle. In the 1940s, Ryle had famously derided Descartes' mind-body duality with the phrase "ghost in the machine." Dennett would later build on this perspective in *Consciousness Explained* (1991), which further denied the existence of what Dennett termed the "Cartesian theater." Supporting himself during graduate school as a jazz musician (Dennett also claims to have introduced the Frisbee to Britain during this period), he completed his DPhil in 1965. He immediately accepted a faculty position at the University of California—Irvine, and remained there until 1971. While in California, Dennett's first book, *Content and Consciousness* (1969), was published, largely based on his graduate work. In 1971, he moved to Tufts University, where he is currently the Austin B. Fletcher Professor of Philosophy and Director of the Center for Cognitive Studies. Dennett and his wife have two grown children, and, in addition to his professional activities, he enjoys sculpture, sailing, and farming.

Much of Dennett's early academic work focused on the mind and consciousness, and in particular, analysis of the mind as an evolved trait. His naturalistic perspective, derived from Quine and Ryles, led him to conclude that natural selection is the only explanatory framework necessary for investigating consciousness. His perspective of the "robotic" nature of conscious thought ("we are each made of mindless robots and nothing else") is extreme among philosophers and conflicts with Western religious perspectives on the soul and free will.

In 1995, Dennett published *Darwin's Dangerous Idea*. The book's title refers to the fear many have about applying Darwinian thinking to understanding consciousness: the ultimate conclusion would be that what is considered to be indicative of the human soul—consciousness—is merely another product of adaptive evolution. (Phillip Johnson, in response, has referred to "Dennett's dangerous idea.") Dennett considers natural selection a "universal acid" that, when invoked as an explanatory framework, dissolves all other possible explanations and lays bare how the universe actually operates. These ideas were expanded in *Kinds of Minds* (1996) and *Freedom Evolves* (2003); the latter book addresses the issue of free will in humans. Dennett maintains that human behavior was at one time completely deterministic, but it no longer needs to be because, with increased scientific understanding of the natural world, we have the freedom to control our own future (human brains can "produce future").

Dennett's ideas have been criticized by some philosophers and evolutionary biologists.

Stephen Gould, whose contributions to evolutionary biology were treated somewhat harshly in *Darwin's Dangerous Idea*, labeled Dennett's ideas "Darwinian fundamentalism." Gould railed against Dennett's strict determinism for failing to account for the role of contingency and nonadaptive change in evolution; that is, human behavior and culture are too complex to be explained solely by the action of natural selection. Not surprisingly, Dennett denied Gould's claims of fundamentalism and bias. This bitter and public exchange (the comments of both were published in the *New York Review of Books*) between two respected figures in evolutionary biology disturbed many biologists.

In 2006, Dennett published *Breaking the Spell: Religion as a Natural Phenomenon*, which considered religious belief from an evolutionary perspective. Dennett professes his atheism (he is an outspoken member of the "Brights" movement), although he considers himself the "good cop" in comparison to arch-atheist Richard Dawkins' "bad cop." However, Dennett has not been particularly conciliatory, having compared the "God meme" to the Mafia; both may have ancillary benefits, but that doesn't justify tolerating either. He acknowledges that it is not surprising that religious belief developed because in prescientific times it would have made an unpredictable and often hostile world comprehensible. But he argues our scientific understanding has surpassed this benefit, and it is time for religion to move aside.

That Dennett has tackled antievolutionism, particularly intelligent design (ID), is not surprising. (It *is* somewhat surprising that he played fire-and-brimstone preacher Jeremiah Brown in his high school's production of *Inherit the Wind*.) He has argued that claims made by the ID movement have been refuted by contemporary biology, and labeled the "teach the controversy" strategy as hollow because ID is devoid of content and, therefore, there is no controversy, only misunderstanding. Antievolutionists have pointed to Dennett's attacks on religion as a clear example of what they claim is the ultimate goal of evolutionary biology: the elimination of religion and its replacement by science.

MICHAEL DENTON (b. 1943)

Michael Denton is Senior Research Fellow at the Department of Biochemistry at the University of Otago where he studies the molecular genetics of retinal disorders. Although Denton accepts that natural selection can cause microevolutionary change, he claims that evidence for speciation is either lacking or supports a typological (unchanging) view of species. Denton believes that there is no known mechanism that could cause speciation, and Darwin's proposal of common descent is unsupported. For Denton, the complexity of organisms precludes that "these things have come about in terms of a gradual accumulation of random changes." Denton views mutation—what he terms the "bedrock of Darwinism"—as being an "implausible" agent for causing adaptive evolution. In papers published in *Nature* and other scientific journals, he and colleagues purport to have discovered "natural laws" that determine the folding of proteins that cannot be accounted for by natural selection. Denton has developed his objections to evolution in *Evolution: A Theory in Crisis* (1986) and *Nature's Destiny: How the Laws of Biology Reveal Purpose in the Universe* (1998).

DEVIL'S CHAPLAIN

Devil's Chaplain was a phrase jokingly coined in 1856 by Charles Darwin in a letter to his friend Joseph Hooker: "What a book a Devil's Chaplain might write on the clumsy, wasteful, blundering low and horridly cruel works of nature." In 2003, biologist Richard Dawkins published a best-selling book titled *A Devil's Chaplain: Reflections on Hope, Lies, Science, and Love*.

HUGO MARIE DE VRIES (1848–1935)

The Mutation theory is opposed to that conception of the theory of selection which is now prevalent.... According to the Mutation theory individual variation has nothing to do with the origin of species.

Hugo Marie de Vries was born February 16, 1848, in Lunteren, Holland, into a family that valued academic achievement: his father was a lawyer and politician, and his mother's family included several university scholars. In 1866, de Vries entered college, and in 1870 he earned his doctorate from the University of Leiden. He then studied in Germany, before becoming a lecturer in plant physiology at the University of Amsterdam in 1877, where he eventually attained the rank of professor and directed the Botanical Institute until his retirement in 1918.

By the 1880s, de Vries was studying inheritance. In 1886, he discovered a diverse local population of the evening primrose (*Oenothera lamarckiana*) that he believed was undergoing speciation. Adopting the primrose as his model organism, de Vries noted that particular characters would occasionally disappear and reappear across generations. These observations led de Vries to the two main concepts with which he is associated: pangenesis as a model of heredity, and mutation as a mechanism of evolution.

De Vries became convinced that the "hereditary factors" within individuals acted as discrete units rather than being mixed together as proposed in the blending model of inheritance. Using Charles Darwin's theory of pangenesis ("gemmules" within an organism determine the characters exhibited by that individual) as a starting point, de Vries developed a model of "intracellular pangenesis." He proposed that the characteristics of organisms are determined by cellular particles he called "pangenes" (the origin of the term *gene*, later introduced by Wilhelm Johannsen), with each specific pangene affecting a particular phenotypic trait. Each cell in an organism bears the same pangenes; however, because pangenes exist in two forms—active (expressed) and latent (not expressed)—cells can differ because they have unique combinations of active and latent pangenes. De Vries developed his theory of pangenesis in *Intracellulare Pangenesis* (1889).

Through his work with evening primrose, de Vries also identified two general classes of variability, "fluctuating" and "specific." "Fluctuating variability" was the minor variation seen within a population, which de Vries ascribed to differences in number of determining pangenes among individuals. He proposed that favorable environmental conditions increase the number of copies of a pangene, and therefore, "selection is in fact the choice of the best-nourished individuals." Fluctuating variability was therefore inconsequential for evolutionary change, and was expected to demonstrate "regression to the mean" of the parental generation, as outlined by Francis Galton. "Specific variability," however, was due to a pangene splitting into multiple forms, each of which could multiply. The creation of these new pangenes allows the development of entirely new characters. These conclusions encouraged de Vries to reject Darwin's model of selection via slow, gradual change in favor of a saltational model driven by mutations.

By the 1890s, de Vries' studies of mutation were based on his model of pangenesis. De Vries explained "hereditary monstrosities"—for example, plants with twisted stems or alternate number of cotyledons—as being due to modifications of the respective pangenes, and he considered these new forms as separate species. De Vries refined his perspective on mutation in his two-volume set *Mutationstheorie* (1901 and 1903), and proposed three types of mutations: progressive, regressive, and degressive. Regressive mutations convert an active pangene into the latent form, and are associated with disappearance of a trait. Degressive mutations explain the presence of intermediate versions of traits because fully active pangenes can be converted into semilatent or semiactive forms. Progressive mutations create new characters, and are therefore the only heritable changes that cause speciation. De Vries did not rule out the action of natural selection (although it was necessarily selection between species), as he believed that only progressive mutations that conferred an advantage on the bearer would persist. De Vries' mutational model of evolution became popular during the early part of the twentieth century, before the basis of heredity was determined.

In 1900, de Vries claimed that he had independently discovered the principles of heredity earlier described by Mendel. At the time, this elevated de Vries to the level of "rediscoverer" of Mendelian inheritance (along with Carl Correns and Erich Tschermak who each, in 1900, also claimed independent discovery of Mendel's principles). However, comparison of de Vries' presentation of his data before and after he is known to have read Mendel's work suggests that he may have modified his data as well as his interpretations in response to having Mendel's conclusions in hand (e.g., a 2:1 phenotypic ratio reported in 1897 became a 3:1 ratio, as found by Mendel, in 1900). Similarly, de Vries' discussion of inheritance prior to 1900 was not in Mendelian terms, but rather was based on his pangenesis model that involved extremely mutable pangenes and fluctuating variability resulting from differing numbers of pangenes among individuals. After 1900, however, de Vries' treatment of these ideas incorporated Mendel's ideas: *latent/active* became *dominant/recessive*, and pangenes were described as being considerably more stable across generations. Consequently, it is now widely suspected that de Vries did not independently derive the laws of heredity discovered by Mendel, but rather attempted to place his work within the framework Mendel had developed. In later publications, de Vries credited Mendel's work, although he never completely harmonized his concepts of heredity with those of Mendel.

De Vries died on May 21, 1935. Although he predicted his synthesis of pangenesis, mutation, and evolution would lead to "wider and wider areas of research in heredity," his ideas were soon eclipsed by those of Mendel. Evidence for the mutational theory of evolution was not found in other species, and even the traits in evening primrose that de Vries had interpreted as mutations were later shown to result from a complex form of genetic recombination that supported Mendel's proposals. De Vries is, however, remembered as an imaginative thinker and dedicated experimentalist, and his work encouraged serious consideration of many aspects of evolutionary biology.

MICHAEL DINI (b. 1954)

Michael Dini was a biology professor at Texas Tech University who based his recommendations for students on his knowledge of the student, their having earned an A in a course he taught, and their acceptance of Darwinian evolution. Dini justified his policy by noting that students who deny the evidence for evolution were committing "malpractice regarding the method of science." In February 2003, student Michael Spradling complained that Dini was using religion as a basis for discriminating against students. When the Liberty Legal Institute complained to the U.S. Justice Department that Dini was denying students' First Amendment right to freedom of religion, the Justice Department launched an investigation. Dini changed his criteria for providing a letter of recommendation, but still required that students provide a scientific explanation for the origin of humans. In 2003, the Justice Department dropped its investigation.

DINOSAUR NATIONAL MONUMENT

Dinosaur National Monument in northeastern Utah is the largest quarry of Jurassic Period dinosaur bones ever discovered and the only national park devoted to dinosaurs. One of the first geologists to visit the area was Earl Douglass (1862–1931), who had been hired by eugenicist and steel magnate Andrew Carnegie to gather specimens for Pittsburgh's new Carnegie Museum. Douglass' discovery in 1909 of eight tailbones of *Apatosaurus* eventually led to the excavation of more than 350 tons of dinosaur skeletons from the site. In 1915, President Woodrow Wilson declared 80 acres around the site a national monument. Since then, the park has expanded to more than 200,000 acres.

DISCOVERY INSTITUTE (Est. 1990)

The Discovery Institute was founded as a Seattle branch of the Hudson Institute, a conservative think-tank based in Indianapolis, Indiana. Bruce Chapman (b.1940), a former Secretary

of State for the State of Washington (who also served as Director of the U.S. Census Bureau in the early 1980s and as Deputy Assistant to President Ronald Reagan from 1983 to 1985), joined the Hudson Institute in the late 1980s and agreed to return to the state of Washington to open the Seattle office.

The Discovery Institute originally focused on regional economic, communications, and transportation policy. However, intelligent design (ID) became a major emphasis of the Institute after Chapman met Stephen Meyer in 1994. At that time, Meyer began working with Phillip Johnson, William Dembski, and Michael Behe to coalesce the ID movement. Chapman likely realized that the perceived conflict between science and religion, in the guise of ID, could attract wealthy donors to his financially struggling organization that was now independent of the Hudson Institute.

Meyer had tutored the children of Howard Ahmanson, a wealthy conservative who had inherited a fortune from his banker father, and he set up a meeting among Ahmanson, Chapman, and himself. Ahmanson, who supported "the total integration of biblical law into our lives," provided $750,000 over three years to found the Center for the Renewal of Science and Culture (CRSC) within the Discovery Institute. The CSRC became the research center for ID, and Meyer, Behe, Dembski, and Jonathan Wells were named fellows of the Center. Meyer became director of the CRSC, and Johnson served as advisor.

Johnson, a Berkeley law professor who concluded in *Darwin on Trial* (1991) that evolution was an unsupported theory, wanted to undermine what he saw as the dominant and pernicious materialism that excluded God as an explanation for nature. Johnson and others developed a multiyear plan—the "wedge strategy"—that used "atheistic evolution" to "replace materialistic explanations with the theistic understanding that nature and human beings are created by God." The strategy was to be kept secret, but the document detailing the plan was discovered when sent out for photocopying and was anonymously posted on the Internet. The CRSC (renamed the Center for Science and Culture to reduce the aura of a religion based agenda) initially denied authorship of the "wedge document," although it eventually acknowledged that it had originated from the Discovery Institute as an "early fundraising proposal."

Out of the controversy surrounding public scrutiny of the wedge document, and coincident with the the Kansas Board of Education's decision to no longer require the teaching of evolution in public schools, the CSC morphed its wedge strategy into "teach the controversy." This approach does not exclude evolution from science curricula, but does portray it as a "theory in crisis." For the teaching of evolution and questions of origins to be impartial, therefore, adequate coverage of the problems and limitations of the theory are required, which in turn allows inclusion of alternative explanations such as intelligent design. The CSC claims that these efforts promote "academic freedom" so teachers can overcome the "Darwinian fundamentalists" who are "waging a malicious campaign to demonize and blacklist anyone who disagrees with them." Meyer also claims that the "teach the controversy" proposal is federally mandated, appealing to *Edwards v. Aguillard* (1987), which he interprets as requiring an equal hearing of all alternative scientific theories. (This strategy actually originated with Wendall Bird of the Institute for Creation Research in the 1980s.) CSC claims that ID is neutral as to who or what the designer is, and therefore ID is a scientific explanation that should be heard. The Discovery Institute claims that it rejects efforts to prevent the teaching of evolution, or to require the teaching of any particular theory, including ID. For example, the Discovery Institute publicly disagreed with the "misguided policies" of the Dover, Pennsylvania, School Board to include ID in ninth-grade biology. However, CSC fellows testified on behalf of the defendants, and the Discovery Institute condemned the conclusions of the presiding judge, John Jones.

CSC helps sympathetic state legislators craft legislation advocating the "teach the controversy" approach. These efforts reached the federal level in 2001 when Johnson wrote what become known as the "Santorum Amendment." This proposed amendment to the 2001 Elementary and Secondary Education Act Authorization Bill (later known as the No Child Left Behind Act) was introduced by Rick Santorum, then a U.S. Senator from Pennsylvania. The amendment proposed that biology instruction should discuss "why [evolution] generates so much continuing controversy," a clear reflection of the CSC's main strategy. The amendment was deleted from the final bill, although the Discovery Institute claims that "teaching the controversy" became federal law.

Although the Discovery Institute promotes itself as the research home for ID (e.g., Chapman announced in 2006 that the Institute had "put over $4 million toward scientific and academic research into evolution and intelligent design" since 1996), critics have noted that the Institute has not produced peer-reviewed publications. In 2001, the Discovery Institute launched "A Scientific Dissent from Darwinism," which asked those holding a PhD in science, engineering, mathematics, or computer science to endorse the statement "We are skeptical of claims for the ability of random mutation and natural selection to account for the complexity of life." As of 2007, more than 600 signatures had been collected. However, the *New York Times* and other organizations reported that most individuals who endorsed this statement are neither biologists nor experts in evolutionary biology. (The National Center for Science Education started the "Project Steve" parody, which asks PhD-level scientists named "Steve"—which should be about 1 percent of all scientists and was chosen in honor of the late Stephen Gould—to sign a letter in support of evolution. As of 2007, more than 800 names had been collected supporting "Project Steve." Also, at the time of the Dover trial in 2005, the "A Scientific Support for Darwinism" project collected 7,733 signatures from scientists worldwide in four days.) The Discovery Institute also provided a viewer's guide titled *Getting the Facts Straight* to accompany PBS's 2002 series on evolution. The guide condemns the documentary as a "misuse of taxpayer money to organize and promote a controversial political action agenda."

The Discovery Institute also studies transportation issues in the Pacific northwest (the Cascadia project), the role of technology in improving the quality of life (the Technology and Democracy Project), and national fiscal and monetary policy. The Institute is well funded, with an annual budget exceeding $4 million (including $1 million a year for ten years from the Bill and Melinda Gates Foundation to support the Cascadia project). ID, however, is viewed by Chapman as his organization's "number one project." The Discovery Institute plans to release a textbook titled *Explore Evolution* that purports to discuss alleged "scientific controversies" associated with evolution. In August 2007, the Institute began promoting *Expelled: No Intelligence Allowed*, a movie scheduled for release in February 2008, which promises to "expose the frightening agenda of the 'Darwinian Machine.'" The movie stars actor, writer, and game-show host, Ben Stein.

AMZI DIXON (1854–1925)

Amzi Clarence Dixon was a prominent Baptist fundamentalist and pastor at Chicago's Moody Church who was hired by Lyman Stewart to edit the first five volumes of *The Fundamentals: A Testimony to the Truth*. Dixon, who felt "a repugnance to the idea that an ape or an orang-outang was my ancestor," denounced Catholicism, liquor, Henry Beecher's liberalism, and evolution, but conceded that Gap evolutionists and theistic evolutionists could be Christians. *The Fundamentals* generated much interest; after publication of the first two volumes, the editors and publisher received 10,000 letters, and after Volume 3, 25,000 letters. Subsequent editors of *The Fundamentals* included Dwight Moody's colleague Reuben Torrey (1856–1928), who claimed that "evolution is a guess pure and simple, without one scientifically observed fact to build upon." Torrey acknowledged "for purely scientific reasons" that someone could "believe thoroughly in the absolute infallibility

of the Bible and still be an evolutionist of a certain type."

THEODOSIUS DOBZHANSKY (1900–1975)

I am a creationist and an evolutionist. Evolution is God's, or Nature's method of creation. Creation is not an event that happened in 4004 BC; it is a process that began some 10 billion years ago and is still under way.

Theodosius Grigorievich Dobzhansky ("Doby") was born January 25, 1900, in Nemirov, Russia (now Ukraine). His family moved to the outskirts of Kiev in 1910. By the age of 12, Doby had decided to become a biologist, and his reading of *On the Origin of Species* at the age of 15 profoundly affected him. Dobzhansky graduated from the University of Kiev with a degree in biology in 1921, and taught zoology and genetics at the Polytechnic Institute in Kiev and the University of Leningrad until 1927. He then started collaborating with Sergei Chetverikov, a geneticist who, essentially unknown outside of Russia, was developing the same theoretical underpinnings of contemporary evolution as the better-known geneticists in Europe and the United States. This early collaboration helped Dobzhansky to later unify genetics and evolution.

In 1924, Dobzhansky married Natasha Sivertzev, also a geneticist. They would have one child, Sophia, and would remain together until Natasha's death in 1969. In 1927, Dobzhansky earned a fellowship from the Rockefeller Foundation in New York to study cytogenetics at Thomas Hunt Morgan's "fly lab" at Columbia University. Morgan and Dobzhansky moved to the California Institute of Technology in 1928. Morgan offered, and Dobzhansky accepted, a position as assistant professor of genetics at Caltech in 1929, advancing to full professor in 1936. In 1940, Dobzhansky returned to Columbia as professor of zoology. He became increasingly uncomfortable with the academic politics at Columbia, and moved to the Rockefeller Institute (renamed Rockefeller University in 1965) in 1962. Dobzhansky retired in 1970, but accepted an adjunct position at the University of California, Davis, in 1971 to collaborate with former student, Francis Ayala.

Dobzhansky was a main architect of the modern synthesis of evolutionary biology of the 1920s and 1930s that integrated Mendel's understanding of inheritance with Darwin's mechanism of evolutionary change. Ronald Fisher, Sewall Wright, and J.B.S. Haldane had developed the mathematical models of evolutionary genetics, but these efforts were primarily theoretical and in a form inaccessible to biologists lacking the necessary mathematical sophistication to use this work.

Under Morgan's tutelage, Dobzhansky pioneered the use of *Drosophila* for experimental genetic studies, both in the lab as well as the field. In contrast to Morgan's predictions, Dobzhansky demonstrated that populations typically harbor a surprising amount of genetic variation, which could allow populations to evolve quickly as environmental conditions change. Dobzhansky melded these experimental findings with the theoretical work of Fisher, Wright, and Haldane in his *Genetics & the Origin of Species* (1937). That book (written in a month as Dobzhansky recovered from a horseback riding accident) was enormously influential on the growth of evolutionary biology because Dobzhansky's synthesis provided observational support of the theoreticians' proposals and addressed a central topic of concern for evolutionary biology, speciation. The book—which Stephen Gould described as "the founding document of the modern synthesis"—has been hailed as second only to *On the Origin of Species* in terms of importance to evolutionary biology.

Dobzhansky continued his studies of *Drosophila* into the 1940s, and transformed genetics research by combining field and lab work. He demonstrated how the frequencies of particular gene variants could be altered by changes in the environment (i.e., temperature), indicating that natural selection can indeed be studied in the field. In 1935, Dobzhansky developed the concept of a species as an "actually or potentially interbreeding array of forms," a

concept later developed as the *biological species concept* by colleague Ernst Mayr. Subsequent work by Dobzhansky demonstrated how the formation of sterile hybrids between *Drosophila* species kept these species separate, leading Dobzhansky to develop the influential concept of "isolating mechanisms" for the process of speciation. In subsequent versions of *Genetics & the Origin of Species*, the importance of nonadaptive evolutionary change (i.e., genetic drift as championed by Wright) was reduced in importance relative to selection. Historians have identified this as a critical event in the evolution of the modern synthesis into the "neo-Darwinian" perspective that emphasized selection as the main force producing evolution.

Because of his personal history (growing up in Imperial Russia; experiencing the rise of Bolshevism) and the era in which he worked (the rise of Fascism in Europe; civil rights movement in the United States), Dobzhansky was eager to apply his understanding of evolution to humans. In *Heredity, Race, & Society* (1946, with L.C. Dunn) and *Genetic Diversity & Human Equality* (1973) he addressed the issue of race. Dobzhansky noted that there is greater genetic variation within human races than between them as evidence that racism and social injustice are cultural constructions. The genetic distinctiveness of individual humans is therefore not a basis for exclusion, but is rather a reason for equality of all people. In his 1962 book, *Mankind Evolving*, Dobzhansky discussed how humans—who he referred to as the "apex of evolution"—are the product of two interacting forces, biology (shared with all other life), and culture (unique to humans). Not surprisingly, Dobzhansky was an outspoken critic of eugenics. He was also troubled by Trofim Lysenko's campaign against genetics (and geneticists) and science in his former homeland, and translated Lysenko's writings into English so they could be available for scrutiny.

In 1972, Dobzhansky addressed the National Association of Biology Teachers (NABT). His theme was "Nothing in Biology Makes Sense Except in the Light of Evolution," and with the publication of his talk in the *American Biology*

Teacher in 1973, this title became a rallying cry for the teaching of evolution for decades to follow. Even if one knew nothing about Dobzhansky's monumental contributions to evolutionary biology, one was well acquainted with this single declarative statement. In his talk, Dobzhansky discussed his religious beliefs, claiming himself to be a "creationist" who believes the Creator works through observable natural forces. Toward the end of his life, Dobzhansky referred frequently to the writings of Jesuit priest Pierre Teilhard de Chardin, and Dobzhansky's statement to the NABT likely has its basis in Teilhard's "Evolution is a light which illuminates all facts, a trajectory which all lines of thought must follow."

Former students and colleagues have referred to Dobzhansky as a "religious man," but organized religion was apparently unimportant to him, although he did belong to the Eastern Orthodox Church. He rejected fundamentalism and possibly the idea of a personal God (although late in life he prayed regularly), but did believe that life had meaning. Dobzhansky became more religious toward the end of his life, with the primary catalysts likely being the death of his wife and his own declining health. Richard Lewontin (b. 1929), a former student of Dobzhansky's and now professor of biology at Harvard, has noted that Dobzhansky "did not seem to believe in a conventional afterlife" and recalls Dobzhansky joking that if Heaven did exist, it would be "a perpetual existence in which experiments would give unambiguous results." Because Dobzhansky believed that natural forces—including evolution—could be guided by divine intervention, Dobzhansky is often cited as an example of the compatibility of religious faith and acceptance of evolution.

In 1969, Natasha, Dobzhansky's wife of nearly forty-five years, died. A year earlier, Dobzhansky himself had been diagnosed with leukemia and given a few months to a few years to live. The disease, however, progressed slowly, and it wasn't until 1975 that his condition became serious. On December 19, 1975, Dobzhansky died as he was being taken to the hospital by his former student, Francisco Ayala (b. 1934).

Dobzhansky received numerous awards and honorary degrees during his long and productive career, including election to the National Academy of Sciences, the American Academy of Arts and Sciences, and the Royal Society. He was awarded the Darwin Medal from the Academia Leopoldina in 1959 and earned the National Medal of Science in 1973. Dobzhansky, a prolific writer, published over 500 papers and a dozen books during his lifetime.

For More Information: Adams, M.B (ed.). 1994. *The Evolution of Theodosius Dobzhansky*. Princeton, NJ: Princeton University.

DOWN HOUSE

Just outside the small village of Downe, Kent—about sixteen miles southwest of London—is Down House, where Charles and Emma Darwin raised their children and lived for more than forty years (Figure 26). (When the village changed its name from Down to Downe, the house kept the old spelling.) The Darwins' large house, which was built late in the eighteenth century, fronted a 15-acre meadow and greenhouse where Charles did many of his experiments. At the western edge of the property was the "Sandwalk" or "thinking path," along which Darwin walked three times per day to digest his thoughts.

Inside Down House, Darwin spent much time in the billiards room, where cause-and-effect relationships were more straightforward than in his theory of natural selection. Across the hall was the dining room and its bay windows that looked out on the Darwins' large backyard. A large portrait of Charles' grandfather, Erasmus, hangs in the dining room; meals in this room were the focal point of the day. Adjacent to the billiards room was Darwin's study, which housed his writing table, microscope, and a few biological specimens—all of which were overlooked by portraits of Josiah Wedgwood, Joseph Hooker, and Charles Lyell. Down the hall, in the drawing room, Emma played piano and backgammon. In his study and in the large armchair by the drawing room's fireplace, Darwin

26. Down House is a sprawling estate in Downe, Kent (16 miles southwest of London), where Charles Darwin and his family lived for more than forty years. Darwin wrote most of his books at Down House, which is now a public museum. (*Randy Moore*)

sat, pen in hand, and discovered his revolutionary idea.

Charles and Emma had ten children:

William Erasmus (1839–1914) was the Darwins' eldest son. William, an amateur photographer, became a banker.

Anne "Annie" Elizabeth (1841–1851), a daughter whose death at the age of 10 changed Darwin's views of Christianity.

Mary Eleanor (1842) died three weeks after birth. She is buried at St. Mary the Virgin Churchyard, not far from Down House.

Henrietta Emma "Etty" (1843–1930), who was the Darwins' only married daughter, helped her father with his work. Etty, who was known around Down House as "Body" and "Budgy," refuted the claims of "Lady Hope" that her father had denounced evolution and converted to Christianity while on his deathbed. In 1915, Etty edited *Emma Darwin: A Century of Family Letters*.

George Howard (1845–1912) graduated from Cambridge, where he was later employed as a mathematician and astronomer. George, who helped his father with proofs of *The Descent of Man*, was knighted in 1905, and was president of the British Association for the Advancement of Science in 1905. When Emma died in 1896, she left Down House to George. In 1903, George Darwin and John Joly were among the first people to claim that radioactivity is at least partly responsible for Earth's heat.

Elizabeth "Bessy" (1847–1926) who never married and had no descendants.

Francis "Frank" (1848–1925) was a Cambridge botany professor. In 1880, Francis and his father wrote *The Power of Movement in Plants*. Francis, who was qualified as a physician but never practiced medicine, served as President of the British Association for the Advancement of Science in 1908, was knighted in 1913, and was elected to the Royal Society in 1882 (the year his father died). In 1887,

Francis edited Thomas Huxley's *On the Responses of the Origin of Species* and *The Autobiography of Charles Darwin*, and later produced *The Life and Letters of Charles Darwin* (1887) and *More Letters of Charles Darwin*. In 1912, Francis received the Royal Society's Darwin Medal, named after his famous father. Francis' son Bernard (1876–1961), who was raised at Down House, became captain of the Royal and Ancient Golf Club in 1934. There was at least one Darwin who was a fellow, without a break, of the Royal Society for 200 years.

Leonard (1850–1943) was a soldier and president of the Royal Geological Society from 1908 to 1911. In 1925, Leonard sent a letter to John Scopes congratulating him for "his courageous effort to maintain the right to teach well established theories...To state that which is true cannot be irreligious...May the son of Charles Darwin send you in his own name one word of warm encouragement." Leonard was a close friend of geneticist and statistician Ronald Fisher, and Fisher dedicated his *The Genetical Theory of Natural Selection* to Leonard. Leonard, who was president of the Eugenics Education Society from 1911 to 1928, was awarded an honorary doctorate from Cambridge in 1912. Leonard was the last surviving child of Charles and Emma Darwin.

Horace (1851–1928) became an engineer who built scientific instruments. Horace, who was knighted in 1918, was the founder and director of the Cambridge Scientific Instrument Company, and mayor of Cambridge from 1896 to 1897. Horace's daughter Nora Darwin Barlow (1885–1989) linked Charles Darwin with twentieth-century scholars—and helped create the "Darwin industry"—when she edited republications of *Diary of the Voyage of the Beagle* (1932), *Darwin and Henslow* (1967), and *Charles Darwin and the Voyage of the Beagle* (1945).

Charles Waring (1856–1858), like Mary Eleanor, was born and died at Down House. Charles Darwin missed the reading of his and Wallace's paper to the Linnean Society on July 1, 1858, because he was mourning his son's death. Charles is buried at St. Mary the Virgin Churchyard.

Several descendants of the Darwins continued to use "Etta," "Emma," and "Charles" as names of their children. Some also continued to intermarry with Wedgwoods, as well as descendants of Thomas Huxley.

Soon after Charles Darwin died in 1882, Emma moved to Cambridge, after which she returned to Down House to spend her summers. When Emma died in 1896, she was buried in the same tomb as Charles' brother Erasmus Alvey Darwin (1804–1881) in the churchyard of nearby St. Mary the Virgin Church, not far from some of her children and Joseph Parslow, the Darwin's servant who worked at Down House for more than thirty-six years.

In 1906, Down House was turned into the Down House School for Girls. Down House was opened as a public museum on June 7, 1929, and today a restored Down House is maintained by English Heritage, and is open for tours most of the year. Down House was scheduled to become a World Heritage Site in 2009.

For More Information: Atkins, H. 1976. *Down: Home of the Darwins.* London: Phillmore.

EUGÈNE DUBOIS (1858–1940)

Embryology and comparative anatomy and atavisms can only provide indirect proof for the existence of the close tie between Man and the animal world, only paleontology provides direct proof.

Eugène Dubois was born into a wealthy family on December 28, 1858, in Eijsden, Holland. Eugène expressed at a young age an interest in natural history, and he was encouraged in these pursuits by his father. When he was 10 years old, newspaper accounts of a public lecture on evolution by biologist Carl Vogt exposed Dubois to the complicated interactions between science and society as he witnessed his village's widespread refusal to accept the evidence for Darwin's theory.

Although Dubois fully embraced evolutionary thinking, it was Ernst Haeckel's *History of Creation* (1868) that had the biggest impact: Haeckel discussed the common evolutionary history of apes and humans and predicted that a transitional form should exist in the fossil record. Haeckel called this—yet undiscovered—transitional form "ape-like man," or *Pithecanthropus*, and identified two characteristics that should mark this "missing link": bipedality and changes in the larynx that in later forms would allow for complex language.

In 1877, Dubois began studying medicine at the University of Amsterdam. During his studies, he completed the transformation to agnosticism began as a child. After finishing his medical studies in 1881, Dubois worked at the University of Amsterdam. By the mid-1880s, Dubois was publishing his anatomical research on the evolution of the larynx, following Haeckel's proposal of the central role of this trait in human evolution. In 1886, Dubois married Anna Lojenga, with whom he had three children.

Early in Dubois' career, sensational reports of Neanderthal fossils being found in caves in Europe rekindled his earlier interest in the evolutionary relationships between apes and humans. A re-reading of Haeckel suggested that if a true "missing link" was to be found—not just the later human-like forms that the Neanderthal fossils represented—one would need to look outside of Europe. Ignoring Darwin's prediction that Africa represented the cradle of human evolution, Dubois, who accepted Haeckel's claim that Asia was where humanity arose, believed that to find the "missing link" he had to look where the great-ape ancestors must have lived—namely, on the Asian side of Wallace's Line. Dubois finally decided that the island of Sumatra was the likely home of the link between apes and humans. Sure of his logic and convinced of the importance of such an undertaking, Dubois resolved to find Haeckel's *Pithecanthropus*. After completing the delicate task of convincing his wife, he resigned his university position (prompting comments that he was throwing away his life for "that crazy book of Darwin's") and enlisted as a doctor in the Dutch Indian Army (the Netherlands then ruled the East Indies). On October 29, 1887, Dubois and family sailed for Sumatra.

Dubois found several fossils of extinct animals (none of them the hoped-for "missing link") and published an article about his discoveries, which prompted the Dutch government to support Dubois with engineering assistance and forced laborers. By 1889, frustrated by his lack of success in Sumatra, Dubois moved his search to Java, and in 1891, after four arduous years, Dubois and his team found a molar of an apelike primate along the Solo River outside the tiny village of Trinil. Within a month they also found a fossil skullcap, and in 1892, a near complete left femur. Based on his training as an anatomist, Dubois determined that the organism had been bipedal with a cranial capacity of around 1,000 cm^3, indicating a brain size intermediate between chimpanzees and humans. In a report summarizing his findings, Dubois concluded that the fossil "is the first known transitional form linking Man more closely with his next of kin among the mammals," and suggested that it supported the contention that "the first step on the road to becoming human taken by our ancestors was acquiring upright posture." Dubois named the find *Pithecanthropus erectus*, combining Haeckel's proposed genus with a specific epithet that emphasized the distinctive bipedal nature of the organism. Dubois referred to the fossil informally as "*P.e.*," but the rest of the world knew it simply as "Java Man." Dubois erected a small monument at the site of his famous discovery; that monument does not include Dubois' name.

Dubois returned to the Netherlands in 1895, but did not receive the acclaim he expected. He published extensively on his discovery, but the fossils were initially dismissed as from a giant ape or a human. The strongest criticism came from the renowned physician and scientist Rudolf Virchow, who believed the fossil to be of a gibbon. Haeckel, however, supported Dubois, which encouraged wider acceptance of Dubois' conclusions. One colleague, Gustav Schwalbe, expressed a particular interest in the fossils, and procured from Dubois a cast of the skullcap as well as an opportunity to examine the fossils themselves. To Dubois' horror, Schwalbe used this access to produce a series of detailed articles on the fossils. Thus betrayed by Schwalbe, and exhausted by his relentless drive to have his findings unequivocally accepted, Dubois began to withdraw from the scientific community.

Because he had invested so personally in uncovering the fossils, Dubois staunchly refused to relinquish control of them, and when traveling among universities to give presentations, he transported the fossils in a battered suitcase. By the turn of the century, his feelings of frustration overwhelmed him and, locking the fossils away, he turned to other scientific endeavors, including study of geology and climate. By 1910, additional finds and continued work by others stimulated renewed interest in *P.e.*, and Dubois was contacted by several anthropologists interested in examining the fossils. Dubois hid his fossils beneath the floor of his house in Holland and refused to let anyone see his now famous fossils. By the 1920s, the scientific community, led by Henry Fairfield Osborn (paleontologist and director of the American Museum of Natural History), wanted Dubois to place the fossils in a museum. Grudgingly, Dubois finally prepared casts of the fossils and sent them to Osborn. These episodes left Dubois even more embittered, and he spent his later years fighting with anthropologists who suggested that *P.e.* was not the "missing link" Dubois had originally claimed. Ultimately, *Pithecanthropus erectus* was reclassified as an example of *Homo erectus*, a widespread species that existed in Asia, Africa, and Europe, 1 to 2 million years ago.

Dubois died of a heart attack on December 16, 1940. His tombstone bears the carving of a skull and crossbones, evocative of a pirate's grave, but also appropriate to one who searched for and found buried treasure on a tropical island.

For More Information: Shipman, P. 2002. *The Man Who Found the Missing Link: Eugene Dubois and His Lifelong Quest to Prove Darwin Right.* Cambridge, MA: Harvard University.

ADRIAN DUPLANTIER (b. 1929)

Whatever "science" may be, "creation," as the term used in the statute, involves religion.

Adrian Guy Duplantier was born in 1929 in New Orleans, Louisiana. Duplantier graduated from the Loyola University School of Law in 1949, after which he began a private law practice. He subsequently taught in the English Department and Dental School at Loyola University, worked as an assistant district attorney in Louisiana from 1954–1956, and served as a state senator in Louisiana from 1960 to 1974. In 1978, Duplantier became U.S. District Judge in the Eastern District of Louisiana.

Duplantier was the presiding judge who, on January 10, 1985, ruled that Louisiana's law requiring "balanced treatment" for evolution and creationism was unconstitutional. Duplantier considered the case against the state so convincing "that whatever that evidence would be, it could not affect the outcome." Duplantier noted that the concepts of creation and a Creator originate in religious conviction, and that Louisiana's law was unconstitutional "because it promotes the beliefs of some theistic sects to the detriment of others." The Fifth U.S. Circuit Court of Appeals upheld Duplantier's ruling, noting that "in truth, this particular case is a simple one, subject to a simple disposal: the Act violates the establishment clause of the First Amendment because the purpose of the statute is to promote a religious belief." In 1987, the U.S. Supreme Court—by a vote of 7–2—affirmed Duplantier's opinion (*Edwards v. Aguillard*).

E

EDIACARA HILLS

The Ediacara Hills of South Australia are where geologist Reginald Sprigg (1919–1994) in 1946 found the earliest known communities of animals. These enigmatic animals, known as Ediacarans, lived during the Precambrian (650–544 million years ago) and comprise a strange collection of mostly sessile creatures. Some Ediacarans may be early relatives of extant phyla; others are extinct mysteries.

LOREN EISELEY (1907–1977)

Charles Darwin saw a vision. It was one of the most tremendous insights a living being ever had.

Loren Corey Eiseley was born on September 3, 1907, in Lincoln, Nebraska. Eiseley had a difficult childhood and, after dropping out of high school, he worked at a variety of jobs before enrolling at the University of Nebraska. When Eiseley contracted tuberculosis in 1933, he dropped out of the university and became a hobo, riding around the country on freight trains. In 1957, Eiseley described his adventures in *All the Strange Hours: An Imaginative Naturalist Explores the Mysteries of Man and Nature*, the most widely read of Eiseley's many books.

When his health improved, Eiseley returned to the University of Nebraska, from which he earned a BS in English and Anthropology. In 1937, he earned a PhD from the University of Pennsylvania, and became a professor at the University of Kansas. After working three years at Oberlin College in Ohio, Eiseley returned to the University of Pennsylvania as head of the Anthropology Department. From 1959 to 1961, Eiseley was provost of the university.

In 1957, Eiseley's *The Immense Journey* described human history. Unlike most other science writers, Eiseley combined science with humanism and poetry. The following year, Eiseley wrote *Darwin's Century* (1958), a popular and award-winning book that described the development of evolutionary theory. *Darwin's Century* became a staple on college campuses for the next two decades. Eiseley spent much time searching for fossils of prehistoric humans, but never made a major discovery.

In 1959, Eiseley claimed that Darwin had taken many of his ideas about natural selection from Edward Blyth (1810–1873), a correspondent of Darwin's who Darwin acknowledged in Chapter 1 of *On the Origin of Species*. In 1979, Eiseley's executors expanded Eiseley's claim into the posthumous *Darwin and the Mysterious Mr. X* (Blyth was "Mr. X"). Eiseley was elected in 1971 to the National Institute of Arts and Letters, a rare honor for a scientist.

Eiseley died on July 9, 1977, and was buried in West Laurel Hill Cemetery in Bala Cynwyd, Pennsylvania. Eiseley's epitaph (shared with his wife Mabel) comes from his poem *The Little Treasures:* "We loved the Earth But Could Not Stay."

For More Information: Christianson, G.E. 1990 (reissued in 2000). *Fox at the Wood's Edge: A Biography of Loren Eiseley*. Lincoln, NE: University of Nebraska.

NILES ELDREDGE (b. 1943)

In a way creationism was good for evolutionary biology... it reminded us that we have, after all is said and done, more in common as evolutionists than we have issues that drive us apart.

Niles Eldredge was born in New York State on August 25, 1943. In ninth grade he was placed in the "slow science class," and remembers hearing "a bit" about evolution in high school. He entered Columbia University in the early 1960s intent on studying Latin. While at Columbia, he met his future wife (with whom he would have two children), who introduced him to anthropologists who, in turn, introduced him to evolution, particularly human evolution. After spending a summer "asking embarrassing personal questions" of Brazilian villagers, he became convinced that cultural anthropology was not the career he should pursue. Instead, he became fascinated by fossils at the village's harbor, and began to consider paleontology as a field of study. After graduating in 1965 with an AB in geology, Eldredge entered graduate school at Columbia to study with Norman Newell, professor of paleontology and curator at the American Museum of Natural History. Eldredge earned his PhD in 1969. He studied trilobites for his dissertation and has continued this research throughout his career.

While in graduate school, Eldredge was surprised by paleontology's negligible impact on evolutionary understanding (it was used primarily to identify patterns in the fossil record). In the 1940s, paleontologist George Gaylord

Simpson had ruefully commented on how the study of evolution had been appropriated by geneticists ever since the modern synthesis began in the 1920s. Eldredge was frustrated that little had changed in the intervening decades, noting that "it was the evolutionary process—and not just the simple facts of evolutionary history—that I longed to study."

In 1971, Eldredge published a paper suggesting the fossil record indicates that most species do *not* gradually and continually change, but rather go through long periods of stasis followed by rapid but infrequent periods of diversification. Eldredge was influenced by the geographic speciation model developed by Theodosius Dobzhansky and elaborated by Ernst Mayr which argued that speciation is most likely at the edges of species' ranges, while the core of the population remains unchanged over long periods. Eldredge's paper generated little interest. However, a year later, Eldredge, along with another of Newell's graduate students, Stephen Gould, developed the idea further, publishing it as "punctuated equilibrium." Gould maintained the idea was generated by Eldredge, but Eldredge credits Gould's "knack for catchy phrases" as being responsible for attracting interest.

Biologists everywhere began discussing the tempo of evolution, often concluding that Gould and Eldredge were wrong (sometimes making their point by referring to punctuated equilibrium as "evolution by jerks"). Eldredge and Gould were labeled as "traitors to the Darwinian tradition." Both spent years trying to demonstrate that punctuated equilibrium is consistent with contemporary understanding of evolution, and that even Darwin suggested that "gradualism" (often portrayed as the mutually exclusive alternative to punctuated equilibrium) is not the only possible pattern of evolution. Creationists used the controversy surrounding punctuated equilibrium as evidence that evolution was a theory in trouble, and Eldredge and Gould were frequently misquoted by antievolutionists.

These events spurred Eldredge to publish *The Monkey Business: A Scientist Looks at Creationism*

(1982) to confront "scientific creationism" and *The Triumph of Evolution and the Failure of Creationism* (2001) in response to intelligent design. He even turned his professional attention to his personal collection of over 400 cornets (he is an amateur jazz trumpeter and avid collector of cornets) to predict the features that an intelligently designed system of objects, like cornets, should possess. His "evolutionary tree" of the cornet has none of the features found in an evolutionary tree of actual organisms (e.g., innovations frequently spread horizontally and simultaneously across lineages in designed objects, but not so in the evolution of actual organisms), as one would expect when comparing designed versus evolved entities.

Eldredge has elaborated the concept of punctuated equilibrium by proposing that local and global environmental change (occurring on different time scales) affect the tempo and scale of evolution. His model, which he calls the "sloshing bucket," invokes ongoing localized environmental change that alters the composition of natural communities without promoting speciation. This standard state is infrequently but catastrophically punctuated by major events (e.g., bolide impacts, volcanism) that lead to mass extinctions that promote diversification (i.e., "most evolution seems to be correlated with extinction events"). These ideas are discussed in Eldredge's 1995 book, *Reinventing Darwin: The Great Debate at the High Table of Evolutionary Theory*.

Eldredge also dismisses the gene-centered view of evolution supported by Richard Dawkins, Daniel Dennett, and others, which proposes that traits of individuals exist only to pass on the "selfish genes" that determine those traits. Eldredge, like Gould, favors a hierarchical model in which evolution results from several forces, including chance and environmental change as well as fitness differences among genetic variants. Applying this perspective to sex, Eldredge argues in *Why We Do It: Rethinking Sex and the Selfish Gene* (2004) that it is unlikely that selfish genes are the sole reason for sex, particularly in humans, because it is such an inefficient process for passing on genes. Instead, Eldredge argues that sex seems to have been decoupled

from reproduction for "economic" reasons (i.e., sex for "fun, profit, and power").

In 2005, the American Museum of Natural History opened the major exhibit *Darwin* that described the development of Darwin's ideas through use of live organisms, videos, and Darwin's manuscripts and possessions. Eldredge was the curator responsible for the exhibit's content. The exhibit was intended to celebrate Darwin's contributions to science, but it also discussed Social Darwinism and creationists' opposition to evolution. *Darwin: Discovering the Tree of Life* (2005), Eldredge's twenty-fourth book, accompanied the exhibit.

ELLESMERE ISLAND

Ellesmere Island is part of the Canadian Arctic Archipelago where, in 2006, University of Chicago paleontologist Neil Shubin discovered fossils of a Paleozoic (375-million-year-old) fish, *Tiktaalik rosea* (in the language of the local Inuit, *tiktaalik* means "large fish in a stream"). Although Ellesmere Island is now only about 900 miles from the North Pole, 375 million years ago—just about the time when the transition of animal life from water to land was suspected to have occurred—Ellesmere Island was a subtropical river delta (continental drift later moved it to its present location). The 1.3-meter-long *Tiktaalik*, which are called "fishapods," had gills, scales, fins, and a snout (like fish), but also—like land animals—had flexible necks, eyes atop broad alligator-like skulls, wrists, and limb-like fins having flexible elbow joints, a primitive wrist, and five fingers. *Tiktaalik* is the sort of transitional fossil predicted by Darwin's theory. As Cambridge University biologist Jennifer Clack noted, "It's one of those things you can point to and say, 'I told you this would exist,' and there it is."

EARL ELLINGTON (1907–1972)

Earl Buford Ellington was the governor of Tennessee who, on May 18, 1967, signed House Bill No. 48 repealing the Butler Law that had been used to prosecute John Scopes. Ellington is buried in Lone Oak Cemetery in Lewisburg, Tennessee.

GEORGE ELLIS

George W. Ellis was a Kentucky state legislator who on January 23, 1922, introduced the nation's first antievolution bill four days after hearing William Jennings Bryan's antievolution speech to a joint meeting of the Kentucky legislature. Ellis' bill was defeated by a vote of 42–41, but Ellis encouraged "open war against Infidel Evolution." During the next ten years, more than forty similar bills were introduced in twenty different states. Seventy-two years after the vote on Ellis' legislation, Kentucky deleted the word *evolution* from its state educational guidelines.

PAUL ELLWANGER

Paul Ellwanger was a respiratory therapist who molded Wendell Bird's proposal for "balanced treatment" into legislation that would introduce "creation science" into public schools. Ellwanger founded Citizens for Fairness in Education, "a citizen group, national in scope, who favor academic freedom and are opposed to suppression of information about evolution and creation." Ellwanger's efforts convinced several state legislatures to consider mandating equal time for creationism. When Judge William Overton ruled in 1982 in *McLean v. Arkansas* that requiring equal time for creationism is unconstitutional and that "creation science" is not science, Ellwanger changed the wording and distributed a new draft of the equal-time act.

EMPEDOCLES (c. 490–430 BC)

Empedocles was a Greek philosopher who around 440 BC claimed that the universe and everything in it has been gradually changing. Empedocles, who promoted spontaneous generation, also suspected that extinctions occur: "Many races of living creatures must have been unable to continue their breed: for in the case of every species that now exists, either craft, or courage, or speed, has from the beginning of its

27. In 1965, Susan Epperson was a high school biology teacher in Little Rock, Arkansas, who sued to overturn the state's ban on the teaching of evolution. Epperson's lawsuit, the first evolution-related lawsuit since the Scopes Trial, produced *Epperson v. Arkansas*, in which the U.S. Supreme Court ruled unanimously that laws banning the teaching of evolution in public schools are unconstitutional. This photo, taken in June 1967, shows Susan discussing her lawsuit in her attorney's office. (*Jerry Tompkins*)

existence protected and preserved it." Empedocles believed that love joined animals' parts together, and hate kept them apart.

SUSAN EPPERSON (b. 1941)

The brief expressed in legal terms everything that I felt and only fear of the controversy was making me hesitate. Now, only cowardice would keep me from signing on as plaintiff in the suit, and that wasn't a good enough reason.

Susan Epperson was born on May 21, 1941, in Little Rock, Arkansas (Figure 27). Her father was a biology professor at the College of the Ozarks in Clarksville, and Epperson often walked to the Science Building to see the plants and animals on display in the classrooms. Epperson attended Clarksville High School, skipping her senior year to attend the College of the Ozarks. In 1962, Epperson earned her BS in biology. She received her MS in zoology from the University of Illinois in Urbana in 1964. In the fall of 1964, Epperson was hired to teach tenth-grade biology at Little Rock Central High School.

In 1965, the Arkansas Education Association (AEA) asked Virginia Minor—a teacher at Little Rock Central High School who directed a kindergarten training program—to help them find a plaintiff to challenge Arkansas' law banning the teaching of human evolution. At about the same time, Susan Epperson, then in her second year as a biology teacher at Little Rock's Central High School, read an article in the *Arkansas Gazette* about the law. In 1965, Epperson had helped choose a new biology textbook—*Modern Biology* by Moon, Otto, and Towle—which claimed that "changes in plants and animals" over time showed that humans and apes "may have had a common, generalized ancestor in the remote past." Although its statements were cautious and qualified, they nevertheless violated the Arkansas antievolution law. (In fact, the law had already been violated, for the school library included books about evolution, including Darwin's *On the Origin of Species*.) Thus, in 1965 Epperson faced a dilemma: she was required to use the prescribed textbook, but doing so would be a crime and subject her to being fired. Most teachers avoided this dilemma by skipping the chapters on evolution; some schools even omitted biology altogether. Epperson, however, decided to challenge the law.

Epperson was an ideal plaintiff for the AEA: She was a public school teacher who had been born, raised, and educated in Arkansas; she had graduated from a church-affiliated college; she was a devout Christian; and Epperson's husband, Jon, was a First Lieutenant in the Strategic Air Command at the Little Rock Air Force Base, thereby neutralizing charges of "outside agitation" that had surfaced during the racial confrontations in Little Rock (as well as during the

Scopes Trial). Jon, as well as Susan's principal at Central High School (Harry Carter), supported Susan's lawsuit. The Eppersons' pastor at Little Rock's Second Presbyterian Church offered to be Susan's bodyguard.

Epperson's challenge of the antievolution law began in mid-September of 1965, when Forrest Rozzell, the Executive Director of the AEA, issued a "personal position" statement outlining the importance of understanding nature and recommending the repeal of the state's 1928 law. Two months later, Rozzell (who had been taught biology at The College of the Ozarks by Epperson's father) met with Epperson, Minor, and Eugene R. Warren (attorney for the AEA) in Minor's kindergarten classroom to discuss the 1928 antievolution law. Epperson agreed to test the law "because of my concept of my responsibilities both as a biology teacher and as an American citizen." The ACLU offered to help Epperson, but she declined; she wanted her lawsuit to be initiated by the people of Arkansas against the Arkansas law. However, the ACLU joined Epperson's lawsuit when it reached the U.S. Supreme Court.

On December 6, 1965, Warren filed the complaint, requested a declaratory judgment about the constitutionality of the antievolution law, and asked that the school board not be allowed to fire Epperson. Hubert Blanchard, Jr., Associate Executive Secretary of the AEA and a parent who wanted his two sons to "be informed of all scientific theories and hypotheses," joined Epperson's action because he believed that his children were being denied a good education. Epperson's lawsuit was the first since the Scopes Trial in 1925 to challenge an antievolution law.

When Epperson's lawsuit was made public, people began taking sides. John Scopes offered encouragement, but admitted that "it's damn near a lost cause." Although the *Arkansas Gazette* praised Epperson's courage, William Tinkle of the Creation Research Society warned Epperson that "evolution is on its way out" and asked, "Can't you delay your action against [the law] while you study the facts?" Epperson released a statement describing her Christian faith and her commitment to "the scientific search for truth," adding that to ignore her responsibilities as a teacher was "the sure path to the perpetuation of ignorance, prejudice, and bigotry." Some Christians asked Epperson why her mother had allowed her to live, and others branded Epperson an atheist bound for hell.

Although Warren wanted to avoid "the carnival atmosphere that went on [at the Scopes' Trial] in Tennessee," others tried to sensationalize the trial. Governor Orval Faubus told his Sunday school class that the antievolution law was good and that he "was not ready to repeal the Bible." Arkansas Attorney General Bruce Bennett—waving a Bible while claiming that "this is the Bible, buddy [and] I intend to defend it"—complained to trial judge Murray Reed that Epperson "was the only person since the law was approved to 'clamor' in favor of teaching that man evolved from monkeys, apes, sharks, porpoises, seaweed, or any other form of animal or vegetable," and that the teaching of evolution would make it possible for "crackpots" to teach the "God is Dead" theory, the theory that "man came from a gorilla," and the "Ham and Eggs Theory of California." Bennett denounced evolution as an atheistic doctrine, and warned that Epperson wanted to convert her students to atheism. Epperson, meanwhile, released her own statement:

> I AM A TEACHER BY PROFESSION. I CHOSE TO BECOME A TEACHER BECAUSE I BELIEVE THAT TEACHING IS THE MOST IMPORTANT PROFESSION TO WHICH A PERSON CAN DEDICATE HIS TALENTS AND ENERGIES. I PURSUED AN EDUCATION COURSE IN COLLEGE TO BECOME A SCIENCE TEACHER AS COMPETENT AS MY CAPABILITIES WILL PERMIT. I RECEIVED A BACHELOR'S DEGREE FROM THE COLLEGE OF THE OZARKS AND A MASTER OF SCIENCE DEGREE FROM THE UNIVERSITY OF ILLINOIS.
>
> MY MOTHER IS A PUBLIC SCHOOL LIBRARIAN AND MY FATHER HAS BEEN A PROFESSOR OF BIOLOGY FOR ALMOST HALF A CENTURY. THEY ARE BOTH DEDICATED CHRISTIANS WHO SEE NO CONFLICT BETWEEN THEIR BELIEF IN GOD AND THE SCIENTIFIC SEARCH FOR TRUTH. I SHARE THIS BELIEF.
>
> AS A NEW TEACHER (THIS IS MY SECOND YEAR), AND, AS A NATIVE ARKANSAN, ARKANSAS'

> ANTI-EVOLUTION LAW HAS DISTURBED ME MORE THAN A LITTLE.
>
> I DO NOT TRY TO TEACH MY STUDENTS WHAT TO THINK. I TRY TO TEACH THEM HOW TO THINK, HOW TO MAKE SOUND JUDGMENTS ABOUT THE VARIOUS RELEVANT ALTERNATIVES. IN DOING THIS, IT IS MY RESPONSIBILITY TO EXPOSE MY STUDENTS TO AND ENCOURAGE THEM TO SEEK AFTER AS MUCH OF THE ACCUMULATED SCIENTIFIC KNOWLEDGE AND THEORIES AS POSSIBLE. RATIONAL KNOWLEDGE, AS ACCURATE AND BALANCED AS HUMANLY POSSIBLE AT ANY GIVEN TIME, IS ESSENTIAL TO THE MAKING OF SOUND VALUE JUDGMENTS.
>
> WHEN I WAS ASKED TO BECOME THE PLAINTIFF IN THIS TEST SUIT I AGREED TO DO SO BECAUSE OF MY CONCEPT OF MY RESPONSIBILITIES BOTH AS A TEACHER OF BIOLOGY AND AS AN AMERICAN CITIZEN. THIS LAW, PROHIBITING ANY TEACHER FROM DISCUSSING IN ANY WAY THE DARWINIAN THEORY OF EVOLUTION, COMPELS ME EITHER TO NEGLECT MY RESPONSIBILITY AS A TEACHER OR TO VIOLATE MY RESPONSIBILITY AS A CITIZEN. AS A RESPONSIBLE BIOLOGY TEACHER, IT IS MY DUTY TO DISCUSS WITH MY STUDENTS AND TO EXPLAIN TO THEM VARIOUS SCIENTIFIC THEORIES AND HYPOTHESES IN ORDER THAT THEY MAY BE AS EDUCATED AND ENLIGHTENED AS POSSIBLE ABOUT MATTERS PERTAINING TO SCIENCE, INCLUDING THE THEORIES OF DARWIN AS SET FORTH IN "ON THE ORIGIN OF SPECIES" AND IN "THE DESCENT OF MAN." HOWEVER, WHEN I DO THIS, I BECOME AN IRRESPONSIBLE CITIZEN – A LAW VIOLATOR – A CRIMINAL SUBJECT TO FINE AND DISMISSAL FROM MY JOB. ON THE OTHER HAND, IF I OBEY THE LAW, I NEGLECT THE OBLIGATIONS OF A RESPONSIBLE TEACHER OF BIOLOGY. THIS IS THE SURE PATH TO THE PERPETUATION OF IGNORANCE, PREJUDICE AND BIGOTRY.
>
> THE ONLY RECOURSE AVAILABLE TO ME WHICH IS CONSISTENT WITH MY CONCEPT OF MY RESPONSIBILITIES AS A TEACHER AND CITIZEN IS THIS TEST SUIT. IT IS MY FERVENT HOPE THAT THIS SUIT WILL RESOLVE THIS DILEMMA NOT ONLY FOR ME BUT ALSO FOR ALL OTHER ARKANSAS TEACHERS.

Epperson's trial occurred on April 1, 1966, in a packed courtroom. Reed had tripled the number of bailiffs; banned photographers, radio coverage, and television coverage; limited the case to constitutional issues; banned questions about the personal beliefs of witnesses; and would not hear arguments about the scientific validity of evolution relative to creationism.

Warren, who hardly questioned the state's four witnesses, questioned Epperson for only ten minutes, during which Epperson focused on the fact that evolution is "a widely accepted theory and I feel it is my responsibility to acquaint my students with it." Bennett took personal charge of Epperson's trial, as he had in the 1957 trial of Daisy Bates—the leader of the state NAACP (The National Association for the Advancement of Colored People)—for refusing to give Bennett the NAACP's membership list. Bennett questioned Epperson for almost an hour, during which time he reminded the Court of the Scopes case and its legacy, as well as the fact that two of Arkansas' neighboring states—Tennessee and Mississippi—had similar statutes. Bennett then attacked Epperson and claimed that the state faced a choice involving "whether we succumb to the atheistic, materialistic concept which is attempting to conquer the world or whether we provide a climate in which our children will be encouraged to retain the spiritual, moral, and human values which our forefathers, in drafting our Constitution, tried so steadfastly to protect." The trial—billed by New York Times as the "New Scopes Trial"—was stopped twice when spectators laughed and became unruly. Reed barred Bennett's fourteen "scientific" witnesses that were to testify on behalf of the law (and, simultaneously, denounce evolution), stating that the issue at hand was a constitutional, not a religious or scientific, issue. Whereas Scopes' trial more than forty years earlier had dragged on for almost two weeks, Epperson's trial involved only six witnesses and ended after only two hours. Bennett "finally gave up" and rested the state's case just thirty minutes into the afternoon session. Warren then rested his case, and Reed—asking for additional briefs—took the case under advisement.

Reed did not reject "creation" in his nine-page per curiam opinion, which he issued two months later. Citing the legacy of the Scopes trial and acknowledging "this Court is not unmindful of the public interest in this case," Reed declared that the Arkansas statute violated the Fourteenth Amendment because it was arbitrary and vague, thereby making interpretation

difficult. Reed also noted that the law "tends to hinder the quest for knowledge, restrict the freedom to learn, and restrain the freedom to teach."

As Bennett had promised, Arkansas appealed Reed's decision to the Arkansas Supreme Court less than one month after Reed's decision. On June 5, 1967, the Arkansas Supreme Court reversed Reed's decision with an unsigned, two-sentence per curiam ruling. Unlike most decisions by the court, this decision was not written by any one justice and did not include the usual published opinion. The first sentence of the decision ruled that the state law "is a valid exercise of the state's power to specify the curriculum in its public schools." The second sentence puzzled everyone: "The court expresses no opinion on the question whether the Act prohibits any explanation of the theory of evolution or merely prohibits teaching that the theory is true."[1] Although the 6-1 decision did not address whether Epperson could assign "the evolution chapter" (i.e., Chapter 39, "The History of Man") of Modern Biology to her students, the appeal did reverse the Chancery Court's ruling, thereby sustaining the statute as an exercise of the state's power to specify the curriculum in public schools. After being legal for just over a year, the teaching of human evolution was again a crime in Arkansas. However, four decades after the Scopes fiasco, there was a case—not in Tennessee, but in Arkansas—to appeal to the U.S. Supreme Court.

On October 16, 1968, Warren addressed the Supreme Court for only ten minutes of the half-hour allotted to him. Warren claimed that teachers were so confused and frightened that "biology is not even taught" in many Arkansas schools. The state's argument was made by Don Langston, a young assistant to new Arkansas Attorney General Joe Purcell (Bennett had been sent back to private practice by Arkansas' voters in 1966). Although Langston would not concede that the law was invalid (to which Justice Abe Fortas replied: "It might not be too

[1]This sentence (and its non-final "decision") was meant to deny the U.S. Supreme Court the jurisdiction of the case (i.e., only a final decision could be appealed).

late, you know"), Justice Hugo Black described Langston's defense of the statute as "unenthusiastic, even apologetic." Langston smiled when Justice Thurgood Marshall asked whether, since the Arkansas Supreme Court disposed of the case in two sentences, he would "object to us disposing of the case in one sentence." At another point, Langston complained that Arkansas "didn't file a written opinion with reasoning." When a justice responded that "maybe they couldn't," Langston responded, "I've heard rumors to that effect."

One month later (on November 12, 1968), the Supreme Court—in an opinion written by Justice Abe Fortas that was strongly influenced by the legacy of the Scopes' Trial—ruled unanimously that banning the teaching of evolution violated the Due Process Clause of the Fourteenth Amendment and the Establishment Clause of the First Amendment to the Constitution. The Court ruled that the Arkansas law was too vague to enforce, attempted to establish religion in a public school, reflected religious dogma, and "sought to prevent its teachers from discussing the theory of evolution because it is contrary to the belief of some that the Book of Genesis must be the exclusive source of doctrine as to the origin of man. It is clear that fundamentalist sectarian conviction was and is the law's reason for existence . . . The law's effort was confined to an attempt to blot out a particular theory because of its supposed conflict with the Biblical account."

Many applauded the Supreme Court's decision; for example, the *Arkansas Gazette* declared that the Court had provided "an overdue rescue from a law of ignorance." As people in Dayton reiterated their claim that they would "convict Scopes still," John Scopes—by this time 68 years old—noted that "this is what I've been working for all along . . . I'm very happy about the decision. I thought all along—ever since 1925—that the law was unconstitutional" (Figure 28).

Epperson is viewed by many legal historians as one of the most important decisions ever issued by the Supreme Court. Indeed, *Epperson* has been applied in a variety of court decisions, including those involving "release time"

from public school for religious training (*Smith v. Smith*, 391 FS 443, 1975), censorship of textbooks (*Daniel v. Waters*, 1975), Bible readings in public schools (*Meltzer v. Board of Public Instruction of Orange County, Florida*, 1977), sex education (*Cornwell v. State Board of Education*, 1969), exhaustion of administrative remedies (*Lopez v. Williams*, 1973), the Establishment Clause of the First Amendment (*Meek v. Pittenger*, 1975), First Amendment rights for students (*Tinker v. Des Moines*, 1969), and the academic freedom of teachers (*Parducci v. Rutland*, 1970). As a result of *Epperson*, all laws banning the teaching of human evolution in public schools were overturned by 1970.

After her day in court, Susan Epperson moved away from Arkansas and never again was involved significantly in the evolution-creationism controversy. Today, Susan and Jon live in Colorado, where Susan teaches biology and chemistry at the University of Colorado in Colorado Springs. She and Jon are members of the Presbyterian Church, United States. Susan remains interested in conveying the compatibility she sees between science and faith.

For More Information: Thorndike, J.L. 1999. *Epperson v. Arkansas: The Evolution-Creationism Debate*. Springfield, NJ: Enslow Publishers.

EUGENICS AND THE EUGENICS RECORD OFFICE

The eugenics movement has its roots in the work of Francis Galton, who coined the term *eugenics* (Figure 29). By examining human pedigrees, Galton accepted the heritability of traits such as intelligence and other personality characteristics. Galton believed that the action of dysegenic selection (decline in the frequency of a desired trait) was already apparent in England in the early 1900s, and he openly advocated positive eugenic measures to improve society. His program of "participatory evolution" was to be mediated through financial and other rewards given to people having favored characteristics. Galton established a research center at University College, London, to support eugenics, with Karl Pearson its first director. Several notable

28. In January 1969, Epperson discussed her lawsuit with John Scopes at a Holiday Inn in Shreveport, Louisiana. (*Jerry Tompkins*)

evolutionary biologists, including Ronald Fisher and J.B.S. Haldane, were affiliated with the Galton Lab.

Galton's proposals were popular, and world leaders from Winston Churchill to Teddy Roosevelt endorsed eugenic efforts. In Europe, Scandinavian countries in particular embraced Galton's proposals, with these efforts unavoidably carrying an undertone of racism (e.g., in Norway, eugenics was seen as a way to prevent undesirable mixing of the Lapp and Norwegian populations). But, except for the extreme case of Nazi Germany, no country implemented Galton's "scientific eugenics" more completely than did the United States.

Well before the start of the twentieth century, U.S. citizens were becoming concerned with the "quality" of their population. The late 1800s was

a time of immense change in America, as the industrial revolution transformed the country from a rural, agricultural society to an urban one. Economic and social stratification of the population, especially in cities, increased, and the economy experienced periodic booms and busts that only caused greater divergence among social classes. Successful businessmen such as John D. Rockefeller and Andrew Carnegie (both of whom supported the eugenics movement in the United States) reinforced the concept of the American dream: those who succeeded *deserved* to succeed, while those who failed lacked the necessary traits and may be doomed to stay at the bottom of society.

Against this backdrop, a book by sociologist Richard Dugdale titled *The Jukes: A Study in Crime, Pauperism, Disease and Heredity* was

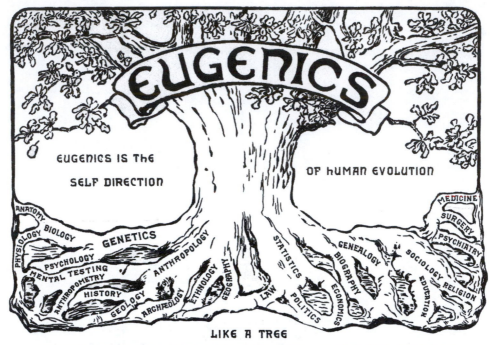

EUGENICS

EUGENICS IS THE
SELF DIRECTION

OF HUMAN EVOLUTION

LIKE A TREE
EUGENICS DRAWS ITS MATERIALS FROM MANY SOURCES AND ORGANIZES
THEM INTO AN HARMONIOUS ENTITY.

29. The Eugenics Records Office promoted improving humanity with selective breeding. This logo, from the Second International Congress of Eugenics in 1921, depicted a tree uniting a variety of disciplines. (*Courtesy of the American Philosophical Society*)

released in 1877. In 1874, Dugdale was a volunteer jail inspector, and at one of his stops, he learned that six family members were simultaneously in jail. Intrigued, he investigated further, and determined that among twenty-nine male "immediate blood relations" within this one family ("Jukes" was a pseudonym), there had been seventeen arrests and fifteen convictions. Dugdale researched the pedigree of the family, determined that the current generation came from a long line of troublemakers, and placed blame on a common ancestor named Margaret, whom he labeled "the Mother of Criminals." The story of the Jukes, combined with the incorporation of Mendel's findings into biology, supported the importance of heredity in human behavior. The possibility of improving the human population by controlling reproduction came to be seen by many as a moral imperative.

In the early 1900s, University of Chicago professor Charles Davenport (1866–1944) persuaded the Carnegie Institution of Washington to found a Station for the Experimental Study of Evolution (SESE) at Cold Spring Harbor Laboratories. At the SESE, established in 1904, Davenport—an engineer and geneticist who embraced the work of Galton and Pearson as well as the findings of Mendel—studied selection in mice, horses, and poultry. By 1907, he turned his attention to humans, and by 1910 had published several papers on the Mendelian inheritance of several human traits. Davenport was focusing on eugenics by 1908, and lobbied for a separate organization to investigate eugenics scientifically. This led to the establishment of the Eugenics Record Office (ERO) at Cold Spring Harbor in 1910, the same year that Davenport became the Lab's director. Davenport's *Heredity in Relation to Eugenics* (1911) was a landmark (and popular college textbook) in American eugenics.

Davenport hired Harry Laughlin (1880–1943), who was teaching agriculture in Missouri, to manage the ERO, and also convened a group of noted scientists (including Alexander Graham Bell) to serve as an advisory board. Laughlin directed the ERO from its inception in 1910 to its closing in 1939 (when the Carnegie Institute withdrew its funding). The goals of the ERO were twofold: (1) study the inheritance of traits (especially traits affecting social behavior) in humans; and (2) disseminate information about eugenics to the public. Davenport's *The Trait Book* (1912) provided a list of human traits (including genetic disorders) of interest. An army of ERO-trained fieldworkers traveled the country interviewing families and collecting pedigree information. These results led to three intertwined efforts, all of which emphasized negative eugenics (the prohibition of reproduction by those with undesirable traits) that characterized the eugenics movement in the United States.

First, data from the ERO, as well as from other entities, indicated that certain negative traits were more frequent among particular ethnic groups. Although several organizations already advocated restricted immigration, and some legislation was already on the books (e.g., the Chinese Exclusion Act of 1882), the ERO provided scientific legitimacy for new anti-immigration proposals. For example, Laughlin urged "purification" of the original Anglo-Saxon lineage of the United States because, he argued, immigrants caused most crime (he was particularly interested in preventing immigration of Jews). Laughlin testified before Congress in support of the Immigration Restriction Act of 1924, which was designed to halt immigration by Italians and eastern European Jews. Upon signing the Act, President Coolidge declared that "America must remain American."

Second, proposals to maintain racial segregation were supported under the guise of eugenics. Famed antievolution crusaders such as Frank Norris and Bob Jones were alarmed by the potential "mongrelizing" of the races through the mixing of "higher" (i.e., white) races with "lower" (i.e., nonwhite) ones. Consequently, by the 1930s, most states had antimiscegenation laws in place. The most infamous was Virginia's Racial Integrity Act of 1924, which was promoted by white supremacists, with assistance by Laughlin. The act banned any white person from marrying anyone of any other race. The law remained in place until 1967 when the U.S. Supreme Court struck it down in consideration of a 1958 case where an interracial couple had been prosecuted under the law. Davenport himself coauthored a study in 1929 that examined the ability of whites, blacks, and those of mixed race to "carry on a white man's civilization." The study declared whites as superior, blacks as inferior, and those of mixed race as "wuzzle-headed."

Third, and most disturbingly, the eugenics movement implemented forced sterilization of individuals deemed unfit due to "feeblemindedness," "shiftlessness," promiscuity, or epilepsy. Laughlin's work with state legislatures in crafting sterilization laws led to his "Model Eugenical Sterilization Law" in 1914, a template to expedite the enactment of such legislation. By the mid-1930s, more than thirty states had such laws in place. (Nazi Germany also drew heavily from Laughlin's proposal, and in 1936 the University of Heidelberg bestowed upon him an honorary degree in recognition of his efforts.) The most famous constitutional test of such legislation was *Buck v. Bell*, based on the forced sterilization of Carrie Buck under Virginia's law. The law was upheld in 1927 by the U.S. Supreme Court, and the majority opinion, written by Justice Oliver Wendell Holmes, concluded that the law was reasonable in accomplishing the public good because "three generations of imbeciles are enough." Tens of thousands of people were forcibly sterilized in the United States and some laws are still on the books. In the 1970s, the U.S. Indian Health Service coerced thousands of Native Americans to accept involuntary sterilization by threatening the loss of benefits.

The ERO and other eugenic organizations promoted the supposed benefits of an active eugenics program. The popular-level *Eugenical News* was published by the ERO for several decades, and exhibits created by Laughlin appeared in many natural history museums, including the American Museum of Natural History, then under the direction of paleontologist and eugenicist Henry Osborn. In advertising the supposed benefits of eugenics, the ERO was aided by several organizations, including the Eugenics Research Association, the Race Betterment Foundation, and the American Breeders Association. The American Eugenics Society established "Fitter Family Contests," introduced at the Kansas Free Fair in 1920, which gave awards to families judged (partly by passing a Wassermann test for syphilis and a psychiatric assessment) to be superior "human stock." Eugenics was also incorporated into the curricula of public schools and universities. In 1917, a Hollywood production called *The Black Stork* illustrated the dire consequences of ignoring eugenic principles: A couple, counseled not to have children, has a baby with birth defects; the mother, now convinced of her error, allows the child to die as Jesus stands waiting.

The eugenics movement was not universally supported, and many biologists opposed it (Alfred Wallace saw it as "the meddlesome interference of an arrogant scientific priestcraft"). Geneticist R.C. Punnett demonstrated that sterilization would only slowly decrease the frequency of an unwanted trait. (R.A. Fisher, who had promoted eugenics in his classic *The Genetical Theory of Natural Selection*, promptly criticized Punnett's analysis.) Nonscientists also saw the dangers, from writer G.K. Chesterson ("eugenicists had discovered how to combine the hardening of the heart with the softening of the head"), to William Jennings Bryan, who noted that science can "perfect machinery, but it adds no moral restraints to prevent the misuse of the machine."

By the 1930s, there was increasing resistance to eugenic programs. The complexity of human behavior was becoming apparent as was the role of the environment as a determining factor. Sterilization laws were seen as infringing on human rights. Finally, the involuntary sterilization of hundreds of thousands of people and the killing of millions more by the Nazis forced a reevaluation of eugenics. Yet, groups propounding the benefits of "human husbandry" have remained active (notables such as Julian Huxley and James Watson have proposed the benefit of eugenics). Bioethicists caution that the ability to select embryos with or without particular characteristics may trigger a new, technology-driven phase in the history of eugenics.

For More Information: Black, E. 2003. *War against the Weak*. New York: Four Walls Eight Windows.

L. WALTER EVANS

L. Walter Evans was a Church of God minister and Mississippi state legislator who in early 1926 introduced legislation to ban the teaching of human evolution in Mississippi. Evans' bill was rejected by the Education Committee by a vote of 10–6. However, after Baptist evangelist T.T. Martin addressed a joint session of the legislature during the first week of February, the House of Representatives approved the bill by a vote of 76–32, the Senate passed the bill by a vote of 29–16, and Governor Henry Whitfield (1868–1927) signed Evans' bill into law on March 12, 1926.

F

HUGH FALCONER (1808–1865)

Hugh Falconer was a Scottish paleontologist, physician, and Indian tea grower who, while working for The British Museum, told Richard Owen about a curious bird-lizard fossil that had been found in Solenhofen, Germany. Owen bought the fossil for £700 (the purchase also included more than 1,000 other fossils) and in 1863 named it *Archaeopteryx* ("ancient feather"). *Archaeopteryx* was one of the first intermediate fossils that supported Darwin's theory of evolution by natural selection. Falconer, along with Huxley, Hooker, and Lyell, was someone whom Darwin admired and from whom he sought advice. Indeed, Darwin sent Falconer a copy of the first edition of *On the Origin of Species* with a letter acknowledging that Falconer would probably "savage" and "crucify" him, but that Falconer would also become "less fixed in your belief in the immutability of species." In 1861, Falconer told Darwin that although he disagreed with Darwin's theory, he and other scientists respected the idea (Falconer accepted evolution, though not by natural selection). In 1864, Falconer and George Busk nominated Darwin for the Copley Medal of the Royal Society. Falconer rests in London's Kensal Green Cemetery.

JERRY FALWELL (1933–2007)

I will continue to stand against the evolutionary and secularist tides by proclaiming that God spoke the heavens and the Earth into existence in six literal days.

Jerry Lamon Falwell, and his twin brother Gene, were born on August 11, 1933, in Lynchburg, Virginia. Falwell's father was a successful Lynchburg businessman who dabbled in a number of enterprises outside the law, including moonshining and dogfights. Falwell's father died when Jerry was 15, leaving behind enough money for his family to be financially comfortable.

In high school, Falwell played sports and edited the school newspaper. He graduated at the head of his class, but was prevented from giving the valedictory address due to unspecified "mischief" he perpetrated during his senior year. He considered playing professional baseball, but eventually narrowed his career options to journalism or engineering, and enrolled in Lynchburg College. In college, Falwell became a Christian and decided to become a minister. In 1952, he transferred from Lynchburg to Baptist Bible College (founded by followers of Frank Norris) in Springfield, Missouri, where he graduated in 1956 with a degree in theology. He married his wife, Macel, in 1958.

After graduation, Falwell returned to Lynchburg to become the founding pastor of Thomas Road Baptist Church. At its inception, the church had only thirty-five members and met in the elementary school Falwell attended as a child. But membership grew quickly, and the church moved into a succession of larger buildings (including, at one point, the former Donald Duck Soft Drink Bottling Company). Today, Thomas Road Baptist Church, with a membership of 22,000, is one of the largest churches in the nation, and in 2006 added a $20 million, 4,800-seat sanctuary. The church espouses a strict Christian fundamentalism that emphasizes a literal interpretation of the Bible. Its doctrinal statement affirms that "human beings were directly created, not evolved, in the very image of God."

Like fundamentalists Billy Sunday, Frank Norris, and William Jennings Bryan before him, Falwell believed that America's societal problems are due to the fact that the country has turned its back on God. He was hostile toward public education because of the alleged secularist bias and moral ambiguity he believed it inculcated into children. In response, Falwell opened the Lynchburg Christian Academy, an accredited K-12 school, in 1967. The school, now known as the Liberty Christian Academy, is located on a 113-acre campus in central Virginia, and enrolls nearly 1,400 students. In 1971, Falwell founded Liberty University, which enrolls more than 9,500 students in thirty-eight undergraduate and fifteen graduate programs. Liberty is openly based upon "an inerrant Bible," "a Christian worldview beginning with belief in biblical Creationism" and "an absolute repudiation of political correctness." In 2006, when Liberty University was searching for a biologist, applicants had to endorse Young-Earth creationism.

Not long after coming to Thomas Road, Falwell began using mass communication to spread his religious and social messages. In 1958, he launched a radio ministry and only six months later began airing *Old Time Gospel Hour* on local Lynchburg television and then eventually to a worldwide audience. Although the popularity of *The Old Time Gospel Hour* declined since its peak in the 1960s and 1970s, Falwell remained one of the most recognizable televangelists. Falwell advanced his vision for America in books such as *Listen, America!* and *America Can Be Saved!* His periodical, *National Liberty Journal*, which focuses on conservative social and political issues of the religious right, began publication in 1995, and he continued to advocate his religious and social agenda through Jerry Falwell Ministries.

In 1979, Falwell formed the Moral Majority, which advocated conservative Christian values in American politics, especially within the Republican Party. Falwell claimed the Moral Majority enabled Ronald Reagan (referred to by Falwell as "my hero") to be elected President, a triumph Falwell viewed as a moral watershed event in American politics. Although Falwell stated that religion and politics don't mix, the Moral Majority infused American society with a conservative, Christian-based agenda that included abolition of abortion, suppression of gay rights, and the inclusion of prayer and the teaching of creationism in public schools. The organization was disbanded in 1989, but was resurrected as the Moral Majority Coalition in 2004.

As a Biblical literalist, Falwell based his church and educational institutions on Young-Earth creationism, and he opposed the teaching of evolution. In 2005, Liberty University hosted and cosponsored, along with the creationist organizations Answers in Genesis and the Creation Research Society, the Creation Mega-Conference. The five-day event featured many of the major names in Young-Earth creationism and included sessions titled "Rocks around the Clock: The Eons That Never Were," "Distant Starlight: Not a Problem for a Young Universe," and "Molecular Evidence for Creation."

In the early 1980s, Liberty University sought state accreditation for its biology department so that its graduates could teach in public schools (or in the words of Falwell, so that Liberty's "hundreds of graduates [could] go out into the classrooms teaching creationism"). Because the

30. Orval Faubus was governor of Arkansas and a vocal opponent of Susan Epperson's attempt to overturn the state's ban on teaching human evolution. Susan taught at Little Rock's Central High School, which became famous when President Dwight Eisenhower used federal troops to ensure its integration (and block Faubus' orders to stop integration). This photo, from 1958, shows Faubus' opposition to integration, which was common among antievolution crusaders in the south. (*Library of Congress*)

course "History of Life" included creationism and was a required part of the science curriculum, the ACLU opposed accreditation. In 1984, after Liberty moved the course to the humanities program and removed creationism from the science curriculum, the program was accredited. The School of Arts and Sciences (dedicated to "an integration of faith and discipline, showing how Biblical presuppositions and a commitment to Christ govern research and teaching") also now houses a Department of Creation Studies.

Falwell, who had a history of heart problems, died in his office on May 15, 2007, of cardiac arhythmia.

ORVAL FAUBUS (1910–1994)

I'm not ready to repeal the Bible.

Orval Eugene Faubus was born in Combs, Arkansas, on January 7, 1910 (Figure 30). His father, "Uncle Sam" Faubus, was a socialist who, in honor of his hero Eugene Debs (one of Clarence Darrow's clients), gave Orval the middle name

of Eugene. Despite having only a high school diploma and a few courses at Commonwealth College, Faubus became a schoolteacher. After serving in World War II with George Patton's Third Army, Faubus—a staunch Democrat—became Arkansas' state highway commissioner. In 1955, Faubus was elected governor on a platform promising improved roads and schools. He desegregated state buses and public transportation, and considered integrating some schools. However, fearing his political demise when attacked by ultraconservative Jim Johnson, Faubus decided to fight the 1954 U.S. Supreme Court ruling, *Brown v. Board of Education*.

On September 2, 1957, despite his upbringing in a racially tolerant home, Faubus ordered the Arkansas National Guard to block nine black children from attending Central High School in Little Rock. After rioting ensued, President Dwight Eisenhower sent U.S. troops to Little Rock and federalized the National Guard to ensure integration of the school. None of the black students got into the school on the appointed day, but they did get in several weeks later after getting a judge's order prohibiting Faubus from interfering. Faubus remained defiant, describing the federal troops as "an army of occupation" and claiming that "blood [would] run in the streets" if black students entered Central High School. Faubus retaliated by closing Little Rock schools for the next two years.

Faubus chose not to run for a seventh term in 1966, and in 1968 Faubus was one of five people considered for the vice presidential position of third-party candidate George Wallace. In 1969, Faubus began managing the Li'l Abner theme park in the Ozarks; that same year he divorced his wife Alta after almost thirty-eight years of marriage and married 30-year-old Elizabeth Westmoreland. Faubus—the longest-serving governor in Arkansas' history—was defeated in attempted comebacks in 1970, 1974, and 1986, the last time losing to future president Bill Clinton.

When Central High School's Susan Epperson decided in 1965 to challenge the Arkansas law banning the teaching of human evolution—the only such law passed by a public referendum—Faubus insisted that the 1928 law was still "the will of the people," and declared the Genesis account of creation "good enough for me." Faubus, who supported the antievolution law "as a safeguard to keep way-out teachers in line," prompted the retired John Scopes to urge Arkansas residents to "figure out what [Faubus] is thinking so you can protect yourself from [him]."

After his divorce, Faubus—almost broke—worked as a bank teller and had to sell his home, which had been built with donations from lobbyists, state employees, and contractors. He also suffered personal tragedy; his only son (Farrell) died of a drug overdose in 1976, and in 1983 his estranged wife Elizabeth was found strangled in the couple's home.

Faubus died of prostate cancer on December 14, 1994, and was buried in Combs Cemetery in Madison County, Arkansas.

For More Information: Reed, R. 1997. *Faubus: The Life and Times of an American Prodigal.* Fayetteville, AR: University of Arkansas.

MIRIAM 'MA" FERGUSON (1875–1961)

I am a Christian mother...I'm not going to let that kind of rot go into Texas textbooks.

Miriam Amanda Wallace was born in Bell County, Texas, on June 13, 1875. In 1899, after attending Salado College and Baylor Female College, Miriam married James Edward "Pa" Ferguson and became the First Lady of Texas when he was elected governor in 1915. As a result of Miriam's dedication to her husband and two daughters, as well as the combination of her initials, she became known as "Ma."

When James Ferguson was impeached during his second term, Miriam announced her candidacy for governor. Her campaign, which was meant to vindicate her husband's reputation, was fiscally conservative, anti-Klan, and antiprohibition; "Ma" also promised to follow the advice of her husband, thereby allowing Texans to get "two governors for the price of one." In November 1924, she handily defeated Republican nominee George Butte. Although Ferguson was the first woman to be elected governor in the United States, she was inaugurated fifteen days after Wyoming's Nellie Ross, and thereby became the second female governor in U.S. history.

Ferguson entered the evolution-creationism controversy in October 1925, when, as head of the state textbook commission, she banned public schools' use of biology textbooks that included evolution. That year, the state adopted Truman Moon's *Biology for Beginners*, but only on the condition that its three evolution-related chapters be removed. Ferguson threatened to fire and prosecute any teacher who used an unapproved book, and justified her edict by reminding Texans that she was "a Christian mother." For the next several decades, the ban imposed by Ferguson forced publishers to produce special editions of their biology books for Texas classrooms.

Ferguson was criticized for granting contracts to her friends and political supporters, as well as for pardoning an average of 100 convicts

31. The First Baptist Church of Minneapolis, Minnesota, was the national headquarters of the fundamentalist movement in the United States in the first half of the 1900s. From this church, antievolution crusader William Bell Riley founded and directed the World's Christian Fundamentals Association, which was responsible for William Jennings Bryan participation in the Scopes Trial. (*First Baptist Church, Minneapolis*)

per month. These controversies, as well as the opposition of antievolution crusader Frank Norris, helped Attorney General Daniel Moody defeat Ferguson in 1926, and in 1930 Ferguson was defeated in another campaign for governor. In 1940, the 65-year-old Ferguson accepted a "popular draft" for the gubernatorial nomination. She lost to incumbent governor W. Lee O'Daniel, and never again ran for public office.

"Ma" Ferguson died of heart failure on June 25, 1961. She rests beside her husband in the Texas State Cemetery in Austin, Texas.

W.F. FERGUSON

W.F. Ferguson was the biology teacher at Rhea County High School for whom John Scopes substituted in 1925. When Ferguson refused to participate in a test of the Butler Law, the trial's instigators asked Scopes if he would agree to be arrested for having taught evolution. Scopes agreed, and the result was the Scopes "Monkey Trial," the most famous event in the history of the evolution-creationism controversy.

FIRST BAPTIST CHURCH OF MINNEAPOLIS (Est. 1853)

First Baptist Church of Minneapolis was the national headquarters of the fundamentalist movement in the early 1900s (Figure 31). Led by William Bell Riley from 1897 to 1947, First Baptist Church housed the World's Christian Fundamentals Association (WCFA) and regularly hosted antievolution crusaders such as William Jennings Bryan, Frank Norris, and Billy Sunday. When Curtis Laws (who coined the term *fundamentalist*) visited Riley's church, he reported that "the whole church edifice seemed to me to be bustling with life like a great department store."

RONALD FISHER (1890–1962)

Natural selection is a mechanism for generating an exceedingly high degree of improbability.

Ronald Aylmer Fisher was born on February 17, 1890, in London, the last of seven surviving siblings (Figure 32). Ronald's twin brother was stillborn. Fisher's intellectual gifts—a colleague would later refer to him as "an outlier, both scientifically and personally"—were recognized at a young age, and he started school early. He also suffered from such poor eyesight that he was discouraged from reading and received most of his instruction verbally and without visual aids. Students and colleagues credited this background for Fisher's preternatural ability to solve difficult mathematical problems quickly in his head. This talent frustrated his teachers, who constantly reminded him to "show his work," as well as, later, readers navigating his research papers.

Fisher entered Cambridge University under scholarship in 1909. Although interested in the biological sciences (Fisher had selected some of Charles Darwin's books as a graduation gift), he

32. Ronald Fisher was a primary architect of the "modern synthesis" and an outspoken eugenicist. Later, Fisher claimed that famed geneticist Gregor Mendel's data were faked. (*Library of Congress*)

concentrated on mathematics, being dissuaded from the study of biology by the prospect of having to memorize the anatomical structures of organisms. Genetics was arising as a new discipline, and while Fisher was an undergraduate, William Bateson (who introduced the term *genetics*), and then R.C. Punnett held the newly created Arthur Balfour Chair of Genetics at Cambridge. This was a period of intense discussion and disagreement about the roles of natural selection and genetics in evolution; Fisher recalls bookshops were then "full of books ... from which one could see how completely many writers ... believed that Darwin's position had been discredited," which contrasted with his own embrace of Darwin's theory.

While at Cambridge, Fisher became an ardent proponent of Francis Galton's views on eugenics. Galton had funded what became the Galton Laboratory for National Eugenics at University College, London in 1909, under the direction of statistician Karl Pearson. Convinced of the benefits of applying selection to the betterment of society, Fisher founded the Cambridge University Eugenics Society in 1911. Fisher also befriended Leonard Darwin, son of Charles Darwin, who had retired from the military and in 1911 had accepted the honorary title of President of the Eugenics Education Society. Darwin remained a trusted advisor to Fisher until Darwin's death in 1942, and Fisher dedicated his opus, *The Genetical Theory of Natural Selection*, to him.

After graduating from Cambridge in 1912, Fisher earned a scholarship for a year of graduate work in physics at the Cavendish Laboratory at Cambridge. By 1913, faced with having to support himself, he accepted a job as statistician with the Mercantile and General Investment Company in London. When World War I broke out, Fisher tried to enlist, but was rejected because of his eyesight, and instead taught school for the next five years, a job he despised. In 1917, he married Ruth Eileen Guinness, with whom he had nine children. Fisher published several papers during this period on statistics, eugenics, and sexual selection.

In 1918, Fisher published an article that laid the foundation of what would become quantitative genetics. The paper demonstrated that the disagreement between the statistical and Mendelian geneticists on the inheritance of continuously variable (quantitative) traits (e.g., height) could be overcome by a model that invoked several loci acting on the trait simultaneously. This work was crucial in unifying the field of genetics. The paper was also significant in that Fisher introduced the term *variance* to describe the statistical dispersion of a distribution as the squared deviations of scores about a mean.

In 1919, Pearson offered Fisher a job at the Galton Laboratory at University College, an ideal location for Fisher, given his twin interests in statistics and eugenics. However, Pearson's supervision would restrict Fisher's intellectual autonomy, and Fisher declined the offer, which started a long-running feud with Pearson. Fisher accepted a second job offer, a temporary position analyzing agricultural data at Rothamsted

Experimental Station, sending his career in an unintended but fruitful direction. This position lasted seventeen years, during which time he single-handedly developed the discipline of experimental design. (The agricultural history of this area of statistics is still reflected in the use of terms like *plot* and *block* in descriptions of experimental design.) Fisher developed sophisticated statistical tools such as analysis of variance and covariance, which today still form the basis of many statistical analyses. By 1925, Fisher's *Statistical Methods for Research Workers* unified his statistical work in a single volume and ensured his place as the most influential statistician of his generation. *Statistical Methods* remained in continuous publication for almost fifty years.

By 1930, Fisher had extended his 1918 paper on the inheritance of quantitative traits into *The Genetical Theory of Natural Selection*. Fisher discussed natural selection in rigorous mathematical terms and effectively united genetics with Darwin's theory. The central theme of the book is Fisher's "fundamental theorem of natural selection," which states that the rate of increase in fitness equals the amount of genetic variance in the population. Fisher then examined the evolution of mimicry, sex ratios, and sexual traits relative to his theorem. Fisher's book supported natural selection as the causative agent for adaptive evolution; in the words of Stephen Gould, it was "the keystone for the architecture of modern Darwinism." Today, Fisher is viewed as one of the main architects, along with J.B.S. Haldane and Sewall Wright, of the modern synthesis of evolutionary biology during the 1920s and 1930s.

Although *The Genetical Theory of Natural Selection* was difficult to read, biologists embraced it and quickly noted its significance. Geneticist E.B. Ford, who collaborated with Fisher for many years, claimed that "no book of the century has had a greater effect upon biology than this one," and Sewall Wright praised it as a "masterpiece." Fisher's fundamental theorem provided the basis for Wright's influential "adaptive landscape" visualization of fitness peaks and valleys introduced in 1932. Ironically, the relationship between Fisher and Wright deteriorated over the years to the point that when Fisher visited Wright's home department at the University of Wisconsin—which he did frequently because Fisher had a daughter who lived in Madison—colleagues arranged his visit so that Fisher and Wright never encountered each other.

The last third of *The Genetical Theory of Natural Selection* advocated eugenics in chapters titled "Reproduction in Relation to Social Class" and "Social Selection of Fertility." Fisher's emphasis of a principle that, after the horrors of World War II, lost its attractiveness may partly explain why Fisher's book was less popular than those of others in the modern synthesis. *The Genetical Theory of Natural Selection* was reissued in 1958, and influenced another generation of biologists, including William Hamilton, who rated the book "only second in importance in evolution theory to Darwin's *Origin*," and considered it "one of the greatest books of the century." Interestingly, Hamilton's original interest in evolutionary biology, like Fisher's, was its potential use for "enhancing the human stock."

As part of his efforts to unite Mendelian genetics with Darwin's ideas, Fisher examined "Mendel's genius in its own terms, and not as it had been perceived" by reanalyzing Mendel's data from his breeding experiments with garden peas. After calculating the expected probability of misclassification of individuals as well as the normal range of variation expected in such data, Fisher concluded that Mendel's data had been "falsified so as to agree closely with ... expectations." Fisher left open the possibility that Mendel may have been deceived by an overzealous assistant, and expressed disbelief that "Mendel himself had any hand in" falsifying the data. Despite these troubling conclusions, Fisher remained an admirer of Mendel's experimental methodology and inferential skills, although Fisher is sometimes considered Mendel's "intellectual assassin."

In 1933, Pearson retired from University College, London, and Fisher left Rothamsted to succeed him as Galton Professor of Eugenics. Fisher began breeding experiments on plants, poultry, dogs, locusts, snails, and mice, and also studied human blood groups to investigate Mendelian inheritance in humans. Fisher moved to Cambridge in 1943 as Arthur Balfour Professor of Genetics, the senior position in the field of genetics in Great Britain. At Cambridge, he cofounded the journal *Heredity* in 1947, and was its editor for ten years. Fisher retired in 1957, and curiously devoted time to disputing the supposed causal link between smoking and lung cancer (Fisher was a longtime pipe-smoker). In 1959, Fisher moved to Australia, accepting a temporary position as Senior Research Fellow at the Commonwealth Scientific and Industrial Research Organization (CSIRO), and then in 1961 moved permanently to the University of Adelaide.

Fisher died in Australia in 1962 after suffering a pulmonary embolism following surgery for colon cancer. Fisher received numerous awards during his lifetime, including Fellow of the Royal Society (1929), membership in the National Academy of Sciences of the United States (1948), the Darwin Medal (1948), and the Copley Medal from the Royal Society (1955). He was knighted in 1952.

For More Information: Box, J.F.R. 1978. *R.A. Fisher: The Life of a Scientist*. New York: Wiley.

ROBERT FITZROY (1805–1865)

The Book! The Book!

Robert FitzRoy was born into a wealthy, aristocratic family in Suffolk, England, on July 5, 1805 (Figure 33). When he was 12 years old, FitzRoy—a fourth great-grandson of Charles II—entered the Royal Naval College at Portsmouth, and in 1819 he joined the Royal Navy. In 1824, he passed his examination for promotion to Lieutenant "with full numbers," a score that had not been achieved before. FitzRoy was appointed to the frigate *Thetis*, which surveyed the coast of South America. Four years later, he was transferred as a Flag Lieutenant to the *Ganges*, which also served in South America. In August 1828, FitzRoy was given his first command, the *Beagle*, a 90′-long brig charged with doing hydrographic survey work off the coasts of Patagonia and Tierra del Fuego (the

33. Robert Fitzroy captained the H.M.S. *Beagle*, which took Charles Darwin around the world (Figure 5). FitzRoy later regretted his role in helping Darwin discover evolution by natural selection. (*National Library of New Zealand*)

Beagle had been without a captain after Pringle Stokes, the *Beagle*'s previous captain, had shot himself in the head). FitzRoy hoped to use the voyage to find scientific proof for the literal truth of Genesis. In 1830, FitzRoy also brought several Fuegans back to England in hopes of civilizing them and teaching them "English . . . the plainer truths of Christianity . . . and the use of common tools." FitzRoy, who helped return them as missionaries to their homeland, paid for the project with personal funds.

On December 27, 1831, the 26-year-old FitzRoy set sail on the *Beagle* from Devonport to survey the Cape Verde Islands, the South American Coast, the Strait of Magellan, the Galápagos Islands, Tahiti, New Zealand, Australia, the Maldives, and Mauritius. The *Beagle*'s voyage established FitzRoy as a world-class navigator. The *Beagle*'s voyage, which lasted four years, nine months, and two days, carried a crew of seventy-four that included three officers, a doctor, and an artist. Also aboard was young Charles Darwin, who shared a cabin at the back of the ship with two officers. Darwin's room was so cramped that he had to remove a drawer each night to make room for his feet.

When the *Beagle* sailed from London, its official naturalist was Robert McKormick, the ship's surgeon. However, FitzRoy had relatively little social contact with the ship's official personnel, so he also needed a companion for his voyage. Fitzroy wanted his companion to also be a naturalist, and his first choice for the job was John Henslow, a friend of Darwin's. Henslow, however, declined the offer, and recommended Darwin for the job. FitzRoy believed that the shape of a person's skull indicated the person's character, and he didn't like the shape of Darwin's nose (FitzRoy believed that it betrayed laziness and hesitancy). Moreover, Darwin's father initially opposed his son's participation in the voyage. Finally, however, FitzRoy invited Darwin to go on the cruise, and Darwin accepted. While at sea, Darwin read Charles Lyell's *Principles of Geology*, which had been given to him by FitzRoy. Lyell refashioned Hutton's uniformitarianism into a coherent scientific theory, documenting that the ancient Earth had been molded by the same slow, directionless forces that mold Earth today—namely, by earthquakes that lift and shift land, by erosion that grinds down rocks and redistributes soil, and by volcanoes that build islands.

Every Sunday, Darwin listened to FitzRoy's sermons. Darwin often disagreed with FitzRoy's biblical literalism—Darwin described their arguments as "bordering on insanity"—and Darwin was once banned from the Captain's table for several days. In Chile, Darwin became seriously ill, and FitzRoy prayed for his recovery. When Darwin survived his illness, FitzRoy concluded that his God was omnipotent and merciful. Darwin questioned FitzRoy's logic, and he later noted that "some parts of [FitzRoy's] brain want mending." Darwin, who became the ship's official naturalist when McKormick was "invalidated out," later described his voyage aboard the *Beagle* as "the most important event in my life."

When the *Beagle* returned to England in 1836, FitzRoy married Maria O'Brien and started a family. Darwin, who had not completely labeled the birds he collected at the Galápagos, asked FitzRoy for his specimens, and FitzRoy graciously gave Darwin his preserved birds. Those birds were important for discovering that each island of the Galápagos housed a unique type of finch.

In June 1837, FitzRoy was awarded the Premium Medal by the Royal Geographical Society. Two years later, FitzRoy published two dense, day-by-day narratives of the *Beagle*'s voyages in the three-volume *Narrative of the Surveying Voyages of His Majesty's Ships* Adventure *and* Beagle, *between the Years 1826 and 1836 Describing their Examination of the Southern Shores of South America and The* Beagle's *Circumnavigation of the Globe in Three Volumes.* (Darwin's contribution was the third volume—and the best-seller—of the set; John Murray later published Darwin's volume with its lasting title, *The Voyage of the Beagle.*) FitzRoy's books described the geology of the regions and the cultures of their inhabitants, as well as his commitment to biblical literalism (e.g., that fossils in rocks atop mountains were proof of Noah's flood). FitzRoy's work was often ridiculed; as Lyell noted, "It beats all the other nonsense I have ever read on the subject." FitzRoy dissociated himself from Darwin's account of the *Beagle*'s voyage, which embraced the new ideas of Lyell.

In 1841, FitzRoy was elected a Tory member of Parliament for Durham. Two years later, FitzRoy was appointed Governor General of New Zealand, a post from which he was later dismissed when he argued that Maori land claims were as valid as those of white settlers.

In 1848, Fitzroy was appointed Superintendent of the Royal Naval Dockyard at Woolich (the same dockyard that had launched the *Beagle* in 1820), and in 1849 was given his last command: the *Arrogant*, a frigate charged with doing meteorological experiments. In 1850, as a result of failing health, FitzRoy was placed on half-pay. Although this ended FitzRoy's naval career, he was promoted to Rear Admiral in 1857, and to Vice Admiral in 1863, by order of seniority. When FitzRoy was elected a Fellow of the Royal Society in 1851, his many supporters included Charles Darwin.

In 1854, FitzRoy was appointed "Meteorological Statist" in a new governmental agency now known as the Meteorological ("Met") Office. To help improve naval navigation, FitzRoy established a series of fifteen stations that telegraphed to London their daily weather conditions, thereby enabling FitzRoy to predict the coming weather. FitzRoy pioneered daily weather forecasts in newspapers (the first of which were published in the London *Times* in 1860) and he invented an inexpensive and serviceable weather station. These barometers were often engraved with "Admiral FitzRoy's Special Remarks" and accompanied by a fifty-page instructional *Barometer Manual*, which contained FitzRoy's popular claims about weather (e.g., "A red sky in the morning is a sailor's warning, but a red sky at night is a sailor's delight"). In 1862, FitzRoy published his meteorological theories and methods in his 400-page *The Weather Book*, in which he repeatedly emphasized that "the state of the air foretells coming weather rather than indicates weather that is present." FitzRoy invented the phrase *weather forecast*, and was the first weather forecaster. Despite these accomplishments, FitzRoy remains overshadowed by the accomplishments of his *Beagle* shipmate, Charles Darwin. Indeed, Darwin's *On the Origin of Species* left FitzRoy feeling betrayed and guilty for his role in Darwin's work.

On June 29, 1860, FitzRoy spoke about British storms at the annual meeting of the British Association for the Advancement of Science at Oxford. The following day, FitzRoy attended the famous debate between Thomas Huxley and Samuel Wilberforce. Fitzroy, dressed in a Rear Admiral's uniform, stood and waved a Bible over his head as he proclaimed "The Book! The Book!" and told listeners that he regretted Darwin's monumental book. Few paid attention to FitzRoy, and he was escorted from the building.

On June 18, 1864, the London *Times* discontinued Fitzroy's forecasts. On April 30, 1865, depressed and in poor health, FitzRoy awoke early, kissed his daughter, and slashed his throat. Today, FitzRoy rests in a roadside grave at All Saints' Church in south London.

When he died, FitzRoy was in debt, and his widow was given the use of a residence by Queen Victoria. There were several fundraisers for FitzRoy's widow; Darwin contributed £100. In 1934, Laura FitzRoy—the Captain's daughter—told Nora Barlow (Charles Darwin's granddaughter) that "Charles Darwin was a great man—a genius—raised up for a special purpose. But he overstepped the mark."

Mount FitzRoy (at the southern tip of South America), the FitzRoy River (in northern Western Australia), and the South American conifer *Fitzroya cupressoides* are among the many things named after FitzRoy. Darwin even named a species of dolphin (*Delphinius fitzroyi*) in honor of his captain. The Met Office's headquarters are on FitzRoy Road, and in 2002 the shipping forecast area Finisterre was renamed FitzRoy. Today, FitzRoy's name rings out in the *BBC*'s daily shipping forecasts.

For More Information: Nichols, P. 2003. *Evolution's Captain: The Dark Fate of the Man Who Sailed Charles Darwin around the World*. New York: HarperCollins.

JOHN FLEMING (1849–1945)

John Ambrose Fleming wrote several influential creationism books, including *The Intersecting Spheres of Religion and Science* and *Evolution or Creation*. Fleming, the first president of the Evolution Protest Movement (the first creationist organization, which was founded in 1932 and is known today as the Creation Science Movement), studied under James Maxwell and is best known for developing the first workable electronic vacuum tube. Fleming is buried in Cambridge, England.

ERNIE FLETCHER (b. 1952)

Ernie Fletcher was a Republican governor of Kentucky who, in 2006, advocated improving science education by teaching "intelligent design." Fletcher claimed that he had a "thorough

understanding of the science and theory of evolution," and grouped "intelligent design" with "2+2=4" as "a self-evident truth." Fletcher—a Primitive Baptist—denounced the Kentucky Academy of Science for its rejection of intelligent design, claiming that it "disappoints and astounds me that the so-called intellectual elite are so concerned about accepting self-evident truths that nearly 90% of the population understands." Fletcher often ridiculed the *Kitzmiller v. Dover* decision while noting that Kentucky already had a law (Kentucky Revised Statute 158.177) that allows the teaching of creationism in Kentucky schools.

Fletcher, the state's first Republican governor in thirty years, was indicted on several misdemeanor charges that were later dismissed after a judge ruled that he could not be tried while in office. The grand jury later issued its findings, claiming that Fletcher had approved a "widespread and coordinated plan" to trade state jobs for support.

On November 5, 2007, Fletcher ordered that the Ten Commandments be displayed alongside other historical documents in the state Capital. On the following day, Fletcher was soundly defeated in his bid for re-election.

THE FLINTSTONES (Est. 1960)

The Flintstones, television's first prime time animated series, portrayed a "modern stone-age family" of humans who lived with dinosaurs, saber-toothed tigers, wooly mammoths, and other extinct animals. The series debuted on ABC on September 30, 1960, and the last of its 166 original episodes aired on September 2, 1966. The series was rebroadcast by NBC from 1967–1970, and today, episodes of *The Flintstones* are shown worldwide in twenty-two languages and in more than eighty countries. *The Flintstones* indoctrinated millions of children and adults with the notion that humans lived contemporaneously with dinosaurs, a foundation of Young-Earth creationists' belief that dinosaurs and humans were created on the sixth day of creation.

In virtually every episode of *The Flintstones*, extinct animals were used as technology;

mammoths were used as vacuum cleaners, and pterodactyls served as airplanes. The series was originally meant for adults (the first sponsor was Winston cigarettes), and was later used to promote countless products, including beer, vitamins, cereals, grape juice, and comic books. Today, you'll find the Flintstones at Flintstones Bedrock City at Custer, South Dakota.

The Flintstones' record as the longest-running prime-time animated series was broken on February 9, 1997, by *The Simpsons*. In that episode, the introductory segment ended with the Simpsons running toward their couch only to find Fred, Wilma, and Pebbles Flintstone already sitting there.

For More Information: Adams, T.R. 1994. *The Flintstones: A Modern Stone Age Phenomenon.* Atlanta, GA: Turner Publishers.

FORDHAM FOUNDATION

The Fordham Foundation reported in 2000 that nineteen of the United States do "a weak-to-reprehensible job of handling evolution in their science standards." The study noted that twelve states' standards did not include the word *evolution*, and urged scientists and science teachers to remember "that most Americans believe that faith in God is the surest way to appreciate the wonder and grandeur of life itself."

ABE FORTAS (1910–1982)

Abe Fortas was the U.S. Supreme Court Justice who in 1968 wrote the *Epperson v. Arkansas* decision. Fortas cited biographies of Clarence Darrow and John Scopes, and compared Arkansas' ban on teaching human evolution with restrictions against teaching that Earth is spherical or revolves on its axis. Fortas, who followed the Scopes Trial when he was fifteen years old and living in Memphis, repeatedly cited "the celebrated Scopes case" and ruled that "the sole reason for the Arkansas law is that a particular religious group considers the evolution theory to conflict with...the Book

of Genesis." The *Epperson* decision was the first opinion assigned to Fortas since President Lyndon Johnson's failed attempt to make him Chief Justice three years earlier. In 1969, Fortas was forced from the U.S. Supreme Court by controversies regarding his personal finances.

WILLIAM FOX (1805–1880)

William Darwin Fox, Charles Darwin's great uncle's grandson, was a friend of Charles Darwin's when Darwin was enrolled at Cambridge. They both collected insects. Fox introduced Darwin to Rev. John Henslow, who later recommended Darwin for the H.M.S. *Beagle*.

FREILER v. TANGIPAHOA PARISH BOARD OF EDUCATION

Freiler v. Tangipahoa Parish Board of Education (1997) established that it is unlawful to require teachers to read aloud a disclaimer stating that the biblical view of creationism is the only concept from which students are not to be dissuaded. Such disclaimers are "intended to protect and maintain a particular religious viewpoint." The court noted that "while encouraging students to maintain their belief in the Bible, or in God, may be a noble aim, it cannot be one in which the public schools participate, no matter how important this goal may be to its supporters."

JOHN FRERE (1740–1807)

John Frere wrote *Account of Flint Weapons Discovered at Hoxne in Suffolk* (1797), which was the first published scientific evidence for prehistoric humans. Frere's work was based on his discovery of flint tools, hand axes and bones from "a very remote period indeed, even beyond that of the present world." Frere's great-great-great-granddaughter was Mary Leakey, who with her husband Louis and son Richard discovered many early hominins and their tools in East Africa.

FUEGIANS

In 1826, the H.M.S. *Beagle*, under the command of Captain Pringle Stokes, left Plymouth Sound for the southern coasts of South America to conduct a survey mission to aid British shipping. On August 1, 1828, Captain Stokes, distraught from the extreme privations experienced during the expedition, shot himself in the head and died eleven days later. Stokes' chief officer, 23-year-old Robert FitzRoy, assumed command of the *Beagle* and completed the survey work, returning to England in October 1830.

While surveying the southern coasts of Tierra del Fuego in 1830, a small whaleboat belonging to the *Beagle* was stolen by the indigenous people ("Fuegians"). FitzRoy responded by capturing three of the local people to serve as hostages in negotiations for the boat's return. The captured group consisted of a 9-year-old girl (named Fuegia Basket by the crew), and two men in their 20s, who were given the names York Minister and Boat Memory. The Fuegians, however, showed no interest in bargaining for the return of their comrades. A chagrined FitzRoy turned this unexpected response into an opportunity to "improve" the Fuegians ("who are now scarcely superior to brute creation") by deciding to "procure for these people a suitable education, and, after two or three years,...send them back to their country." Before leaving the area, a 12-year-old boy was also brought on board by trading his father a large mother-of-pearl button for his son, earning this fourth recruit the name "Jemmy Button."

Soon after landing in Plymouth, Boat Memory succumbed to small pox, but the remaining three, after being successfully vaccinated, were transferred to the care of the Church Missionary Society. Fuegia, York, and Jemmy were taught some English, the "truths of Christianity," and practical skills like gardening. They became celebrities and met the King and Queen. Jemmy, in particular, embraced European culture, and enjoyed dressing in the starched collars and polished boots of the day. FitzRoy, however, decided to move up the return of the Fuegians to their homeland when suspicions arose that York

had a sexual interest in young Fuegia. FitzRoy implored the Admiralty to schedule another surveying expedition to Tierra del Fuego, but his requests were denied, and he eventually planned to fund the trip himself. However, influential relatives of FitzRoy were finally able to secure support for the planned voyage.

On December 27, 1831, the *Beagle* again sailed for South America with FitzRoy as Captain, and Charles Darwin and the three Fuegians aboard. Also aboard was Richard Matthews, a young missionary who would land with Fuegia, York, and Jemmy to help convert their fellow Fuegians into civilized Christians. During the voyage, Darwin grew fond of Jemmy, especially as he helped Darwin, who was new to ocean travel, through his bouts of seasickness. Aware of Jemmy's recent history, Darwin was "incessantly struck [by] how similar [the Fuegians'] minds were to ours." Encountering the Fuegians in their native land, however, also impressed Darwin, and he described them as resembling "the representations of devils on the stage" and exclaimed that "I could not have believed how wide was the difference between savage and civilised man; it is greater than between a wild and domesticated animal."

In January 1833, the three Fuegians were put ashore, along with Matthews and supplies (consisting mostly of items such as soup tureens, wine glasses, and tea trays). The *Beagle* continued surveying the local area, and returned several days later to find that the supplies had been plundered, prompting Jemmy to remonstrate that "my people very bad; great fool; know nothing at all." Darwin described with sorrow the plight of Jemmy, but recorded in his diary that he believed he would retain his newly acquired European manners; FitzRoy was less optimistic. Darwin's response to the Fuegians has been interpreted as racist. However, his observation that "three years has been sufficient to change savages into, as far as habits go, complete and voluntary Europeans" and the sadness he felt at leaving Jemmy behind, did leave open the possibility for Darwin that a privileged Englishman and a Fuegian might not be so different.

In 1848, Allen Gardiner, a retired sea captain, resurrected FitzRoy's plan to civilize the Fuegians and established the Patagonian Mission Society. He traveled to Tierra del Fuego in 1850, but he and his entire crew died in 1851. The Mission Society made several more attempts, and eventually established themselves in the Falkland Islands. By the late 1850s, the organization was resettling Fuegians on Keppel Island in the Falklands. They had, amazingly, found Jemmy Button (now with two wives and several children) more than two decades after the *Beagle* had left. Button initially aided the missionaries, but refused to move to Keppel Island. A mission was established in Tierra del Fuego in 1859, but the missionaries were killed by the Fuegians. The one survivor, the cook, was sheltered by Jemmy Button until help arrived. A final attempt at civilizing the Feugians was undertaken in the 1860s, when a single missionary started living among them according to their customs. Unfortunately, the introduction of diseases had by this time decimated the Fuegian population from around 8,000 in 1833 to fewer than fifty a hundred years later. Jemmy Button died of infection in 1864, although Fuegia Basket was known to be alive as late as 1883.

For More Information: Keynes, R.D. 2003. *Fossils, Finches, and Fuegians*. Oxford: Oxford University.

G

NORMA GABLER (1923–2007) AND MEL GABLER (1915–2004)

Mel Gabler and his wife Norma Rhodes Gabler founded the influential Educational Research Analysts in 1973 in Longview, Texas, to finance their campaign against "secular humanism" and textbooks they considered anti-American and which "eliminate coming to Christ for forgiveness of sin." In 1969, after the Gablers claimed that the evolution-based Biological Sciences Curriculum Study (BSCS) biology textbooks had been written by communists, the Texas Board of Education removed two BSCS books from the list of textbooks approved for use in the state's public schools, and in 1971 proclaimed that all biology textbooks must identify evolution as a theory. Because the textbook market was so large in Texas, the changes promoted by the Gablers affected textbooks nationwide. The Gablers' *Scientific Creationism* (1985) included contributions by Henry Morris and Duane Gish. Mel Gabler was shy, so Norma usually testified at hearings. During the 1970s and the 1980s, the Gablers were the most effective textbook-censors in America.

GALÁPAGOS ISLANDS

The Galápagos Islands are an archipelago of thirteen major volcanic islands (plus six small islands and more than 100 rocks and islets) that straddle the equator in the Pacific Ocean approximately 620 miles west of Ecuador (Figure 34). The oldest islands formed approximately five million years ago, and the youngest are still being formed by volcanic activity. The islands, which derive their name from *galápago* (the Spanish word for *tortoise*), are part of Ecuador.

The Galápagos Islands were discovered in 1535 when Dominican Fray Tomás Berlanga, the Bishop of Panama, and his crew were instructed by King Charles V to sail to Peru to settle a dispute between Francisco Pizarro and his officers after they had conquered the Incas. When winds died, currents pushed Berlanga to the Galápagos. Berlanga later wrote about the islands' tame tortoises (Figure 35), iguanas, sea lions, and birds, but his overall depiction of the archipelago was damning; for example, he described the soil as "worthless, because it has not the power of raising a little grass, but only some thistles."

In 1574, Abraham Otelius, a Flemish cartographer, first placed the "Islas de Galápagas" on maps. The islands were more fully described by buccaneers (for which the archipelago became a refuge), but their reports were like earlier ones—namely, that the archipelago was uninhabitable. In the early 1700s, the islands were visited by

34. The Galápagos Archipelago was made famous by Charles Darwin's visit in 1835 aboard the *Beagle*. This photo shows Pinnacle Rock and part of the Bartolomé (Bartolomew Island) isthmus, looking toward the Sullivan Bay lava flows on Isla Santiago (James Island). This is one of the most famous views in the Galápagos. (*Randy Moore*)

35. The Galápagos Islands are home to, and named for, the giant Galápagos tortoise (*Geochelone elephantopus*). (*Randy Moore*)

Alexander Selkirk, who later became famous as Robinson Crusoe. Several colonies were established, but almost all collapsed. Except for buccaneers' crude maps, the Galápagos remained uncharted.

Ecuador annexed the Galápagos Islands in 1832, and in 1892—the 400th anniversary of Christopher Columbus' voyage to the Americas—the islands were named Archipelago de Colon by the government of Ecuador. The islands provided sanctuary to pirates, and between 1780 and 1860 they became home to British and American whalers. Whaling was highly profitable in the first half of the nineteenth century, and whales were abundant in the Galápagos, where the confluence of ocean currents produced ideal sites for whales. In 1812, the United States sent navy hero, David Porter, and the *Essex* to protect American whalers from harassing British ships. Porter was in the Galápagos for five months in 1813, and he condemned the islands as uninhabitable. Whalers and other visitors to the islands decimated the populations of whales and tortoises, which were excellent sources of food.

In 1841, the 22-year-old Herman Melville visited the Galápagos while aboard the New England whaler, *Acushnet*. Melville, who published ten sketches of the archipelago, rejected—and at times parodied—Darwin's earlier observations of the islands. In his classic *Moby-Dick*, Melville quoted from Darwin's *Journal of Researches*.

When Darwin landed at the Galápagos on September 15, 1835, he noted that the archipelago "seems to be a little world within itself; the greater number of its inhabitants, both vegetable and animal, being found nowhere else." Darwin was in the Galápagos for five weeks, nineteen days of which he spent ashore. Darwin visited Chatham (San Cristobel), Charles (Floreana), Albemarle (now Isabella), and James (Santiago) Islands; all of the specimens he collected came from these islands. Darwin collected almost fifty tortoises, but the shells of only four made it back to England's museums. FitzRoy's map of the Galápagos remained in use until the U.S.S. *Bowditch* recharted the archipelago in 1942.

In 1845, Darwin published a general account of his voyage; that book eventually became known as *The Voyage of the Beagle*, and it remains one of history's greatest travel books. Darwin noted that: "The natural history of these islands is eminently curious, and well deserves attention ... The different islands to a considerable extent are inhabited by a different set of beings." Some of the specimens collected during expeditions to the Galápagos ended up in the hands of Joseph Hooker, who included them in his 1847 publication "Enumeration of the Plants of the Galápagos Archipelago."

Although many biology textbooks cite the Galápagos as a type of catechism about Darwin's theory of evolution, *On the Origin of Species* (1859) mentions the Galápagos only six times while making the following points:

> In isolated places such as the Galápagos, one group of plants or animals is sometimes replaced by another group of plants or animals. For example, "in the Galápagos Islands reptiles ... take the place of mammals." Darwin reasoned that some groups of plants and animals could not reach such places, and others evolved to fill the niche.
>
> The Galápagos contain many closely related species found nowhere else, a "fact [that] might have been expected on my own theory." These species often resemble those of the nearest mainland. As Darwin asked, "Why should the species which are supposed to have been created in the Galápagos Archipelago, and nowhere else, bear so plain a stamp of affinity to those created in America?"
>
> In Darwin's world, a deity did not create species adapted to the Galápagos. Instead, chance arrivals evolved to live there.

In 1872, antievolutionist Louis Agassiz and his wife Elizabeth (who later became the founding president of Radcliffe College) visited the Galápagos, and used their observations to undermine Darwin. The Agassizs described happy, playful porpoises and portrayed the Galápagos as a harmonious place far from the Malthusian world of Darwin. When Agassiz returned from his voyage, he wrote in *Nature* that "Darwin's hypothesis of gradual variation of species, and the natural selection for preservation of those whose variations were favorable to them in the struggle for life, seems to me to have few facts to sustain it, and very many to oppose it." Agassiz, who died a year after visiting the Galápagos, was the last great American biologist to proclaim the special creation of species.

Construction of the Panama Canal in 1914 made the Galápagos more accessible, and the Galápagos were changed profoundly by World War II. On December 11, 1941—just four days after Pearl Harbor was attacked by the Japanese—Ecuador gave U.S. troops permission to occupy the Galápagos. Early the next year, construction began on an airbase on Baltra (just north of Santa Cruz), and several thousand Americans came to live on the island. When Americans left the Galápagos after the war, all of the islands' residents were allowed to remove one of the wooden buildings from the base. For the next several decades, these buildings were common on the inhabited islands.

In the 1930s, Julian Huxley—grandson of Thomas Huxley and then Secretary of the Zoological Society of London—asked English schoolteacher and amateur ornithologist David Lack to observe Galápagos finches for an entire breeding season. Lack and five colleagues documented natural selection through interspecific competition, after which they examined specimens of Galápagos finches in museums in California and at the American Museum of Natural History in New York, where Lack roomed with German émigré Ernst Mayr. Lack, who measured more than 8,000 finches, later described his work in *Darwin's Finches*, which forever designated the new name of the Galápagos birds. Although Darwin had used Galápagos mockingbirds to make his points in *On the Origin of Species*, textbooks followed Lack's lead and discussed "Darwin's finches."

More recently, Peter and Mary Grant's studies of the Galápagos finches have shown that populations of these birds evolve rapidly in response to environmental conditions. The Grants have studied the Galápagos finches for decades, and their work is some of the most impressive fieldwork ever reported. Thanks to the Grants,

Lack, and others, "Darwin's finches"—like the Galápagos Islands themselves—are an icon of evolution. The dark-colored birds, which are about the size of a sparrow and live on all the islands, comprise a subfamily called Geospizae, which is found only in the Galápagos Islands and on Cocos Island to the northeast. Thirteen species are endemic to the archipelago; all of these species originated from an original species, *Melanospiza richardsonii*, which lives on Sainte Lucie Island in the Caribbean.

The first legislative efforts to protect the Galápagos Islands was enacted in 1934, near the time that two groups on Floreana Island—a German dentist and a "neurotic" Austrian baroness, both with lovers in tow—became the "Adam and Eve of the Galápagos." The following year, Ecuador issued commemorative stamps recognizing the centenary of Darwin's visit; the stamps featured a map, a marine iguana, a giant tortoise, and the head of Darwin. In 1959, the centenary year of Darwin's *On the Origin of Species*, Ecuador declared more than 97 percent of the archipelago to be the country's first national park. The Charles Darwin Research Station on Santa Cruz Island was established in 1960. Noteworthy species that inhabit the Galápagos include Galápagos sea lions (*Zalophus californianus*), Galápagos penguin (*Spheniscus mendiculus*, which reached the Galápagos on the Antarctic Humboldt Current), Galápagos tortoise (*Geochelone elephantopus*), and the Galápagos marine iguana (*Amblyrhynchus crisatus*), the only iguana that feeds at sea.

In 1971, a park ranger discovered a tortoise on Pinta Island, the first tortoise to be seen there since explorers from the California Academy of Science (who made Charles Darwin an Honorary Member in 1872) killed three there sixty-five years earlier. That tortoise, a 90-kg male named "Lonesome George" (after 1950s comedian "Lonesome" George Gobel), now lives at the Darwin Station, and is the last survivor of a doomed race. There's a reward of $10,000 for a mate, but the reward has gone unclaimed.

Today, the Galápagos National Park Service oversees the many visitors to the islands, and the Charles Darwin Research Station administers research and conservation programs (e.g., by breeding and releasing tortoises and iguanas). In 1986, the surrounding 43,500 square miles of ocean were declared a marine reserve; this reserve is second in size only to Australia's Great Barrier Reef. Although only about 300 people lived on the Galápagos when Darwin visited the archipelago in 1835, the islands are now inhabited by more than 30,000 people, and more than 100,000 tourists and visitors visit each year. The Galápagos have also been the backdrop of movies such as *Master and Commander* (2003), in which naturalist Stephen Maturin discovered new animal species. Kurt Vonnegut's 1985 novel *Galápagos* described how the isolated survivors of "the Nature Cruise of the Century" evolved into a peaceful race of humans after war, disease, and starvation had killed everyone else.

Today, the archipelago is one of Ecuador's national parks, a whale sanctuary, and a founding "Man in the Biosphere" preserve. In 1978, the Galápagos became the first World Heritage Site chosen by the United Nations Educational, Scientific and Cultural Organization (UNESCO), and in June 2007, UNESCO added the caveat "in danger" to the designation to draw attention to the threat posed by tourism and immigration from Ecuador's mainland (since 2001, the increased flights and boat traffic have contributed to a 60 percent increase in introduced species). The Galápagos' bustling capital includes Melville Avenue and Charles Darwin Avenue, which intersect near the water's edge. As biologist Robert Bowman noted, "No area on Earth of comparable size has inspired more fundamental changes in Man's perspectives of himself and his environment than the Galápagos Islands."

For More Information: Constant, P. 2004. *The Galápagos Islands: A Natural History Guide.* New York: W.W. Norton; Larson, E.J. 2001. *Evolution's Workshop: God and Science on the Galápagos Islands.* New York: Basic Books.

GEORGE GALLUP (1901–1984)

George Horace Gallup was an American statistician who invented the Gallup poll to sample public opinion. In 1982, Gallup reported that "debate over the origin of man is as alive

36. Francis Galton, a cousin of Charles Darwin, was a famous mathematician and the founder of eugenics. (*Library of Congress*)

today as it was at the time of the famous Scopes trial in 1925, with the public now about evenly divided between those who believe in the biblical account of creation and those who believe in either a strict interpretation of evolution or an evolutionary process involving God." In 2004, a Gallup poll reported that 35 percent of Americans believed that Darwin's theory of evolution by natural selection is well-supported by evidence. Gallup is buried in Princeton, New Jersey.

FRANCIS GALTON (1822–1911)

I take Eugenics very seriously, feeling that its principles ought to become one of the dominant motives in a civilised nation, much as if they were one of its religious tenets.

Francis Galton was born February 16, 1822, in Birmingham, England (Figure 36). Galton's father was a wealthy banker who encouraged his son to consider a career in medicine. A cousin of Galton's, Charles Darwin (Galton's mother was the half-sister of Charles' father, Robert), was similarly encouraged to enter medicine by his own father. Although neither Darwin nor Galton became physicians, both were influential figures in science.

By all accounts, Galton was a child prodigy. Under the tutelage of an invalid sister, he learned the alphabet by the age of eighteen months, could read by three, and was reciting and discussing poetry by the time he was five. These abilities, however, did not translate into good grades in school. When he was sixteen, Galton was removed from King Edward's School partially due to disciplinary issues. His father then arranged for him to learn medicine at Birmingham General Hospital, still advocating a medical career for his bright but wayward son. A year later, Galton moved to the medical school at King's College, London, but by 1840, he had decided to transfer to Cambridge to study mathematics. Underachievement and emotional problems continued to plague Galton, and after his third year at Cambridge, followed by another stint at studying medicine, he decided to leave school altogether. The final event precipitating this decision was the death of his father in 1844, which provided, in Galton's words, "a sufficient fortune to make me independent of the medical profession."

In 1850, freed of having to build a career, Galton led an expedition by the Royal Geographic Society to southwest Africa to explore, in the footsteps of David Livingstone, the Lake Ngami region of Namibia. This trek produced the book *Tropical South Africa* (1852), which achieved popular acclaim and for which Galton was awarded the Founders Medal from the Royal Geographic Society. Galton was soon thereafter made a fellow of the Royal Geographic Society. About this same time, he married Louisa Butler, to whom he would remain married until her death in 1897.

In 1855, Galton published his second book, *The Art of Travel*, which provided survival tips for those traveling under primitive conditions (and included such chapters as "Bush Remedies" and "Management of Savages"). Although Galton was interested in a variety of topics (including the efficacy of prayer), his name is forever linked to heredity and eugenics, topics he studied during the second half of his life.

Galton's lifelong obsession of counting and recording information—he followed the motto of "Whenever you can, count" and he published data on the number of people fidgeting at scientific meetings and the number of attractive women he passed on walks through various cities—would serve him well in quantifying the heritability of traits. His cousin Charles had recently published *On the Origin of Species*, which Galton claimed "made a marked epoch in my own mental development." Galton embraced the possibility of evolutionary change, although he disagreed with Darwin's ideas on the mechanism of inheritance. (Galton conducted an experiment where he transfused blood between differently colored rabbits to disprove Darwin's proposal of pangenesis that invoked Lamarckian inheritance through "gemmules.")

Galton studied the inheritance of intelligence and personality in humans by using biographical records to compare the frequency of eminent descendants being produced by eminent versus common men. Galton concluded that "talent and peculiarities of character are found in the children, when they have existed in either of the parents, to an extent beyond all question greater than in the children of ordinary persons.... The feeble nations of the world are necessarily giving way before the nobler varieties of mankind.... We are living in a sort of intellectual anarchy, for want of master minds."

In 1869, Galton's *Hereditary Genius* expanded his examination of pedigrees and introduced the use of the normal distribution ("bell curve") as an analysis tool. This was followed in 1883 by *Inquiries into Human Faculty and Development*, a collection of papers, which advocated for what he termed *eugenics*, an inquiry into "questions bearing on what is termed in Greek, eugenes, namely, good stock, hereditarily endowed with noble qualities." In *Inquiries*, Galton discussed studies of identical twins (he had earlier used the phrase "nature and nurture" and realized that twin studies could examine the relative contribution of each component) as evidence for the strength of hereditary characters.

Convinced that human traits had a strong hereditary basis, Galton developed a theory of inheritance to explain the patterns he observed. Ultimately, he proposed the existence of "germs" within individuals, with the "stirp" (genome) of the individual being the collection of germs that determine the separate characteristics of an individual. Germs existed as "patent" and "latent" forms: the patent forms are expressed in the individual bearer, while the latent forms are passed to offspring, which could explain reversion of traits to ancestral states. Hence, Galton took Darwin's concept of "gemmules," eliminated the Lamarckian aspect, and in broad outline developed the germ-line theory of inheritance soon to be proposed by August Weismann.

Galton tested his model of heredity by analyzing inheritance of seed size in sweet pea plants. In this study, Galton developed a powerful statistical technique when he plotted the average seed size of offspring versus average seed size of parents; he called the slope of the line through these points the "coefficient of regression." He later also developed the statistical technique of correlation, and Galton is, therefore, credited with establishing the foundations of biometry. Galton also concluded that there is an overall tendency of "regression to the mean" in each generation, which meant that Darwin's model of gradual selection operating on continuous variation would be ineffective; instead, speciation could only proceed through large-scale genetic changes ("sports") as advocated by "saltationist" geneticists of the early 1900s.

Anxious to apply his ideas to humans, Galton devised an "anthropogenic laboratory" at the International Health Exhibition in 1884. During a thirty-minute period, a visitor would pass through the laboratory and undergo a number of strength and sensory tests. By the end of the exhibition, Galton had measured over 9,000 people. Analysis of this large dataset produced (in addition to several new statistical methods) the book, *Natural Inheritance* (1889). Galton, who worried that natural selection was being hindered by misguided programs that encouraged "unfit" people to have children, advocated policies that would avert a "reversion toward mediocrity."

Convinced of the heritability of human traits, Galton promoted eugenics. In 1901, he used the Huxley Lecture at the Royal Anthropological Institute to advocate the use of athletic and scholastic tests (for men) and medical tests (for women) to identify the "best" of each; those that passed would receive certificates that documented their desirable status. By 1907, Galton had established the Laboratory for National Eugenics at University College, London, whose first director was Galton disciple and statistician, Karl Pearson. Although Galton viewed enhancing reproduction of those in the upper tail of the distribution as preferable to "repressing the reproductivity of the worst," he also noted, "What nature does blindly, slowly, and ruthlessly, man may do providently, quickly, and kindly." During the final decade of his life, Galton's ideas became popular in Europe and the United States. Galton envisioned his proposals would ultimately produce a eugenic utopia, an idea he explored in an unfinished novel, *Kantsaywhere*.

In recognition of a lifetime of work, Galton was awarded the Darwin Medal from the Royal Society (1902), made a Fellow of Trinity College, Cambridge (1902), and was knighted in 1909. On January 17, 1911, following a bronchial attack, Galton died of heart failure, and was buried at St. Michael's and All Angels' Church in Warwickshire, England. Despite his accomplishments and eminence, and therefore the high probability of passing on desirable traits, he died without having fathered any children.

For More Information: Gillham, N.W. 2001. *A Life of Sir Francis Galton*. New York: Oxford University.

GARDEN OF EDEN

The Garden of Eden is described in Genesis as the place where the first man and woman (i.e., Adam and Eve) lived after they were created by God. God made Adam from dirt, and then made Eve from one of Adam's ribs to be Adam's helper. When Adam and Eve ate from the Tree of the Knowledge of Good and Evil, they were expelled from the Garden. Many theologians believe that the Garden of Eden was near Mesopotamia. Eden was the setting of much of Milton's *Paradise Lost*, and was depicted by Michelangelo on the ceiling of the Sistine Chapel.

NORMAN GEISLER (b. 1932)

Norman Geisler was a popular preacher, writer, and debater who testified during *McLean v. Arkansas Board of Education* that UFOs are "the Devil's major, in fact, final attack on the Earth," that he knew "at least 12 persons who were clearly possessed by the Devil," and that UFOs exist because he "read it in the *Readers Digest*." Geisler, a long-time member of the American Scientific Affiliation, who endorsed several claims by the Bible-Science Association, also claimed that the Arkansas law mandating equal time for creationism and evolution did not introduce religion into schools because God is not a religious concept—as Geisler noted, "It is possible to believe that God exists without necessarily believing in God." Judge William Overton dismissed Geisler's claim "as contrary to common understanding." Geisler, whose *The Creator in the Courtroom—Scopes II* described his experiences during *McLean*, later cofounded and served as dean of the Southern Evangelical Seminary in Charlotte, North Carolina. In the mid-1980s, Geisler was one of the first people to revive intelligent design as an alternative to evolution.

ÉTIENNE GEOFFROY SAINT-HILAIRE (1772–1844)

There is, philosophically speaking, only a single animal.

Étienne Geoffroy Saint-Hilaire was born April 15, 1772, in Étampes, just outside Paris. His father, a judge, advised his son to become a lawyer. Étienne considered entering the priesthood, but followed his father's advice and earned a law degree from the University of Navarre in 1788. While in college, Geoffroy

decided to become a physician, and entered the College of Cardinal Lemoine. Soon, however, Geoffroy's main interest became natural history.

During the Reign of Terror of the French Revolution, one of Geoffroy's university mentors, René Just Haüy, was arrested. Geoffroy, risking imprisonment himself, advocated for Haüy's release, which he finally secured. A grateful Haüy was instrumental in having Geoffroy appointed assistant curator at the Jardin des Plantes in 1793. The Revolution forced a reorganization of scientific institutions in Paris, and in 1793 when the Jardin des Plantes became the Museum of Natural History, Geoffroy was named professor of zoology.

The following year, Geoffroy was contacted by a former mentor who recommended Georges Cuvier for employment at the Museum. Cuvier was hired as an assistant, and was soon named professor of comparative anatomy. Cuvier and Geoffroy became friends (although they were of different personalities), and coauthored four scientific papers during their first year as colleagues. Geoffroy tried to identify the single body plan upon which he believed all animals were built (his "unity of plan"), while Cuvier classified organisms according to a few distinct body-plans (*embranchments*). Although Cuvier and Geoffroy had minor disagreements, their perspectives were somewhat congruent: Cuvier, like Geoffroy, believed that body plans were shared among organisms, but that there were simply multiple plans.

In 1798, Geoffroy accompanied Napoleon to Egypt, and returned in 1801 with mummified cats and birds (which proved of great interest to Cuvier). Geoffroy married Pauline Brière in 1804, and they would have three children, including Isodore, who assumed his father's position at the Museum upon Geoffroy's death. While in Egypt, Geoffroy developed a nervous disorder that caused wild mood swings that plagued him for the rest of his life. In 1807, he was elected to the Academie des Sciences, and was appointed professor of zoology at the University of Paris in 1810.

Geoffroy's goal of elucidating the common archetypal body-plan led him to compare the anatomies of a large number of species, at both adult and embryonic stages. This work resulted in the two-volume set, *Philosophie Anatomique* (1818–1822). In the first volume, Geoffroy compared the structural similarities of a wide range of vertebrates (i.e., from fish to primates) and, applying the concept of recapitulation, proposed that the developmental sequence of embryos reflects the history of changes that have occurred to the basic body plan in that species. In the second volume, Geoffroy examined "monstrosities"—that is, individuals that show unusual, sometimes extreme traits—to demonstrate that these are the exceptions that prove the rule: the unusual characteristics of these monsters are simply modified versions of body parts found in the typical version of the organism. Based on this work, Geoffrey proposed three principles that guide how the basic body plan can be modified: (1) the *principle of development* (organs appear or disappear gradually; this explains the presence of vestigial organs); (2) the *principle of compensation* (the elaboration of one organ requires a corresponding reduction in another organ); and (3) the *principle of connections* (structures can become differentially modified in different organs, but they always maintain the same relative position). Geoffroy used his principle of connection to identify what he called "analogies," or what would now be labeled homologous structures.

Initially, Cuvier was not distressed by these proposals. However, Geoffrey's *Mémories sur l'organization des Insectes* (1820), published in the period between release of the two volumes of *Philosophie Anatomique*, finally split Cuvier and Geoffroy. In *Mémories*, Geoffrey extended his argument to propose that arthropods are built upon the same body plan as vertebrates (i.e., the exoskeleton of insects is merely a single vertebrae used on the outside of the body). Cuvier could not accept this, and he attacked Geoffrey's proposal, dismissing it as "lack[ing] logic from beginning to end."

Geoffroy became convinced that, through modification of the basic body plan, transformation of organisms occurs. Geoffroy rejected Lamarck's proposals for evolutionary change,

and did not suggest an unguided process, as Darwin would later do. Rather, Geoffroy viewed transformation as following a deterministic, pre-ordained plan. Cuvier, however, firmly believed in the immutability of species, and organic change was a contentious issue between the two colleagues.

In 1830, a paper was submitted for consideration by the Academie des Sciences. The authors applied Geoffroy's principle of connections to suggest that mollusks had the same body plan as vertebrates. Geoffroy responded enthusiastically to the paper (and may have even rewritten parts of it) because it integrated yet another of Cuvier's embranchments (the Mollusca) into Geoffroy's unified body-plan. Cuvier interpreted this as a public repudiation of his work, and forced a pubic confrontation with Geoffroy. Over the course of eight public meetings, the two men debated the merits of their respective positions. Cuvier is usually declared the winner of the debates, but this conclusion probably reflects his tremendous knowledge of comparative anatomy and his clear and organized speaking skills. In contrast, Geoffroy's halting style did not work in his favor. The debates were interrupted by the overthrow of Charles X, and Cuvier died less than two years later.

In 1840, Geoffroy lost his sight and was forced to resign from teaching and research in 1841. He died on June 19, 1844.

For More Information: Le Guyader, H. 2004. *Étienne Geoffroy Saint-Hilaire: A Visionary Naturalist*. Chicago, IL: University of Chicago.

GEOLOGICAL SOCIETY OF LONDON (Est. 1807)

The Geological Society of London elected Charles Darwin to membership in 1836 after he was nominated by Adam Sedgwick and John Henslow. Darwin was the Society's secretary from 1838 to 1841. In 1859, the Society awarded Darwin its highest honor, the Wollaston Medal. In 2007, the society celebrated its bicentenary by opening The Lyell Centre, a virtual library. The Geological Society of London is the oldest national geological society in the world.

GERTIE THE DINOSAUR (Est. 1914)

Gertie the Dinosaur was the first animal that starred in a cartoon. Gertie was created by Winsor McCay (1871–1934), a cartoonist who had introduced the friendly dinosaur in the comic strip, *Little Nemo in Slumberland*. McCay later developed a vaudeville act featuring Gertie. Gertie was the first time the public saw an animated dinosaur; in subsequent years, many such animations would follow (e.g., *Barney*, *The Flintstones*). Gertie's frequent association with steamboats, locomotives, and cars reinforced the association of dinosaurs and machines with fossil fuels and the giant reserve of buried energy that can be released by technology.

DUANE GISH (b. 1921)

Creationism is in every sense as scientific as evolution.

Duane Tolbert Gish was born February 17, 1921, in White City, Kansas (Figure 37). Gish served in the South Pacific during World War II, and then attended the University of California, Los Angeles and earned a BS in chemistry in 1949, and a PhD in biochemistry from the University of California at Berkeley in 1953. Gish moved to Cornell University, after which he returned to Berkeley to study tobacco mosaic virus with Nobel laureate Wendell Stanley. He then worked in Kalamazoo, Michigan for eleven years as a research associate at the Upjohn Company. From 1952 through 1970, Gish published forty articles in scientific journals on various aspects of biochemistry.

Gish's father was not religious, but his mother was a devout Methodist who helped her son "find a personal relationship with Christ," and he eventually became a fundamentalist Baptist. Gish traces his antievolution activism to John R. Howitt's pamphlet *Evolution: Science Falsely So-Called* during the late 1950s. Howitt, a Canadian psychiatrist and staunch antievolutionist, was a member of the American Scientific

37. Duane Gish was a famed creationist who participated in more than 300 debates with evolutionary biologists throughout the world. Gish worked for the Institute for Creation Research and was a founder of the Creation Research Society. (*ICR*)

Affiliation (ASA) during the 1940s and 1950s, but left when the organization drifted toward theistic evolution. After reading Howitt's work, Gish also joined the ASA, but soon left for the same reason cited by Howitt.

In 1963, Gish joined nine other creationist scientists (including Henry Morris) to form the Creation Research Society (CRS). CRS espoused a literal interpretation of the Bible—including the reality of a worldwide Noachian flood—and emphasized "publication and research which impinge on creation as an alternate view of origins." Gish did not initially stress Young-Earth creationism, considering the Flood to be "a matter of indifference." He suggested that any common ground with Old-Earth creationists could be put to good use, a perspective that proved controversial during the early years of the CRS. However, by the time Gish began publishing his antievolution books through the Institute for Creation Research (ICR), he believed flood geology was central to creationism.

In 1970, Gish began to work full-time for the advocacy of creationism. Henry Morris had recently joined with Tim LaHaye to found Christian Heritage College (now San Diego Christian College). By 1971, Gish had joined the faculty of the school as Professor of Natural Science. When Morris established ICR in 1972, Gish moved with him to serve as Associate Director. When ICR started offering graduate degrees, Gish became Senior Vice President.

When Gish began fighting evolution, he attacked the nation's educational system because "the dogmatic fashion in which evolution is usually taught in our schools and universities amounts to indoctrination." In 1970, while still at Upjohn, he published an essay in a recurring section (called, ironically, "The Devil's Advocate") of the *American Biology Teacher*, a journal published by the National Association of Biology Teachers (NABT). Gish's essay stressed standard creationist objections (i.e., lack of transitional forms) about the fossil record. In 1972, when the NABT annual convention bore the theme "Biology and Evolution," Gish (now at ICR) presented, and later published in the *American Biology Teacher* (1973), his alternative interpretation of the fossil record. (This was the same convention at which Theodosius Dobzhansky gave his famous "Nothing in Biology Makes Sense Except in the Light of Evolution" talk.) The presentation outlined the major themes Gish would emphasize repeatedly over the next few decades: evolution demotes humans to just another species (evolution is the "molecules-to-man theory"); evolution is not science ("creation is . . . no more religious or less scientific than evolution"), which in turn requires a "balanced" approach in the classroom; and the evolution-creationism conflict is ultimately one of religion versus materialism ("the authoritarianism of the medieval church has been replaced by the authoritarianism of rationalistic materialism").

Gish claims that the Bible is "the only rational explanation of origins."

Gish's message and tactics were well received within the creationist community; even the theistic-leaning ASA reported favorably on his NABT presentation. Gish started debating evolutionary biologists in 1972 when, as Gish was giving a public lecture at the University of California at Davis, he was confronted by biologist G.L. Stebbins. Gish earned a reputation among fellow creationists as a worthy debate opponent, earning him the moniker "Creation's Bulldog," playing upon the informal title, "Darwin's Bulldog," by which Thomas Huxley had been known. By 2005, Gish had participated in almost 300 debates in all fifty U.S. states (except Maine) and more than forty foreign countries. Gish avoided discussions of the age of the Earth in his debates; although an advocate of Young-Earth creationism, he tried to prevent his opponents from using what he calls the "if the Earth is old, that proves evolution" argument to undermine his position.

Gish's many books and pamphlets advocating his creationist perspective, published through ICR, include *Evidence against Evolution* (1972), *Evolution: The Fossils Say No!* (1986), *Evolution: The Fossils Still Say No!* (1995), *Teaching Creation Science in Public Schools* (2000), and a children's book, *Dinosaurs: Those Terrible Lizards* (1978). In 2005, Gish, at the age of 84, retired and assumed the title of "Emeritus Vice President" of ICR. He continues to author books and remains a spokesman for Young-Earth creationism.

For More Information: Gish, D.T. 1995. *Evolution: The Fossils Still Say No!* El Cajon, CA: Institute for Creation Research.

JOHN GODSEY (1874–1932)

John Luke Godsey was one of John Scopes' original attorneys. On June 13, 1925, Godsey filed a motion seeking to have the charge against Scopes dropped because the Butler Act was unconstitutional. The motion was rejected, and Godsey bowed out of the case before the trial began. Godsey was replaced by F.B. McElwee, a former student of John Neal.

RICHARD GOLDSCHMIDT (1878–1958)

Richard Goldschmidt was a geneticist who coined the phrase "hopeful monsters" in 1933 to describe the products of sudden evolutionary jumps he felt were required to account for speciation. "Monster" referred to macromutations or mutants that might arise in a single generation, thereby producing a selective advantage in the changing environment; "hopeful" referred to Goldschmidt's belief that a macromutation might be so adaptive in the new environment that it would be selected as the new norm. "Hopeful monsters" remains an appealing, but discredited, idea for evolutionary change.

HENDRIK GOOSEN

Hendrik Goosen was a fisherman who, in 1938, caught a coelacanth, a fish thought to have become extinct more than 65 million years ago. Goosen's catch caused a sensation, and a £100 reward for other coelacanths produced more than 200 specimens. Today, coelacanths, which are similar to their 350-million-year-old ancestors, are depicted on the 5-franc coin of the Comoros Islands.

For More Information: Thompson, K. 2001. *Living Fossil: The Story of the Coelacanth.* New York: W.W. Norton.

PHILIP GOSSE (1810–1888)

When the catastrophic act of creation took place, the world presented instantly the structural appearance of a planet in which life had long existed.

Philip Henry Gosse was born on April 6, 1810, in Worcester, England. Gosse's *Entomologia Terrae Novae* was the first scientific study of Canada's insects. Gosse also invented the seawater aquarium, the first of which he exhibited at the London Zoo. He wrote several books about biblical chronology, including *The 6000 Years of the World's History Now Closing: A Table of Scripture Chronology* (1846).

Gosse, who corresponded with Charles Darwin, was disturbed by the conflict between science and religion. His reconciliation of science and religion began with an age-old question—because Adam and Eve were created directly by God and not born like other people, did they have navels? If Adam had a navel, it would have implied his nonexistent birth from a nonexistent mother. Gosse answered this and related questions in his 1857 book, *Omphalos: An Attempt to Untie the Geological Knot*, in which Gosse claimed that God had used fossils and geology to suggest a history that did not actually exist. In *Omphalos* (Greek for "navel"), Gosse claimed that Adam and Eve had navels, and argued that all geological evidence for an ancient Earth had been created about 6,000 years ago when God created the universe as if it had an ancient history (Gosse called the events that did not occur "prochronic," meaning "outside time"). Gosse claimed that because nature is cyclical, creation had to start from some part in the cycle, and if one assumed a creation from nothing, there had to be traces of a previous existence that never existed. This meant that Adam's navel was evidence of a natural birth that did not occur. Similarly, Gosse claimed that trees in the Garden of Eden were created with tree rings so that they could be part of nature's cycle, and that geologic sediments were created with fossils already in them (i.e., the apparently extinct organisms had never existed).

Gosse's claim that God had created evidence of a past that never existed was not popular; many critics equated it with divine fraud because it implied that humans cannot trust what God created. Similarly, Gosse's idea was nicknamed "Last Thursdayism" by British philosopher Bertrand Russell (1872-1970) as in "the world might have been created last Thursday—how would we know the difference?" Young-Earth creationists who accept Gosse's claim that God created a "mature" Earth are called "Mature-Earth creationists." Today, *Answers in Genesis* claims that Adam and Eve lacked navels because they were created directly by God.

The public's rejection of his idea prompted Gosse to largely abandon natural history, and he devoted the rest of his life to religion and crime stories. Gosse died on August 23, 1888. Today, two portraits of Gosse hang in London's National Portrait Gallery. In *Creation According to God's Word* (1966), John Whitcomb—the coauthor with Henry Morris of *The Genesis Flood*—claimed that creation included Gosse's "superficial appearance of history or age." Gosse's irreligious son, Edmund, wrote in *Father and Son* (1907) that his father was "a strictly scientific observer who was also a slave to revelation."

JOHN GOULD (1804–1881)

[If it were not for] the constancy of species, ornithology would no longer be a science.

John Gould was born in Dorset, England, on September 14, 1804, and was the son of a successful gardener. Gould learned his father's trade, and at the age of 14 he apprenticed at the Royal Gardens of Windsor, where he was responsible for picking dandelions for Queen Charlotte's tea. While at the Royal Gardens, Gould learned taxidermy. In 1824, Gould became a full-time taxidermist, and in 1827 he became the first Curator and Preserver at the museum of the Zoological Society of London. In 1930, Gould received a large collection of birds from the Himalayas, which he described in *A Century of Birds from the Himalayas* (1832). Most of the hand-colored illustrations in Gould's books were life-size. Gould's fans included Queen Victoria and Prince Albert.

On January 4, 1837, Charles Darwin gave Gould the birds that he and others aboard the *Beagle* had collected at the Galápagos Islands. Darwin believed the birds were blackbirds, "gross-bills," and finches, but one week later Gould reported that the birds were "a series of ground Finches which are so peculiar" because they formed "an entirely new group, containing 12 species." Two months later, Darwin met with Gould, who reported that the Galápagos "wren" was another species of finch, and the mockingbirds that Darwin had collected on different islands were separate species. Darwin did not label his specimens carefully, so he gathered specimens from Captain FitzRoy and others who were aboard the *Beagle*. Gould discovered that the birds were all finches, and that their primary difference was the shape of their beaks. Gould described Darwin's birds in 1837 in the *Proceedings of the Zoological Society of London*, and designed fifty hand-colored plates for *Zoology of the Beagle*. Darwin later realized that immigrants to the Galápagos Islands had somehow been modified and that each species was adapted to a particular island.

Although Gould had no formal university training, he was elected a Fellow of the Royal Society in 1843. Gould published a variety of illustrated books about birds, and his lithographs—especially those of hummingbirds and toucans—are prized by collectors.

Gould died in London on February 3, 1881. His final wish was that his epitaph be, "Here Lies John Gould, Bird Man."

STEPHEN JAY GOULD (1941–2002)

Humans are not the end result of predictable evolutionary progress, but rather a fortuitous cosmic afterthought, a tiny little twig on the enormously arborescent bush of life, which if replanted from seed, would almost surely not grow this twig again.

Stephen Jay Gould was born September 10, 1941, in Queens, New York. His career path was determined at the age of five when, awestruck at the sight of a *T. rex* skeleton at the American Museum of Natural History (AMNH), he decided to become a paleontologist. As a boy, Gould read everything he could about dinosaurs, and his penchant for collecting fossils earned him the name "Fossilface" from his peers. His parents encouraged his scientific pursuits, especially his father whom Gould described as "an intellectual without official credentials." (Gould's father was also a Marxist, which may explain the younger Gould's leftward political leanings sometimes evident in his writing.) Upon entering Jamaica High School in Queens, he eagerly anticipated learning more about evolution (Charles Darwin and

Joe DiMaggio—Gould was a lifelong New York Yankees fan—were his childhood idols). But Gould was quickly dismayed at how little the subject was discussed, a situation he later attributed to the lasting effects of the Scopes Trial.

Gould earned a BA in geology from Antioch College (Ohio) in 1963, after which he entered graduate school at Columbia University to study with Norman Newell, professor of paleontology and curator at the AMNH. He married Deborah Lee in 1965 and the following year accepted a faculty position in the Geology Department at his undergraduate alma mater. He earned a PhD in 1967 from Columbia University, and the same year he joined the faculty of Harvard University, eventually becoming curator of invertebrate paleontology and Alexander Agassiz Professor of Zoology, and where he would remain until his death in 2002. Gould had two sons, Jesse and Ethan. In 1995, Gould—by then divorced—married Rhonda Shearer, a sculptor and art historian.

Gould's professional career encompassed two overlapping yet distinct roles: academic scientist and popular science writer. He began his long-term study of land snails of the genus *Cerion* in the West Indies while still an undergraduate and continued this work for years after going to Harvard, producing hundreds of scientific papers. Early on, Gould was also interested in the relationship between the development of individuals (ontogeny) and they way in which the species to which they belong evolves. His book *Ontogeny and Phylogeny* (1977) explored this subject in detail, and in general outline predicted the discipline known as "evo-devo," the interface between developmental and evolutionary biology. But it was his broader views about evolution that produced the greatest effect on the academic world, views that spilled over into the public arena.

The scientific idea most closely associated with Stephen Jay Gould is punctuated equilibrium, first introduced in print with fellow paleontologist Niles Eldredge in 1972. (Gould was a strong proponent of punctuated equilibrium, but gave Eldredge primary credit as the main developer of the idea.) Gould and Eldredge were struck by how the fossil record suggests that species often remain unchanged for millions of years, but that these periods of stasis can be interrupted by bursts of diversification. Building upon Ernst Mayr's idea that rapid speciation is favored in peripheral, isolated populations, Gould and Eldredge argued that the fossil record demonstrates that evolutionary change is generally nonexistent in species composed of large populations due to the homogenizing effect of interbreeding among individuals. Instead, significant evolutionary change is more likely where populations are fragmented, as on the edges of species' ranges, due to the increased strength of evolutionary forces in small, isolated populations. According to Gould and Eldredge, the lack of smooth transitional sequences in the fossil record does not, therefore, represent the unavoidably imperfect (due to the chancy process of fossilization) chronicle of slow gradual change, but rather accurately reflects the actual operation of the evolutionary process.

The standard model for evolutionary change, phyletic gradualism, proposes that evolution is a slow, gradual process apparent across an entire species as that species evolves in response to changes in its environment. Because Gould and Eldredge referred to punctuated equilibrium in their 1972 paper as "an alternative to phyletic gradualism," the rapid and vocal response by the scientific community to an idea that seemingly upended conventional thinking was predictable. It is now generally accepted that both phyletic gradualism and punctuated equilibrium are represented in the fossil record. Gould and Eldredge, however, continued to maintain that punctuated equilibrium was the dominant mode for evolutionary change.

Creationists used the debate about punctuated equilibrium to question the status of evolution within science: if scientists themselves can't agree about how evolution works, then evolution cannot be the well-established concept as has been claimed. Creationists also argued that punctuated equilibrium was an example of scientists' flip-flopping (evolution is first proposed to work slowly, now it's proposed to work quickly) when confronted with contradictory evidence (i.e., lack of transitional forms in the fossil record). In particular, creationists attempted to conflate punctuated equilibrium with the discredited idea of macromutations producing large evolutionary leaps ("saltations") to question evolution.

This was, predictably, infuriating to scientists, especially to Gould, who saw his work distorted to support antievolutionism. In response, he frequently attacked creationism in his popular books and monthly "This View of Life" column in *Natural History* magazine. Collections of these columns in book form (e.g., *Ever Since Darwin, The Panda's Thumb, Hen's Teeth and Horse's Toes*) reached a wider audience, establishing Gould as a recognizable and outspoken proponent of evolutionary biology and critic of antievolutionism. In the 1970s and 1980s, Gould demonstrated how the main form of antievolutionism at that time, "creation science," was, in fact, an oxymoron. Ironically, the antievolution forces tried to interpret this active response as an indication of the thinness of scientists' arguments.

Gould testified for the plaintiffs in *McLean v. Arkansas Board of Education*, a lawsuit that challenged Arkansas' Act 590 mandating equal time for evolution and creationism. Even though Judge William Overton ruled Act 590 to be unconstitutional and credited Gould's testimony as being influential, Gould's tendency to provide responses that could be misinterpreted except by trained biologists ("I don't believe that mutation and natural selection are sufficient...") was used by creationists to portray evolution as a weak concept with little explanatory power. Inadvertently abetting the creationist agenda was a charge Gould faced during much of his career.

The triumph in Arkansas and soon after that of *Edwards v. Aguillard* (another case involving "equal time" for creation science, this time in Louisiana) encouraged Gould to mistakenly declare in 1987 that the fight against creationism was over. However, the rise of intelligent design in the 1990s prompted Gould to renew his defense of the teaching of evolution. When, in the late 1990s, the Kansas Board of Education eliminated evolution from that state's science

standards, Gould testified on the need for evolution in the public schools (e.g., "to teach biology without evolution is like teaching English without grammar"). Gould saw creationism as "politics, pure and simple" and part of the wider effort of religious conservatives to push that overall agenda.

The continued conflict between some religious groups (biblical literalists) and science fascinated Gould. Although a declared agnostic and a secular humanist, he had long admired religion and had stated that he regretted not having been raised within a particular religious tradition. Gould explored these topics in his 2002 book, *Rock of Ages*, in which he claimed that science and religion are independent ways of knowing about the universe ("non-overlapping magesteria" or NOMA), a position embraced by some, but rejected by others. In particular, Oxford University biologist Richard Dawkins (with whom Gould had frequent disagreements about various aspects of evolutionary biology) was scathing in his indictment of Gould's acceptance of religion, chastising Gould for "a cowardly flabbiness of the intellect."

Gould's development of NOMA was consistent with his contention that morality is independent of evolution. Gould bristled at the suggestion of biological determinism of human behavior, and had long disagreed with Harvard biologist E. O. Wilson about the genetic basis of human behavior. Wilson had founded modern sociobiology, which holds that behavior (in particular social behavior) results from the reproductive advantage of genetically determined behavioral traits (i.e., via natural selection). Gould's 1981 book, *The Mismeasure of Man*, dealt primarily with intelligence, a trait that had been proposed as being in large measure biologically determined, and examined the negative implications for individuals and society of such a perspective. The most recent manifestation of sociobiology, evolutionary psychology, was viewed by Gould as pseudoscience that had sinister political and social implications that amounted to a resurrection of the social Darwinism of the late nineteenth century and the eugenics programs of the early twentieth century.

Gould argued throughout his career for the crucial role of "contingency" in the evolutionary process. He often noted that humans likely exist only because of chance events: restart the process of biological evolution on Earth and our species would probably not arise. This position was frequently misstated to suggest that evolution was a random process. Instead, Gould's point was that evolution is not goal-directed (any particular species or even an increase in complexity is not inevitable) and can be influenced by chance events (e.g., an asteroid impact may have allowed mammals to diversify quickly after the demise of the dinosaurs). Gould also used this position to attack those he termed "Darwinian fundamentalists" (most notably Richard Dawkins and philosopher Daniel Dennett), who he thought wrongly viewed natural selection as the only significant evolutionary force.

Gould's ability to intrigue a reader with discussions of even arcane aspects of science attracted many students to careers in science. Gould was also an accomplished historian and philosopher of science, cited by Ronald Numbers as second only to Thomas Kuhn in influence. But many have also questioned Gould's merits as a scientist; for example, eminent evolutionary biologist John Maynard Smith noted that "the evolutionary biologists with whom I have discussed his work tend to see him as a man whose ideas are so confused as to be hardly worth bothering with." Regardless, Gould received numerous awards during his career, including a Schubert Award for excellence in paleontological research (1975), an American Book Award (1981), and a MacArthur Foundation Award (1981). He also became fellow of the American Association for the Advancement of Science (1983; serving as President in 2000) and was elected to the National Academy of Sciences (1989).

In 1982, Gould was diagnosed with abdominal mesothelioma, a rare form of cancer, usually due to asbestos exposure. At the time of his diagnosis, the median expected length of survival was eight months; however, he underwent a new experimental treatment, and lived another twenty years. But cancer eventually returned, although in a different form, and Gould died in 2002.

For More Information: Gould, S.J. 1977. *Ever Since Darwin*. New York: Penguin; Gould, S.J. 1981. *The Mismeasure of Man*. New York: Norton; Gould, S.J. 1989. *It's a Wonderful Life: The Burgess Shale and the Nature of History*. New York: W.W. Norton.

THE GRAND CANYON

The Grand Canyon is a spectacular steep-sided gorge carved by the Colorado River in Arizona. The Canyon was originally inhabited by Native Americans who built settlements in the Canyon's walls. The first scientific expedition to study the Canyon was conducted in 1869 by John Wesley Powell, who reported that sediments forming the Canyon's walls were "leaves in a great storybook." The Grand Canyon—one of the natural wonders of the world—was given federal protection in 1893; it later was made a national monument, and became a National Park in 1919 (three years after the National Park Service was formed). The Canyon is 277 miles long, 0.3–15 miles wide, and attains a depth of more than a mile. Geological evidence indicates that the Canyon is approximately 6 million years old.

Many Young-Earth creationists believe that the Grand Canyon was formed by the retreating waters of Noah's flood approximately 4,500 years ago, and therefore claim that the Canyon is a "monument to the world-wide Flood" (Figure 38). Institute for Creation Research (ICR) geologist Steven Austin (b. 1948) claims in *Grand Canyon: Monument to Catastrophe* (1994) that "the real battle in regard to understanding the Grand Canyon is founded not just upon Creation and Noah's flood, but upon Christianity versus humanism." Similarly, Colorado River guide Tom Vail's coffee-table book *Grand Canyon: A Different View* (2003), which was published by Master Books (formerly Creation-Life Books, the publishing arm of the ICR), includes contributions by Young-Earth creationists Henry Morris, Duane Gish, Ken Ham, John Whitcomb, and Kurt Wise. Vail, who has led "Christ-centered rafting trips through the canyon," noted that "For years,

Grand Canyon
Creation Tour

Join ICR
for the tour
of your life!

38. The Grand Canyon is cited by many creationists as evidence of a worldwide flood. Antievolution groups such as the Institute for Creation Research often sponsor creationism-based tours of the canyon. (*ICR*)

as a Colorado River guide I told people how the Grand Canyon was formed over the evolutionary time scale of millions of years. Then I met the Lord. Now, I have 'a different view' of the Canyon, which, according to a biblical time scale, can't possibly be more than about a few thousand years old." Vail, like many creationists, bases his conclusions about nature on the Bible, not scientific evidence.

When the National Park Service began selling *Grand Canyon: A Different View* at its bookstore at the rim of the Canyon, hundreds of scientists— including the presidents of seven geological societies—protested, claiming that the National

Park Service was promoting "the anti-science movement known as Young-Earth creationism." The Public Employees for Environmental Responsibility (PEER) described the situation as a "Christian fundamentalist" takeover of the National Park Service, adding that "the Park Service leadership now caters exclusively to conservative Christian fundamentalist groups. The Bush Administration appears to be sponsoring a program of Faith-Based Parks." PEER also noted that religious plaques had been restored in the Park, and that rangers at Grand Canyon National Park could not mention that Earth is more than a few thousand years old.

In 1919, the Canyon hosted 44,173 visitors; in 2006, more than 5 million visitors came to the Grand Canyon.

For More Information: Vail, T. 2003. *Grand Canyon: A Different View*. Green Forest, AR: Master Books.

PETER GRANT (b. 1936)
AND ROSEMARY GRANT (b. 1936)

Peter and Rosemary Grant are a husband-and-wife team that has studied the finches of the Galápagos Islands for more than thirty years. Peter (the younger Grant by eighteen days) was born in London, England, and earned a BA at Cambridge University (1960) and a PhD in zoology at the University of British Columbia (1964). After a fellowship at Yale University, he was a faculty member at McGill University for twelve years, and then at the University of Michigan for eight years. In 1985, he moved to Princeton University, where he is currently Professor of Ecology and Evolutionary Biology. Rosemary Grant, born in Arnside, England, earned a BA at the University of Edinburgh in 1960, and a PhD in zoology at the University of Uppsala in 1986. She has held research positions at the University of British Columbia, Yale, McGill, the University of Michigan, and is currently Lecturer and Senior Research Biologist at Princeton. The couple married in 1962, and they have two children.

In the early 1970s, the Grants were interested in studying evolution occurring in real time in actual populations. Although the thirteen species of Galápagos finches had become famous because of their association with Darwin's visit to the islands, as well as through David Lack's 1947 book, *Darwin's Finches*, little research had actually been done on the islands (even Lack only spent four months in the islands). When the Grants first visited the Galápagos in 1973, they quickly decided they had found a "gold mine" of a natural laboratory. The birds exist in discrete populations that exhibit variation in a number of relevant characteristics (most notably, beak size). And, as is often true for species on remote islands, the birds are extremely tame, which makes marking and observing them easy. Living conditions for the researchers are, however, primitive, and just getting to the islands is a major undertaking.

During their decades of research, the Grants have witnessed Darwin's mechanism of natural selection firsthand. The most famous case, now cited in many textbooks, is the change in beak size that occurred within the population of the medium ground finch (*Geospiza fortis*) on the island of Daphne Major. The Grants had been studying this population for a couple of years when, in 1977, a severe drought occurred, which greatly reduced the seed supply the birds use for food. Once the preferred, smaller seeds had been consumed, birds that had larger bills had a competitive advantage because they could open the larger, tougher seeds that remained. Larger-billed birds were more likely to survive—which meant they also tended to reproduce at a higher rate—than birds with smaller bills, and the large-billed trait was passed to the next generation at a higher frequency. The Grants demonstrated that the distribution of bill sizes in the population shifted toward larger sizes as a result of the drought, a clear case of differential reproductive success causing a population to evolve.

Long-term monitoring of natural populations is rare because of funding and logistical challenges. The Grants' work is therefore of great significance because of the continuous record they and their graduate students have generated, having studied more than twenty-five generations and 20,000 birds. The Grants are currently focusing on speciation, extinction, and character displacement due to interspecific competition.

For More Information: Grant, P.R. 1999. *Ecology and Evolution of Darwin's Finches*. Princeton, NJ: Princeton University; Weiner, J. 1994. *The Beak of the Finch: A Story of Evolution in Our Time*. New York: Alfred Knopf.

ROBERT GRANT (1793–1874)

Robert Grant was a physician, marine biologist, admirer of Lamarck, and friend of Charles Darwin during their time together at Edinburgh University. Grant offered to examine the corals that Darwin brought back from his voyage aboard the *Beagle*, but Darwin declined his offer.

ASA GRAY (1810–1888)

God himself is the very last, irreducible causal factor and, hence, the source of all evolutionary change.

Asa Gray was born on November 18, 1810, into a farming family in Sauquot, New York (Figure 39). After attending Clinton Grammar School and Fairfield Academy, Gray enrolled at the College of Physicians and Surgeons in Fairfield, New York, where he received an MD degree in 1831. After teaching for three years at Bartlett's High School in Utica, New York, Gray began practicing medicine in Bridgewater, New York.

Although the diminutive Gray was trained as a physician, his passion was plants. Gray served as Curator at the New York Lyceum of Natural History from 1835 to 1838, and in 1836 published *Elements of Botany*, the first of his several textbooks. Gray then became the first professor at the newly formed University of Michigan in 1838, after which he went to Europe to buy $5,000 worth of books for the university. After being appointed Fisher Professor of Natural History at Harvard University in 1842, Gray published several botany books, including *How Plants Grow* (1858) and *How Plants Behave* (1872). Gray collected plants throughout Europe and North America, and with John Torrey was the first scientist to attempt to classify and describe work supported Darwin's theory, for Gray suspected that the similarities of plants in eastern Asia and eastern North America resulted from the fact that they were descendants of a single species that lived across the Northern Hemisphere before the Ice Age.

In 1854, botanist Joseph Hooker showed Darwin Gray's review of Joseph Hooker's *Flora of New Zealand*, in which Gray questioned the stability of species. Soon thereafter, Darwin urged Gray to update Gray's *Manual of the Botany of the Northern United States* (which had first appeared in 1848), and in September 1857 Gray became only the third scientist (after Hooker and Charles Lyell) to whom Darwin revealed his theory of evolution by natural selection. A letter from Darwin to Gray in 1857 was part of the Darwin–Wallace paper (i.e., the first public announcement of evolution by natural selection) presented to the Linnean Society on July 1, 1858.

Gray received his copy of *On the Origin of Species* in late December 1859. As he read it, Gray wrote exclamation points and notes such as "Yes" and "Well put" in the margins of Darwin's book. Although Gray questioned some of Darwin's conclusions, he recognized the importance of Darwin's work, and understood that "different species represent different adaptations to conditions." Gray, the most important botanist of the nineteenth century, became Darwin's most vocal scientific ally in the United States.

After *On the Origin of Species* appeared, Gray—an orthodox Christian—launched his public defense of Darwin's theory with a strong, insightful review of Darwin's book in the March 1860 issue of *The American Journal of Science*. Gray described Darwin's views about variation as "general, and even universal," and Darwin's evidence as "fair and natural." Gray concluded his review by noting that any theory—not just Darwin's—could be used to support atheism. Although Gray tried to reconcile his religious views with Darwin's idea, Gray rejected the lack of purpose that Darwin's theory seemed to imply. This lack of purpose, and not a literal reading of Genesis (which most Christians in Gray's time rejected), was the teleological issue that Gray discussed in his review, and about

39. Asa Gray was a famous biologist and theistic evolutionist. Gray was America's foremost Darwinist until his death in 1888. (*Library of Congress*)

all of the known plants in North America. By the early 1860s, Gray's personal collection of plants included almost 200,000 specimens.

Gray's first edition of *First Lessons in Botany and Vegetable Physiology* (1857) claimed that "the Creator established a definite number of species at the beginning, which have continued by propagation, each after its kind." However, the 1887 edition of the book noted instead that "nearly related species probably came from a common stock in earlier times." Although Gray's book did not include the word *evolution* or describe natural selection, evolution replaced the fixity of species.

During the 1850s, Gray and Darwin often exchanged letters, and Darwin used Gray as a source of information about the distribution and variation of nondomesticated plants. Gray's which he corresponded with Darwin for several years.

Although Gray was reluctant to accept all of Darwin's theory of natural selection, Darwin liked Gray's review, and hoped that it would be used as a preface to the second American edition of *On the Origin of Species*. As Darwin noted, "No one person understands my views & has defended them so well as A. Gray;...though he does not by any means go all the way with me." Darwin did not accept Gray's teleology; as he wrote to Gray, "It has always seemed to me that for an Omnipotent & Omniscient Creator to foresee is the same as to preordain." Darwin later added that "I had no intention to write atheistically. But I own that I cannot see, as plainly as others do, & as I [should] wish to do, evidence of design & beneficence on all sides of us. There seems to me to be too much misery in the world....I am inclined to look at everything as resulting from designed laws, with the details, whether good or bad, left to the working out of what we may call chance." Darwin closed a subsequent letter to Gray with this postscript: "I cannot possibly believe that a false theory would explain so many classes of facts."

In 1860, Gray published three widely read articles about evolution in *The Atlantic Monthly*; those articles later comprised a chapter entitled "Natural Selection not Inconsistent with Natural Theology" in Gray's *Darwiniana* (1876). Darwin complimented Gray on his articles, but again noted that "I grieve to say that I cannot honestly go as far as you do about Design." Darwin rejected Gray's belief that evolution was guided by God. For Darwin, a design-based doctrine that included suffering and evil was not worthy of consideration.

Gray was America's foremost Darwinist, and a colleague of Gray's at Harvard—Louis Agassiz, a professor of zoology and geology—was Darwin's biggest scientific foe. Gray promoted Darwin's theory and accepted natural selection, noting that "natural selection is not the wind which propels the vessel, but the rudder which, by friction...shapes the cause." Agassiz—a staunch supporter of special creation—attacked Darwin's idea. According to Agassiz, species had been created by God, and were therefore immutable. Agassiz believed that catastrophes had periodically destroyed life on Earth. Gray and Agassiz were cordial, but distant, colleagues.

After 1861, Gray returned most of his attention to botany, and his Harvard laboratory became a clearinghouse for the identification of plants from newly explored areas of North America. In 1864, Gray's donation of his books and plant specimens to Harvard University became the nucleus of the Botany Department and the Gray Herbarium. In 1872, Gray retired from teaching to work on his *North American Flora*, and the following year he became a Charter Member of the National Academy of Sciences. Gray traveled extensively and continued to correspond with Darwin until Darwin's death in 1882.

Asa Gray died on January 30, 1888, and is buried in Mount Auburn Cemetery in Cambridge, Massachusetts. Gray's wife, Jane, edited his *Letters*, which was published in 1893. Today, Gray's home in Cambridge is a National Historic Site. The University of Michigan Library honors Gray with its Asa Gray Society, and in 1984 the American Society of Plant Taxonomists established the Asa Gray Award to honor living botanists for their career achievements.

For More Information: Dupree, A.H. 1988. *Asa Gray*. Baltimore, MD: Johns Hopkins.

VERNON GROSE (b. 1928)

Vernon Grose was an engineer and Pentecostal creationist who in 1969 told members of the California State Board of Education—many of whom had been appointed by Governor Ronald Reagan—that science classes should include evolution and creationism. Grose believed that "creation in scientific terms is not a religious belief." Grose's claims represented a change in creationists' tactics—instead of asking states to ban evolution, they now asked that teachers also teach creationism. Grose, who invoked arguments about "the last days" and claimed that his "citizenship is in heaven," said that he was "a little Christ...[who] felt something like Jesus did when he overthrew the tables and money changers in the temple." Grose's complaints, which he said would "interfere with Satan's well-laid plans," resulted in evolution being treated as a theory, not as a fact, and described by conditional statements in California's public schools. Grose's complaints also triggered years of debate and legal wrangling about evolution and creationism in California.

JAMES GULLY (1808–1883)

James Manby Gully was born in 1808. After studying medicine in Paris and Scotland, he began a career as a surgeon in London in 1830. In 1842, Gully read an article by Austrian surgeon Vincent Priessnitz advocating "water therapy" (hydrotherapy), which is the use of water to treat illnesses. Soon thereafter, Gully gave up his surgical career and established (with James Wilson, another hydrotherapist) the Water Cure Establishment in Malvern, a small city northwest of London famous since the 1600s for its bottled, low-mineral spring water. Advocates of hydrotherapy claimed that immersing one's body in cold water drew blood away from the inflamed nerves of the stomach, thereby calming the nerves. Gully's popular clinic attracted patients such as Charles Dickens, Lord Tennyson, Florence Nightingale, and Samuel Wilberforce.

In March 1849, Sullivan Bartholomew—one of Darwin's shipmates aboard the *Beagle*—recommended that Darwin seek treatment at Gully's Water Cure Establishment. Darwin was skeptical, but nevertheless went (with his entire family) to Malvern for Gully's treatment that included a regimen of wrapping, scrubbing, and bathing with cold water, exercising, drinking water, homeopathic medicines, and a prescribed diet. Four months later, Darwin left Malvern feeling better, and returned home to continue his research. At Down House, Darwin arose every day at 7:00 AM and, after immersing himself in cold water, was scrubbed by his butler. In 1849, during Darwin's first visit to Malvern, Gully made Darwin give up snuff.

In 1850, 9-year-old Annie Darwin became ill, and Darwin took her to Gully's center for treatment. In early 1851 she rallied, but by late March she had to be returned to Malvern for more treatments. In April, Annie's condition worsened, and on April 23, 1851, she died, with Charles at her bedside. Annie's death destroyed the remnants of Charles' faith in a beneficent God. Annie was buried beside Priory Church, not far from where she and her father received Gully's treatments. Charles later gave up water-cures and relied on diets and resting his "nerves."

Gully, a popular speaker and writer, believed that many women's psychological problems (e.g., depression, anxiety) resulted from pressures in Victorian England for women to be ambitionless, chaste providers. In 1872, Gully retired and had an affair with Florence Ricardo, and Gully aborted her pregnancy the following year. Soon thereafter, Gully left Malvern. However, when Ricardo married barrister Charles Bravo in 1875, Gully was enraged. On April 18, 1876, when Bravo died of poisoning, Gully was suspected of the murder. However, Gully was never arrested, and the murder was never solved.

Gully died in 1883. Today, the Malvern Hills Brewery produces "Doctor Gully's Winter Ale" (promoted as "Just What the Doctor Ordered!"), and Schweppes & Malvern bottles the world famous "Malvern Water" used by the Queen of England. The water's high purity is the basis for it being known for "containing nothing at all."

ARNOLD GUYOT (1807–1884)

Any length of time that Darwin might desire for his transformations, would never suffice to make of the monkey a civilizable man.

Arnold Henri Guyot was born on September 28, 1807, in Boudevilliers, Switzerland. With plans on becoming a minister, Guyot graduated from the College of Neuchâtel in 1825. Guyot then moved to Germany, where he met fellow countryman Louis Agassiz (then a student at the University of Heidelberg); Agassiz became a lifelong friend and colleague of Guyot's. Agassiz and Guyot spent most of their vacations studying nature, kindling Guyot's interest in science. While in Germany, Guyot interacted with philosopher Georg Wilhelm Friedrich Hegel and naturalist/explorer Alexander von Humboldt, both of whom influenced Guyot's interests. But Guyot's most influential teacher was geographer Carl Ritter, who approached geography with the underlying assumption that the entire Earth is a well-tuned entity composed of parts that are designed by the Creator. Ritter and others convinced Guyot that equal dedication to the study of nature and his theological studies would not be possible, and he finally decided to concentrate on geography. Guyot earned his PhD in 1835, having written his thesis on "The Natural Classification of Lakes."

When Agassiz proposed in 1838 his idea of a universal ice age, Guyot's interest in glaciers was piqued, and Agassiz encouraged his friend to study the glaciers of the Alps. Guyot followed this advice, and made fundamental discoveries about the structure and movement of glaciers. Guyot announced these findings at a meeting of the Geological Society of France, but did not publish them until 1883—well after Agassiz's death—to avoid infringing on his friend's intellectual ownership of this area of inquiry.

In 1839, Guyot chaired both the History and Physical Geography Departments at the newly established Academy of Neuchâtel. Agassiz was also on the faculty there, and the friends enjoyed collaborating (although Agassiz moved to the United States in 1847). Guyot, who distinguished himself as a gifted, popular teacher, also studied glaciology, meteorology, geology, and hydrography, including mapping of Lake Neuchâtel. However, in the wake of the Civil War of 1845 and the national reorganization of Switzerland in 1848, the Academy of Neuchâtel was closed. Agassiz encouraged Guyot to move to the United States, and in late 1848, Guyot followed his friend to Massachusetts.

Soon after his arrival, Guyot lectured about geography at the Lowell Institute in Boston. Although given in French, the lectures were well received and were translated and published as *Earth and Man* (1849). Guyot was a devout Christian, and he emphasized that natural processes lead to increased order or harmony. Guyot further applied the concept of increasing harmony in interpreting the "geographical march" of human history. He proposed that civilization passed through three distinct stages tied to particular geographic regions: (1) subjection, in Asia; (2) individual freedom, in Greece; and (3) association, in Rome. Having passed through these youthful phases, Guyot proposed that mature civilization was now best exemplified by his adopted home, America.

Due to the success of his Lowell lectures and his reputation as a talented teacher, Guyot was asked by the Massachusetts Board of Education to teach geography and instructional methods to the state's teachers. This work spurred him to develop atlases and wall maps for use in teaching, and Guyot's materials set the standard for such resources. In 1854, Guyot accepted an endowed chair of Physical Geography and Geology at the College of New Jersey (renamed Princeton University in 1896), where he remained for the rest of his career. He also served as lecturer at the State Normal School at Trenton, and as Lecturer Extraordinary at the Princeton Theological Seminary, where he lectured on "the connection of revealed religion and physical and ethnological science." Guyot also used his personal collection to establish a natural history museum, now housed in Guyot Hall, at Princeton. In 1867, Guyot married Sarah Haines, daughter of a former governor of New Jersey.

Although Guyot believed that the Bible is the word of God, he also felt that "the chief design of the Bible . . . is to give us light upon the great truths needed for our spiritual life; all the rest serves only as a means to that end, and is merely incidental." Based on his studies of geology, Guyot accepted the antiquity of the Earth, and he dismissed a literal interpretation of Genesis as a "pretended history of six solar days, founded upon imaginary facts of which geology has no knowledge." In his last published work, *Creation*, or the *Biblical Cosmogony in the Light of Modern Science* (1884), Guyot instead interpreted the six "days" of creation as separate epochs of time, which included three instances of special creation: creation of matter, creation of life, and creation of humans. All other major changes (e.g., creation of land animals on the fifth day) were merely "transformations...or a continuation of the same kind of creation," and therefore "the question of evolution...remains still open." However, Guyot refused to extend this possibility to the appearance of humans, claiming that "man is the crowning act of the Creator" and is a "being of a new and superior order, and therefore must be kept distinct."

Guyot integrated his understanding of the natural world with his devout Christian beliefs to advocate a perspective on creation that became known as day-age creationism (each biblical "day" was an "epoch" of unrestricted length). Day-age creationism became popular after the turn of the century as increases in scientific understanding required harmonization with Scripture, and became associated with William Jennings Bryan and William Bell Riley, among others.

Arnold Guyot died in Princeton on February 8, 1884, and is buried in Princeton Cemetery. His legacy includes his scientific approach to the study of geography, and this impact is reflected by an Alaskan glacier and two mountains in the eastern United States bearing his name. He was one of the last prominent scientists of the late 1800s to not fully embrace evolution.

H

HADRIAN'S WALL

James Hutton was intrigued by Hadrian's Wall, a seventy-three-mile barrier that formed the northwest frontier of the Roman Empire for most of the period 122–410 AD. The wall was built by Roman legionnaires from 122 to 128 AD "to separate the Romans from the Barbarians" (Figure 40). The wall, named after Roman emperor Hadrian (who ordered its construction), stretches from Wallsend in the east to Bowness-on-Solway in the west, and marked the northern limits of the Roman Empire. Hadrian's Wall was originally an active military zone manned by Roman soldiers for almost 300 years to restrain marauding Celts. Hutton knew that rocks had been in the wall for almost 1,700 years, yet they looked like that had not weathered at all. Nearby mountains, however, had eroded a great deal. Hutton concluded from this and other observations that the Earth must be very old.

When it was built, Hadrian's Wall stood 12′ high and 6′ wide. Over the centuries, however, people dismantled Hadrian's Wall and used its stones to build homes, churches, and other buildings. Today, the best-preserved segments of Hadrian's Wall—a UNESCO World Heritage Site—are about 5′ high. Along its winding length are museums and beautiful countryside.

For More Information: Breeze, D.J., and B. Dobson. 2000. *Hadrian's Wall*. New York: Penguin.

ERNST HAECKEL (1834–1919)

The real cause of personal existence is not the favor of the Almighty, but the sexual love of one's Earthly parents.

Ernst Heinrich Philipp August Haeckel was born February 16, 1834, in Potsdam, Prussia (now Germany). At his father's urging, Haeckel pursued a medical degree, first at the University of Berlin, and later at the University of Würzburg, where he assisted famous pathologist Rudolf Ludwig Karl Virchow. He completed his thesis, on the histology of crayfish, and graduated in 1859. Although still ambivalent about being a doctor, he set up a practice in Berlin. Haeckel eventually decided that medicine was not to be his life's work, and he traveled to Sicily to study the surrounding sea life, an interest of his since his university days. While casting about for how to support himself (he even considered becoming a landscape painter), Haeckel discovered several unknown species of radiolarians, an experience that fueled his already intense interest in natural history. In 1862, after working as a lecturer at the University of

40. The rocks of Hadrian's Wall, which was built almost 1,900 years ago in Scotland, intrigued James Hutton. Hutton noted that the wall's rocks had barely weathered, but the nearby mountains had weathered significantly. This helped convince Hutton that Earth was ancient. (*David Cantrell*)

Jena, Haeckel was appointed associate professor; the same year, he published a monograph about his radiolarians. He became a full professor and remained at Jena for the rest of his career.

Haeckel admired Darwin's efforts in *On the Origin of Species* to "introduce comprehensible laws of nature in place of the incomprehensible miracles," and immediately embraced the concept of descent with modification. (He was, however, less accepting of Darwin's proposed mechanism, natural selection.) Haeckel corresponded with Darwin frequently, and by 1876, he had visited Down House twice. Although Emma Darwin could not stand to be in the same room as Haeckel, Haeckel later said meeting Charles Darwin "was as if some exalted sage of Hellenic antiquity ... stood in the flesh before me."

For Haeckel, evolution was a unifying principle not only for biology, but also for philosophy. He believed that progress had finally moved human understanding beyond the superstition and mysticism that had ruled previously; teleology and the dominance of religion were now

undermined. Haeckel explored these ideas in his *General Morphology of the Organism* (1866), which contained evolutionary trees (where organisms were placed not on simple branching diagrams, but on drawings of actual trees) for all known organisms, including humans. Using the growth of crystals as his metaphor, Haeckel proposed that increasing morphological complexity arose through evolutionary time due to increasingly complex (yet symmetrical) relationships between tissues and body parts. Hence, just as complex crystals result from symmetrically arranged additions, so too do complex biological morphologies.

Haeckel used Karl von Baer's laws of development, that united degree of taxonomic relationship with similarity in development, as a mechanism for evolutionary change: alterations in the development of an embryo, especially in later developmental sequences, would produce exactly the morphological relationships manifested in Haeckel's phylogenetic trees. Haeckel codified this relationship as his "biogenic law":

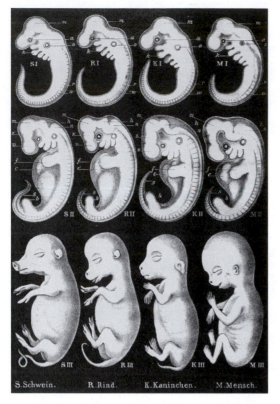

41. Ernst Haeckel was a German biologist who became famous for his biogenic law, namely, that ontogeny recapitulates phylogeny. Although he was a gifted artist, Haeckel's famous drawings of embryos were inaccurate. Today, many creationists cite these drawings as evidence that evolution is based on fraudulent data. Haeckel's claim that "politics is applied biology" was later used by Nazis to justify their militant racism and naturalism.

ontogeny recapitulates phylogeny. In support of his law, Haeckel studied and drew embryos of a variety of species (e.g., fish to humans). Now famous, these drawings (Figure 41) purported to demonstrate the commonality of early developmental stages among organisms, indicating that, for example, mammals passed through a "fish" stage, an "amphibian" stage, and so on. Haeckel concluded that "the connection between them is not of an external or superficial, but of a profound, intrinsic, and causal nature," and he marshaled his evidence in *A Natural History of Creation* (1868), and *Anthropogeny, or the Development of Man* (1874).

Criticisms arose about Haeckel's drawings. The stylized nature of the drawings overemphasized the similarity among particular stages of the embryos, and some details were conveniently omitted. Haeckel claimed that he was simply trying to capture the overall morphology of each embryo in a general way that may or may not be present in any particular specimen, and that the drawings were meant to provide "schematic" information. Whether Haeckel purposely produced fraudulent drawings is unknown, and scholars continue to debate this point. Perhaps the most damning evidence is Haeckel's own abilities as an artist. His skill and attention to detail were remarkable throughout his vast body of work; it is interesting that the accuracy of Haeckel's drawings was lowest when so much was depending on what those drawings demonstrated.

Creationists have cited Haeckel's drawings as either evidence for a supposed materialist conspiracy to undermine religion, or as proof against evolution itself. However, within thirty years after Haeckel had produced the embryo drawings, science had already cast aside his biogenetic law. (Unfortunately, textbooks did not do the same as quickly.) There are identifiable developmental similarities among organisms, and these similarities reflect some aspects of common ancestry. However, the full-fledged recapitulation of evolutionary history does not occur, as von Baer had noted as early as the 1830s. Although Darwin cited von Baer's work in the *Origin*, Haeckel's work had no influence on Darwin because Haeckel's infamous drawings would not exist for several more years. Regardless, creationists continue to use Haeckel's embryos to refute evolution (e.g., as in Jonathon Wells' *Icons of Evolution: Science or Myth? Why Much of What We Teach about Evolution Is Wrong*).

Haeckel proselytized for Darwin, and gave frequent public lectures, replete with drawings and specimens, emphasizing the enlightened worldview a Darwinian perspective provides. Haeckel's views became increasingly monistic, and his struggle against religion intensified. He presented his clearest exposition of an evolution-based philosophy in his book *Welträtsel* (1899), which sold more than 500,000 copies in Germany and was translated into nearly thirty languages. Haeckel helped found a German Monists' Association at Jena in 1906, which among other activities, held antichurch rallies that sometimes attracted thousands of supporters. As he aged, Haeckel's positions became increasingly extreme and by the 1910s, he was espousing racism and violence, including the suppression of "less advanced races that were relics of earlier stages of human evolution."

Haeckel died in Jena, Germany, on August 9, 1919. The Ernst Haeckel Museum ("Villa Medusa") in Jena houses many of his records and personal items. Mount Haeckel (13,418') in the eastern Sierra Nevada overlooks Evolution Basin.

For More Information: Breidbach, O. 2006. *Visions of Nature: The Art and Science of Ernst Haeckel*. Munich: Germany.

A.P. HAGGARD (1862–1932)

A.P. Haggard was born Pleasant Andrew Haggard in 1862. Later, Haggard changed his name to Andrew Pleasant, and became known as "A.P." At the time of the Scopes Trial, Haggard was a popular lawyer in Dayton, Tennessee; he was also Dayton's wealthiest resident and was involved in virtually all important enterprises in the area. During the Scopes Trial, Haggard handled several of the logistical issues associated with the trial, especially those regarding security. Haggard referred to John Scopes as "Mr. Dayton." Haggard's second wife was the sister of Scopes prosecutors, Sue and Herbert Hicks.

After the Scopes Trial, Haggard lost his bank, home, and money, and became seriously ill with diabetes. He moved into the home of Fred "The Hustling Druggist" Robinson, which Haggard had built near the courthouse. In 1932, in an upstairs bedroom of the Robinson home, Haggard ended his life with a razor. Today, Haggard rests atop a hill in Dayton's Buttram Cemetery. Haggard's son Wallace helped prosecute John Scopes. Wallace, the brother-in-law of Fred Robinson and a founder of Bryan College, died in 1971.

J.B.S. HALDANE (1892–1964)

There are still a number of people who do not believe in the theory of evolution. Scientists believe in it, not because it is an attractive theory, but because it enables them to make predictions which come true.

John Burdon Sanderson Haldane was born into an aristocratic family on November 5, 1892, in Oxford, England. When he was 8 years old, Haldane's father took him to hear a lecture on the "rediscovery" of Mendelian genetics, an event that had a lasting impact on the younger Haldane. Haldane won a scholarship in mathematics to Oxford University, and graduated with First Class degrees in Mathematics and Classic Studies in 1911. His formal education ended at this point (i.e., he earned no graduate degrees), and the only biology course he ever had was in vertebrate anatomy. Haldane had intended to follow his father into physiology, but the outbreak of World War I ended this plan. He joined the Scottish Black Watch, serving four and half years as a bombing officer. Haldane was wounded twice (but claimed he still enjoyed the experience of war), and was discharged in 1919. During the war, Haldane published his first paper (1915) that combined a re-analysis of data by A.D. Darbishire (who had given the talk of Mendelism he heard when a boy) with experimental studies of mice that he had conducted in 1912 with his sister Naomi (who later attained fame as author Naomi Mitchison) and friend A.D. Sprunt. (Sprunt was killed in World War I). This work was the first to identify genetic linkage (then referred to as "reduplication") in mammals.

After the war, Haldane became a fellow in physiology at Oxford University, sometimes collaborating with his father. In 1918, Haldane published a theoretical paper (based primarily on thinking done while recovering in a hospital in India during the war) that constructed a linkage map based on the frequency of recombination. This work, plus his 1922 paper describing what came to be known as "Haldane's Rule" (when two species hybridize, it is generally the heterogametic sex that is inviable or sterile) provided him with increased recognition. Haldane also studied the regulation of blood pH, and often used himself and colleagues as research subjects. In one set of experiments, he consumed various quantities of ammonium chloride, which lead to the discovery that ammonium chloride can reduce fits associated with excess alkalinity of infants' blood. Haldane also learned how to speak while exhaling and inhaling, a technique he sometimes used during arguments.

In 1923, Haldane accepted a readership position in biochemistry at Cambridge University, which began the most productive period of his career. His studies of enzyme kinetics ultimately produced the book *Enzymes* (1930). Haldane also analyzed the quantitative effects of selection, and from 1924 to 1932, published a series of mathematical models of the action of selection. This work was synthesized in the 1932 book, *The Causes of Evolution*, which, along with R.A. Fisher's and Sewall Wright's independent and complementary work of the same time, unified Darwinian evolution with Mendelian genetics— a unification now known as the modern synthesis. Haldane's pioneering work influenced evolutionary biology for decades (e.g., his insights provided the foundation for Motoo Kimura's development of the neutral theory of molecular evolution). Haldane also had a public romance while at Cambridge with the already-married Charlotte Burghes. After Burghes divorced, she married Haldane in 1926, but the affair had already prompted Cambridge's Sex Viri Committee (charged with protecting the morals of the university) to dismiss Haldane. Not shy about controversy, Haldane appealed his dismissal and was reinstated.

Because of his studies of enzymes, Haldane was in a unique position to appreciate the relationship between proteins and genes, and he began to suspect that genes carry the information to produce enzymes. A part-time position at the John Innes Horticultural Research Station from 1927 to 1932 enabled Haldane to investigate this proposal. Studies of the inheritance of pigmentation in *Primula* showed that certain alleles were correlated with particular patterns of pigmentation. In 1942, these and other findings were joined in *New Paths in Genetics*, which, among other things, speculated on the semiconservative nature of DNA synthesis.

Haldane left Cambridge in 1933 to become Professor of Genetics at University College, London, the same year that Ronald Fisher became Galton Professor of Eugenics at the same institution. During World War II, both men moved temporarily to the Rothamsted Experimental Station, where Fisher had worked before moving to University College. In 1937, Haldane became Weldon Professor of Biometry at University College. Haldane's research during these years focused on human genetics and mutation, but he also worked on improving the safety of submarines. Spurred by the sinking in 1939 of the *Thetis* (a submarine of the Royal Navy), Haldane subjected himself and others to treatments of varying pressure and mixtures of gases, and in the process seriously injured his spine.

With the ongoing political upheaval in Europe of the 1930s, Haldane feared that fascism and Nazism would spread, and he was dissatisfied with his own government's response. By 1942, both Haldane and his wife Charlotte had become members of the Communist Party of Great Britain; he edited the *Daily Worker*, the party's official publication, and was a member of the party's governing body for several years. Haldane had earlier embraced dialectical materialism, and had explored the relationship between science and Marx's teachings in the book *The Marxist Philosophy and the Sciences* (1939). Charlotte eventually left the communist party, and she and Haldane drifted apart; they were divorced in 1945. Soon thereafter, Haldane married Helen Spurway, a former student to whom he would remain married until his death. Haldane had no children from either marriage. During the 1940s, Haldane's politics and science attracted the attention of John Maynard Smith, then an engineer designing warplanes for Britain. Smith, who became a student of Haldane's in 1948, later was an influential figure in evolutionary biology.

After World War II, controversy surrounded Haldane's response to Trofim Lysenko's suppression of science (and scientists, especially geneticists) in the Soviet Union. Even though he rejected Lysenko's claims (e.g., support for Lamarckian inheritance), Haldane believed that something positive could ultimately arise, particularly in the area of agriculture. Haldane's views were colored by his political affiliation; however, by 1950, he had become disillusioned with the Soviet system and with communism, and he quietly left the communist party.

Haldane wrote dozens of popular essays about science for a general audience. In 1922, he published the book *Daedulus*, or the *Future of Science*, which presented Haldane's views on what science might provide in the future. Of these predictions, most notable was the possibility of producing what are now called "test-tube babies." Aldous Huxley (grandson of Thomas Huxley, "Darwin's bulldog") and Haldane were close friends, and some of Haldane's ideas appeared in Huxley's writing, such as the "bottle-babies" in *Brave New World* (1932). (Huxley also included a none-too-flattering Haldane-like character in the novel *Antic Hay*, published in 1923.) Haldane also authored a collection of children's stories, *My Friend Mr. Leakey* (1937).

By the late 1950s, Haldane was irritated with European, especially British, politics. In 1957, Haldane neared retirement, and he and Helen Spurway moved to India, working first at the Indian Statistical Institute in Calcutta, and then at Bhubaneswar where he established a government-funded laboratory of genetics and biometry in 1962. Both he and Spurway became Indian citizens in 1961.

On December 1, 1964, J.B.S Haldane died from a protracted battle with rectal cancer. The numerous awards granted Haldane include being made Chevalier of the Legion d'Honneur of France (1937), and the Kimber Award from the National Academy of Sciences (1961).

For More Information: Clark, R. 1968. *J.B.S.: The Life and Work of J.B.S. Haldane*. London: Hodder and Stoughton.

MARSHALL AND SANDRA HALL

Marshall and Sandra Hall published a variety of books in the 1970s linking evolution with topics such as Catholicism, astrology, Masonry, Mormons, the United Nations, NASA, homosexuality, and equal rights for women. Their 1977 book, *The Connection between Evolutionary Theory and the Metric Swindle*, blamed Charles Darwin for the metric system. The Halls, who claimed that the John Birch Society's failure to stop communism was due to its soft stance on evolution, wanted anyone who accepted evolution to be tried and convicted of treason.

KEN HAM (b. 1951)

No apparent, perceived, or claimed evidence in any field, including history and chronology, can be valid if it contradicts the Scriptural record.... If God did not mean what He said in Genesis, then how could one trust Him in the rest of the Scriptures?

Kenneth Alfred Ham was born on October 20, 1951, in Queensland, Australia (Figure 42). Ham earned a bachelor's degree in applied science from the University of Queensland and a diploma of education at the University of Queensland in Brisbane, after which in 1979 he cofounded The Creation Science Foundation. In 1987, Ham moved to the United States and began working for the Institute for Creation Research, and in 1994 he moved to northern Kentucky to establish Answers in Genesis, a ministry that claims the Bible is historically correct, literally true, and scientifically accurate. In 1993, Ham received an honorary Doctor of Divinity from Temple Baptist College, and in 2004 he was awarded an honorary Doctor of Letters by Jerry Falwell's Liberty University.

Ham, who has started over 100 "Creation Clubs" in American schools, is a Young-Earth Creationist who believes that dinosaurs ("missionary lizards") and modern humans lived together, and that the entire universe was created approximately 6,000 years ago, and that Noah's flood occurred about 1,500 years later. Although

42. Ken Ham is the founder and CEO of Answers in Genesis, an evangelical organization that condemns evolution while promoting Young-Earth creationism. (*Courtesy of www.creationmuseum.org*)

Ham acknowledges that natural selection can produce new species of the same "kind," he rejects modern evolution and geologists' claims of an old Earth because he believes that they undermine the authority of the Bible. As Ham has noted, "if something disagrees with the Bible, it is wrong, regardless of the evidence."

Like many other antievolutionists before him, Ham believes that evolution is responsible for many societal ills, including racism, pornography, homosexuality, abortion, humanism, euthanasia, and eugenics. Ham believes that "there is a war going on in society—a very real battle ... that, at the foundational level, [is] really creation versus evolution." Ham—who claims that rainbows should remind us that God judges sin—believes that God will soon destroy the world again, and labels evolution the "biggest stumbling block to people being receptive to the gospel of Jesus Christ." Ham encourages children to "always trust God" and reject science when scientific findings contradict a literal reading of the

Bible. Ham has written many books, including *The Lie: Evolution* and *Dinosaurs of Eden*, in which he claims that Jesus was a Young-Earth creationist, Old-Earth creationism undermines God, and Adam and Eve "were perfect."

Ham is the CEO of Answers in Genesis USA, and joint CEO of Answers in Genesis International. Answers in Genesis is influential and well funded, and in 2007 opened a $27 million Creation Museum and Family Discovery Center near Cincinnati, Ohio, to spread its message. Ham's radio program, *Answers ... with Ken Ham*, is broadcast on over 1,300 stations worldwide. Ham is one of the world's most popular speakers about creationism.

For More Information: Ham, K. 1987. *The Lie: Evolution.* Green Forest, AR: Master Books; Ham, K. 2006. *The New Answers Book.* Green Forest, AR: Master Books.

WILLIAM HAMILTON (1936–2000)

A gene may receive positive selection ... if it confers sufficiently large advantages on relatives.

William Donald Hamilton was born in Cairo, Egypt, on August 1, 1936. Hamilton earned a BA in 1960 from Cambridge University, and while there, became a disciple of biologist Ronald Fisher. In 1968, Hamilton earned a doctorate in genetics at the University of London. He then began teaching at London's Imperial College of Science and Technology, where he remained until 1977.

In 1964, while a graduate student, Hamilton published his groundbreaking theory of kin selection. Although kin selection was hinted at by J.B.S. Haldane and others in the 1930s, Hamilton made it explicit in his two "gene's eye view" papers titled "The Genetical Evolution of Social Behavior," which were published in 1964 in *The Journal of Theoretical Biology*. According to Hamilton's Rule, an altruistic gene can spread by natural selection, so long as the cost incurred by the altruist is offset by a sufficient amount of benefit to relatives. That is, animals are more likely to behave altruistically toward their relatives than toward unrelated members of their species, and

Hamilton's Rule predicts that the degree of altruism will be greater if the relationships are closer. Hamilton's theory has been confirmed in a variety of species, including birds and social insects.

Hamilton, who considered geneticist R.A. Fisher to be his "hero of twentieth-century evolution," revolutionized our understanding of nature and evolution by showing how natural selection acts on social behavior. Hamilton's papers became some of the most cited papers in all of science. In recognition of his work, Hamilton received many awards, including the Kyoto Prize, the Fyssen Prize, the Wander Prize, the Darwin Medal of the Royal Society, and the Crafoord Prize from the Swedish Academy of Sciences. Hamilton revered the work of Fisher, Sewall Wright, and J.B.S. Haldane—"the grand triumvirate," as he called them—throughout his life; the only four references in his "The Evolution of Altruism" paper in *American Naturalist*— the paper that in 1963 introduced biologists to Hamilton's Rule—were two to Haldane, one to Fisher, and one to Wright. Hamilton made many important contributions to biology, including those involving population sex ratios, why some animals flock together when threatened by predators, why organisms invest energy in dispersing their offspring, and how cooperation can evolve between unrelated individuals in the fundamentally selfish world of Darwin.

Hamilton had a long feud with population geneticist John Maynard Smith, a former student of Haldane. Smith reviewed Hamilton's manuscripts for *Journal of Theoretical Biology*, and he demanded that Hamilton revise the papers before they were published. This delayed the publication of Hamilton's work by almost a year. In the meantime, Smith published a paper entitled "Group Selection and Kin Selection" in *Nature,* in which Smith attacked "group selection" and suggested that "kin selection" was the primary force driving the evolution of altruism. In doing so, Smith created the label for the process that Hamilton had outlined. Hamilton was infuriated that Smith—who had "inside knowledge" of, and delayed the publication of, Hamilton's work—would preempt his work.

Although Smith apologized repeatedly for his actions, Hamilton never forgave Smith.

In 1977, Hamilton—disappointed by his lack of recognition at Imperial College—accepted a position at the University of Michigan. There, his collaboration with political scientist Robert Axelrod produced the classic paper "The Evolution of Cooperation" in *Science* in 1981, which won the Newcomb-Cleveland Prize of the AAAS as the best publication of the year. Whereas Hamilton's Rule explained the cooperation and altruism between relatives, the paper by Axelrod and Hamilton focused on the evolution of cooperation in the absence of genetic relatedness. The Axelrod-Hamilton paper elevated Hamilton to a science superstar, and in 1984 Hamilton became a Royal Society Research Professor at Oxford University, one of the most prestigious positions in British academia.

In 2000, while on a biological expedition to war-torn Congo to study whether HIV originated with the polio vaccine, Hamilton contracted malaria. He responded well to treatment in Africa, and decided to return to England. When his flight landed in London, he went to the hospital, but left when he discovered the line for treatment was too long. The next morning, Hamilton passed out in a hospital bathroom and never regained consciousness. After several weeks in intensive care, the 63-year-old Hamilton died on March 7, 2000. He was buried in Wytham Wood in Southern England.

Hamilton's influence on evolutionary biology has been immense. For example, E.O. Wilson and Richard Dawkins credited their conversion to gene-centered thinking to Hamilton's work. Obituaries referred to Hamilton as "the primary theoretical innovator in modern Darwinian biology, responsible for the shape of the subject today," "the most influential evolutionary biologist of his generation," and "more original and more profound than that of any other biologist since Darwin." The London *Times* described Hamilton as "one of the towering figures of modern biology," and Richard Dawkins' obituary for Hamilton opened with: "W.D. Hamilton is a good candidate for the title of most distinguished Darwinian since Darwin."

For More Information: Hamilton, W.D. 1997. *Narrow Roads of Gene Land*. New York: Oxford University.

JOHN HAMPDEN (1819–1881)

John Hampden was an advocate of biblical flat-Earthism who claimed that teaching that Earth is spherical was Satanic and would destroy morality. Antievolutionists such as George Price and Henry Morris later made similar claims about the teaching of evolution. When Hampden offered £500 to anyone who could prove Earth is spherical, Alfred Wallace was encouraged by Charles Lyell to accept Hampden's challenge because doing so might "stop these foolish people." On March 5, 1870, newspaper editor J.H. Walsh—Wallace's choice as an umpire—ruled that Wallace had proven the point, and gave the £500 prize to Wallace. For the next sixteen years, Hampden harassed Wallace, Walsh, and their friends with letters condemning them as liars and cheats. When Walsh sued Hampden for libel, the court ordered Hampden to stop his letter writing campaign and "keep the peace" for a year. When Hampden sent Wallace's wife a letter describing her as "a miserable wretch" and her husband as an "infernal thief" and "convicted felon" who would have "every bone in his head smashed to a pulp," Wallace sued Hampden. Hampden was fined and spent a week in jail, but he continued to harass Walsh and Wallace. After declaring bankruptcy, Hampden sued, and a court ordered Walsh to return Hampden's money. Wallace paid Walsh's expenses in the lawsuit, and lost much money during the venture. Wallace later described the Hampden incident as "the most regrettable incident in my life."

HAPPY FAMILY

Happy Family was a large assemblage of animals—many of which are "natural enemies"—that appeared in a tranquil Garden of Eden atmosphere in *Guidebook to the Consolidated P.T. Barnum Greatest Show on Earth* (1896). The Happy Family, which was often displayed at county fairs, sharply contrasted with the Darwinian struggle for existence.

BENJAMIN HAWKINS (1807–1889)

Benjamin Waterhouse Hawkins was a British sculptor who teamed with Richard Owen to create the first life-size models of dinosaurs. Hawkins' dinosaurs—which were made of brick, iron, and reinforced concrete, and reflected the best scientific knowledge of their time—were a major attraction at the Crystal Palace Exhibition in 1854. It was inside the belly of Hawkins' *Iguanodon* that Owen hosted a dinner on New Year's Eve, 1853, celebrating the opening of the Crystal Palace Exhibit. Hawkins also used specimens collected by Charles Darwin to draw the plates for *Fish and Reptiles* that was part of the *Zoology of the Voyage of the HMS Beagle* (part 1 of which was written by Richard Owen, part 2 by George Waterhouse, part 3 by John Gould, part 4 by Leonard Jenyns, and part 5 by Thomas Bell). When Hawkins came to the United States, he set up a studio near what is now the American Museum of Natural History. Associates of William "Boss" Tweed later smashed Hawkins' models, the remains of which were later buried in the south end of Central Park. Hawkins then became a recluse at Princeton, and later returned to England. Hawkins worked with Thomas Huxley and Edward Cope, and some of Hawkins' paintings of dinosaurs are still displayed at Princeton Museum. Two of Hawkins' *Iguanodon* stand in Sydenham Park south of London. Hawkins—the first great dinosaur reconstruction artist—was the first artist to stimulate the public's interest in ancient life.

ARTHUR HAYS (1881–1954)

Is there anything in Anglo-Saxon law that insists that the determination of either court or jury must be made in ignorance? We plead for the freedom of education, for the liberty to teach, and the liberty to learn.

Arthur Garfield Hays was born on December 12, 1881, in Rochester, New York, and was educated at Columbia University (BA, 1902; MA and LLB, 1905). Hays was admitted to the New York bar, and from 1914 to 1915 he practiced international law in London. In New York, Hays became a successful corporate lawyer and provided counsel for the ACLU. He enjoyed his work with the ACLU because it brought him into contact "with a variety of circles, usually poor, defenseless, and unpopular, always the dissenters and persecuted."

At the Scopes Trial, Hays—a short and stocky man named by his Republican father after a string of conservative presidents—managed the defense team and developed its overall strategy. Hays, an expert on technical aspects of law, was responsible for keeping the court record in shape for the appeal. Hays was also the ACLU's chief counsel at the Scopes Trial, which he viewed as an opportunity to "educate the people and the newspapers."

On Monday, July 20, 1925, Hays announced the most dramatic event of the trial: As Dudley Malone whispered to John Scopes that "hell is going to pop now," Hays sprang Darrow's trap by announcing to the court that "The defense desires to call Mr. Bryan as a witness." Darrow's questioning of Bryan was a turning point in the public's view of science and fundamentalism, as well as in the many re-tellings of the trial (e.g., *Inherit the Wind*). Hays considered the trial a success, noting that people "learned more about evolution through the Dayton exhibition than they could have in any other way." Toward the end of the trial, Hays earned the court's last collective chuckle when he offered Judge Raulston copies of Darwin's *On the Origin of Species* and *The Descent of Man*. Raulston accepted both.

Hays began Scopes' appeal confident that the conviction would be overturned. However, in 1927, when the Tennessee Supreme Court set aside Scopes' conviction—thereby ending the chance of an appeal to the U.S. Supreme Court—Hays admitted that Scopes' defense team "did not credit [the judges] with the astuteness that they really possessed."

After the Scopes Trial, Hays continued to practice law in New York, but was increasingly drawn to society's underdogs. Hays accompanied columnist H.L. Mencken to Boston to sell banned books, and helped Clarence Darrow obtain an acquittal for Ossian Sweet, an African-American charged with murdering a member of an angry mob that had assembled outside his house. Hays also participated in the legal appeals of Nicola Sacco and Bartolomeo Vanzetti (Italian anarchists in Boston who were executed in 1927 for a crime they denied committing) and the Scottsboro case (in which a group of black men from Alabama were charged with raping two white women). Hays treasured his association with Darrow, noting that "nothing has been more inspiring or humanly helpful than his counsel, his example, and his friendship."

Like Malone and Darrow, Hays attacked Scopes' prosecutors and antievolution laws that were "born in ignorance—ignorance of the Bible, of religion, of history, and of science." Hays, who rejected all limitations on free speech, wrote several books, including *Let Freedom Ring* (an account of civil liberties cases of the 1920s; 1928) and *Democracy Works* (a defense of the U.S. system of government; 1939), and in 1942 summarized his long career as a civil rights attorney in *City Lawyer: The Autobiography of a Law Practice*.

After more than forty years of fighting for civil rights, Hays died of a heart attack in New York City on December 14, 1954.

HENDREN v. CAMPBELL

Hendren v. Campbell (1977) ended when the County Court in Marion, Indiana, ruled that it is unconstitutional for a public school to adopt creationism-based biology books because these books advance a specific religious point of view. Glendon Campbell was a member of the Indiana Textbook Commission, and Jon Hendren was a ninth-grader at West Clark Community Schools who, with the aid of the ACLU, sued the schools when they adopted the creationist textbook *Biology: A Search for Order in Complexity*. Judge Michael T. Dugan ruled that the use of *A Search for Order* was unconstitutional, either alone or in conjunction with evolution-based materials. Dugan's decision was not appealed, and the Indiana Textbook Commission dropped *A Search for Order* from its list of approved textbooks.

SUE HENDRICKSON (b. 1949)

Sue Hendrickson was a self-taught fossil-hunter who made headlines worldwide when she discovered the largest and most complete fossil (*Tyrannosaurus rex*) ever found. That fossil, named "Sue," was purchased for $8.8 million and is displayed at the Field Museum in Chicago, Illinois.

JOHN HENSLOW (1796–1861)

I have long since seen that the noble expedition upon which you are entering would have been no way fitted for L. Jenyns. With a little self-denial on your part I am quite satisfied you must reap an abundant harvest of future satisfaction.

John Stevens Henslow was born February 6, 1796, in Rochester, England. In 1814, Henslow entered Cambridge University and studied mathematics, although he was primarily interested in natural history (degrees in such subjects were not yet available). After earning a BA in 1818, Henslow became a fellow of the Linnean Society, and then, along with geologist Adam Sedgwick, founded the Cambridge Philosophical Society in 1819, to promote science at Cambridge. Henslow's influence on science instruction and research at Cambridge increased when he was appointed chair of mineralogy in 1822, and then chair of botany in 1824, a position in which he excelled (although he later admitted that when appointed he "knew very little indeed about botany"). Henslow became an ordained minister in 1824, and immediately started serving at St. Mary's Church in Cambridge. He married Harriet Jenyns in 1823, with whom he had three daughters (the oldest of whom married botanist Joseph Hooker) and three sons (one of whom died in infancy).

At Cambridge, Henslow made the Botanic Garden a first-rate teaching resource, and expanded the university herbarium. Henslow's herbarium collections were unique in that he often included multiple individuals (sometimes more than thirty) of a single species per herbarium sheet, a process he called "collation." Henslow believed in the fixity of species, but understood that enormous variation can occur within any particular species. Through collation, he tried to demarcate the many species he collected. Henslow's most popular work is *The Principles of Descriptive and Physiological Botany* (1835).

Henslow stressed the importance of independent observation by his students, and had them collect and examine plants themselves rather than simply review herbarium specimens. His teaching abilities ultimately gained the attention of Prince Albert, who had Henslow teach his children. Another famous student was Charles Darwin, who attended Henslow's classes from 1829 to 1831, and who later noted Henslow's "power of making the young feel completely at ease." The Henslows frequently hosted students and faculty at their home, and Darwin often took long walks with Henslow to discuss natural theology. Darwin became so close to his botany professor that he became known as "the man who walks with Henslow." Henslow's engaging teaching style may have encouraged Darwin's emphasis on science to the exclusion of his theological studies.

In August of 1831, Henslow received a letter from Cambridge mathematician and astronomer George Peacock describing an opening for a ship's naturalist on the upcoming voyage of the H.M.S. *Beagle* to South America and beyond. After considering the offer himself, Henslow approached his brother-in-law, Leonard Jenyns (like Henslow, a vicar with a background in natural history), about his interest. Jenyns declined the offer, after which Henslow contacted his former student, Charles Darwin, telling Darwin that although he was not "a finished naturalist," he was certainly "qualified for collecting, observing, and noting anything worthy to be noted in Natural History." Darwin accepted, and sailed on the *Beagle* on December 27, 1831. Before Darwin left, Henslow encouraged him to read Charles Lyell's *Principles of Geology*, but included the admonition to not "believe any of the wild theories." Henslow, who had long dreamed of an expedition to an exotic locale, lived vicariously through Darwin, who sent him specimens collected during the *Beagle's* voyage. Henslow recorded the contents of each crate, and directed the specimens to appropriate experts. Henslow, however, never accepted Darwin's conclusions about evolution.

In 1837, Henslow moved to a new congregation in Suffolk. Although he officially maintained his position at Cambridge for several more years, he concentrated his efforts on his new parish. Henslow implemented several changes aimed toward "the common good," including establishment of schools, charity benefits, horticultural shows, and day trips to London. He also founded the Ipswich Museum in 1847.

In 1860, Henslow chaired the meeting of the British Association for the Advancement of Science at Oxford at which Bishop Samuel Wilberforce and Thomas Huxley met in their famous "debate." A year later, Henslow died of a heart condition exacerbated by a bronchial attack. He is buried near his wife in his parish cemetery at Hitcham in Suffolk. Upon learning of Henslow's death, Darwin—who named Anne, George, and Leonard Darwin after Henslow's children—remarked that "a better man never walked this Earth."

For More Information: Walters, S.M., & E.A. Stow, 2001. *Darwin's Mentor: John Stevens Henslow*. New York: Cambridge University.

SUE HICKS (1895–1980)

We will consider it a great honor to have William Jennings Bryan with us in the prosecution.

Sue Kerr Hicks was born in 1895 in Madisonville, Tennessee, into a legally minded family; his father was a successful attorney in Madisonville, and his uncle had written the first manual of Chancery law in Tennessee. Hicks was named after his mother Susan, who had

died during his birth. Hicks graduated from Hiawassee College and the University of Kentucky, and began practicing law in Dayton, Tennessee, with his older brother Herbert Hicks, who had been appointed acting Rhea County Attorney.

In March 1925, Hicks talked with Dayton school board superintendent Walter White about the Butler Act, which banned the teaching of human evolution in Tennessee's public schools. When Hicks and a group of Dayton's businessmen decided to test the Butler Law as a way of stimulating the local economy, Hicks—a friend of John Scopes—volunteered to prosecute Scopes if Scopes would cooperate. During the Scopes Trial, most of Hicks' announcements were procedural.

When Bryan died five days after the Scopes Trial, Hicks delivered a stirring eulogy at Bryan's memorial service in Dayton. Hicks accompanied Bryan's body as far as Chattanooga on its way to Arlington Cemetery in Virginia. During the trip to Chattanooga, Hicks met Bryan's nephew Jennings Bryan, who promised Hicks that moving to Florida would enable him "to make a lot of money." Hicks accepted the advice, turned his law practice over to his brother Herbert, and moved to Miami, where he practiced law for four years. He then returned to Dayton, where he served as a judge and was active in various civic organizations. In 1935, Hicks defeated Walter White for a seat in the state legislature. Fifty years after the trial, Hicks claimed that the Scopes Trial was conceived not by George Rappleyea at Robinson's Drug Store, but instead in his Dayton law office by him and Walter White.

Hicks considered *Inherit the Wind* to be heresy and had to be dissuaded by his family from buying television time to set the record straight. Hicks later became a folk hero when he inspired Chicago writer Shel Silverstein to write *A Boy Named Sue*, a song popularized in 1969 by Johnny Cash's live recording *At San Quentin*.

Sue Hicks died on June 17, 1980, and was buried in Haven Hill Memorial Gardens in Madisonville, Tennessee.

EDWARD HITCHCOCK (1793–1864)

Edward Hitchcock was a famous cleric-geologist and president of Amherst College whose best-selling book, *Elementary Geology* (1860), reconciled geology with Genesis. Hitchcock, who proclaimed that "scientific truth is religious truth," accepted an ancient Earth and argued that the Bible and science must be related. Hitchcock's *Utility of Natural History* (1823) discussed how geology shows the attributes of God. Hitchcock often noted that "the real question is not whether these hypotheses accord with our religious views, but whether they are true." Hitchcock collected more than 20,000 fossilized footprints of dinosaurs, and believed that they were left by huge, extinct birds. Hitchcock, who was America's leading advocate of catastrophism-based gap creationism, was a Charter Member of the National Academy of Sciences.

CHARLES HODGE (1797–1878)

Charles Hodge was a famous and influential Princeton theologian whose 1874 book, *What Is Darwinism?* declared that "it is atheism" because it denies divine design in nature. Hodge, who often quoted Louis Agassiz and described Darwin's theory as "interesting," also claimed that the "denial of [God's] design in nature is virtually the denial of God." Hodge—an Old-Earth creationist—was one of the first prominent theologians to declare that Darwinism threatens religion.

ARTHUR HOLMES (1890–1965)

It is perhaps a little indelicate to ask our Mother Earth her age, but Science acknowledges no shame.

In 1911, Arthur Holmes used Bertram Boltwood's findings about radioactive decay to determine—in his first scientific paper—that rocks from the Devonian were 370 million years old. (At the time, this was the oldest claimed age of a rock.) Holmes' research ultimately produced an absolute timescale for geological history. In *The Age of the Earth* (1913), Holmes estimated Earth to be 1,600 million years old, and in 1944 his classic *Principles of Physical Geology*—whose title honored Lyell's *Principles of Geology*—influenced generations of geologists. (Interestingly, the final chapter of Holmes' book addressed continental drift.) In 1947, Holmes calculated that Earth was 3.35 billion years old, and in 1956 he extended that date to 3.45 billion years. By 1960, Holmes' geological timescale was similar to that used today. Holmes, who received the Wollaston Medal in 1956, is considered to be the "Father of Geological Timescales." As Holmes noted in 1964, "Earth has grown older much more rapidly than I have—from about 6,000 years when I was 10, to 4 or 5 billion years by the time I reached sixty."

Holmes died of bronchial pneumonia on September 20, 1965, in London's Bolingbrook Hospital, and was cremated a few days later. The Arthur Holmes Medal of the European Geosciences Union is named after Holmes.

For More Information: Lewis, C. 2000. *The Dating Game: One Man's Search for the Age of the Earth.* New York: Cambridge University.

ROBERT HOOKE (1635–1703)

Robert Hooke was an English biologist famous for first describing the cell walls of cork as "these pores, or cells" in Observation XVIII of *Micrographia* (1665), one of the few science books of its time written in English. Although Isaac Newton wrote to Hooke in 1676 his famous line, "If I have seen further it is by standing on the shoulders of giants," the two scientists feuded. Hooke was the first person to use a microscope to examine fossils, including petrified wood and fossil shells. In *Micrographia*, Hooke concluded that shell-like fossils are the remains of once-living organisms. Hooke's *Lectures and Discourse on Earthquakes and Subterranean Eruptions* (1705), which influenced geologist James Hutton, argued that shells atop mountains could not be accounted for by a flood. Hooke's conclusion that past events could be recognized and dated with fossils forecast the science of biostratigraphy. Hooke—who spent inordinate amounts of time in coffeeshops—went blind and died on March

3, 1703. After an elaborate funeral, Hooke was buried at St. Helen's Bishopsgate, after which he was exhumed and reburied in "North London." His image was memorialized in a stained glass window at London's St. Helen's Church, which was destroyed by Irish Republican Army bombs in 1992. No paintings of Hooke survive, and most of his personal belongings have vanished. Hooke, Edmond Halley, and Newton established the preeminence of British science at the end of the seventeenth century.

For More Information: Inwood, S. 2005. *The Forgotten Genius: Biography of Robert Hooke*. San Francisco, CA: MacAdam/Cage; Jardine, L. 2004. *The Curious Life of Robert Hooke*. New York: HarperCollins.

JOSEPH HOOKER (1817–1911)

Plants, in a state of nature, are always warring with one another, contending for the monopoly of the soil—the stronger ejecting the weaker—the more vigorous overgrowing and killing the more delicate.

Joseph Dalton Hooker was born June 30, 1817, in Halesworth, England, to William and Maria Hooker. From an early age, Joseph was surrounded by botanical specimens (his father's private collection later occupied thirteen rooms in the family's house), and as a boy he was passionate about science. In preparation for a career in natural history, Joseph received the standard scientific training for the period, an education in medicine, and graduated from the University of Glasgow in 1839.

As a boy, Hooker had been entranced by tales of Captain Cook sailing the globe, and he dreamed of traveling to exotic locales. After earning his degree, he also realized that entry into the scientific community required additional experience. Because training in natural history was not then available as part of a university education, the primary way to build such a background was to follow in the footsteps of Hooker's hero, Charles Darwin, and accompany one of the survey expeditions launched by the British government. In 1839, Hooker, eager for adventure, joined an expedition to the southern hemisphere led by famous Arctic explorer

James Clark Ross. Hooker was appointed assistant surgeon on the H.M.S. *Erebus* (a sister ship, the *Terror*, would accompany the *Erebus*), a position that was a disappointment for Hooker, as he would have a variety of duties besides just making scientific collections and observations. Hooker had to support himself (unlike Darwin on his voyage), and serving in the Royal Navy was the only way to be part of the expedition. The success Darwin had achieved from his travels on the *Beagle* was uppermost in Hooker's mind; Hooker kept proofs of Darwin's *Journal and Remarks* (later known as *Voyage of the Beagle*) under his pillow while he slept.

The *Erebus* and the *Terror* visited Australia, the southern tip of South America, New Zealand, and Antarctica during its four-year voyage, and Hooker amassed a large collection of plants from these locales. Fascinated by the regional patterns he observed, Hooker hoped to elucidate the overarching laws governing plant distribution. In 1843, the expedition returned to England. Hooker's father was now director of the Royal Botanic Gardens at Kew, which meant that the younger Hooker had access to the Kew's extensive herbarium collections in addition to his own collections. Hooker's first volume of *Botany of the Antarctic Voyage* was published but generated no income, and Hooker was without permanent employment during the 1840s. He worked at various short-term positions, including as an assistant with the Geological Society, and as temporary lecturer at the University of Edinburgh.

Despite his financial hardships, Hooker's involvement in science continued uninterrupted. Darwin was aware of Hooker's work even before the *Erebus* had returned; Hooker's letters to his father were passed to Charles Darwin through the Lyell family. Upon Hooker's return, Darwin asked Hooker to classify the plants Darwin had collected in the Galápagos. Hooker agreed, and the two scientists soon became close friends. Hooker was subsequently the first person in whom Darwin confided his ideas on evolution. In a letter dated January 14, 1844 (over fifteen years before the publication of *On the Origin of Species*), Darwin wrote that "I am almost convinced . . . that species are not . . . immutable." He eluded to a proposed mechanism, and famously acknowledged the implications of his views—"it is like confessing a murder." By all accounts, Hooker was open-minded and expressed his own suspicions in this regard. In 1859, Hooker became the first biologist to state in print his support for evolution by natural selection; in the introduction to his *Flora Tasmaniae*—published just one month after the *Origin* appeared—Hooker interpreted the distribution of the Tasmanian flora through specific reference to the proposals of "Mr. Darwin and Mr. Wallace."

Hooker's studies of the geographic distribution of plants was aided by the global network of collectors his father had cultivated. In the late 1840s, while visiting the Himalayas, Hooker was imprisoned for a few weeks after an ill-advised border crossing. Hooker wrote about these experiences in the travel book, *Himalayan Journals* (1854), his only work to garner attention outside the scientific community. Not long after his return from Asia, Hooker married Frances Henslow, daughter of Darwin's mentor, cleric and Cambridge botanist, John Henslow. (Frances would die in 1874, and Hooker would marry Hyacinth Jardine in 1876.) In 1855, Hooker finally gained steady employment when he became assistant to his father.

In June of 1858, Darwin received the fateful letter and paper from Alfred Russel Wallace that detailed Wallace's independent discovery of natural selection. Distraught at the prospect of losing credit for an idea he had developed many years earlier, and having to contend with a seriously ill child, Darwin put the matter into the hands of his two closest friends, Joseph Hooker and Charles Lyell. Hooker and Lyell proposed that at a meeting of the Linnean Society, three documents would be read: extracts from a manuscript of Darwin's that outlined his conceptualization of natural selection; a letter from Darwin to Asa Gray that established Darwin as the original discoverer of natural selection; and Wallace's paper. There was little immediate fallout from the meeting, but Hooker expressed relief at no longer having to keep

Darwin's secret. Hooker also attended the famous Huxley-Wilberforce "debate" at the 1860 meeting of the British Association for the Advancement of Science. Huxley's comments have become legendary, and have overshadowed that Hooker provided the most eloquent and forceful defense of Darwin. Hooker, who often visited Darwin at Down House, arranged for Charles Lyell to be buried in Westminster Abbey.

Hooker became director of the Royal Botanic Gardens when his father died in 1865. (When Joseph Hooker retired, he was succeeded by his son-in-law.) He expanded the research focus of the institution and reduced the use of the Gardens as a place for public recreation. In 1885, Hooker retired from his position at the Royal Botanic Gardens, and continued to publish, although at a more leisurely pace, and participated in expeditions to Africa, Asia, and North America. From 1892 to 1895, Hooker supervised publication of *Index Kewensis*, a two-volume description of 400,000 genera of plants. *Index* was paid for "by the members of the family of the late Charles Darwin."

By 1911, Hooker's health was starting to decline, and on December 10, 1911, he died in his sleep. An offer of burial in Westminster Abbey was made (so as to lie next to Lyell and Darwin) but, honoring his own wishes, Hooker was buried next to his father in the family plot at St. Anne's Church in Kew. Honors bestowed on Hooker included election to the Royal Society in 1842, and service as president of the Society from 1873 to 1878. The Royal Society awarded him its highest honors, including the Royal (1854), Copley (1887), and Darwin (1892) medals. He was knighted in 1877.

For More Information: Allen, M. 1967. *The Hookers of Kew, 1785–1911*. London: Joseph.

EARNEST HOOTON (1887–1954)

Earnest Albert Hooton was an anthropologist whose influential *Up from the Ape* (1946) advocated learning about humans by studying humans' evolutionary lineage, thereby expanding anthropology from an anatomical to a behavioral science. While a professor at Harvard from 1914 to 1954, Hooton trained the first generation of professional physical anthropologists in the United States.

ELIZABETH REID HOPE (1842–1922)

Elizabeth Reid "Lady" Hope was a British evangelist who, soon after Charles Darwin's death, announced at Dwight Moody's school in Northfield, Massachusetts that she visited Darwin just before he died, and that during that visit Darwin denounced evolution and became a Christian. Moody encouraged Lady Hope to publish her story, which appeared in the August 15, 1915, issue of the Baptist newspaper *Watchman Examiner*. Lady Hope's story was debunked by Darwin's children; for example, daughter Henrietta Darwin Litchfield wrote in the February 23, 1922 issue of *The Christian* that "Lady Hope was not present during [Charles Darwin's] last illness, or any illness…He never recanted any of his scientific views, either then or earlier…The whole story has no foundation whatever." Similarly, Francis Darwin noted in 1918 that "Lady Hope's account of my father's views on religion is quite untrue. I have publicly accused her of falsehood, but have not seen any reply. My father's agnostic point of view is given in my *Life and Letters of Charles Darwin*, Vol. I, pp. 304–317. You are at liberty to publish the above statement. Indeed, I shall be glad if you will do so."

Lady Hope died in 1922 and was buried in the Anglican cemetery at Rookwood in suburban Sydney, Australia.

JOHN R. "JACK" HORNER (b. 1946)

John R. "Jack" Horner was the Curator of the Museum of the Rockies in Bozeman, Montana, and a famous popularizer of dinosaurs in the late twentieth century. The largely self-taught Horner named several dinosaurs (e.g., *Maiasaura*, *Orodromeus*), wrote several books (e.g., *Digging Dinosaurs*, *Complete T. rex*, *Dinosaur Lives*), discovered a fossilized nesting colony of the hadrosaur *Maiasaura*, and was a technical advisor for the movies *Jurassic Park* and *The Lost World*.

KATHARINE HORNER (1817–1915)

Katharine Murray Horner wrote *Life, Letters and Journals of Sir Charles Lyell*, which was published in two volumes in 1881. In 1875, Horner—who married Charles Lyell's brother—asked Charles Darwin to be a pallbearer at Lyell's funeral. Darwin was too sick to attend.

WILLIAM HOUGHTON (1887–1947)

Your chief business on Earth is to be a witness. In fact, it's your only excuse for living. If you are less than that you are nothing.

William Henry Houghton was born in 1887 in South Boston. Houghton was active in his local church as a boy, and frequently participated in religious plays, which gave him his first taste of life in front of an audience. He decided to become an actor (and would later draw upon these experiences in his famous "From Stage to Pulpit" sermon), and Houghton performed on the vaudeville circuit until he was in his mid-twenties. In 1909, he attended a Nazarene tabernacle and immediately felt a call to the ministry. Houghton married Elizabeth Andrews in December 1918, and the couple had a son late in 1919.

Houghton enrolled at Eastern Nazarene College in Rhode Island, but quit after six months, and started working with a traveling evangelist. By 1915, he was ordained and had served at a number of churches in Pennsylvania. Houghton started publishing religious tracts that he distributed free of charge. His writings garnered attention, and in 1929, Houghton replaced the late John Straton as pastor at Calvary Baptist Church in New York City. In 1934, Houghton moved to Chicago to serve as president of the Moody Bible Institute. In 1938, he expanded the influence of the Institute through a series of national radio addresses titled, *Let's Go Back to the Bible*. The text from these sermons was published as a book of the same name. Another series of broadcasts ensued the following year, producing another book, *Back to the Bible*.

Around this time, Rev. Irwin A. Moon was becoming a well-known evangelist through his "Sermons to Science" presentations that used electricity and other phenomena to illustrate "the visible evidence of a Divine plan." Although dismissed by many as "gadget evangelism," Moon's efforts gained Houghton's attention, and he invited Moon to join the Moody Institute. Moon agreed, but only if his emphasis could remain on reaching "those who would never be reached by ordinary methods, particularly high school and college young people" instead of "over-fed Christians in Bible conferences." Houghton agreed and they devised a plan to bring together Christian scientists to discuss the theological dangers of contemporary science, especially evolution. (In 1945 Moon became director of the new Moody Institute of Science in Los Angeles.)

Houghton proposed an annual conference to "help ministers...who are dealing with scientific subjects," and in September 1941, five Christian scientists met at the Moody Bible Institute. This meeting led to the founding of the American Scientific Affiliation (ASA), an organization formed to provide information that interpreted science "accurately" relative to Scripture. ASA members had to sign a statement attesting to the inerrancy of the Bible that included "I cannot conceive of discrepancies between statements in the Bible and the real facts of science." Once established, Houghton did not have an active role in the ASA.

Houghton's sermons, radio programs, and Moody's publication, *Moody Monthly*, promoted his perspectives on religion and society. He bemoaned the loss of strict adherence to Scriptural doctrine and suggested that many social ills exist because "instead of sending for the revivalist [people] call in the psychologist." The active role of what he saw as godless universities did not escape Houghton's scorn, and he blamed evolution for the "great mechanical monster" of Nazism.

Houghton died on June 13, 1947. His body lay in state at the Moody Institute, allowing thousands of mourners to pay final respects, and his funeral service was attended by more than 4,000 people. Houghton was buried in New Bethlehem, Pennsylvania.

For More Information: Smith, W.M. 1951. *Watchman on the Wall: The Life of Will H. Houghton.* Grand Rapids, MI: Eerdmans.

KENT HOVIND (b. 1953)

If evolution is true, abortion, euthanasia, pornography, genocide, homosexuality, adultery, incest, etc., are all permissible.

Kent "Dr. Dino" Hovind was born on January 15, 1953. He graduated from East Peoria High School (Peoria, Illinois) in 1971, after which he received a bachelor's, masters, and doctoral degree in Christian Education from unaccredited institutions. Hovind's doctoral dissertation concluded with, "I believe Jesus was right."

In 1975, Hovind began preaching and teaching at several Baptist-related schools. By 1989, Hovind—a Young-Earth creationist—founded Creation Science Evangelism in Pensacola, Florida, and two years later opened an antievolution amusement park known as Dinosaur Adventure Land ("where dinosaurs and the Bible meet"). The 7-acre park generated more than $1 million per year, and included rides, a science center, and a museum showing dinosaurs and humans coexisting in the last 4,000–6,000 years. The park also described a worldwide flood and claimed that *T. rex* was vegetarian until Eve sinned.

According to the flamboyant Hovind, Noah's family and two of every animal boarded the Ark, after which Earth and other planets were hit by an ice meteor. These impacts produced "supercold snow" and the Grand Canyon in two weeks, and caused "the Earth [to] wobble around for a few thousand years." In 2001, Arkansas state Representative Jim Holt asked Hovind to testify as an expert before the State Agencies and Governmental Affairs Committee to support Holt's bill requiring "that when public schools refer to evolution that it be identified as an unproven theory." Holt later admitted that much of the information for his proposed legislation came from Jonathan Wells' *Icons of Evolution*.

In his sermons and speeches, Hovind often asked, "Do you know chimpanzees are still having babies? Why don't they make another human?" Hovind also claimed that fossils are "all fake. They're coming from China. These Chinese guys who make 40 cents a year.... They're faking them. They spend years forging these things." Just as creationist George Price had offered $1,000 to anyone who could "prove that one kind of fossil is older than another," in 1990 Hovind began offering $2,000 to anyone who could prove that dogs and bananas have a common ancestor, and $10,000 to anyone who can "prove the theory of evolution." In 1999, Hovind raised the offer to $250,000 (winning required challengers to prove the Big Bang by producing an actual Big Bang). This caused some creationists to distance themselves from Hovind; for example, Answers in Genesis noted that it "would prefer that creationists refrained from gimmicks like this."

Hovind claimed that the U.S. government was responsible for the Oklahoma City and 9/11 attacks; that the 1969 moon landings never occurred; that UFOs are "apparitions of Satan"; that the Smithsonian Institution "has 33,000 sets of human remains in their basement ... many of them taken while the people were still alive," and that "Adam and Eve probably had hundreds of children." Hovind—who linked evolution with the shootings at Columbine High School, Joseph Stalin's executions, and Nazi horrors—produced radio programs and hosted conferences promoting creation science. Hovind appeared on *The Ali G Show*, and in 2000 was given the "P.T. Barnum: One Born Every Minute" award from the New Mexicans for Science and Reason.

Hovind was charged with a variety of crimes, including felony assault and misdemeanor battery (these charges were later dropped by the alleged victim, who was Hovind's secretary). In 1996, Hovind unsuccessfully filed for bankruptcy to avoid paying federal income tax, claiming that everything he owned belonged to God and that he did not have to pay taxes because he does God's work. The judge called Hovind's arguments "patently absurd" and

ordered him to pay the delinquent taxes. In 2002, when Hovind was again delinquent in paying his taxes, he unsuccessfully sued the IRS for harassment. In 2004, IRS agents confiscated financial records at Hovind's home and business, and on July 7, 2006, the U.S. Tax Court ruled that Hovind owed more than $470,000 in taxes. The Court also noted that Hovind paid his employees in cash and labeled them "missionaries" to avoid paying taxes.

On November 2, 2006, Hovind was convicted of fifty-eight federal charges, including tax evasion. Hovind's wife Jo was convicted of forty-four of the charges. Kent Hovind was remanded into custody because U.S. District Judge Casey Rodgers deemed him a flight risk and "a danger to the community" (agents raiding Hovind's home found $42,000 in cash and a variety of guns, including a semiautomatic rifle). On January 19, 2007, in a courtroom packed with Hovind's praying followers, Kent Hovind appeared for sentencing. Although Hovind had ridiculed the government's ability to sentence him to prison (claiming that he could "make their lives miserable"), at the sentencing Hovind cried and begged for leniency, claiming that his followers would pay his fines. After reminding attendees that Hovind's case "is not and has never been about religion," Judge Rodgers fined Hovind $611,954 ($604,876 in restitution to the Internal Revenue Service, and $7,078 to cover the prosecution costs for Hovind's trial) and sentenced him to ten years in federal prison. Hovind's appeal to the U.S. Court of Appeals for the 11th Circuit was denied. On June 29, 2007, Jo Hovind was sentenced to one year plus one day in prison and fined $8,000.

FRED HOYLE (1915–2001)

Fred Hoyle was a British astronomer who in 1978 suggested that the seeds of life might have come from outer space. In 1989, Hoyle's claim was supported by the discovery of amino acids within meteor craters. Hoyle, who advocated intelligent design and claimed that *Archaeopteryx* was a fake, coined the term *big bang* in a sarcastic

43. Ken Hubert was a teacher at Faribault (Minnesota) High School who confronted Rod LeVake about LeVake's refusal to teach evolution in his biology course. This confrontation ultimately produced *LeVake v. Independent School District #656*, in which the court ruled that a teacher's right to free speech does not permit the teacher to circumvent the prescribed curriculum. (*Randy Moore*)

comment about Lamaître's theory that the universe is expanding.

EDWIN HUBBLE (1889–1953)

Edwin Hubble, in 1929, substantiated Belgian priest Georges-Henri Lamaître's (1894–1966) suggestion that the universe is expanding. Hubble's observation—that light from other galaxies is redshifted—became known as Hubble's Law. Extrapolated to the past, these observations suggest that the universe began as a "primeval atom," now known as the Big Bang. The term *big bang* was coined in 1949 by Fred Hoyle in a sarcastic comment about Lamaître's theory ("this big bang idea").

KEN HUBERT (b. 1955)

I couldn't imagine deleting evolution from an introductory biology course.

Ken Hubert was born on June 10, 1955, in Duluth, Minnesota (Figure 43). He graduated in 1978 from the University of Minnesota in Duluth, after which he began teaching high school biology in Beloit, Wisconsin.

In 1984, Hubert began teaching at Faribault High School in Faribault, Minnesota. All of the school's biology teachers devoted two or three weeks to evolution, and incorporated evolution throughout their courses. However, soon after Rod LeVake began teaching biology at Faribault High School in 1997, Hubert and other science instructors learned that LeVake was not including evolution in his course (as prescribed in the school's curriculum); that LeVake rejected evolution; and that LeVake was awarding extra credit to students for summarizing articles from creationist magazines. Just before a meeting of the science faculty, Hubert asked LeVake how he was teaching evolution. After LeVake responded that "I cannot teach evolution" and that "I will not teach something that is not true," school administrators reassigned LeVake to teach physical science. LeVake then asked Pat Robertson's American Center for Law and Justice to file a lawsuit on his behalf. When LeVake lost the case, he appealed the verdict. LeVake's case ended on January 7, 2002, when the U.S. Supreme Court refused, without comment, to hear his appeal (*LeVake v. Independent School District #656*).

Hubert was criticized for his role in LeVake's reassignment. Letters claimed that Hubert was a Nazi; that the Bible records all of the basic laws of science; that evolution is a lie; that teachers who teach evolution should be fired; and that the teaching of evolution is responsible for premarital sex, abortion, homosexuality, drugs, gangs, Satanism, and suicides. Today, Hubert is the Activities Director and an award-winning swimming coach at Faribault High School.

CHARLES HUGHES (1862–1948)

Charles Evans Hughes was a governor of New York, associate justice of the U.S. Supreme Court, and unsuccessful Republican candidate for president. Hughes was the ACLU's choice to defend John Scopes at his Dayton trial, but Scopes chose Clarence Darrow and Dudley Malone. Hughes agreed to defend Scopes if Scopes' case ever reached the U.S. Supreme Court, but it never did. Hughes rests in Woodlawn Cemetery in Bronx, New York.

DAVID HUME (1711–1776)

Our most holy religion is founded on Faith, not on reason; and it is a sure method of exposing it to put it to such a trial as it is, by no means, fitted to endure.

David Hume (born Home, he changed his name when in his 20s) was born in Edinburgh, Scotland, on April 26, 1711. Hume entered the University of Edinburgh when he was 11 or 12 years old, but left in 1725, not having earned a degree. A prolonged period of lack of direction in his life ensued, lasting well into the 1730s. In 1734, resolved to support himself, Hume began working as a clerk with a sugar merchant in Bristol, a foray into business that was short-lived, and which began many years of temporary employment. Hume never married.

In 1739, Hume published anonymously, in two volumes, a masterpiece of philosophical discourse, *A Treatise of Human Nature*. (A third volume followed in 1740.) Hume, building upon the empiricism of John Locke, outlined a perspective where understanding flows only from direct experience. Central to the goal of philosophy, then, is an understanding of human nature and how it affects our perceptions and conclusions (along with a casting aside of superstition and metaphysics). For Hume, understanding is a product of our senses alone, which creates beliefs, not objective knowledge. These beliefs in how the world works then lead to habits. Knowledge, therefore, is nonrational because we are not justified in expecting the future to be like the past, yet we are forced to act as if this were so (i.e., we act out of habit). Hume urged science to employ "mitigated skepticism" that is "sensible of the strange infirmities of human understanding," especially when interpreting cause-and-effect relationships.

Treatise was not successful; as Hume noted, "it fell dead-born from the press, without reaching such distinction as even to excite a murmur among the zealots." A shortened and somewhat altered version of the book, *An Enquiry Concerning Human Understanding* (1748), also failed to excite interest. However, the ideas in these works ultimately proved enormously influential. Given the poor sales of *Treatise*, Hume had to find a way to support himself. After unsuccessfully (and in spite of help by his friend Adam Smith) pursuing university positions, in 1745, Hume began tutoring the Marquess of Annandale. Hume's prospects improved, however, when he was asked to serve as secretary to General James St. Clair, who was undertaking an expedition through Europe. This led to a two-year position at the British embassy in Paris in 1763. During this time, Hume continued writing, especially on politics and history (*History of England, Political Discourses*), and he was noticed by the French intelligentsia, which established him as a noteworthy writer and thinker. For the remainder of his life, Hume supported himself as an author.

Hume was not an atheist, claiming that the level of evidence (or faith) needed for such a position equals that required of a believer, and would be classified with Thomas Huxley, the originator of the term *agnostic* (and the author of a book about Hume). However, Hume felt organized religions that claimed special knowledge of the universe were promulgating superstitions, as questions about the existence of a deity are unanswerable based on input from our senses.

In his posthumously published *Dialogues Concerning Natural Religion* (1779), Hume tackled the "argument from design" for asserting the existence of a creator. In the form of a conversation, the protagonists Philo (antidesign advocate; usually considered to be Hume himself) and Cleanthes (prodesign advocate) discuss inferring the existence of a deity by observing an apparently complex and designed world. Filtered through the limitations of eighteenth-century scientific understanding, Hume rejected the design-based argument by invoking the existence of evil, by noting that human creations are subject to later improvements, and by questioning the assumption of the existence of complexity. Hume's analysis has formed the basis for subsequent refutations of the design argument, including those countering claims of the modern intelligent design movement.

While in France, Hume befriended fellow philosopher Jean-Jacques Rousseau. Rousseau's books were controversial, and as he was fleeing persecution in France, Hume accompanied him to England. However, Rousseau's paranoia overcame him, and he incorrectly and publicly accused Hume of treachery. Public opinion mostly sided with Rousseau, and Hume's reputation suffered.

In the 1770s, Hume retired to Edinburgh, where he, Joseph Black, and economist Adam Smith founded Edinburgh's Oyster Club. There, they and others met to eat, drink, and discuss the scientific and philosophical issues of their age.

Hume died on January 4, 1776. He is buried in a large Roman mausoleum in Edinburgh's Old Calton Burial Ground.

For More Information: Graham, R. 2004. *The Great Infidel: A Life of David Hume.* Edinburgh: John Donald.

CARL HUNT

Carl Hunt was a constituent of Arkansas State Senator James Holsted who sent to Holsted a copy of Wendell Bird's resolution for equal time for evolution and creation science in public schools. After being lobbied by Duane Gish, Kelly Segraves, Henry Morris, the Creation Science Research Center, and other creationist organizations, Holsted molded the resolution into a bill, which became law. The "equal time" law was overturned by *McLean v. Arkansas Board of Education*.

GEORGE HUNTER

George William Hunter taught at New York's DeWitt Clinton High School and wrote *A Civic Biology: Presented in Problems* (1914), a best-selling textbook that John Scopes used when he taught biology in Dayton, Tennessee. Hunter's book, which William Jennings Bryan said, "could not be more objectionable," claimed that "the civilized white inhabitants of Europe and America" were "the highest type of all." At the Scopes Trial, Bryan was especially upset by a diagram in Hunter's book grouping humans with mammals instead of being in a group of their own. After the Scopes Trial, Tennessee

abandoned Hunter's book, which was later retitled *New Civic Biology* to distinguish it from the book associated with Scopes' sensational trial.

HURST v. NEWMAN

Hurst v. Newman began in January 2006, when geologist Kenneth Hurst—the father of two students at California's Frazier Mountain High School—sued the local school district, claiming that a course taught by a soccer coach promoting intelligent design and Young-Earth creationism as an alternative to evolution was unconstitutional (Steve Newman was a member of the El Tejon Unified School District Board of Trustees). The case ended two weeks later when the parties agreed that courses entitled "Philosophy of Design" or "Philosophy of Intelligent Design" would be terminated within ten days and never offered again. When the decision was announced, evangelist Pat Robertson proclaimed it "terribly wrong," adding that "there are inexplicable gaps in the so-called evolution theory."

JAMES HUTTON (1726–1797)

The result, therefore, of our present inquiry is that we find no vestige of a beginning, -no prospect of an end.

James Hutton was born into a wealthy family in Edinburgh, Scotland, on June 3, 1726, during the Scottish Enlightenment. After graduating from college in Edinburgh in 1743, Hutton became interested in chemistry, law, and medicine. In 1747, Hutton moved to Paris to study chemistry and anatomy, and in September 1749 received his medical degree. By 1750, however, Hutton had abandoned medicine for agriculture and began farming in northeastern Berwickshire on land that he had inherited. While traveling throughout Europe to study agriculture, Hutton became interested in geology; he was "very fond of studying the surface of the earth," possibly as a result of reading Robert Hooke's *Micrographia* and *Dissertation on Earthquakes*. Although Hutton described farming as "the story of my life," around 1768 Hutton quit farming and

44. This site in Edinburgh's Salisbury Crags, today known as Hutton's Section, is where James Hutton understood that molten rock (magma) under pressure can intrude between sedimentary strata. Hutton concluded that igneous rocks originated as magma. This conclusion was incompatible with the then popular view that igneous rocks crystalized like salt from sea water. (*Clifford Ford*)

returned to Edinburgh, where he lived with his three sisters until his death. Hutton was a lifelong bachelor, but he had one illegitimate son (also named James Hutton) to whom he occasionally sent money. Hutton was friends with philosopher David Hume and economist Adam Smith, both of whom lived in Edinburgh.

"A question naturally occurs with regard to time," said Hutton. "What had been the space of time necessary for accomplishing this great work?" Geologists known as Neptunists, led by German geologist Abraham Werner (1750–1817), claimed that rocks had precipitated from the water of a great flood that had once covered Earth. However, in 1785 in rocks at Glen Tilt of the Scottish Highlands, Hutton found granite extending into metamorphic schists in a way indicating that the granite had been molten. Hutton concluded that granite formed from cooling (and not by precipitation from water) and that the granite was therefore younger than the schists

in which it was found (Figure 44). Hutton believed that this was evidence of intense subterranean heat, and that this heat was the engine that created new rock. In 1855, Charles Lyell paid homage to Hutton by including a sketch of rocks at Glen Tilt in his *Manual of Elementary Geology*.

On March 7 and April 4, 1785, Hutton's study of Earth's history was reported to the recently formed Royal Society of Edinburgh in a thirty-page dissertation titled *Concerning the System of the Earth, Its Duration, and Stability*. Hutton claimed that rocks now at the Earth's surface were formed from pressurized sediments under the ocean, raised to the Earth's surface by pressure and heat, and eroded to the sea, where the process began anew. During the upheavals, molten rock was injected into cracks of dislocated strata. Copies of the abstract of Hutton's dissertation are rare, and it wasn't until 1947 that scholars documented that the anonymous papers were, in fact, Hutton's (Figure 45).

177

In 1788, Hutton visited Siccar Point, the third and most famous of the unconformities of which he wrote. There, he saw that the uppermost sediments were horizontal, but those beneath them were vertical. Hutton concluded that the lowermost layers had been deposited in water, but were then tilted and raised above sea level by an earthquake, after which they were again covered by water and a new set of horizontal layers of sediment. That same year, Hutton published his ideas in the inaugural volume of *Transactions of the Royal Society of Edinburgh*. His claims were ridiculed, and Hutton was accused of atheism and poor logic. In 1795, Hutton published his most famous and influential book, the two-volume, 1,100-page *Theory of Earth*, in which he claimed that great, endless cycles are constantly yet imperceptibly changing Earth, thereby challenging the claim that Earth is only 6,000 years old. In doing so, Hutton opened geology to scientific observation, removed it from the influence of Bible-based chronologies, and gave geology its most transforming idea—namely, that Earth is ancient. Although Hutton's writings were full of deistic metaphysics and teleology ("The glove of this earth is evidently made for man"), his conclusions meant that long periods of time were available for evolution to occur.

To Hutton, it was change, and not stasis, which explained geology, and he argued that no great catastrophes (e.g., a worldwide flood) were needed to explain Earth's geology. Instead, Hutton claimed that "in examining things present we have data from which to reason with regard to what has been.... The ruins of an older world are visible in the present structure of our planet." Hutton believed that the age of the Earth was "indefinite" and beyond comprehension, as suggested in the poetic and now-famous sentence that concluded *Theory of Earth*, "we find no vestige of a beginning, -no prospect of an end." In context, the importance of this sentence was not the magnitude of its inferred timescale, but instead an eternal equilibrium (which Hutton suggested was analogous with the Newtonian "system" of orbiting planets). Hutton did not

A B S T R A C T

O F A

D I S S E R T A T I O N

READ IN THE

ROYAL SOCIETY OF EDINBURGH,

UPON THE

SEVENTH of MARCH, and FOURTH of APRIL,

M,DCC,LXXXV,

CONCERNING THE

SYSTEM OF THE EARTH,

ITS DURATION, AND STABILITY.

45. James Hutton, a Scottish farmer and geologist who believed that earth is ancient, argued that geology is best explained by change, not stasis, over vast periods of time. Shown here is the 1785 abstract announcing Hutton's earliest ideas about an ancient Earth.

speculate on how the Earth originated or how it might end.

Theory of Earth attracted little attention during Hutton's life because it was long and difficult to read. Indeed, Hutton's idea about metamorphism is described in a sentence that is 136 words arranged in several ponderous clauses. However, Hutton's friend John Playfair—who had accompanied Hutton to Siccar Point—popularized Hutton's ideas in 1802 (five years after Hutton's death) when he published *Illustrations of the Huttonian Theory of the Earth*. Many of the most notable people in church history—for example, the prophet Elijah, St. Augustine, St. Thomas Aquinas, and Martin Luther—had estimated the age of the Earth, and all concluded that Earth is about 6,000 years old. Hutton never made such an estimate. Hutton concluded that Earth is an ancient planet in a perpetual state of decay and rebirth.

Hutton, a theist, also had interesting ideas about animal breeding. Buried in the more than 2,000 verbose pages from Volume 2 of Hutton's *Investigation of the Principles of Knowledge* is this: "...if an organised body is not in the situation and circumstances best adapted to its sustenance and propagation, then, in conceiving an indefinite variety among the individuals of that species, we must be assured, that, on the one hand, those which depart most from the best adapted constitution, will be the most liable to perish, while, on the other hand, those organised bodies, which most approach to the best constitution for the present circumstances, will be best adapted to continue, in preserving themselves and multiplying the individuals of their race." Hutton supported his suggestion by citing his studies of animal breeding, noting that dogs survived by "swiftness of foot and quickness of sight...the most defective in respect of those necessary qualities, would be the most subject to perish, and that those who employed them in greatest perfection...would be those who would remain, to preserve themselves, and to continue the race." Similarly, if an acute sense of smell was "more necessary to the sustenance of the animal...the same principle [would] change the qualities of the animal, and...produce a race of well scented hounds, instead of those who catch their prey by swiftness." The same "principle of variation" would influence "every species of plant, whether growing in a forest or a meadow." However, Hutton provided no data to support his suggestion, and his idea about life's diversity faded away.

Hutton's critics included Jean-André de Luc (1727–1817), who coined the term *geology*, and the editors of *Encyclopaedia Britannica*, whose twelve-page critique of Hutton's work in 1797 noted that the world is full of evidence of the Flood. *The Analytical Review* dismissed Hutton's work as "philosophical romance," and geologist John Williams claimed in *The Natural History of the Mineral Kingdom* (1789) that Hutton's "wild and unnatural notion" would lead to "skepticism, and at last to downright infidelity and atheism." Hutton's most vocal critic was the eccentric Richard Kirwan (1733–1812), whose *Elements of Mineralogy* (1784) was the first English-language text on the subject. Kirwan believed that rocks had precipitated from a biblical flood, and his *Geological Essays* (1799) was a book length attack on Hutton. When Kirwan died in 1812 at age 79, his copy of Hutton's *Theory of Earth* was found with many of its pages uncut (i.e., unread). Hutton's *Natural Philosophy* (1792) included essays about meteorology and the nature of matter, and his 3,200-page *Principles of Knowledge* (1794) discussed perception, reason, matter, motion, and religion.

Hutton's geological ideas became known as *uniformitarianism*, the doctrine that the distant past is best explained by common forces that operate today ("the present is the key to the past"). Hutton's uniformitarianism, which founded modern geology, opposed Cuvier's catastrophism and inspired the thinking of Lyell and Darwin. To geologists, rocks are pages in Earth's autobiography, and Hutton showed how to read them.

Hutton died in his home in Edinburgh on the evening of Saturday, March 26, 1797, at the age of 70. For the next 150 years his body rested in an unmarked grave below the south side of Edinburgh Castle in the South Yard of historic Greyfriars Kirk churchyard, the largest cemetery in Edinburgh. The third volume of *Theory of Earth* was published by the Geological Society of London in 1899, and on November 3, 1947, a memorial to Hutton was placed by the Royal Society of Edinburgh on the east wall of the part of the churchyard known as the Covenanters Prison, in the lair of the Balfour family to whom Hutton was related (through his mother, Sarah Balfour). The memorial is simple, "The Father of Modern Geology, James Hutton, 1726–1797." A portrait of Hutton by famed painter Henry Raeburn hangs in the Scottish National Portrait Gallery in Edinburgh; that painting shows Hutton surrounded by a few rocks and the manuscript of *Theory of the Earth*.

In 1947, historians discovered two copies of Hutton's paper that had appeared in *Transactions*. Those papers, which were still in their original blue wrappers and signed "From the Author Dr Hutton," sold for £90 and £110. Hutton's personal collection of rocks was stored at Edinburgh University's Museum of Natural History, where they eventually were entrusted to Robert Jameson, a Wernerian geologist who opposed Hutton's ideas. Jameson refused to catalogue the collection, and the collection slowly disappeared.

Today, the James Hutton Trail celebrates Hutton and the geological significance of the Scottish Borders, and includes an exhibition about Hutton's life, Hutton's farm, and Siccar Point. The northeast tower of the Scottish Portrait Gallery is decorated with a statue of Hutton holding a hammer in his right hand and a rock in his left.

In 2002, the James Hutton Memorial Garden opened in Edinburgh at the site of Hutton's original house. The garden includes boulders from Glen Tilt having granite veins and a plaque on which is carved Hutton's famous claim, "We find no vestige of a beginning, -no prospect of an end."

For More Information: Bailey, E.B. 1967. *James Hutton: The Founder of Modern Geology*. Amsterdam: Elsevier; Baxter, S. 2004. *Ages in Chaos: James Hutton and the True Age of the World*. New York: Forge; Repcheck, J. 2003. *The Man Who Found Time: James Hutton and the Discovery of the Earth's Antiquity*. Cambridge, MA: Perseus.

HUTTON'S SECTION

Hutton's Section is a prominent slab of sedimentary rock injected with magma that now forms the remains of glaciated rocks called Salisbury Crags that overlook Edinburgh, Scotland. The strata were pushed upward as the magma forced its way into the sediments. When Hutton saw this in the late eighteenth century, he concluded that igneous rocks originated as molten magma, thereby refuting the claim by Abraham Werner and others that all rocks had precipitated as salts from a supposed primordial sea. Although Hutton never discussed this part of Salisbury Crags in his published works, insights there were an important part of his ideas presented in *Theory of the Earth*.

Near Hutton's Section is an isolated outcrop known as "Hutton's Rock," which includes a vein of iron ore formed as a molten solution that percolated through fissures in the cooling rock. Hutton requested that this outcrop be saved from quarrying, thereby making it the earliest known example of geoconservation in the world. Hutton's Section, which Charles Darwin referred to as "that classical ground," is visited by thousands of geology pilgrims each year.

JULIAN HUXLEY (1887–1975)

Operationally, God is beginning to resemble not a ruler but the last fading smile of a cosmic Cheshire cat.

Julian Sorell Huxley was born on June 22, 1887, in London, the eldest of four children of Leonard and Julia Huxley. Leonard Huxley was the son of T.H. Huxley, "Darwin's bulldog." The youngest of Julian's brothers, Aldous, became a famous author, writing such well-known books as *Brave New World* (1932) and *Island* (1962). Julian's half-brother (his father remarried after the death of Julia) was Andrew Fielding Huxley, who won a Nobel Prize in 1963 for his studies of action potentials in nerves. Julian Huxley graduated from Oxford in 1909, and then studied at the Naples Biological Station in Italy.

In 1910, Huxley returned to Oxford as Lecturer in Biology and University Demonstrator in Zoology, and in 1912 produced his first book, *The Individual in the Animal Kingdom*. He left Oxford in 1913 to accept a faculty position and establish a biology department at the Rice Institute (now Rice University) in Houston, Texas. In 1916, Huxley joined the British Army, and served as an intelligence officer in Italy until 1919. He then returned to Oxford as a Senior Demonstrator in Zoology, and in 1921 led a research expedition to the island of Spitsbergen. Huxley married Juliette Baillot in 1919, with whom he had two sons. In 1925, Huxley became the Chair of Zoology at King's College, London.

Huxley's experimental manipulations of developing axolotls (salamanders) demonstrated hormonal inducement of artificial metamorphosis. Surprisingly, the research attracted media attention as newspapers reported that the "elixir of life" had been discovered. Huxley enjoyed the attention, presaging his role as a gifted spokesperson for science. His later research examined polarity in developing embryos, and, working with geneticist E.B. Ford, Huxley demonstrated the Mendelian inheritance of developmental rate. This work culminated in Huxley's 1932 book, *Problems of Relative Growth*.

Huxley, who had been an avid birdwatcher since childhood, also studied avian courtship displays and mating behavior as tests of Darwin's proposals on sexual selection. His *Courtship Habits of the Great Crested Grebe* (1914) was a landmark contribution to the developing field of ethology. He later wrote *Ants* (1929) and *Bird-Watching and Bird Behaviour* (1930), and helped found the Institute for Animal Behaviour in 1936, now known as the Association for the Study of Animal Behaviour. He was similarly involved in starting the Society for the Study of Evolution.

Only two years after accepting the position at King's College, Huxley resigned to concentrate on writing. In 1931, he published the popular *The Science of Life*, cowritten with the father-and-son team, H.G. and G.P. Wells. Huxley's role as a popularizer of science expanded to include public lectures and radio shows (e.g., he was a regular panelist on the BBC's popular quiz show *The Brains Trust* for several years). His film *The Private Life of the Gannets* (1934) was a milestone in nature documentaries.

Huxley contributed to the modern synthesis, the unification of Darwin and Mendel's ideas, through his *Evolution: The Modern Synthesis* (1942). This book was a thorough and up-to-date overview, and proposed that, with the successful integration of evolution, biology "no longer presents the spectacle of a number of semi-independent and largely contradictory sub-sciences but is coming to rival the unity of older sciences like physics." The centrality of evolution was, by this point in time, not particularly contentious in biology. However, Huxley's application of evolutionary thought, in three related areas, was controversial, both inside science and with the general public.

First, Huxley (who once described the concept of God as "not only unnecessary but intellectually dubious") considered evolutionary biology the logical basis for a science-based philosophy of life that would supplant traditional religion (i.e., where religious expression could be redirected from a nonexistent God and toward humanity). In 1927, Huxley outlined his "evolutionary humanism" in *Religion Without Revelation*, which was followed by a series of books, including *Towards a New Humanism* (1957), *New Bottles for New Wine* (1958), and *The Coming New Religion of Humanism* (1962). Huxley argued that "Darwinism removes the whole idea of God as the Creator of organisms from the sphere of rational discussion."

Second, Huxley concluded that evolutionary history documents progressive change, and he cited some species' autonomy and control over the environment as evidence. He tried to avoid teleology, but his admiration of the ideas of Pierre Teilhard de Chardin undercut his proposals. Huxley's ideas about evolutionary progress were not widely accepted, and were criticized for conflating complexity and progress, and for invoking a particularly anthropomorphic view of evolutionary change.

Third, the unique position humans hold on Earth was viewed by Huxley as providing the means for "consciously controlling [our] own destinies and the destinies of all life," and he was a strong supporter of eugenics. In the 1930s, he advocated universal education and medical care so that the "best" individuals could be identified. By the 1960s, Huxley proposed financial inducements or penalties (depending on the traits, good or bad, an individual had), artificial insemination, and the long-term storage of sperm from particularly fit donors to achieve these eugenic goals.

Huxley had a high-profile career, serving as secretary for the Zoological Society of London

(1935–1942), which put him in charge of the London Zoo, and as the first secretary-general of UNESCO (1946–1948). In recognition of a lifetime of accomplishments across a wide range of disciplines, Huxley received many awards, including election to the Royal Society (1938), the Darwin Medal of the Royal Society (1956), and a Gold Medal from the International Union for the Conservation of Nature (1970). He was knighted in 1958. In 1973, Julian Huxley suffered a stroke from which he never fully recovered, and died on February 14, 1975. He is buried at Compton Chapel, Surrey.

For More Information: Baker, J.R. 1978. *Julian Huxley, Scientist and World Citizen, 1887–1975: A Biographical Memoir*. Paris: UNESCO.

THOMAS HUXLEY (1825–1895)

I am Darwin's bulldog.

Thomas Henry Huxley was born on May 4, 1825, in Ealing, just outside of London, the youngest of his parents' six surviving children. As a teenager, Huxley read widely in science (e.g., he read James Hutton's *Theory of Earth* when he was 12 years old), but he also studied the writings of David Hume and Thomas Carlyle, which at a young age caused him to consider the relationship between science and religion. Although originally hoping to become an engineer, when Huxley was 14, his sister Ellen married Dr. John Cooke, who accepted Huxley as an apprentice. (Another sister also married a doctor, John Salt, who later served as a surgeon in the Confederate Army during the American Civil War.) Cooke was a hard drinker—Huxley described him as "a mass of beer and opium"—and the introduction he gave Huxley to the medical field was unpleasant (especially his first observation of an autopsy).

By 1841, Huxley was enrolled in Sydenham College, where Cooke taught. In 1842, both Huxley and his brother James earned scholarships to study medicine at Charing Cross Hospital, where Thomas published his first research paper (on a layer of cells that surround hair follicles and that is still known as "Huxley's layer"). However, the scholarship ended in 1846, and Huxley left school without earning a degree, and to support himself, he joined the Royal Navy and became assistant surgeon on the H.M.S. *Rattlesnake*. The *Rattlesnake* explored northern Australia and New Guinea, and during its four-year voyage, Huxley attended to day-to-day medical matters and assisted the ship's naturalist, and published several papers on the anatomy and classification of the organisms they collected. While stationed in Australia, Huxley also discovered that his position as a naval officer conferred invitations to parties given by the upper class of Sydney society. At one such party he met Henrietta ("Nettie") Heathorn, and he soon proposed marriage. The proposal was accepted, but the marriage was postponed, with Nettie remaining in Australia until the *Rattlesnake* had returned to England. With a wife and perhaps a family in the near future, Huxley became serious about his scientific investigations, seeing them as a route to a career after completing his military service.

Huxley returned to England in October of 1850, and with the backing of Richard Owen of the British Museum, convinced the Admiralty to provide him with a leave and partial pay to complete a book about his research. His scientific accomplishments earned Huxley a fellowship (1851) and a Gold Medal (1852) from the Royal Society, and election to the Society's Council in 1853. (Huxley later sold the Gold Medal to cover his mounting bills.) The Society, however, would not pay the cost of publishing his book because he was still in the Navy. By 1854, Huxley was still writing and still on leave from military service; he was also actively, but unsuccessfully, seeking an academic position. When the impatient Royal Navy finally ordered him back to active duty, Huxley refused, and he was promptly dismissed. With neither income nor prospects, Huxley lost faith in his dream of a scientific career; as he wrote to Nettie, "a man of science can earn great distinction, but not bread."

However, in July 1854, Huxley secured a lectureship, which eventually became a professorship, in natural history at the School of Mines, a position he would hold for thirty years. He also gained a part-time position with the Geological Survey the same year. Flush with success, he married Nettie in 1855. (Huxley used his honeymoon to collect specimens for his research, and to study the coast for the Geological Survey.) Huxley advocated the need for a scientifically literate citizenry, and in 1855 he began a series of open lectures, which not only helped disseminate information to the public, but also built Huxley's reputation as a dynamic speaker. Financial support for Huxley's book was finally available, and *The Oceanic Hydrozoa* appeared in 1859, reduced somewhat in importance due to research by others in the intervening years. Huxley's book was also overshadowed by another book published that year, Charles Darwin's *On the Origin of Species*.

Huxley established himself as a materialist, and labeled as "creationist" anyone who invoked supernatural explanations. (When Huxley joined London's Metaphysical Society in 1869, he was asked to categorize his metaphysical philosophy. He coined the term "agnostic" to describe his perspective, which he explained is "not a creed but a method" of understanding based on observation.) Huxley hoped science would replace religion as the dominant authority in society. Relative to evolution (a topic much discussed even prior to 1859), Huxley left open the possibility of "transmutation," but initially considered such change possible only within the constraints of natural groups. However, he considered such discussion "fruitless" because there was no identifiable mechanism to explain such changes, and such understanding may never be possible. (Lack of a mechanism, outside of supernatural intervention, was the basis for Huxley's denunciation of Robert Chambers' 1844 book, *Vestiges of the Natural History of Creation*.)

Publication of *On the Origin of Species* in 1859 changed everything for Huxley: "The *Origin* provided us with the working hypothesis we sought." After reading *Origin*, Huxley remarked "How stupid of me not to have thought of that!" He wrote an anonymous, 5,000-word review of Darwin's book for *The Times* of London, which Darwin predicted would cause the "old fogies to think the world will come to end."

Darwin was not disappointed. One of the most forceful responses came from a protége of Georges Cuvier, Richard Owen, superintendent of the natural history departments of the British Museum and one of the most influential scientists of the mid-1800s. In 1848, Owen (reportedly the only man the normally congenial Darwin openly hated) used his book *On the Archetype and Homologies of the Vertebrate Skeleton* to claim that anatomical similarities among organisms reflect design by a creator. Owen wrote a scathing review of *On the Origin of Species*, and even though it was published anonymously, the identity of the author was obvious. Although Huxley did not embrace all aspects of Darwin's thesis (he rejected the possibility of selection acting in a gradual fashion, advocating a role for larger-scale, "saltational" changes), he did embrace the larger message and was eager to serve as the "bulldog" for the retiring Darwin; Huxley assured him that "As for your doctrines I am prepared to go to the stake if requisite...I am sharpening up my claws and beak in readiness." Huxley implored scientists to take up Darwin's cause "if we are to maintain our position as the heirs of Bacon and the acquitters of Galileo."

Owen and Huxley had already clashed over the relationship between humans and apes. By the late 1850s, Owen had acknowledged that he could no longer "shut my eyes to the significance of that all-pervading similitude of structure" between humans and apes, but he found refuge in the one final trait, the brain, which he felt separated the two. Owen searched for significant differences in brain anatomy between humans and apes, and concluded that he had finally found one, the hippocampus minor. Huxley, thus challenged, set about demonstrating him wrong, which he eventually did.

The conflict over evolution came to a showdown at the weeklong annual meeting of the British Association for the Advancement of Science at Oxford in 1860, which attracted the major figures in British science. Darwin's recently announced theory was widely discussed, and Owen and Huxley openly clashed about its implications for human evolution. By the end of a long week of disputation, Huxley (a renowned workaholic due to financial necessity) was tiring and had decided to skip the last part of the meetings. It was only upon encountering Chambers that Huxley learned of the Anglican Bishop Samuel Wilberforce's plans to discuss evolution. Wilberforce (who had been coached by Owen) delivered a long speech denouncing Darwin's theory that Joseph Hooker described as brimming with "ugliness and emptiness and unfairness." This proved too much for Huxley; not only was science being undercut, but the Church was wielding the axe. According to legend (no transcripts exist of the exchange), Wilberforce asked Huxley if he was related to apes on his grandmother's or grandfather's side. Amid applause and laughter, Huxley purportedly whispered "the Lord hath delivered him into mine hands," and rose to declare that:

> If then, said I, the question is put to me would I rather have a miserable ape for a grandfather or a man highly endowed by nature and possessed of great means of influence and yet who employs these faculties and that influence for the mere purpose of introducing ridicule into a grave scientific discussion, I unhesitatingly affirm my preference for the ape.

Huxley, who proclaimed *On the Origin of Species* to be "the most potent instrument for the extension of the realm of knowledge which has come into man's hands since Newton's *Principa*," then proceeded to defend Darwin. Although not an actual debate, the exchange between Wilberforce and Huxley has been hailed as a turning point in the acceptance of evolution by the scientific community, and certainly challenged the authority of the Church. By most accounts, the meeting's most eloquent and detailed statement in support of evolution was, however, provided by Joseph Hooker, but his words were overshadowed by Huxley's sensational exchange with Wilberforce.

Soon after the meetings at Oxford, Huxley's 4-year old son, Noel, died of scarlet fever. This was a devastating loss, and caused Huxley to completely abandon religion. After traveling to Switzerland to contend with the loss, Huxley wrote the influential *Evidence as to Man's Place in Nature* (1863), which was the first book to explicitly claim that humans have an apelike ancestor. Huxley followed the book with numerous public lectures on the topic, and broadened his discussion to include the origin of life from inanimate matter.

In 1870, Huxley was elected to the London School Board, from which he advocated improved science curricula in schools. Having once said that "If I am to be remembered at all, I would rather it should be as a man who did his best to help the people than by any other title," Huxley stressed the importance of a laboratory-based curriculum for learning science. (To achieve his larger goals, he also compromised by endorsing reading of the Bible in school.) Huxley wrote science textbooks, maintained his teaching and research program (now at the recently reorganized Royal College of Science), and continued to present public lectures. Presumably due to overwork, Huxley suffered a breakdown in 1871, and was forced to resign from the School Board, but the changes he had initiated had already been set in place.

Huxley, who did more than anyone to advance Charles Darwin's ideas, toured the United States in the late 1870s, lecturing and interacting with American scientists. In 1881, he became dean of his school, and in 1883 he was elected president of the Royal Society. Decades of overwork and personal loss (his daughter Marian died of pneumonia in 1887) hardened Huxley's perspective later in life. He began to despair that the world is "doomed to forever be at war," and the rise of the industrial revolution, which created struggle among nations, encouraged him to press harder for effective science instruction. He published articles on his agnosticism, including one in response to Prime Minister William Gladstone's support of the revealed and inerrant truth of the Bible. Huxley used a specific story from the Bible—Jesus driving demons out of a possessed man and into a herd of pigs—to demonstrate what he thought to be the absurdity of the Prime Minister's proposition. Exchanges in the press between Huxley and Gladstone—who had visited Darwin at Down House—continued for months.

In 1890, Thomas and Nettie left London to live in the countryside of Eastbourne. In March of 1895, Huxley was stricken with influenza. He weakened considerably over the next few weeks, and died at his home on the afternoon of June 29, 1895. He was buried—without a eulogy—on July 4 in St. Marylebone Cemetery in East Finchley, next to his son, Noel. Huxley's burial, which was attended by Joseph Hooker, Edwin Lankester, Lord Kelvin, Charles Darwin's sons, and other science dignitaries, was described as "the last and greatest gathering of Victorian scientists to gather on one spot." Huxley's headstone bears an inscription written by his wife:

> Be not afraid, ye waiting hearts that weep;
> For still He giveth His beloved sleep,
> And if an endless sleep He wills, so best.

These lines were later used by Huxley's student, H.G. Wells, in *The New Machiavelli* (1911).

In 1925—the centenary of Huxley's birth—famed journalist H.L. Mencken eulogized Huxley as "the greatest Englishman of the 19th century—perhaps the greatest Englishman of all time." Huxley's grandsons, Julian Huxley and Andrew Huxley, became famous biologists (Andrew won a Nobel Prize in 1963), and their brother Aldous became a famous writer (e.g., *Brave New World*). Today, a life-sized statue of Huxley (Figure 46) adorns the British Museum

46. Thomas Huxley—known as "Darwin's Bulldog" for his ferocious defense of evolution by natural selection—is memorialized by this statue in the British Museum of Natural History. Nearby is a similar statue of Charles Darwin (Figure 24). (*Randy Moore*)

of Natural History, near a similar one honoring Charles Darwin.

For More Information: Desmond, A.J. 1997. *Huxley: Evolution's High Priest.* New York: Viking Penguin.

I

INHERIT THE WIND

He that troubleth his own house shall inherit the wind—Proverbs 11:29

Inherit the Wind (Figure 47) is a play written in 1950 by Jerome Lawrence and Robert E. Lee that used a fictionalized account of the Scopes Trial to expose the threat to intellectual freedom associated with the anti-Communist hysteria of the McCarthy era. *Inherit the Wind* opened on January 10, 1955, to favorable reviews in Dallas, Texas. Three months later it opened at Broadway's National Theatre, where its three-year run included 805 performances. The play described "the famous 'Monkey Trial' that rocked America" that was "the most explosive trial of the century." Although *Inherit the Wind* includes some historically accurate moments (e.g., Darrow's condemnations of anti-intellectualism), the play's many historical inaccuracies are noted prominently in the playwrights' note:

> *Inherit the Wind* is not history....Only a handful of phrases have been taken from the actual transcript of the famous Scopes Trial....*Inherit the Wind* does not pretend to be journalism. It is theater. It is not 1925.

Although the closing rant by the character modeled after William Jennings Bryan had no counterpart in the Scopes Trial, it resembled Senator Joe McCarthy's behavior on June 17, 1954, when McCarthy's House un-American Activities Committee came to an abrupt end.

The original Broadway play starred Paul Muni as the character representing Clarence Darrow, Ed Begley as the character representing William Jennings Bryan, and Tony Randall as the character representing caustic journalist H.L. Mencken. When its Broadway run ended in 1957, *Inherit the Wind* was the most successful and longest-running drama in Broadway's history. The success of *Inherit the Wind* prompted the *Encyclopedia Britannica* to first include the Scopes Trial in 1957.

The movie version of *Inherit the Wind* had its world premiere in early July 1960 at the Berlin Film Festival, and its American premier at the Dayton Drive-In Theatre on July 21, 1960, at a celebration commemorating the thirty-five-year anniversary of the Scopes Trial (Figure 48). The 127-minute movie, which was directed by Stanley Kramer, starred two Oscar-winning Hollywood legends, Spencer Tracy (playing Henry Drummond, the character representing Darrow) and Fredric March (playing Matthew Harrison Brady, the character representing

47. *Inherit the Wind* is a play and movie that uses a fictitious retelling of the Scopes Trial to examine the dangers of zealotry. The movie version of *Inherit the Wind*, which appeared in 1960, remains the most influential film associated with the evolution-creationism controversy. (*Randy Moore*)

IN HONOR OF
SCOPES TRIAL DAY
YOU ARE CORDIALLY INVITED TO THE
FIRST PUBLIC U. S. PRESENTATION
OF
STANLEY KRAMER'S
"INHERIT THE WIND"
STARRING
SPENCER TRACY
FREDRIC MARCH
GENE KELLY
WITH
DICK YORK — DONNA ANDERSON
AND FLORENCE ELDRIDGE

PRODUCED AND DIRECTED BY STANLEY KRAMER

A UNITED ARTISTS RELEASE

————————————

DAYTON DRIVE-IN THEATRE
THURSDAY, JULY 21, 1960 — 8:30 P.M.

ADMISSION BY
RESERVATION ONLY
R.S.V.P.

J. J. RODGERS, M.D.
MAYOR
DAYTON, TENNESSEE

48. *Inherit the Wind* had its American premier at the Dayton Drive-In Theater on July 21, 1960. This invitation was sent to Scopes' prosecutor and friend, Sue Hicks. (*Randy Moore*)

Bryan). *Inherit the Wind* was nominated for four Academy Awards—Best Actor in a Leading Role (Tracy), Best Adapted Screenplay, Best Black-and-White Cinematography (Ernest Laszio), and Best Film Editing (Frederic Knudtson)—but won no Oscars. *Inherit the Wind* also featured Gene Kelly in a rare dramatic role as E.K. Hornbeck (the character representing Mencken). The movie's screenplay was produced by Nedrick Young and Harold Smith. Young had been blacklisted by McCarthy and wrote under the name Nathan Douglas; Young's credits weren't restored until 1997. Late in 1960, *Inherit the Wind* became the world's first in-flight movie when TransWorld Airlines used it to lure first-class passengers. The film's many highlights include Tracy's eleven-minute summation, which was filmed in a single take to heighten the tension of the movie's climax.

Scopes, who promoted the movie at the studio's request, acknowledged that the movie was not historically accurate, but that it "captured the emotions in the battle of words between Bryan and Darrow." When the movie version of *Inherit the Wind* premiered, Scopes—at the urging of his wife—accepted Kramer's invitation to attend the film's North American premiere at the Dayton Drive-In Theatre. Scopes was the only surviving principal of the trial who attended; most of the trial's principals (e.g., Darrow, Malone, Hays, Neal, Robinson) were dead.

Since 1960, *Inherit the Wind* was been remade several times for television. The 1965 remake, which removed much of the humor and several important passages, starred Melvyn Douglas as Drummond, and Ed Begley as Brady. The 1988 version starred Jason Robards as Drummond, Kirk Douglas as Brady, and Darren McGavin as Hornbeck; and the 1999 version starred Jack Lemmon as Drummond, George C. Scott as Brady, and Beau Bridges as Hornbeck. For their performances, Lemmon won a Golden Globe

49. The Institute for Creation Research (ICR) is one of the most influential and well-funded antievolution organizations. ICR operates the Museum of Creation and Earth History in Santee, California. (ICR)

Award, and Robards won an Emmy. Although the 1988 remake won an Emmy for Outstanding Drama-Comedy Special, none of the remakes achieved the popularity or impact of the original movie. In 2007, *Inherit the Wind* again opened on Broadway.

For More Information: Lawrence, J., & Lee, R.E. 1955. *Inherit the Wind.* New York: Bantam.

INSTITUTE FOR CREATION RESEARCH (Est. 1970)

There seems to be no possible way to avoid the conclusion that, if the Bible and Christianity are true at all, the geological ages must be rejected altogether.

The Institute for Creation Research (ICR) was founded by Henry Morris and Tim LaHaye to promote "creation science, biblical creationism, and related fields." ICR was originally the research division of LaHaye's Christian Heritage College, but became autonomous in 1981 (Figure 49).

In 1974, ICR published two versions of the textbook, *Scientific Creationism*: one for public schools, containing no references to the Bible, and another for Christian schools that included a chapter titled "Creation According to Scriptures." In the late 1970s, ICR launched a massive publicity campaign that included advertisements in mainstream publications such as *TV Guide*.

In 1979, geologist Steve Austin (b. 1948) joined ICR full-time; for many years, Austin was the only employee who did nonliterary research. Within a decade, ICR published more than fifty books, as well as its newsletter *Acts & Facts*. In 1981, Morris estimated that more than 500,000 ICR books were in circulation.

ICR's first graduate courses were taught in 1981, and all of ICR's graduate programs (e.g., astronomy, biology, geology, and science education) were approved by California that same year. In 1990, Bill Honig (b. 1937)—the California Superintendent of Public Instruction—revoked ICR's license to award degrees, but two years later it was reinstated by a federal judge. ICR's faculty included Duane Gish, a famed debater for creationism. In 2007, ICR concluded its eight-year Radioisotopes and the Age of the Earth (RATE) study that concluded that accelerated radioactive decay "such as during the Genesis flood, the Fall of Adam, or early Creation week" explains the apparent conflict between the Bible-based conclusion that Earth is young and the science-based conclusion that Earth is ancient. In early 2007, ICR announced that "details regarding the Flood are beginning to unfold" and that "scientific discoveries continue to erode Darwinism." ICR tells its members that "there is now abundant evidence that man and dinosaurs lived at the same time."

ICR claims that "all things in the universe were created and made by God in the six literal days of the creation week described in Genesis 1:1–2:3"; that the Bible is "factual, historical, and perspicuous"; and that "all theories of origins or development which involve evolution in any way are false." ICR also claims that evolutionary thinking produces abortion, promiscuity, drug abuse, and homosexuality. ICR sells books, "Bibleland Cruises" with ICR's Museum of Creation and Earth History in Santee, California, includes...creationists, and videos with titles such as *The Dangers of Evolution* and *Dinosaurs in the Bible*. The president of ICR is John Morris (b. 1946), the son of Henry Morris (who stepped down as ICR's president in 1996). One of ICR's radio programs (*Science, Scripture, and Salvation*) is aired on more than 600 stations. ICR has a variety of outreach programs and offers "scientific and legal consultation, service as expert witness, etc." ICR promotes research, but requires students and faculty to promise that publications will reject evolution and agree with a literal interpretation of the Bible. ICR pledges to "tear down the strongholds of evolutionary thinking."

ICR is based in Dallas, Texas, where in 2006 it established the Henry M. Morris Center for Christian Leadership. ICR's Museum of Creation and Earth History in Santee, California, includes a Noah's Ark diorama that documents how animals were packed aboard the Ark. The museum, which is visited by more than 25,000 people per year, explains the age of the Earth, the meaning of life, what happens when we die, and why there is suffering in the world. In 2007, ICR moved its administrative offices to Dallas, Texas.

J

FORREST HOOD "FOB" JAMES (b. 1934)

Forrest Hood "Fob" James, Jr., was a former professional football player who succeeded George Wallace as governor of Alabama from 1979 to 1983 (as a Democrat) and from 1995 to 1999 (as a Republican). During a 1995 appearance before the Alabama Board of Education, James—supported by antievolution crusaders such as Jerry Falwell and Pat Robertson—ridiculed evolution by slowly crossing the stage, beginning in a crouch and then ending erect. The following year, James used discretionary funds to send all of Alabama's biology teachers a copy of Phillip Johnson's *Darwin on Trial*. James vowed to use state troopers and the National Guard to keep the Ten Commandments in an Alabama courtroom. James, a flamboyant fundamentalist who advocated states rights, helped pass a law allowing teachers to lead students in prayer. The law was declared unconstitutional in May 1983.

THOMAS JEFFERSON (1743–1826)

Thomas Jefferson, the third president of the United States and founder of the University of Virginia, published one of the first scientific descriptions of a fossil mastodon skeleton found in the United States. Jefferson's fossils inspired Georges Cuvier to begin studying elephants and other large animals. In 1796, Jefferson hired William Clark (later of Lewis and Clark fame) to search for remains of extinct mastodons in Kentucky. President Jefferson encouraged the excavation of fossils, and shipped a giant intact skeleton of a mastodon to Europe to counteract claims that "all animals are smaller in North America than Europe." When Congress authorized the Lewis and Clark expedition of 1803, Jefferson helped pay for the expedition, and he asked the explorers to send bones of large animals back to Washington. At one point, the East Room of the White House was filled with bones of ancient animals, and Jefferson's belief that extinct creatures were alive elsewhere resurfaced for centuries (e.g., the Loch Ness plesiosaur, Conan Doyle's classic *Lost World*). While enroute to his inauguration as Vice President in 1797, Jefferson delivered a wagonload of fossils to the American Philosophical Society.

In an 1802 letter to the Baptist Association of Danbury, Connecticut, Jefferson—who named the giant ground sloth (*Megalonyx*; it was later named *M. jeffersoni* in his honor)—praised the Establishment Clause of the U.S. Constitution for "building a wall of separation between church and state." Jefferson, for whom rational inquiry was an inalienable right, is buried at Monticello in Charlottesville, Virginia.

LEONARD JENYNS (1800–1893)

Leonard Jenyns was a pastor and the brother-in-law of John Henslow, who was offered the job as naturalist aboard the *Beagle*, but turned the offer down to work at a church just outside Cambridge. While attending Cambridge University, Darwin became friends with Jenyns, who—like Darwin—collected insects. Darwin gave the 137 species of fish he collected while aboard the *Beagle* to Jenyns. In 1871, Jenyns changed his name to Leonard Blomefield to obtain an inheritance.

WILHELM JOHANNSEN (1857–1927)

The word gene is completely free of any hypothesis; it expresses only the evident fact that, in any case, many characteristics of the organism are specified in the germ cells by means of special conditions, foundations, and determiners which are present in unique, separate, and thereby independent ways—in short, precisely what we wish to call genes.

Wilhelm Johannsen was born in Copenhagen, Denmark, in 1857. His family moved to the island of Zealand in eastern Denmark when Wilhelm's father, an army officer, was transferred. Johannsen, unable to attend college due to his parents' limited finances, was apprenticed at the age of 16 to a pharmacist. By 1879, Johannsen completed his pharmacist training, but also received instruction in botany and chemistry. He then worked with a pharmacist who indulged Johannsen's interest in science by outfitting him with a small laboratory.

In 1881, Johannsen became an assistant in the Chemical Department at the new Carlsberg Laboratorium in Copenhagen. He studied morphological variation in wheat and barley and the physiological determinants of dormancy in plants. Johannsen became a lecturer in botany at the Royal Veterinary and Agricultural College in 1892, where he began investigating variation in the progeny of self-fertilized plants.

Johannsen recognized that self-fertilizing plants were an ideal system for examining the role of variability in evolution, and he created experimental lines of beans descended from self-fertilized individuals. He was impressed with Francis Galton's statistical approach to studying variation, and was intrigued by Galton's demonstration of partial regression to the mean during selection experiments: when breeders were selected from one extreme of the morphological distribution, the average value of the trait in the next generation fell between the average of the previous generation and the average of the parents. Johannsen expected this outcome in his selection experiments with beans. However, he found that the morphological distributions (e.g., for seed size or weight) of successive generations remained unchanged. The only way Johannsen could replicate Galton's results was to mix descendants from more than one self-fertilizing individual prior to selecting breeders.

To Johannsen, these results indicated that individuals descended from a single ancestor are genetically identical, and in 1903 he called these lineages *pure lines*. Johannsen concluded that morphological variation within a pure line must be due to the effects of the environment. This variability (referred to as "fluctuating variability" by Hugo de Vries) is not heritable, which allowed dismissal of Lamarckian evolution. In 1909, Johannsen's *Elements of Heredity* integrated these findings with Mendel's results. In *Elements*, the terms *genotype* and *phenotype* are introduced, and Johannsen coined the term *gene* to describe the unit of heredity that influences a phenotypic character (although Johannsen did not consider genes as material entities). Johannsen's work contributed to the elucidation of the nature of heredity and set the stage for the integration of genetics with evolution during the first third of the twentieth century.

In 1905, Johannsen was appointed professor of plant physiology at the University of Copenhagen and was promoted to rector of the university in 1917. Johannsen—who never earned a university degree—maintained that his "scientific unsophistication" allowed him to consider topics from unique perspectives. After his work with pure lines was completed, he retired from experimental science and spent his time writing. Johannsen died in 1927.

50. Phillip Johnson is a former law professor whose best-selling *Darwin on Trial* condemned evolution and helped found the modern "intelligent design" movement. (*InterVarsity Press*)

PHILLIP JOHNSON (b. 1940)

This isn't really, and never has been a debate about science. It's about religion and philosophy.

Phillip E. Johnson was born in 1940 in Aurora, Illinois (Figure 50). His family was only nominally religious, and in college he was agnostic. Johnson entered Harvard University when he was 16 years old, and graduated with a degree in English in 1961. After graduation, Johnson taught English in East Africa.

When he returned home, Johnson found that his father had submitted an application for him to the University of Chicago Law School. He was accepted, and graduated in 1965 at the head of his class. He clerked for Chief Justice Robert Traynor of the California Supreme Court, and then for Chief Justice Earl Warren of the United States Supreme Court. Johnson accepted a faculty position at the Boalt School of Law at the University of California at Berkeley in 1967, and remained at Berkeley for the rest of his career (although he also served as Deputy District Attorney for Ventura County while on leave from teaching). He retired in 2000, and currently is Jefferson E. Peyser Professor of Law, Emeritus at Berkeley. In the mid-1980s, Johnson was considered to fill a vacancy on the Ninth Circuit Court of Appeals.

Although he had achieved professional success, in the 1970s Johnson perceived a lack of meaning in his life. His marriage dissolved around this time and he was unfulfilled by what he labeled "superficial" interactions he had with colleagues. While visiting his 11-year-old daughter at a bible camp, Johnson became an evangelical Christian. It was through his new church that he met his second wife.

While on sabbatical in 1987 at University College, London, Johnson read Richard Dawkins' *The Blind Watchmaker*. Johnson admired Dawkins' rhetorical abilities, but found Dawkins' argument for the power of unguided evolution unconvincing. As Johnson saw it, evolutionary biologists did not consider all the evidence (i.e., especially that for design), and instead were confusing assumptions and conclusions. Johnson also claimed that scientists accept evolution only because they have a dogmatic, *a priori* commitment to naturalism. According to Johnson, science has therefore not shown that God does not exist, it merely assumes there is no God because to do otherwise is not "science"; accepting evolution then necessarily means adhering to an atheistic worldview, because one cannot simultaneously accept evolution and believe in God.

Johnson began writing a manuscript that considered evolution from a lawyer's perspective: What does the evidence indicate as to the likelihood of this concept explaining life as we know it? *Darwin on Trial* (1991) concluded that evolution fails to explain the observed patterns we see in nature. Not surprisingly, the book was rejected by scientists. The review by Harvard paleontologist Stephen Jay Gould in *Scientific American* was typical: "full of errors, badly argued, based on false criteria, and abysmally written." (Johnson submitted a rebuttal to Gould's review, but *Scientific American* declined to publish it.) Scientists noted that acceptance of evolution does not endorse atheism (e.g., Gould's well-known "non-overlapping magesteria") and claimed that Johnson was confusing the methodological materialism required of science with the philosophical materialism that some scientists (including Dawkins) propose flows from scientific understanding.

Johnson attacked naturalism as science's "creation myth" in his next book, *Reason in the Balance* (1995), and introduced the mutually exclusive options of "methodological naturalism" (i.e., contemporary science) and "theistic realism." Johnson's theistic realism assumed that "the universe and all its creatures were brought into existence for a purpose by God." This perspective is not scientific, nor does it accommodate belief systems that reconcile religion and evolution, such as theistic naturalism, which claims that God has created and works through natural forces, including evolution.

Johnson also began developing a strategy to undermine evolution. In 1992, members of the nascent intelligent design (ID) movement met at Southern Methodist University (SMU). The meeting included Johnson, Michael Behe, Stephen Meyer, and William Dembski, all of whom became leading proponents of ID. The meeting represented a pivotal moment in the development of ID because it was there that the "wedge strategy" was first proposed.

The wedge strategy ("a log is a seeming solid object, but a wedge can eventually split it by penetrating a crack and gradually widening the split. In this case the ideology of scientific materialism is the apparently solid log"), honed over the next few years at meetings similar to the one at SMU (but kept secret from public scrutiny), was well developed by the mid-1990s. A home for the strategy was created in 1996 when the Discovery Institute, a Seattle-based Christian think-tank founded by Bruce Chapman (former Secretary of State for the state of Washington), announced the establishment of the Center for the Renewal of Science and Culture, later renamed the Center for Science and Culture (CSC). Johnson served as Program Advisor, and Stephen Meyer as Director, with most other notable ID proponents as Fellows of the Center. Johnson was the leading edge of the wedge, or in his words, "I'm like an offensive lineman in pro football... the idea is to open up a hole that a running back can go through." Although Johnson has at times emphasized keeping the discussion purposely vague about who or what the "designer" is (even claiming that it could be an alien that brought life to Earth), at other times he has claimed that he wants to "get the issue of intelligent design, which really means the reality of God, before the academic world and into the schools."

The wedge strategy was encapsulated in the "wedge document" (whose existence the CSC initially denied), which outlined five-year and twenty-year goals for the "overthrow of materialism and its cultural legacies." The plan included three phases: scientific research, writing, and publication; publicity and opinion-making; and cultural confrontation and renewal. Despite claims to the contrary by the CSC, no peer-reviewed research relative to ID has appeared in the scientific literature. However, an article by CSC director Stephen Meyer was published in the *Proceedings of the Biological Society of Washington* in 2004. This article reviewed existing information—no new scientific data were presented—and the managing editor alone decided to publish it (i.e., there was no peer review). The publishers of the journal issued a statement stating that, given the opportunity, they "would have deemed the paper inappropriate" for publication.

Johnson and colleagues have implemented the other two phases of the wedge strategy, publicity and public confrontation. Johnson has written seven books discussing evolution and, until recently (he suffered a series of strokes in 2002 and 2005), traveled frequently to speak about ID and the struggle against scientific naturalism. In 1999, in the wake of the Kansas State Board of Education's decision to abandon the teaching of evolution, Johnson helped craft a modified tactic—"teach the controversy." This strategy shifted the focus from teaching ID to portraying evolution as a controversial theory in trouble; any discussion of evolution therefore should include its (allegedly) serious problems and an examination of alternatives. "Teach the controversy" is often included in legislation to promote ID in public school science curricula.

Johnson also authored the "Santorum Amendment" that former U.S. Senator Rick Santorum of Pennsylvania tried to attach to the 2001 Elementary and Secondary Education Act Authorization Bill (later renamed the No Child Left Behind Act). The amendment invoked standard "teach the controversy" ideals about science education, and specified evolution as a topic needing such discussion. Although the Discovery Institute supported the amendment, it was not included in the final version of the bill enacted into law.

For More Information: Johnson, P.E. 1997. *Defeating Darwinism by Opening Minds*. Downers Grove, IL: InterVarsity.

JOHN JOLY (1857–1933)

In 1899 Irish geologist John Joly speculated that when the oceans originally formed, they contained freshwater, and that the ocean's present salinity was due to salts leaching from rocks. Joly assumed that the rate of leaching had been constant, and therefore the age of Earth's first ocean equaled the mass of sodium in the oceans divided by the annual rate of sodium input into the oceans; this calculation—using a previously ignored method originally proposed by English astronomer Edmond Halley (1656–1742)—produced an age of 89 million years (Halley had believed that his method would confirm "the evidence of the Sacred Writ [that] Mankind has dwelt about 6,000 years"). Joly's calculation was embraced by some scientists, but rejected by others, the most notable of which was Arthur Holmes. Joly later argued that radioactive decay was an unreliable clock because decay rates were not constant throughout geological history. (Today, many creationists make similar claims.) In 1914, Joly used sediment accumulation data to

argue that Earth is 47–188 million years old. Joly, with Ernest Rutherford, used "pleochronic halos" (the commonly observed dark rings in micas) to conclude that the Devonian was not less than 400 million years, and that radiation (and radiogenic heat) is ubiquitous in rocks. Later in his life, Joly's Radium Institute began the use of radioactivity in medicine.

ROBERT 'BOB" JONES (1883–1968)

Robert Reynolds "Bob" Jones, Sr., was born on October 30, 1883, in Dale County, Alabama. Jones graduated from Mallalieu Seminary in 1900, and the following year he entered Southern College (later Birmingham-Southern College) in Greensboro, Alabama. Jones supported his studies by preaching, but was soon such a successful preacher that he quit college. Jones' first wife (Bernice Sheffield) died within ten months of their 1905 marriage, and in 1908 Jones married Mary Stollenwerck. The couple had one child, Bob Jones, Jr.

When he was 13, Jones organized his first congregation of fifty-four members, and by the time he was 16, Jones was attracting enormous crowds to his worship services. Jones' fame continued to grow, and in the 1920s he was eclipsed only by Billy Sunday and Aimee Semple McPherson. By the time Jones was 40, he had preached to more than 15 million people. Jones' sermons, like those of Sunday, were often theatrical.

In 1924, William Jennings Bryan told Jones that "if schools and colleges do not quit teaching evolution as a fact, we are going to become a nation of atheists." After months of preaching about the evils of evolution and other aspects of modernism, Jones—an avowed segregationist—founded Bob Jones University the year after the Scopes trial. The university, which was originally located in Florida, banned the admission of blacks until 1970 (because of its widespread discrimination against blacks, Bob Jones University lost its tax-exempt status in 1983). Jones claimed that the Civil Rights Movement was "of the devil" and that modernists were eradicating racial boundaries that God had set; the college even awarded an honorary doctorate degree to self-proclaimed racist and Alabama governor George Wallace. Until a public-relations problem prompted the university to change its policy in 2000, the university banned interracial dating, which was viewed as "playing into the hand of the Antichrist" by defying God's will regarding the God-made differences among the races. Like fundamentalists such as Sunday and Frank Norris, Jones also condemned Catholicism, claiming it "is a satanic counterfeit" and that popes "are demon possessed." Jones' son and grandsons, who succeeded him as President of Bob Jones University, have described Catholics and Mormons as "cults which call themselves Christian."

In the 1920s, Jones became one of the first preachers to broadcast his sermons on radio. His sermons in 1925—many of which denounced the corrupting influence of evolution—were among the first to originate from an evangelistic crusade. During his lifetime, Jones delivered approximately 10,000 radio messages, many of them condemning evolution and other societal ills such as dancing and drinking. He urged listeners to believe that "whatever the Bible says is so."

Like Sunday, Norris, and others, Jones supported—and accepted money from—the Ku Klux Klan. Jones claimed that God supported racial segregation, and that opponents of segregation were opposing God. Jones also supported the Klan's endorsement of religious fundamentalism, prohibition, and opposition to the teaching of evolution.

Bob Jones University remains a citadel of creationism. Instructors at Bob Jones University produce procreationism textbooks and other materials that denounce evolution because it produces immorality. The biology department supports Young-Earth Creationism by training "Christian biologists who see the living world indelibly marked with the fingerprints of a God of limitless wisdom and power." Bob Jones University Press is the largest textbook supplier to home-school families in America.

Today, Bob Jones University is in Greenville, South Carolina. The university is the world's largest fundamentalist Christian school; it enrolls more than 5,000 students representing every state and more than forty foreign countries. Students who have attended Bob Jones University include Tim LaHaye and Billy Graham, who received an honorary degree in 1948. However, Jones later attacked Graham, prompting Graham to distance himself from the university. The Jones-Graham feud was often nasty (for example, Bob Jones University students were forbidden from attending Graham's 1966 revival in Greenville). The university has also awarded honorary degrees to Henry Morris and Billy Sunday.

Jones died of heart failure on January 16, 1968. He is buried on campus at Bob Jones University.

JOHN E. JONES III (b. 1955)

ID cannot uncouple itself from its creationist, and thus religious, antecedents.

John E. Jones III was born in Pottsville, Pennsylvania, in 1955. After law school, he joined the firm Dolbin & Cori as a trial attorney, and in 1986 he formed his own law firm, John Jones & Associates. From 1994 to 1996 he also served as solicitor for the City of Pottsville. Jones has been married to his wife, Beth Ann, since 1982, and they have two children. He is also part-owner of a five-course golf business, Distinct Golf.

In 1994, Jones, a Republican, ran for the United States House of Representatives seat in the Sixth Congressional District in Pennsylvania, but narrowly lost to Democratic incumbent Tim Holden. That same year, Jones cochaired Governor-elect Tom Ridge's transition team. The following year, Ridge named Jones to the Pennsylvania Liquor Control Board, the largest state liquor monopoly in the United States. In 1997, Jones attempted but failed to privatize some of the state-run liquor stores, and he later banned the sale of Bad Frog Beer because the labels depicted a frog making an obscene gesture. Jones considered running for Governor in 2001.

Jones was appointed by George W. Bush to the United States District Court for the Middle District of Pennsylvania in February 2002; unanimous Senate confirmation followed in July 2002. Jones developed a reputation for fairness, humor, and intellect. He presided over several free-speech cases, including barring Shippensburg University from enforcing a speech code that would have prevented students from distributing anti-Osama bin Laden posters.

In December 2004, Jones heard about a federal lawsuit filed in Dover, PA, regarding the teaching of intelligent design (ID), a topic about which he knew little. The next morning he discovered that he had been assigned the *Kitzmiller v. Dover Area School District* trial. This lawsuit was brought by eleven parents of students in the Dover school system when the school board required science teachers to read to students a statement proposing ID as an alternative to biological evolution. The trial began on September 26, 2005, and included extended testimony from well-known individuals on both sides of the ID debate. On November 4, 2005, the trial ended, and on December 20, 2005, Jones issued his decision.

Jones' 139-page decision was unequivocal: "We find that the secular purposes claimed by the Board amount to a pretext for the Board's real purpose, which was to promote religion in the public school classroom, in violation of the Establishment Clause." Jones further proposed that application of both the *Lemon* and the endorsement tests allowed direct analysis of "the seminal question of whether ID is science. We have concluded that it is not." In summary, Jones determined that "the citizens of the Dover area were poorly served by the members of the Board who voted for the ID Policy. It is ironic that several of these individuals, who so staunchly and proudly touted their religious convictions in public, would time and again lie to cover their tracks and disguise the real purpose behind the ID Policy."

Anticipating his critics, Jones also noted that "those who disagree with our holding will likely mark it as the product of an activist judge. If so, they will have erred as this is manifestly not an activist Court. Rather, this case came to us as the result of the activism of an ill-informed faction on a school board, aided by a national public interest law firm (Thomas More Law Center) eager to find a constitutional test case on ID, who in combination drove the Board to adopt an imprudent and ultimately unconstitutional policy. The breathtaking inanity of the Board's decision is evident when considered against the factual backdrop, which has now been fully revealed through this trial." Despite this, Jones was accused of judicial activism, and received death threats after announcing his decision.

Jones was listed in 2006 by *Time* magazine as one of the "Time 100: The People Who Shape Our World." *Wired* magazine also granted him the dubious distinction of being one of "2005's 10 Sexiest Geeks" as well as giving him a Rave Award in 2006 for innovation and leadership.

DAVID STARR JORDAN (1851–1931)

When organizations in the name of religion strive to resist the progress of knowledge and to punish or ostracize men and women who think for themselves and by the truth are made free, their influence is evil.

David Starr Jordan was born on January 19, 1851, in Gainesville, New York. In 1869, he enrolled in the newly established Cornell University, where in 1872 he received an MS (he never received a bachelor's degree). After teaching science in Indiana, he received an MD from Indiana Medical College. In 1875, Jordan became a professor of natural history at Butler University (from which he would receive a PhD in 1878), after which he taught at—and in 1885 became president of—Indiana University. There, he wrote *Darwinism* (1888) and *Footnotes to Evolution* (1898), and taught what he claimed was the first evolution course offered anywhere in the world. In 1891, Jordan became the first president of Leland Stanford Junior University (which later came to be known as Stanford University), and while in that position coauthored with Vernon Kellogg *Evolution and Animal Life*. Jordan was influenced by famed biologist Louis Agassiz, but was never converted to Agassiz's antievolutionary views. Throughout his life, Jordan was distant from religion, often claiming that "much of what we have called religion is merely the debris of our grandfather's science."

After serving as president of the American Association for the Advancement of Science from 1909 to 1910, Jordan—a founding member of the Sierra Club—served as president of the National Education Association in 1915, and from 1906 to 1916 was on the first board of trustees of the Carnegie Foundation for the Improvement of Teaching. Jordan, who directed the World Peace Foundation from 1910 to 1914, often approached war and peace from a biological perspective, noting that war was detrimental to the health of species because it removed the strongest individuals from the gene pool. In 1925, Jordan chaired the Tennessee Evolution Case Defense Fund Committee for the ACLU. Jordan gathered more than $11,000 for the fund and, when the Scopes Trial ended, he returned $561.29 to contributors and transferred the remaining $1,868 to the ACLU to support other fights for academic freedom. The Fund also contributed to John Scopes' graduate education at the University of Chicago. Jordan, who dismissed Protestant revivalism as "simply a form

of drunkenness no more worthy of respect than the drunkenness that lies in the gutter," then retired from public life.

Jordan died on September 19, 1931, in his home in Palo Alto, California. The *David Starr Jordan*, a National Oceanic and Atmospheric Administration (NOAA) research ship, was commissioned in his honor in 1966. Cornell University, Indiana University, and Stanford University jointly sponsor the David Starr Jordan Prize, an international award given every three years to honor a young scientist's innovations in evolution, ecology, and population biology.

CHARLES JUDD (1873–1946)

Charles Hubbard Judd was a defense witness at the Scopes Trial whose affidavit stressed that banning evolution from the curricula of public schools would damage public education. Hubbard, head of the Department of Education at the University of Chicago and a past-president of the American Psychological Association, urged that evolution be taught as the "fundamental basis for the understanding of all human institutions." Judd's testimony was read into the court record by Arthur Hays.

K

PAUL KAMMERER (1880–1926)

Paul Kammerer was born on August 17, 1880, in Vienna, Austria. He studied music at the Vienna Academy, but eventually graduated with a degree in biology. Kammerer studied Lamarckian inheritance, and claimed that many genetic traits had been suppressed by animals' lifestyles. Kammerer studied this by investigating how the breeding of amphibians in unusual habitats affected their offspring. Kammerer claimed many successes, and biologists from throughout Europe visited his lab. Kammerer delivered popular lectures about how humans might become superhumans.

Kammerer's most famous claim involved midwife toads (*Alytes obstetricians*), a group so-named because males carry fertilized eggs on their backs and hind legs. Most toads mate in water and have black, scaly nuptial pads on their hindlimbs that help them cling to each other while they mate in their slippery environment. However, midwife toads mate on land and lack these pads. When Kammerer forced midwife toads to mate in water, he reported in 1919 that they laid fewer eggs and developed the black pads. Kammerer's apparent confirmation of Lamarckian inheritance made front-page news throughout the world. Because Kammerer's claims supported socialist ideals and were consistent with Trofim Lysenko's Lamarckian version of genetics, Kammerer accepted a job in Moscow.

Kammerer was hailed as a successor to Charles Darwin, and Kammerer's work was described as "the greatest biological discovery of the century." One newspaper noted, "Professor Kammerer's work on the inheritance of acquired characteristics has startled the world.... His results are in the forefront of discussion today in biological circles. We all want to believe his facts if they are true. It means so much to the educator, to society in general, if they are true."

Kammerer's facts weren't true; other biologists could not replicate his work. In the August 7, 1926, issue of *Nature*, G. Kingsley Noble—a biologist at the American Museum of Natural History who had examined Kammerer's frogs—claimed that Kammerer's data were fake. William Bateson agreed, claiming that the alleged nuptial pad "was no nuptial pad at all, but just a spot of black pigment." Subsequent examination of one of Kammerer's pickled toads showed that the black pads—that is, the acquired trait that had allegedly been inherited—were nothing more than black ink that had been injected into the toad's foot. Kammerer's reputation was destroyed.

Kammerer was in serious financial trouble, and much of his research was lost during World

War I. Kammerer claimed to be astonished by Noble's accusation, and denied any wrongdoing. However, six weeks later—on September 23, 1926, just before he was to begin work in Moscow—Kammerer committed suicide. While on a walk in the Theresien Hills, he shot himself in the head.

After Kammerer's death, the Soviet Union produced a film titled *Salamandra* that ended with Kammerer's triumphant arrival in the Soviet Union. In 1971, Arthur Koestler's *The Case of the Midwife Toad* speculated that Kammerer's results might have been planted by Nazi sympathizers or Darwin supporters intent on discrediting Lamarckian inheritance.

For More Information: Koestler, A. 1971. *The Case of the Midwife Toad.* London: Hutchinson.

KANSAS STATE BOARD OF EDUCATION

On August 11, 1999, the Republican-dominated Kansas State Board of Education endorsed by a vote of 6–4 a set of science education standards for the state's 305 public school districts. These standards, which had been developed by the Creation Science Association for Mid-America, included no mention of biological macroevolution, the age of the Earth, or the origin and early development of the universe. Earlier, the Board had rejected standards developed by a twenty-seven-member panel of scientists and science educators, and had deadlocked over standards developed by board member Steve Abrams and Tom Willis of the Creation Science Association for Mid-America. *Science* reported the story as "Kansas Dumps Darwin," and *The New York Times* described the vote with a headline reading "Board for Kansas Deletes Evolution from Curriculum—A Creationist Victory." Conservative columnist and political commentator George Will noted that "Every [political] party at any given time has a certain set of issues on its fringe that can make it look strange, and this is one that can make the Republicans look strange."

The Kansas decision prompted many people to ridicule Kansas, and triggered debates about science standards throughout the nation. A few months after the Kansas vote, Kentucky—which already had a law allowing the teaching of creationism in its public schools—replaced the word *evolution* with *change over time* in its state science standards. The National Academy of Sciences, the American Association for the Advancement of Science, and the National Science Teachers Association began denying the Kansas State Board of Education's requests to use their documents in the new Kansas science education standards. An editorial in *Scientific American* suggested that college admissions boards question the qualifications of applicants from Kansas, but antievolution crusader Phillip Johnson (who had given money and public support to a creationist candidate) called the board's decision "courageous," and Michael Behe called it "heartening." Duane Gish declared that his hero was Board chairwoman Linda Holloway, who had raised enough money to buy the first television commercial in the history of Kansas school-board elections. When Holloway lost her bid for reelection, Johnson blamed the results on "very heavy-handed intimidation."

On February 14, 2001, the Kansas State Board of Education, now having some newly elected members, voted 7–3 to reinstate the teaching of biological evolution and Earth's history into the state's science standards. In February 2007, another newly elected Board removed the antievolution guidelines and again made the state the subject of much ridicule. This change was the fourth major change in the treatment of evolution in the previous eight years.

In 2005, Kansas' decision to include intelligent design in the science curriculum prompted physicist Bobby Henderson (b. 1980) to begin promoting the Flying Spaghetti Monster as the designer of the universe. Henderson's *The Gospel of the Flying Spaghetti Monster* (2006) was a bestseller.

ARTHUR KEITH (1866–1955)

Religious leaders and men of science have the same ideals; they want to understand and explain the universe of which they are part; they both earnestly desire to solve, if a solution be ever possible, that great riddle: Why are we here?

Arthur Keith was born on February 5, 1866, in Aberdeen, Scotland, and graduated from the University of Aberdeen in 1888. When Keith topped other students in his anatomy classes, he was awarded a copy of Darwin's *On the Origin of Species* as a prize. By 1894, Keith was a Fellow of the Royal College of Surgeons of London, and his *Human Embryology and Morphology* became a standard textbook. Keith was elected a Fellow of the Royal Society in 1913 and was knighted in 1921.

Keith was especially interested in reconstructing prehistoric humans based on fossils collected in Africa and Europe. Keith rejected the claim that African australopithecine fossils were human ancestors because, Keith argued, human qualities appear only when the volume of the brain is at least 750 cm^3, a delineation named "Keith's Rubicon." (Keith used 750 cm^3 because it was the smallest functioning modern human brain that anatomists had seen in Keith's era.) Consequently, *Australopithecus africanus* did not qualify as a human because its brain volume was only 450 cm^3. Today, Keith's Rubicon is not regarded as a valid measurement for defining a "human" primate.

Keith often touted the importance of Piltdown Man, and was bitter when it was discovered to be a fraud. Some people suspected Keith was involved with the hoax, but he was acquitted of any involvement. For many years, Keith participated in public debates about evolution and agnosticism. He also helped purchase Down House for the British Association and the Royal College of Surgeons. In 1932, Keith established a research institute in Kent, where he lived the rest of his life.

Keith died on January 7, 1955, in Downe, not far from Down House.

BILL KEITH

Bill Keith was a state Senator from Louisiana who in 1980 introduced a bill requiring the inclusion of creationism in the state's public schools

in 1980. When his bill died in committee, Keith—in the name of "academic freedom"—introduced a new bill that he had received from Paul Ellwanger (that had been drafted by creationist Wendell Bird) requiring equal time for evolution and creationism in the state's public schools. After Keith distributed materials from the Institute of Creation Research proclaiming that creationism is "pure science" having "no missing links," Keith's modified bill passed the state House of Representatives by a vote of 71–19. After Keith told his colleagues that "evolution is no more than a fairy tale about a frog that turns into a prince," the Senate also passed the legislation. When Governor David Treen signed the bill, it became law. Keith's bill was later overturned by the U.S. Supreme Court in *Edwards v. Aguillard*. Keith claimed that God delayed *Edwards* so President Ronald Reagan could appoint more federal judges who supported "creation science."

VERNON KELLOGG (1867–1937)

Vernon L. Kellogg was a renowned Stanford entomologist who was one of the first biologists to promote a genetics-based view of evolution. Kellogg, a founder of the National Research Council, published a variety of evolution-related books, including *Evolution and Animal Life* (1907, with David Starr Jordan) and *Darwinism Today* (1907). Kellogg's 1915 high school textbook noted that "Although there is much discussion of the causes of evolution there is practically none any longer of evolution itself. Organic evolution is a fact, demonstrated and accepted." In 1917, Kellogg's *Headquarters Nights* convinced William Jennings Bryan that German militarists had used evolutionary theory to justify their actions and lead people away from Christianity. Kellogg, whose research provided a foundation for the modern synthesis, died on August 8, 1937, in Hartford, Connecticut.

DENNIS JAMES KENNEDY (1929–2007)

In 1974, Dennis James Kennedy founded Coral Ridge Ministries, a $37 million evangelical church, school, seminary, and business empire in Florida that reached several million people. Kennedy opposed same-sex marriage, wanted to impeach any judge who did not believe in God, wanted teachers to promote Christianity, and urged his followers to "reclaim America for Christ, whatever the cost." Kennedy's *Darwin's Deadly Legacy* (2006) cited British biologist Arthur Keith to link evolution to Hitler (Kennedy claimed that Darwin's theory "fueled Hitler's ovens"), and denounced evolution as "the big lie" that "resulted in the deaths of more people than have been killed in all of the wars in the history of mankind." Kennedy—a Young-Earth creationist—claimed that there are "only two religions...Christianity [and] evolution," and that "anyone who does not realize that evolution is a religion does not know much about evolution."

Kennedy, who founded Knox Theological Seminary (in 1989) and the Center for Reclaiming America for Christ (in 1996, but disbanded in 2007), believed that there is a conspiracy among scientists to "repress" antievolution facts, and produced antievolution publications, including *The Crumbling of Evolution* (1983), *The Collapse of Evolution* (1981), and *Evolution's Bloopers and Blunders* (1986). Kennedy claimed that life has no meaning to evolutionists, and urged his followers to fight evolution. As Kennedy noted during his *Truths that Transform* radio program, "Why is the Creation account in Genesis so important? Because if it didn't happen, then we don't need Jesus today! That is what is at stake in the creation/evolution debate."

Kennedy was a prominent member of Jerry Falwell's Moral Majority, and Kennedy's Center for Christian Statesmanship (founded in 1995) awarded its "Distinguished Christian Statesman Award" to Tom DeLay and other conservative Christians. Kennedy was the honorary chairman of Answers in Genesis' Creation Museum, and his radio program featured intelligent design advocates Phillip Johnson and William Dembski. Kennedy has promoted the work of numerous other antievolutionists, including Ann Coulter's (b. 1961) bestseller *Godless: The Church of Liberalism*, which rejected Darwinian evolution.

In late 2006, Kennedy suffered a massive heart attack. He died in Ft. Lauderdale, Florida on September 5, 2007.

KENTUCKY LEGISLATURE

The Kentucky legislature made effective on July 13, 1990, the Kentucky Revised Statute 158.177, which allowed teachers to teach biblical creationism and give students credit on exams for answers based on biblical creationism. This law defied the 1987 U.S. Supreme Court decision in *Edwards v. Aguillard*, but has not been challenged. In 1996, the Superintendent of Schools in Marshall County, Kentucky, ordered that evolution-related pages in the schools' science textbooks be glued together.

DEAN KENYON (b. 1939)

Dean H. Kenyon was a biology professor at San Francisco State University who in 1969 wrote *Biochemical Predestination*, a book that advocated biochemical evolution (the preface of the Russian edition was written by A.I. Oparin). Although Kenyon was inspired to study evolution by Julian Huxley (the grandson of Thomas Huxley), in 1976 Kenyon read *The Genesis Flood* and other neocreationism books and became a creationist. Kenyon began teaching creationism, and in 1992 was censured for teaching religion in his biology courses. Kenyon was scheduled to testify in *McLean v. Arkansas Board of Education*, but was convinced by Wendell Bird—the general counsel for the Institute for Creation Research—to leave town before testifying. Later, Kenyon's deposition defending creationism was read into evidence at *Edwards v. Aguillard* and *Kitzmiller v. Dover Area School District*.

Kenyon, who once described Charles Darwin as "one of the greatest naturalists who ever lived," is most famous for writing with Percival Davis the 166-page *Of Pandas and People: The Central Question of Biological Origins* (1989), the first textbook promoting intelligent design. *Pandas*, which ignored the age of the Earth and promoted the alleged lack of transitional fossils, was denounced by scientists as "incompetent" and a "wholesale distortion of modern science." As Judge John Jones noted in the *Kitzmiller* decision, "the definition for creation science in early drafts [of *Pandas*] is identical to the definition of [intelligent design]." After *Edwards*, the publisher replaced "creationism" with "intelligent design," and "creationists" with "intelligent design proponents," at one point changing "creationists" to "cdesign proponentsists." After the *Kitzmiller* decision, the publisher changed the name of the sequel from *Pandas* to *The Design of Life*. That book—written by Kenyon, Michael Behe, Jonathan Wells, William Dembski, and other creationists—cited *Inherit the Wind* as an example of how the book's claims have been distorted. After retiring in 2001, Kenyon became a Fellow at the Discovery Institute. Despite being exposed in *Kitzmiller* as a creationist book, the Foundation for Thought and Ethics continued to claim that *Pandas* had "improved the way origins is taught."

For More Information: Davis, P. & D.H., Kenyon 1993. *Of Pandas and People*, 2nd edition. Dallas, TX: Haughton Publishing.

HENRY BERNARD DAVIS KETTLEWELL (1907–1979)

Henry Bernard Davis Kettlewell was a British physician, population geneticist, and moth enthusiast who became famous for his evolutionary studies in the 1950s of the influence of industrial melanism on natural selection in peppered moths (*Biston betularia*). In the 1920s, British entomologist J.W. Heslop Harrison (1881–1967) claimed that pigmentation of these moths was a result of Lamarckism. However, scientists could not repeat Harrison's work, and evolutionary biologists such as Ronald Fisher argued that Harrison's explanation required much higher rates of mutation than had been reported. Despite this criticism, Harrison continued to argue for Lamarckian inheritance.

In 1951, Kettlewell suspected that wing coloration was an important adaptation in peppered moths. Kettlewell's lab work showed that dark-winged moths tended to choose dark backgrounds (and light-winged moths tended to choose lighter backgrounds). In field experiments, Kettlewell demonstrated that birds could function as selective agents, but predation was a learned behavior (i.e., birds had to learn to recognize a type of prey before they could exploit it). Kettlewell confirmed his claims with mark-release-recapture experiments in polluted forests near Birmingham and in pristine forests near Dorset. Regardless of the environment, camouflaged moths were twice as likely to be recaptured as conspicuous moths.

Kettlewell's work showed that lightly colored moths were more conspicuous than darkly colored moths in polluted areas, and in those areas were more susceptible to predation by birds. Nobel laureate Niko Tinbergen's (1907–1988) movies of the differing predation rates of the moths by birds were shown at science meetings throughout the world, and Kettlewell's work appeared in a *Scientific American* article titled "Darwin's Missing Evidence." Kettlewell's work was described in 1978 by Sewall Wright as "the clearest case in which a conspicuous evolutionary process has actually been observed."

For many years, Kettlewell's work was cited in biology textbooks as the best example of natural selection in action. However, a 1987 article in *New Scientist* announced the "exploding" of the "myth of the melanic moths." Although no one questioned that natural selection had made dark moths more abundant in polluted areas and light moths more abundant after pollution was controlled, some biologists questioned the role of bird predation as the cause of the change. The article raised a variety of questions: Were dark-colored moths better suited to handle pollution? If so, was selection operating at the caterpillar stage rather than the adult stage? Moth coloration is influenced by what moths eat; was the different coloration in different habitats due simply to moths eating different types of food rather than predation by birds? Other questions involved pollution killing lichens (which provided camouflage for the moths) and predation by bats at night (when color would be less important). Although Kettlewell's research represented a monumental amount of work, and has been supported by more recent studies, it has been replaced in most textbooks by the research of Peter and Rosemary Grant, who have monitored populations of Galápagos finches for several decades.

Kettlewell also studied how to use radioactivity to tag insects, and in 1974 coauthored (with Julian Huxley) *Charles Darwin and His World*. Kettlewell died of a drug overdose in 1979.

BENJAMIN KIDD (1858–1916)

Benjamin Kidd was a social philosopher who wrote *The Science of Power* (1918), a book that helped convince William Jennings Bryan that German militarists had used the evolutionary theory to justify their actions and lead people away from Christianity. *The Science of Power* also examined Darwin's influence on German philosopher Friedrich Nietzsche (1844–1900), who in 1882 famously noted that "God is dead." In *Social Evolution* (1894), Kidd claimed that natural selection favors the preservation of "nonrational" institutions such as religion.

MOTOO KIMURA (1924–1994)

Motoo Kimura was a Japanese biologist who in 1968 proposed the neutral theory of molecular evolution, in which genetic drift is the main force that changes allelic frequencies. Kimura argued that most molecular differences are selectively neutral (i.e., do not influence fitness), and his ideas became a basis for the "molecular clock" that is used to measure the time passed since species diverged from a common ancestor. Kimura's theory, which remains controversial, was initially claimed by some creationists to refute Darwinian evolution, but this claim was unfounded.

CHARLES KINGSLEY (1819–1875)

Charles Kingsley was a highly respected English writer and theologian who was Chaplain to Queen Victoria. Charles Darwin suspected that Kingsley would support his theory of evolution, and sent Kingsley an advance copy of *On the Origin of Species*. After reading *Origin*, Kingsley

wrote to Darwin that, "[a]ll that I have seen of it awes me...if you [are] right, I must give up much that I have believed." In 1863, Kingsley published *The Water Babies*, a book that creationists denounced as evolutionary propaganda for children. *The Water Babies* notes that Owen, Huxley, and Darwin are "very wise men [and] you should listen respectfully to all they say." *The Water Babies* became a classic, and was reprinted for more than a century. Kinsgley died in 1875 and was buried in St. Mary's Church in Eversley, England.

RUDYARD KIPLING

Rudyard Kipling (1865–1936) wrote the children's book *Just-So Stories* (1902) that included fanciful stories about how animals get their traits (e.g., "How the Giraffe Got Its Long Neck," "How the Camel Got Its Hump"). Similar stories were often invoked by biologists claiming that structures developed "for the good of the species." (Stephen Gould and Richard Lewontin used Voltaire's fictional Dr. Pangloss to counter this attitude.) Kipling's *Jungle Books* were posthumously made into several movies by the Walt Disney Company. Kipling is buried in Poet's Corner of Westminster Abbey.

TAMMY KITZMILLER (b. 1966)

I feel they brought a religious idea into the classroom, and I object to that.

In 2004, Tammy Kitzmiller was a 38-year-old divorced mother of two teenage girls living in Dover, Pennsylvania, and working as an office manager for a landscaping company. In the fall of that year, the local school board decided that ninth-grade biology students would be required to hear a one-minute statement that "The Theory [developed by Charles Darwin] is not a fact" and "Intelligent design is an explanation of the origin of life that differs from Darwin's view." Upon reading about the school board's actions in the local paper, Kitzmiller was uneasy because this seemed to violate the First Amendment of the Constitution. Kitzmiller, a self-described shy person, decided that she was "going to sue because I can't believe they want to bring up religion."

Kitzmiller was apprehensive but was reassured to learn that other parents shared her concerns. Eventually Kitzmiller and ten other parents sued the Dover Area School District in U.S. District Court for the Middle District of Pennsylvania on December 14, 2004. Kitzmiller was selected as the lead plaintiff because her last name had a solid Pennsylvania Dutch sound to it and she was the only plaintiff with a child in ninth grade at the time.

The proceedings started on September 26, 2005, and lasted six weeks. Kitzmiller and the other plaintiffs were called to testify, as were several expert witnesses on both sides of the issue. Kitzmiller's testimony lasted for just a few minutes, and consisted mainly of being asked what she knew of the Board's plans and how she learned of them. She also indicated that her 14-year-old daughter invoked her "opt out" option and left the room when the Board-approved statement was read in her biology class. The many witnesses against intelligent design (ID) included John Haught, a professor of theology at Georgetown University who believes that evolution challenges Christians to develop new understandings of God and how God creates.

On December 20, 2005, presiding judge John Jones ruled that the Board's actions violated the First Amendment. Having won, Kitzmiller and the other plaintiffs returned to their lives, although Tammy Kitzmiller's name, like John Scopes', has remained associated with the controversy surrounding the teaching of evolution. In November 2007, PBS broadcast "Judgement Day: Intelligent Design on Trial," a two-hour documentary about the *Kitzmiller* trial.

For More Information: Humes, E. 2007. *Monkey Girl: Evolution, Religion, and the Battle for America's Soul.* New York: HarperCollins.

STANLEY KRAMER (1913–2001)

Stanley Earl Kramer was born in Brooklyn, New York, on September 29, 1913, and was raised by his mother Mildred and his maternal grandparents in the Hell's Kitchen section of New York. Mildred worked for Paramount Pictures, and Kramer's uncle was an agent in Hollywood. Kramer attended DeWitt Clinton High School and New York University, and intended to go to law school. However, during his senior year of college, Kramer became a writer for 20th Century Fox. Kramer also built sets at MGM Studios.

In 1943, Kramer was drafted into the military, during which time he worked for an army film crew in New York. He then founded a production company called Screen Plays Inc., whose first movie (*So This Is New York*) was a failure. However, its second film (*Champion*)—starring Kirk Douglas as a hard-nosed boxer—received six nominations for Academy Awards. In 1950, Kramer produced *The Men*, a drama about paraplegic war-veterans, which included the screen debut of Marlon Brando. Kramer's 1952 western *High Noon*—which starred an aging Gary Cooper as a lone man standing up for his beliefs—won four Oscars and launched the career of Grace Kelly. In 1954, Kramer's hit *The Caine Mutiny* starred Humphrey Bogart in his final movie.

While most other movies during Kramer's era stressed escapism, Kramer's movies were often "message" films, including *Inherit the Wind* (1960) and *Judgment at Nuremberg* (1961). Kramer's 1959 movie *On the Beach*, which premiered in Moscow and seventeen other cities around the world, depicted a world facing nuclear annihilation, and prompted Nobel Laureate Linus Pauling to speculate that "*On the Beach* [may have] saved the world." Kramer, who attributed his empathy for underdogs and moral causes to his own hardships, became known as "Hollywood's Conscience." Kramer's *Inherit the Wind* remains the most influential film or documentary associated with the evolution-creationism controversy.

In 1963, Kramer produced and directed the all-star comedy, *It's a Mad, Mad, Mad, Mad World*, and in 1967 Kramer's *Guess Who's Coming to Dinner* included the final screen pairing of Katharine Hepburn and Spencer Tracy (Tracy starred in four of Kramer's films, the first of which was *Inherit the Wind*). *Guess Who's Coming to Dinner* was nominated for eight Academy Awards, including a posthumous nod for Tracy. In all, Kramer's thirty-five motion pictures earned eighty-five Academy Award nominations and won fifteen Academy Awards, none of which went to Kramer. Kramer's only Oscar was his 1961 Irving Thalberg Memorial Award for overall excellence.

In 1977 Kramer retired from filmmaking and left Hollywood to live in Seattle, Washington. There, Kramer taught at local colleges, wrote columns for newspapers, and hosted radio shows. He returned to California in the late 1980s, and in 1997 wrote his autobiography, *A Mad Mad Mad Mad World: A Life in Hollywood*.

In the late 1990s, Kramer—a diabetic—became increasingly frail. After contracting pneumonia, Kramer died in his sleep in Woodland Hills, California, on February 19, 2001.

KU KLUX KLAN (Est. 1866)

The Ku Klux Klan was formed in Pulaski, Tennessee, in 1866 by Confederate hero Nathan Bedford Forrest to protect widows and orphans of the Confederate dead. Soon, however, many white Southerners, frustrated by federal Reconstruction policies, used the Klan to lash out against occupying troops and/or blacks who were benefiting from Reconstruction. Unable to control the increasing violence, Forrest dissolved the Klan in 1869. However, most of its chapters continued to function, and the Klan's influence continued to spread. By 1877, the Klan and its clones had helped overthrow Republican rule in every Southern state.

On Thanksgiving night in 1915, the Klan was reborn at a cross-burning ceremony at Stone Mountain, Georgia, the world's largest monolith and the Klan's most sacred site. Like virtually all racist organizations, the Klan based its beliefs on special creation; a separate origin for whites and blacks—as could be gleaned from creationism but not evolution—enabled the Klan and similar groups to claim that blacks and other ethnic minorities were not entitled to the same rights as whites. The Klan gave the

antievolution movement powerful support, and in 1925 became the first organization to urge that creationism and evolution be given "equal time" in public schools. Decades later, this argument was resurrected by other creationists and ruled unconstitutional by the U.S. Supreme Court (*Edwards v. Aguillard*).

The Klan supported, and was supported by, several leaders of the antievolution movement in the early 1900s (Figure 51). For example, Bob Jones and William Jennings Bryan endorsed Klansmen in political races and, in return, the Klan openly supported their campaigns against evolution and other social ills. At the 1924 Democratic National Convention in New York City, Bryan—who was not a member of the Klan—spoke against an amendment denouncing the Klan. When Bryan died five days after the Scopes Trial, the Klan proclaimed that they would "take up the torch [that has fallen] from the hand of William Jennings Bryan" and burned crosses at memorial services honoring Bryan. After one such ceremony, a large cross was raised bearing the inscription, "In memory of William Jennings Bryan, the greatest Klansman of our time, this cross is burned; he stood at Armageddon and battled for the Lord." At a rally of 30,000 Klansmen one month after Bryan's death, the Klan laid a wreath at Bryan's grave in Arlington National Cemetery.

The Klan's philosophy overlapped significantly with that of many fundamentalists of the early twentieth century, such as Bob Jones and Frank Norris, both of whom claimed that Genesis prohibits race-mixing; that God destroyed Man in a worldwide flood because Man had mixed races; and that evolution is responsible for virtually all societal ills. Similarly, both groups believed in the literal truth of the Bible, wanted to mandate prayer and ban the teaching of evolution in public schools, supported racial segregation, had a similar social agenda, and believed that "human reason bows" before the Bible. Antievolution groups such as The Supreme Kingdom and The Bible Crusaders, which were sympathetic to the Klan's causes, recruited members by stressing to the public

51. The Ku Klux Klan is a militant white-supremacy group that supported, and was supported by, several prominent antievolution crusaders in the 1920s. This advertisement appeared in the *Winston-Salem Journal* during one of Billy Sunday's crusades in North Carolina in the early 1920s.

that accepting evolution made blacks as good as whites.

The Klan was a vital organization of the religion of the South, and Southern ministers often attacked evolution while defending the Klan and white supremacy. Southern Baptists, who had split from other Baptists over the slavery issue in 1845, repeatedly denounced Darwinism while simultaneously opposing integration. As other religions denounced the Klan, Southern Baptists remained silent.

Many fundamentalist evangelists openly supported, and were openly supported by, the Klan. For example, Jones accepted $1,568 from the Klan after a three-week revival in Alabama, and Frank Norris's church often hosted Klan meetings. Billy Sunday accepted the Klan's money and endorsed the Klan Kreed that promoted white supremacy and a literal interpretation of the Bible. Sunday pleaded for a return to the Klan's version of "the old time religion of our fathers," just as Klansmen promoted Sunday's "100% American" social and religious

attitudes (e.g., the evils of evolution, booze, and other ideas associated with modernism). Except for H.L. Mencken, few social commentators deplored the Klan's seduction of religious fundamentalism.

Antievolution crusader Norris claimed that the teaching of evolution would lead to the breakdown of White Supremacy. Norris denounced Catholics by claiming that Catholics favored equality for blacks. Although Norris was not a member of the Klan, Norris often praised—and was supported by—the Klan (e.g., his acquittal in 1927 for the murder of Fort Worth lumberman D.E. Chipps was aided by the Grand Dragon of the local Klan). Norris, who published articles by Klansmen in his *The Fundamentalist* magazine, linked racial equality with the teaching of evolution and with communism, and said that communists, like Catholics, advocated interracial marriage to form a "mongrel race." Just two weeks before his death in 1952, Norris denounced interracial marriage, condemned the teaching of evolution, and advocated White Supremacy by claiming that "God didn't make [blacks equal]. It's hard to go against God's laws."

At the peak of its popularity (near the time of the Scopes Trial), Klan membership exceeded 4,000,000. In 2007, Klan groups were active in thirty-six states and had a combined membership of approximately 8,000. Today, the Imperial Klans of America describes itself as "a private Christian organization."

For More Information: Wade, W.C. 1987. *The Fiery Cross: The Ku Klux Klan in America*. New York: Simon and Schuster.

WILLIAM KUNSTLER (1919–1995)

William Moses Kunstler was a flamboyant defense attorney who in 1967 defended Gary Scott when Scott was fired in Jacksboro, Tennessee, for teaching evolution. Scott's lawsuit ultimately led the Tennessee legislature to repeal the state's Butler Law banning the teaching of human evolution in public schools. Kunstler, a self-described "radical lawyer," also defended the Chicago Seven, Lenny Bruce, Malcolm X, Martin Luther King, Leonard Peltier, and other controversial individuals. Kunstler died on Labor Day in 1995.

L

DAVID LACK (1910–1973)

Darwin's finches, the Geospizinae of the Galápagos, form a little world of their own...the factors involved in their miniature adaptive radiation seem to be the same as those involved in the larger world, so that they help to demonstrate evolutionary and ecological principles with unusual clarity.

David Lambert Lack was born on July 16, 1910, in London, and in 1936 earned a degree in natural history from Cambridge. He taught school for most of the 1930s, and in 1935 visited the United States, where he met biologist Ernst Mayr at the American Museum of Natural History. Three years later, Lack—at the request of Julian Huxley, the grandson of "Darwin's Bulldog" Thomas Huxley—took a year off from teaching to go to the Galápagos Islands to study birds for an entire breeding season. While at the Galápagos, Lack witnessed natural selection at work among the islands' finches. He then returned to New York and spent the next year writing a report of the expedition. Mayr was Lack's roommate.

From 1940 to 1945, Lack served in the British Army, where he developed the radar program. Lack later used this experience to study patterns of bird migration. In 1943, Lack published *The Life of the Robin*, a popular book that garnered much publicity. Two years later, he became director of the Edward Grey Institute of Field Ornithology at Oxford, a position he held until 1973.

In 1947, Lack published his classic, *Darwin's Finches*. Although Charles Darwin had mentioned Galápagos mockingbirds (but not finches) in *On the Origin of Species* (e.g., in Chapter 12, Darwin notes "closely allied species of mocking-thrush, each confined to its own island"), Lack's book made Darwin's name synonymous with the Galápagos finches. Soon thereafter, textbooks began including the finches as evidence for Darwin's theory, and the finches became an icon in evolution. After reading Lack's book, evolutionary biologist E. O. Wilson proclaimed that Darwin's finches "shout the truth of evolution." Following Lack's lead, ornithologists—such as Peter and Rosemary Grant—have flocked to the Galápagos ever since.

In 1951, Lack was made a fellow of the Royal Society. In subsequent years, he published a variety of books (e.g., *Evolutionary Theory and Christian Belief* in 1957 after his conversion from agnosticism to Anglicanism; *Population Studies of Birds* in 1966) and won several prestigious awards, including the Darwin Medal of the Royal Society. Unperturbed that a benevolent deity could reign over nature's struggle for existence, Lack claimed that "man is surely unqualified to judge whether this [natural] ordering is in any way evil, or contrary to divine plan."

Lack died on March 12, 1973, in Oxford, England.

For More Information: Lack, D. 1947. *Darwin's Finches*. Cambridge: Cambridge University.

TIM LAHAYE (b. 1926)

It's the lie of evolution that all men are just evolved and that they're all equal, and that all creatures are equal.

Timothy F. LaHaye graduated from Bob Jones University in 1950, and in 1958 he moved to San Diego and became pastor of Scott Memorial Church. In 1971, LaHaye founded Christian Heritage College (now San Diego Christian College), a school "unreservedly committed to strict creationism." Soon thereafter, LaHaye and fellow fundamentalist Henry Morris expanded the College by opening the Institute for Creation Research (ICR). From 1970 to 1978, Morris was Christian Heritage College's Vice President for Academic Affairs, where he required "all curricula to be founded on creationism and full Biblical authority." ICR later separated from the college, but it remains a dominant force in the evolution-creationism controversy in the United States.

In the 1960s and 1970s, LaHaye ran training seminars for the John Birch Society, and when Jerry Falwell formed the Moral Majority in 1979, LaHaye was its California chair. In the late 1970s and early 1980s, LaHaye helped found a variety of conservative political organizations, including the American Coalition for Traditional Values and the secretive Council for National Policy, which linked conservative Christian leaders with like-minded politicians and financiers. In 1988, LaHaye initially supported Pat Robertson for president, but left Robertson's organization to be the national cochair of Jack Kemp's presidential campaign. One week later, LaHaye was forced to resign when an article in the Baltimore *Sun* reported that LaHaye viewed Roman Catholicism as "a false religion."

LaHaye's wife Beverly headed Concerned Women for America, a conservative group of 500,000 members that filed a brief in *Edwards v. Aguillard* supporting the teaching of creation science. Concerned Women for America, which opposes evolution and sex education, is "built on prayer" and dedicated to "bringing biblical principles into all levels of public policy." In 1992, Beverly LaHaye was given an honorary Doctorate of Humanities by Jerry Falwell's Liberty University, and *Time* magazine described Tim and Beverly as "the Christian Power Couple."

In *The Battle for the Mind* (1980), LaHaye claimed that "no humanist is qualified to hold any governmental office," and in *The Ark on Ararat* (1976), LaHaye and John Morris claimed that demons have prevented Christian investigators from proving the Bible right and evolution wrong. LaHaye and Morris conclude *The Ark on Ararat* by telling people to "climb on board the Ark then. Receive Jesus Christ as your personal Savior."

LaHaye blamed evolution for communism, socialism, and a variety of social ills. LaHaye wrote the introduction to Henry Morris' *The Troubled Waters of Evolution*, and LaHaye's bestselling *Mind Siege* opens with students being arrested by armed SWAT teams for distributing antievolution tracts at a public school. To LaHaye, anything not biblical literalism was "evolutionism."

In the 1980s, LaHaye and Jerry Jenkins began publishing the *Left Behind* series of novels. These books popularized biblical end-times prophecy, much like Morris's books popularized creation science. The *Left Behind* books, whose sales exceed 60 million copies, spawned an industry of movies, television specials, violent video games, and associated materials, all proclaiming premillenial dispensationalism. In 2001, LaHaye donated $4.5 million to Liberty University to open the Tim LaHaye School of Prophecy. LaHaye's *The Rapture* was released on June 6, 2006, presumably to emphasize the connection with 666. The *Left Behind* series is the fastest-selling adult fiction series in history.

JEAN-BAPTISTE LAMARCK (1744–1829)

Do we not therefore perceive that by the action of the laws of organization . . . nature has in favorable times, places, and climates multiplied her first germs of animality, given place to developments of their organizations and increased and diversified their organs? Then ...aided by much time and by a slow but constant diversity of circumstances, she has gradually brought about in this respect the state of things which we now observe. How grand is this consideration, and especially how remote is it from all that is generally thought on this subject?

Jean-Baptiste Pierre Antoine de Monet, Chevalier de Lamarck was born into a military family on August 1, 1744, at Bazentin-le-Petit, Picardy, in rural northern France. He was the youngest of eleven children. Phillipe, Lamarck's father, expected Jean to have a church-related career, and in 1756 Lamarck enrolled in a Jesuit seminary. However, when Phillipe died in 1760, Lamarck quit the seminary, bought a horse, and joined the French army. In his first battle, Lamarck distinguished himself for bravery and was made an officer. However, an injury inflicted by a comrade forced Lamarck out of the military, after which he worked as a bank clerk in Paris. Lamarck then began to study botany and medicine.

At age 34, Lamarck published *Flore Francaise*, an acclaimed book about the plants of France. The following year—with the help of Buffon, whose son Lamarck tutored—Lamarck was elected to the French Academy of Science. In 1793 (the same year that Louis XVI and Marie Antoinette went to the guillotine), Lamarck helped reorganize the French Museum of Natural History, and was appointed a professor there the following year. The museum was to be run by twelve professors in twelve scientific fields, and Lamarck was in charge of studying "insects, worms, and microscopic animals." Lamarck knew little about insects and worms, but he later coined the word *invertebrate* to describe them. Today, a plaque at the museum's entrance notes that Lamarck lived there from 1795 until his death in 1829.

Lamarck first presented his ideas about evolution in a lecture on May 11, 1800. In 1809, Lamarck—who was a protégé of Georges-Louis Buffon and botanist to King Louis XVI—published his ideas about evolution in his most famous work, *Philosophie Zoologique*. In this purely theoretical book, which Darwin read while aboard the *Beagle*, Lamarck discussed life's "tendency to progression" and "tendency to perfection," and claimed that life is in a constant state of advancement and improvement that is too slow to be perceived except with the fossil record. Lamarck's claim required spontaneous generation of new species to replace those transformed to more-advanced species. Whereas contemporaries such as Buffon had hinted at evolution, Lamarck was its champion: "...species have only a limited or temporary constancy in their characters...there is no species which is absolutely constant." Lamarck was confident of his conclusions; as he once noted, "I am not submitting an opinion, but announcing a fact."

Lamarck, who coined the term *biology*, was perhaps the world's premier invertebrate zoologist. He argued that organisms contained a "nervous fluid" that enabled them to adapt to their local environments. Lamarck believed that organisms evolve to become more complex over time; that these purposeful changes are brought about by the use or disuse of acquired traits; and that these changes made the organisms better able to survive in new environments and conditions. Simply put, Lamarck believed that what an animal did during its lifetime was passed on to its offspring.

According to Lamarck, environmental changes alter the needs of organisms living in that environment. In turn, the organisms' altered needs change the organisms' behaviors, and these altered behaviors then lead to the greater or lesser use of different structures. The more an organism used a part of its body, the more developed that part would become (similarly, the disuse of a part would result in its decay). Lamarck referred to this idea—namely, that the use or disuse of a structure would cause the structure to develop or shrink—as his "First Law." This was followed by Lamarck's "Second Law," which proposed that the changes acquired as a result of the First Law would be inherited by the organism's offspring. As a result, species would gradually change as they became adapted to their environment. For example,

Lamarck argued that wading birds evolved long legs as they stretched them to keep high and dry. Similarly, giraffes evolved long necks as they stretched their necks to reach leaves high in trees. When they stretched, Lamarck claimed, their "nervous fluid" would flow into their necks and, over successive generations, cause their necks to grow longer. This inevitable need-based change (i.e., necks getting longer to get food) would give giraffes permanently longer necks, and these long necks would be passed to the giraffes' offspring. Lamarck's idea, which came to be known as "inheritance of acquired characteristics," suggested that there was a drive toward perfection and complexity, analogous to species climbing a ladder. Lamarck rejected extinction, instead claiming that organisms evolved into different, more perfect species via a process he called "transmutation." New, primitive organisms constantly formed from inorganic matter at the bottom of the scale.

Lamarck—an ardent materialist—was his era's most renowned advocate of evolution, and his model for evolution was the first testable hypothesis to explain how a species could change over time. Many people—Erasmus Darwin among them—endorsed Lamarck's idea. However, it was rejected (and sometimes ridiculed) by the leading scientists of his time (e.g., his revered colleague Cuvier), and was later dismissed by other scientists. Nevertheless, Lamarck's idea was popular with the public—so much so that Charles Darwin alluded to it in later editions of his *On the Origin of Species*. Darwin wrote to Joseph Hooker in 1844 that "Heaven forfend me from Lamarck nonsense of a 'tendency to progression,' 'adaptations from the slow willing of animals,' etc.! But the conclusions I am led to are not widely different from his; though the means of change are wholly so. I think I have found out (here's presumption!) the simple way by which species become exquisitely adapted to various ends."

Although Lamarck's name is most often associated with his discredited "inheritance of acquired traits," Darwin and many other scientists acknowledged him as a great zoologist and one of evolution's early thinkers. Although

Darwin described Lamarck's book as "veritable rubbish," Darwin conceded in 1861 that "Lamarck was the first man whose conclusions on the subject excited much attention. This justly celebrated naturalist . . . first did the eminent service of arousing attention to the probability of all changes in the organic, as well as in the inorganic world, being the result of law, and not of miraculous interposition." Lamarck's speculative suggestions regarding the origin of new traits continue to overshadow his otherwise important contributions to biology.

Lamarck married Marie Delaporte, the mother of his first six children, on her deathbed in 1792. In 1795 he then married Charlotte (who died in 1797), and in 1798 he married his third wife, Julie Mallet, who died in 1819. He is rumored to have married a fourth time, but no documents support this claim. When in his 70s, Lamarck's eyesight began to deteriorate. Despite his great contributions to biology, Lamarck spent his last decade a blind, penniless man living in obscurity and cared for by his unemployed daughters.

Lamarck died in Paris on December 18, 1829. Lamarck's papers, books, and belongings were auctioned, and Lamarck was buried in a lime pit in Montparnasse cemetery with other paupers. Cuvier, who respected Lamarck's studies of invertebrates but rejected his theory of evolution, used his eulogy to ridicule and discredit Lamarck. Similarly, Lamarck's obituary in the London *Times* did not mention his many contributions to biology, instead focusing on the politics of finding his replacement at the museum. Although Lamarck's daughter claimed that "posterity will remember you," Lamarck's corpse was later excavated and piled with other nameless remains in the Paris catacombs. Today, Lamarck is memorialized with a large statue inscribed "Founder of the Doctrine of Evolution" at Jardin des Plantes in Paris (Figure 52). Far across the city is *rue Lamarck*, alongside which is the 86-meter, *rue Darwin*.

For More Information: Burkhardt, R.W. 1995. *The Spirit of the System: Lamarck and Evolutionary Biology.* Cambridge, MA: Harvard University.

52. Although Lamarck made enormous contributions to invertebrate biology, he is best known for proposing the inheritance of acquired traits. This statue at the Jardin des Plantes in Paris describes Lamarck as the "Founder of the Doctrine of Evolution."

WALTER LAMMERTS (1904–1996)

Walter E. Lammerts was a Lutheran biologist and an influential disciple of George McCready Price who helped found the Creation Research Society. Lammerts was one of the first Flood geologists to earn a doctorate in biology (in 1930 from the University of California), after which he began a successful career as a plant breeder. Lammerts joined the Deluge Geology Society in 1941, and his work was cited in *The Genesis Flood*. Lammerts—who believed that there could be "no discrepancies" between nature and the Bible—rejected theistic evolution and insisted on "the absolute fixity of species."

Unlike most creationists of his era, Lammerts supported civil rights and conservation, abhorred far-right extremists (e.g., John Birch Society), and rejected the claim that communism resulted from evolutionary concepts. His approach to the evolution controversy was simple: "If a man is such a stupid fool he can't see that evolution is wrong, I'm not going to try to convince him."

WALTER LANG (1913–2004)

Walter Lang was a Lutheran preacher from Nebraska who in 1964 founded the Bible-Science Association (BSA) as a lay counterpart to the Creation Research Society to promote scientific creationism. BSA publications endorsed "the Bible as the inerrant word of God, true in every subject which it touches, whether the plan be salvation, science, or history," and noted that "talking about dinosaurs, Noah's Ark...and UFOs can lead to...salvation through Jesus Christ." Lang attracted much attention with claims that nature has been "contaminated" by sin; that "very primitive people still exist today" in Bolivia and the Philippines; and that accepting Jesus will protect people from UFOs and other "Satanic deceptions." According to Lang, "the government needs the creation approach—the alternative is a Hitler or Stalin." Lang often worked with fellow fundamentalists Clifford Burdick, Duane Gish, and Henry Morris. BSA has flirted with geocentrism, and several of its members claim that they would "not be geocentrists if it were not for the Scriptures." BSA, along with the Creation Research Society, claimed that people did not have to accept evolution to be scientific and modern. Despite its name, most of the Association's 15,000 members were not scientists. Critics of Lang's biblical literalism argued that its proponents cite only evolution-related scriptures, not those involving meteorology (Job refers to "storehouses of the wind"), public health (Samuel II refers to plagues caused by the Angel of Death), or geology (I Kings refers to how "the Earth was split").

EDWIN LANKESTER (1846–1929)

Edwin Ray Lankester was an invertebrate biologist and protégé of Thomas Huxley who strongly supported Darwin. When he was a teenager, Lankester exposed the famous American "Spirit-Medium" Henry Slade, who had claimed that he could communicate with dead people. When Lankester replaced Richard Owen as Director of The British Museum of Natural History, he shifted the emphasis from classification to Darwinian evolution; this change was manifest in Lankester's moving the life-size statue of Darwin to the Museum's most prominent position, atop a stairwell overlooking the Great Hall. Lankester, who was knighted in 1907 and awarded the Copley Medal in 1913, published *Degeneration, A Chapter in Darwinism* (1880). He also wrote a long-running newspaper column titled "Science from an Easy Chair," and his book *Extinct Animals* (1905) was the source of the creatures used by Sir Arthur Conan Doyle in *The Lost World*. (Doyle also created Sherlock Holmes, and described Lankester as the hero's "gifted friend" in that book.) Lankester promoted population genetics, which helped revitalize Darwin's theory of evolution by natural selection.

EDWARD LARSON (b. 1953)

Darwin wrote his Origin of Species in 1859. At that time Queen Victoria was on the throne in England and James Buchanan was President of the United States. Now who has a greater impact on us today?

Edward John Larson was born in 1953 in Mansfield in north-central Ohio. After earning a BA from Williams College (1974), and an MA from the University of Wisconsin (1976), Larson graduated from Harvard University with a J.D. degree (1979). After Harvard, Larson joined the Seattle law firm Davis, Wright & Tremaine in 1979, leaving in 1983 to become associate counsel for the U.S. House of Representatives Committee on Education and Labor.

Larson became increasingly interested in the interaction between law and science, and entered the graduate program in the History of Science at the University of Wisconsin. He worked with well-known historian of the evolution-creationism conflict Ronald Numbers (author of *The Creationists*), earning his PhD in 1984. His graduate work eventually produced the book *Trial and Error: The American Controversy Over Creation and Evolution* in 1985.

Larson accepted the positions of Richard B. Russell Professor of History and Talmadge Professor of Law at the University of Georgia in 1987. His specialties as a law scholar include expertise in health care, science and technology, and bioethics, but his overarching interest is in historical analysis of law in the United States. In 1997, Larson published the book *Sex, Race, and Science: Eugenics in the Deep South*. This was followed by *Summer for the Gods: The Scopes Trial and America's Continuing Debate Over Science and Religion* in 1997, which won the 1998 Pulitzer Prize for history. The latter book examined the Scopes Trial from a legal perspective, an analysis Larson felt was lacking among the large number of publications devoted to this seminal event in the history of antievolutionism. He has continued to study evolution, and recent books include *Evolution's Workshop: God and Science on the Galápagos Islands* (2001) and *Evolution: The Remarkable History of a Scientific Theory* (2004).

Following up on psychologist James Leuba's landmark studies of religious beliefs among American scientists in the early part of the twentieth century, Larson and Larry Witham (b. 1952, whose books include *Where Darwin Meets the Bible: Creationists and Evolutionists in America* and *The Measure of God: Our Century-Long Struggle to Reconcile Science & Religion*) repeated Leuba's survey in the late 1990s. They found that the rate of belief among U.S. scientists had not changed since 1933 (still averaging around 40 percent across those surveyed). Larson pointed out that sixty-five years earlier this number was "shocking" to many because of how low it was. This included William Jennings Bryan, who interpreted Leuba's findings as demonstrating a need for a crusade against evolution and scientific materialism. Today, however, this same number surprises many because of how large it is

within a discipline often viewed as opposing religion.

Larson was honored with the George Sarton Award in 2000 from the American Association for the Advancement of Science, which recognizes a historian of science for a combined body of work. He also served on the panel examining the history of the Human Genome Project. In 2006, Larson accepted the Hugh and Hazel Darling Chair at the School of Law at Pepperdine University. He is married with two children. Larson was an unpaid consultant on the Discovery Institute's Cascadia Project that studies the transportation system of the Cascade region of the Pacific Northwest. He severed this relationship as the Discovery Institute became involved with the intelligent design movement.

JEROME LAWRENCE (1915–2004)

Almost if not all of our plays share the theme of the dignity of every individual mind, and that mind's lifelong battle against limitation and censorship.

Jerome Lawrence Schwartz was born in Cleveland, Ohio, on July 14, 1915. After graduating Phi Beta Kappa from The Ohio State University, Lawrence worked as a reporter and editor for several newspapers. In 1937 he began graduate school at UCLA and worked as a writer for a CBS radio station in Beverly Hills.

Lawrence met Robert Lee in a Madison Avenue bar in New York City, and the writers decided to form a freelancing writing partnership. They finished their first play—*Laugh, God!*—in 1939, and in 1942 Lawrence and Lee joined the Army after the U.S. entered World War II. While in the Army, Lawrence and Lee helped establish Armed Forces Radio Services, which aired patriotic programs on hundreds of radio stations. When Lawrence and Lee left the army in 1945, they resumed their writing, and in 1948 produced their first Broadway show, *Look Ma, I'm Dancing*.

Lawrence and Lee's best-known play was *Inherit the Wind*, which appeared in 1955. Harold Freedman, Lawrence's agent, stopped the play for almost a year, and finally Margo Jones opened *Inherit the Wind* in Dallas, Texas, on January 10, 1955. Lawrence and Lee also produced a radio version of *Inherit the Wind* in 1965. *Inherit the Wind* won numerous awards and has been translated into more than thirty languages. Far more than anything that occurred in Dayton, *Inherit the Wind* strongly shaped how future generations viewed the Scopes Trial and the evolution-creationism controversy.

Lawrence and Lee collaborated on thirty-nine works, including *Auntie Mame* and *The Night Thoreau Spent in Jail*, a story of Thoreau's imprisonment for "moral" tax evasion. These collaborations earned Lawrence a variety of awards (e.g., New York Drama Critics Poll Award) and honorary degrees, and The Ohio State University named its theatre archive to honor Lawrence and Lee (The Jerome Lawrence and Robert E. Lee Theatre Research Institute). Lawrence, who was elected to the Theatre Hall of Fame in 1990, subsequently taught playwriting at the University of Southern California (USC).

Lawrence died from complications of a stroke on February 29, 2004, in Los Angeles, California.

CURTIS LAWS (1868–1946)

Those who still cling to the great fundamentals and who mean to do battle royal for the fundamentals shall be called "Fundamentalists."

Curtis Lee Laws was born in Loudon County, Virginia, and educated at Crozer Theological Seminar. He then became a Baptist pastor. In 1913, the conservative Laws became editor of *The Watchman Examiner*, a newspaper that had the largest circulation of any Baptist periodical in the North. Laws edited the paper for twenty-five years, during which time he became a respected voice of Baptist ideology.

In June 1920, Laws—with the help of William Riley, John Straton, and Frank Norris—convened a "Fundamentals of Our Baptist Faith" conference just before the annual Northern Baptist Convention in Buffalo, New York. At that conference, Laws and his colleagues expressed "increasing alarm" about "the havoc which rationalism is working in churches," and urged their followers to "restate, reaffirm, and reemphasize the fundamentals" of their faith. At the same conference, Riley reminded attendees that "this is a war from which there is no discharge."

In the July 1, 1920 issue of *The Watchman Examiner*, Laws coined the term *fundamentalist*, which he defined as someone willing to "cling to the great fundamentals" and "do battle royal" for the faith. Laws hoped to promote theological orthodoxy and biblical Christianity, and his broad definition of *fundamentalist* required neither inerrancy nor dispensationalism. Laws emphasized verification of the Bible's truth, not its value as a scientific document, and believed that these truths would be known by common sense.

Laws died in 1946.

NICHOLAS LAWSON

Nicholas Lawson was vicegovernor of the prison colony Charles Island in the Galápagos Islands, who told Charles Darwin on September 24, 1835 that he could "pronounce with certainty from which island [in the Galápagos] any tortoise had been brought from the shape of its shell." At the time, Darwin did not grasp the significance of Lawson's claim, thus possibly explaining why Darwin did not collect shells from all of the islands.

THE LEAKEYS

Louis Seymour Bazett Leakey was born August 7, 1903, in Kabete, Kenya, just outside of Nairobi. His parents were British missionaries with the Kikuyu tribe, and as a child, Leakey was fluent in the Kikuyu language as well as English. When young, the Leakey children were taught by private tutors, augmented by schooling in England. It was planned that Louis would return to England for several more years of formal education, but World War I prevented such travel, and Louis remained in Africa for the rest of his childhood. As a teenager, he discovered stone tools near his home, and abandoning plans to become an ornithologist, decided to pursue anthropology.

In 1922, Leakey entered Cambridge. A rugby accident the next year, however, temporarily halted his education, and he returned to Africa as part of an archaeological expedition in 1924. He returned to Cambridge in 1925, and graduated with a degree in anthropology in 1926. Immediately after graduation, Leakey joined an anthropological expedition to East Africa, where he met his first wife, Frida. The couple married in 1928, and their first child was born in England in 1931. Leakey, who earned his PhD in 1930, developed a reputation as a good field anthropologist, although some have claimed his success was also due to good fortune—the so-called "Leakey luck." Leakey often made extreme claims, especially about the antiquity and importance of his finds, with little supporting evidence. This risky strategy backfired in 1932 when he reported finding the oldest *Homo sapiens* fossils then known. The claim was convincingly disputed, leaving Leakey's reputation tarnished.

In addition to his professional reputation, Leakey's public persona was also damaged in 1933, when he had an affair with Mary Nicol, a scientific illustrator he had hired. At the time, Frida and Louis were still married, and Frida was pregnant with the Leakeys' second child. When Louis returned to Africa for the next expedition, Mary accompanied him. In 1936, Frida divorced Louis, and Mary and Louis married later that year. Together, these events made it difficult for Louis to secure funding for his work, and he struggled for decades to support his research. Although Mary and Louis Leakey had productive professional careers, they later became estranged.

Mary Nicol was born February 6, 1913, in London, England. Her parents, Erskine and Cecilia, were both artistic, and her father introduced Mary to archeology through the cave paintings of Europe. Mary was inquisitive but disliked formal education, and she never finished high school (being expelled from school several times). By the age of 17, she was employed as an illustrator at archeological excavations in England.

In 1937, Mary and Louis Leakey returned to East Africa. Louis began an ethnographic study of the Kikuyu, whose culture he feared would be obliterated by European expansion into Africa. Mary began her own archeological studies. By 1939, the impending war made return to England impossible, and Louis started intelligence work for the Allied powers under the guise of a traveling salesman. Jonathan, the couple's first child, was born in 1940, followed by Richard in 1944, and Phillip in 1948. The Leakeys' successful Pan-African Congress of Prehistory and Paleontology in 1947 recaptured some of the respectability that was earlier lost.

In 1948, while working on Rusinga Island in eastern Lake Victoria, Mary found a fragmented skull of *Proconsul africanus*, an extinct primate extant about 20 million years ago. The discovery attracted much attention, but did not garner support from funding agencies. At the same time, the Mau Mau uprising (an insurgency against British rule started by the Kikuyu people) made conditions unsafe for the Leakeys, who were now living in Nairobi while Louis worked as a curator at the Coryndon Museum. Louis was convinced that Africa was where humans originated (many other anthropologists stressed Asia as the cradle of humanity), and the site the Leakeys believed to be the most fertile for investigation was Olduvai Gorge. The political turmoil in Kenya made the early 1950s an opportune time to move to Tanganyika (now Tanzania).

Mary made important discoveries at Olduvai, while Louis spent an increasing amount of time on other projects that required significant travel. (These projects included the well-known primatological studies led by Jane Goodall, Dian Fossey, and Biruté Galdikas. Louis also actively pursued a project, unsupported by evidence, aimed at demonstrating the presence of humans in North America over 80,000 years ago.) In 1959, Mary found a 1.8-million-year-old fossil of *Zinjanthropus biosei* ("Zinj"), later reclassified as *Australopithecus boisei*. This discovery made the Leakeys famous. Adequate funding was finally available to continue their work, which led to the discovery of a *Homo habilis* ("handy man") fossil in 1964. Stone tools found at the same site suggested tool use, and Louis declared *H. habilis* to be the earliest member of the genus *Homo*.

In 1976, Mary started working at Laetoli, a site near Olduvai, where she discovered fossilized animal footprints. Two years later, she found, preserved in solidified volcanic dust, three sets of fossilized footprints left by individuals who walked upright. The surrounding rock was 3.6 million years old, indicating that hominins walked upright before they evolved large brains. In 1974, Donald Johanson and colleagues discovered "Lucy," a 40 percent complete skeleton of a new hominin species. Johanson and Mary Leakey concluded that the Laetoli footprints (plus a jawbone also found at the site) and the fossils found by Johanson's team represented the same species, and a joint paper announcing the species was prepared. Prior to its release, Johanson announced at a conference, with Mary in attendance, the name assigned to the new species: *Australopithecus afarensis*. Mary never considered the australopithecines to be related to the genus *Homo* (a minority opinion in anthropology), but she did believe the footprints she found had been made by human ancestors; Johanson's pronouncement was therefore unacceptable. She removed her name from the paper describing the species, and a long-lasting feud developed between Mary and Johanson. However, the type specimen for *A. afarensis* remains the Laetoli find.

By the 1970s, Louis' health was deteriorating, and on October 1, 1972—while in London finishing work on the second volume of his autobiography—he died of a heart attack. Mary continued field work until her retirement to Nairobi in 1983. She died on December 9, 1996. Their son Richard was initially reluctant to emulate his parents (although, like his mother, he also never finished high school), and was more interested in working on photographic safaris and studying live organisms. However, in 1964, on a flight to Nairobi, he spotted from the air a promising location for excavations. The site, on the shore of Lake Rudolf (now Lake Turkana), was rich in fossils: in 1984, his team found a near-complete skeleton of a young *Homo erectus* ("Turkana boy"), and the following year, they found the first skull (the "black skull") of the species *Australopithecus aethiopicus*.

In 1969, Richard divorced his first wife, Margaret, and married Maeve Epps (b. 1942), a primatologist with a PhD in zoology. They had two children, Louise and Samira. Richard's work shifted into wildlife conservation in the 1980s, first as director of the Kenya Wildlife Service and then as a private consultant on conservation issues. In 1997, he was elected to the Kenyan Parliament. These accomplishments have been achieved despite major adversity, including a kidney transplant (from his brother Phillip) in 1979, a plane crash in 1993 that resulted in the loss of both legs below the knee, and a public beating in 1995 during political unrest in Kenya. Maeve Leakey and daughter Louise now lead the research in the Turkana Lake basin.

For More Information: Morell, V. 1995. *Ancestral Passions: The Leakey Family and the Quest for Humankind's Beginnings*. New York: Simon Schuster.

JOSEPH LECONTE (1823–1901)

Joseph LeConte was raised in a devout Southern family. The death of both of his parents—especially his unbelieving father—before LeConte turned 15 plunged LeConte into a spiritual crisis that ultimately resulted in LeConte becoming a Christian. He graduated from what is now the University of Georgia in 1841, from the New York College of Physicians and Surgeons in 1845, and—after practicing medicine in Macon, Georgia for a few years—enrolled at Harvard, where he apprenticed with Louis Agassiz. (LeConte later claimed—inaccurately—that Agassiz had founded the "biogenic law," that ontogeny recapitulates phylogeny.) LeConte then taught at the University of Georgia and what came to be the University of South Carolina. During the Civil War, LeConte continued to teach at the University of South Carolina and oversaw the production of medicine and explosives for the Confederacy. During this time, the death of his 2-year-old daughter Josie prompted another spiritual crisis that affected LeConte for the rest of his life.

By the early 1870s, LeConte had moved to California, where he helped organize the

University of California; he was also appointed the university's first professor of geology, a position he held until he died. In California, he became a "reluctant evolutionist," but years later enthusiastically endorsed evolution as being "entirely consistent with a rational theism." During this period, LeConte was one of the most famous reconcilers of evolution and religion, and his definition of evolution—continuous progressive change resulting from resident forces according to certain laws—was a standard for many years. His popular books included *Religion and Science* (1874) and *Evolution: its History, its Evidence, and its Relation to Religious Thought* (1888). In 1874, LeConte was made a member of the National Academy of Sciences, and in 1892 was president of AAAS. By the 1890s, however, LeConte was a pantheist who—largely because of the death of Josie—believed in immortality but rejected the efficacy of intercessory prayer, the existence of heaven and hell, and the fall of man.

LeConte—a charter member of the Sierra Club—died of a heart attack in Yosemite Valley, California, on June 6, 1901. LeConte is buried in Mountain View Cemetery in Oakland, California, beneath a tombstone carved from a piece of granite collected during his final trip to Yosemite.

ROBERT LEE (1918–1994)

Theatre is the universal means of expression. It embraces all of the arts through which human minds seek to reach one another.

Robert Edwin Lee was born on October 15, 1918, into a working-class family in Elyria, Ohio. After graduating from Elyria High School in 1935, Lee earned a degree in astronomy from Ohio Wesleyan University. He then worked as a writer at Cleveland radio stations and at an ad agency in New York.

In 1955, Lee and Jerome Lawrence produced the play *Inherit the Wind*, which used a fictitious account of the Scopes Trial to examine the perils of zealotry associated with McCarthyism. *Inherit the Wind* won numerous awards, and it remains the most popular and influential retelling of the Scopes Trial. After *Inherit the Wind*, Lawrence and Lee collaborated on more than thirty other major works, the most famous of which was *Auntie Mame*.

Lee subsequently taught playwriting at UCLA, received several honorary doctorate degrees, and in 1990 was elected into the Theater Hall of Fame. His final collaboration with Lawrence was *Whisper in the Mind* (1990).

Lee died of cancer in Malibu, California, on July 8, 1994, and is buried in Forest Lawn Cemetery in Hollywood Hills, California. His gravestone features a line from *Inherit the Wind*: "An idea is a greater monument than a cathedral."

LEMON v. KURTZMAN

Lemon v. Kurtzman was a case in which the U.S. Supreme Court in 1971 established the so-called "Lemon Test" detailing the requirements for legislation concerning religion. The Lemon Test has been cited repeatedly in court cases associated with the evolution-creationism controversy, including *McLean v. Arkansas Board of Education*. The Lemon Test's three "prongs" require that a government's action (1) have a legitimate secular purpose; (2) must not have the primary effect of either advancing or inhibiting religion; and (3) must not foster excessive entanglement of the government with religion. If any of these three prongs is violated, the government's action is unconstitutional under the Establishment Clause of the First Amendment to the United States Constitution. These three prongs resulted from three separate cases (*Lemon v. Kurtzman, Earley v. DiCenso*, and *Robinson v. DiCenso*), which were joined because they involved similar issues. The final decision became known by the first case of the list.

LEONARDO DA VINCI (1452–1519)

If you wish to say that it was the Deluge which carried these shells hundreds of miles from the sea, that cannot have happened.

Leonardo da Vinci was born on April 15, 1452, in the town of Vinci, in the Tuscan region of Italy. In 1469, Leonardo apprenticed with the sculptor Andrea del Verrocchio, from whom he learned sculpture and technical drawing. By 1481, Leonardo was working independently and moved to Milan to accept the title of "painter and engineer for the duke" (Duke Ludovico Sforza). Leonardo's seventeen years in Milan produced some of his most famous paintings, including *The Virgin of the Rocks* and *The Last Supper*. Leonardo left Milan around 1500, after the city was captured by the French. He spent a year as surveyor and military engineer for Cesare Borgia, who commanded the papal army charged with controlling the Papal States of Romagna (in modern-day northern Italy) for his father, Pope Alexander VI.

In 1503, Leonardo returned to Florence to begin a period of work marked by active scientific inquiry. He adhered to the philosophy of *saper vedere* ("to see is to know") and believed that artists were specially gifted to observe nature and were, therefore, best qualified for understanding the natural world. While in Milan, Leonardo recorded his scientific observations in notebooks, or codices, which included extensive observations and sketches of nature. During his second stay in Florence, Leonardo dissected dozens of human corpses and investigated plant and animal structure and flight in birds. Leonardo painted his masterpiece, *The Mona Lisa*, while living in Florence at this time.

Leonardo's codices cover a staggering range of information from technical drawings (including human anatomy, flying machines, and submarines) to basic scientific observations. The Leicester Codex (named for Lord Leicester, who bought the codex after it was discovered in the late 1690s but is now owned by Bill Gates) is primarily about the properties and action of water, but also discusses Earth's evolutionary history. Leonardo believed that the human body and the Earth are interconnected and parallel entities, what he referred to as "microcosm" and "macrocosm" respectively. He believed that processes in the human body, like movement of fluids, similarly occur within the entire planet.

One of these shared processes, the circulation of fluids, was proposed to distribute landmasses. Consequently, parts of the Earth were constantly rising and falling to achieve an overall planetary equilibrium as water circulated globally. Leonardo interpreted marine fossils on mountaintops as evidence for this vertical movement of landmasses.

Through his frequent engineering projects, Leonardo had ample opportunity to observe geological strata and fossils, and he may have identified the principle of superposition—"stratified stones of the mountains are all layers of clay, deposited one above the other by various floods"—before it was developed by Nicolaus Steno in the 1600s. But Leonardo was especially struck by fossils of marine organisms atop mountains. Leonardo rejected the two dominant explanations for the location of such fossils (i.e., the action of the Noachian flood and the spontaneous creation of such fossils *in situ*). He reasoned that, if due to the flood, the fossils should be jumbled, but instead they are found in groups like the colonies of similar living organisms. Furthermore, dead, heavy-shelled organisms should not rise in the water column, but should sink to the bottom of the flood waters. Even live organisms, especially those with low mobility, would not have had enough time to migrate hundreds of miles inland from the original ocean during the biblical Flood. Of the second explanation, Leonardo responded that only "ignoramuses" would accept the idea that such fossils were created in such a place: How would they grow if encased in rock? Why aren't they found in all types of strata? Why does the arrangement of fossils so closely resemble how extant marine organisms live?

Although Leonardo could not unify his concepts of the microcosm and the macrocosm with his observations of the functioning of the human body and Earth, he accepted the concept of an ancient Earth. He invoked the idea of a changing planet that experienced mountain building and erosion, and presaged the uniformitarianism of Hutton and Lyell. However, Leonardo's interpretation of fossils was cast aside during the

Reformation when a literal interpretation of the Bible, including the Flood, would prevail, supported by Cuvier's catastrophism.

Leonardo returned to Milan in 1506 at the behest of the French governor. He focused on sculpture and scientific studies, working particularly on his thesis that the forces of nature were orderly and understandable. Leonardo moved to Rome in 1513, and permanently left Italy in 1516 when King Francis I invited him to serve France. He took up residence at the Chateau of Cloux, the summer residence of the King, located in Ambiose. In 1519, Leonardo died and was buried at the church at the royal residence's compound. During the French Revolution, the church was destroyed and the exact location of Leonardo's grave is now unknown.

PETER LESLEY (1819–1903)

Peter Lesley was an influential American geologist and occasional Congregationalist minister. Lesley graduated from the University of Pennsylvania in 1838 and from Princeton Theological Seminary in 1844. Lesley disliked the Christian name "Peter," so he added a "J" (for "Junior," since Peter was named after his father and grandfather) before his name. Lesley was subsequently known as J. Peter Lesley and John Peter Lesley.

Although Lesley accepted humans' origin from apes, he rejected Darwinism, claiming that it was "the prevalent epidemic scientific superstition of the day." Lesley, who believed that orthodox Christianity hinders science, placed science ahead of church doctrine. In 1849, after becoming a Unitarian and declaring that Genesis is "a poem, not a textbook," the Presbytery removed Lesley's license to preach. Late in his life, Lesley—who was president of the American Association for the Advancement of Science in 1884—became a pantheist, believing that "God is nature, nature is God."

In 1893, Lesley—while preparing a major report for the Pennsylvania Geological Survey—had a breakdown from which he never recovered. He died in 1903.

JAMES LEUBA (1868–1946)

James Henry Leuba was a psychology professor at Bryn Mawr College who reported that over half of the *American Men of Science* doubted or rejected a personal God and personal immortality. In 1916 (and again in 1920) Leuba published his findings in *Belief in God and Immortality*, and noted that many students had lost their religious faith during college after being exposed to modern ideas, including evolution. William Jennings Bryan, who viewed Leuba's results as proof that evolution was destroying moral standards, was convinced by his book to become involved in the antievolution campaign. In 1997, Edward Larson repeated Leuba's survey, and found that the rate of belief in a personal god (approximately 40 percent) remained unchanged since Leuba's original survey. Both surveys showed, however, that belief in a personal god was substantially lower among members of the National Academy of Sciences than among less renowned scientists.

RODNEY LEVAKE (b. 1954)

The process of evolution itself is not only impossible from a biochemical, anatomical, and physiological standpoint, but the theory of evolution has no evidence to show that it actually occurred.

Rodney LeVake was born on November 16, 1954, in Colorado Springs, Colorado. After attending St. John's University and double-majoring in Natural Science and Social Science, LeVake enrolled at Minnesota State University at Mankato, where he earned an MA in Teaching in Life Science. In 1984, LeVake began teaching math and science in Faribault Junior High School in Faribault, Minnesota, and in 1997, he began teaching biology at Faribault High School. The course syllabus, course registration guide, and the curriculum adopted by the school board listed evolution as part of the biology curriculum. When other science teachers suspected that LeVake had not taught evolution, LeVake was confronted by his friend and colleague Ken Hubert about his teaching of evolution. When

LeVake responded on April 15, 1998, that he rejected evolution because it is not scientific, he was reassigned to a ninth-grade physical science course.

LeVake, a self-described fundamentalist, felt that he had been treated unfairly. Seeking advice, LeVake contacted The Rutherford Institute, Focus on the Family, and The American Center for Law and Justice (ACLJ). The ACLJ, an organization founded by televangelist Pat Robertson to defend "the rights of believers," helped LeVake sue the school and its administrators, claiming that LeVake was reassigned because his religious beliefs opposed evolution. LeVake—a member of the Institute for Creation Research—asked the Court to give him $50,000 (plus court costs) and declare "the district's policy, of excluding from biology teaching positions persons whose religious beliefs conflict with acceptance of evolution as an unquestionable fact, to be unconstitutional and illegal under the U.S. and Minnesota Constitutions." The District Court ruled against LeVake, noting that a teacher's right to free speech does not permit the teacher to circumvent the prescribed curriculum. LeVake began a lengthy appeal. The Minnesota Court of Appeals supported the original ruling, and LeVake's case ended on January 7, 2002, when the U.S. Supreme Court refused, without comment, to hear his case (*LeVake v. Independent School District #656*).

Today, LeVake teaches at Faribault High School in Faribault, Minnesota.

SINCLAIR LEWIS (1885–1951)

Harry Sinclair Lewis was an American novelist and playwright who wrote *Elmer Gantry*, a best-selling satire in 1927. The main character in *Elmer Gantry*—an opportunistic evangelist—was based partly on antievolution crusader John Straton. *Elmer Gantry* was banned in several American cities. In 1930, Lewis became the first American to win a Nobel Prize in Literature and the first to turn down a Pulitzer Prize. Lewis, who claimed that "when fascism comes to America it will be wrapped in the flag and carrying a cross," died in Rome in 1951 of complications

from alcoholism. Lewis is buried in Greenwood Cemetery in Stearns County, Minnesota.

WILLARD LIBBY (1908–1980)

Willard F. Libby was a chemist who in 1950 developed radioactive carbon dating. Because ^{14}C has a half-life of 5,730 years, it is only useful for dating objects less than 50,000 years old; that is, ^{14}C is useful for establishing archaeological chronologies (such as the Turin Shroud, which dates to 1275–1381 AD), but not useful for most geological applications. Libby received a Nobel Prize in 1960.

JOHN LIGHTFOOT (1602–1675)

Man was created by the Trinity about the third houre of the day, or nine of the clocke in the morning.

John Lightfoot was born into an ecclesiastical family on March 29, 1602, in Staffordshire, England. After graduating from the University of Cambridge, Lightfoot taught for two years at Repton, Derbyshire. Lightfoot then became a chaplain at Bellaport. In 1629, Lightfoot published his first work, entitled *Miscellanies, Christian and Judaical, penned for Recreation at Vacant Hours*, which—as the title suggests—he wrote in his spare time. He then became minister of Saint Bartholomew's Church, and moved closer to London to supervise the publication of his next book, the wonderfully titled *A Few, and New Observations, upon the Booke of Genesis, the most of them certain; the rest, probable; all, harmless, strange, and rarely heard of before*, which was published in 1642. In 1654, Lightfoot became Vice Chancellor of Cambridge University, and in 1658 published the first volume of his best-known work, *From the Talmud and Hebraica*. Lightfoot, an original member of the Westminster Assembly, was a leading Hebrew scholar and was the first Christian scholar to call attention to the importance of the Talmud. Lightfoot was also an accomplished botanist; he wrote the first flora of Scotland.

Lightfoot is often credited with claiming that creation occurred at 9:00 AM on October 23,

4004 BC (a date similar to that claimed eight years later by Irish prelate James Ussher). However, in Verse 26 on page 4 of the small, twenty-page booklet *A Few and New Observations*, Lightfoot wrote that "Man was created by the *Trinity* about the third houre of the day, or nine of the clocke in the morning." Nowhere did Lightfoot mention the creation of the Earth; his "nine of the clocke in the morning" referred only to the creation of humans. In fact, Lightfoot never wrote that creation occurred on October 23; that was added by subsequent writers, most notably Andrew Dickson White (1832–1918) in his *A History of the Warfare of Science with Theology in Christendom* (1896), a book in which White told readers that science and religion were in perpetual conflict. In 1644, Lightfoot's *The Harmony of the Foure Evangelists: Among themselves, and with the Old Testament*, set a date for creation: "From the beginning of time to this fullness of it, hath laid this great, wondrous, and happy occurrence of the birth of the Redeemer in the yeere of the world, three thousand nine hundred twenty eight." Lightfoot even gave the actual day—the September equinox, September 12. Nevertheless, Lightfoot continues to be (inaccurately) associated with the claim that creation occurred at 9:00 AM on October 23, 4004 BC.

Lightfoot died of pneumonia on December 6, 1675, in Ely, Cambridgeshire, England. He was buried in Much Munden.

For More Information: Bowden, J.K. 1989. *John Lightfoot: His Work and Travels*. London: Bentham-Moxon Trust at the Royal Botanic Gardens, Kew.

CARL LINNAEUS (1707–1778)

If you do not know the names of things, the knowledge of them is lost too.

Carl Linnaeus was born May 23, 1707, in Råshult, Sweden, to parents Nils and Christina (Figure 53). The family name, Linnaeus, was created by Nils when he started his university education. Inherited surnames were uncommon among Swedish peasants, such as Nils, but were required by the University of Lund where Nils went to study. Recalling the linden trees on his family's farm, Nils invented a name, "Linnaeus," based on *lind* (the Swedish word for "linden trees"). Nils Linnaeus was a Lutheran minister and recreational gardener, and hoped that his eldest son would follow him into the same vocation. Carl, however, was more influenced by his father's avocation (he was tending his own garden by the age of 5), and at a young age was fascinated with learning the names of the various plants he and his father grew. His father, disappointed by his son's lack of interest in the ministry, considered apprenticing Carl to local artisans. Before that occurred, Nils was told by a local physician that his son showed some aptitude for medicine. In 1727, Carl entered the University of Lund to study medicine, transferring to the University of Sweden the next year.

Linnaeus was not interested in medicine, and spent most of his time at the University's botanical garden. He was encouraged to study plants by several faculty, and was soon helping teach botany at the university. In 1730, he proposed a survey of Lapland (the wild northern region of Sweden) to assay the mineral (and, of course, botanical) resources of this relatively unknown area. In 1732, the 24-year-old Linnaeus left alone on an expedition that took five months and encompassed 3,000 miles of travel. Linnaeus identified more than 100 new species of plant, and made ethnographic observations of the Sami, the indigenous people of the region.

Another expedition, this time to central Sweden, began in 1734. It was on this trek that Linnaeus met his future wife, Sara Moraeus, to whom he proposed within two weeks of their meeting. Her father, concerned about the earning potential of his daughter's future husband, requested that Linnaeus complete his medical degree before he would agree to an engagement. Linnaeus consented, and traveled to the University of Harderwijk to finish his studies, earning his degree in 1735 after producing a thesis about malaria. He then studied at the University of Leiden, and visited eminent botanists of Europe. The same year, he published the first version of his *Systema Naturae*, which outlined his proposals for the classification of life.

In 1738, Linnaeus returned to Sweden, and the following year he married Sara. Linnaeus had planned to work as a physician, but because he had been gone for three years, he discovered that "nobody dared entrust their dear life in an untried Doctor's hands, not even his dog." Before long, however, Linnaeus was teaching botany at the University of Sweden, where he joined the faculty in 1741. (He and another professor switched positions after each was hired, to better fit the qualifications of each.) Linnaeus immediately began improving the university's botanical garden, and recruited students to travel the world to collect undiscovered species. In 1739, Linnaeus helped found the Swedish Academy of Sciences, the body that now awards Nobel Prizes. Carl and Sara had several children during this period, including two that died in childhood.

Like many people in his era and since, Linnaeus believed that God created life in the Garden of Eden. Linnaeus accepted the biblical account of the Flood, but did not believe that it could have moved organisms very far inland and covered them in sediments in the time available. As he noted, "He who attributes all this to the Flood, which suddenly came and as suddenly passed, is verily a stranger to science and himself blind, seeing only through the eyes of others, as far as he sees anything at all."

Linnaeus—who believed that classifying organisms would reveal the pattern of creation—was appointed the King's physician in 1747, a position of great prestige. In 1761, the King granted him nobility, whereupon Linnaeus took the name of Carl von Linné. He published significant work during this time, including *Species Plantarum* (first published as a fourteen-page booklet in 1735), which established the Latin binomial system for naming plants and is considered the beginning of contemporary botanical nomenclature. (Less successful were his attempts to make Sweden more self-sufficient by trying to grow tropical crops like cacao, coffee, and bananas in the harsh climate of Sweden.) Linnaeus' son, also named Carl, succeeded his father in his university position when the elder Linnaeus retired in 1763.

53. Carl Linnaeus, the first person to include humans in a biological classification scheme, hoped to uncover God's plan for creation by classifying organisms. His system of naming, ranking, and classifying organisms influenced generations of biologists and remains in use today. (*Library of Congress*)

In the 1770s, Linnaeus' health declined, restricting his ability to work. In 1778, he suffered a series of strokes and died on January 10, 1778, during a ceremony in Uppsala Cathedral. When his son Carl also died five years later, and without heirs, the family sold the elder Linnaeus' collection to Sir James Smith, who then founded the Linnean Society of London. Linnaeus is buried with his son in Uppsala Cathedral. Linnaeus' home is now a museum.

The system Linnaeus proposed in *Systema Naturae* revolutionized classification and taxonomy. He identified common characteristics—in particular, reproductive structures—that can be used to categorize life, but understood that his system would create artificial groups (e.g., the Cryptogamia, plants which do not have obvious reproductive structures.) Today, the single unifying theme in systematics, evolution, allows organisms to be classified based on the degree of shared ancestry. But having a systematic method that hierarchically cataloged species (Linnaeus used a system of kingdoms, classes, orders, and genera of increasing specificity) was a great advance, and even today, his system underlies all classification.

Although Linnaeus was the first person to include humans in a biological classification system, he was uncertain about whether humans should be classified into a separate genus. As he noted in a letter to a friend in 1747, "I ask you and the whole world for a generic difference between man and ape which conforms to the principle of natural history. I certainly know of none…If I were to call man ape or vice versa, I should bring down all the theologians on my head." Linnaeus named and placed humans, daringly, in the same group of mammals that included monkeys and apes. He even named two of the humans: *Homo sapiens* (Man the Thinker) and *Homo troglodytes* (Cave Dwelling Man).

Because open discussion of sex in the eighteenth century was somewhat taboo, the central role of sexual reproduction in Linnaeus' classification led to interesting consequences. Linnaeus usually alluded obliquely to the function of the structures he was measuring, which tended to enliven his writing so that it sometimes read like a romance novel: Linnaeus once described flowers as "bridal beds which the Creator has so gloriously arranged, adorned with such noble bed curtains, and perfumed with so many soft scents that the bridegroom with his bride might celebrate their nuptials with so much greater solemnity." Not surprisingly, some botanists were hesitant to employ Linnaeus' seemingly prurient system. (When Johann Siegesbeck referred to Linnaeus' system as "loathsome harlotry," Linnaeus named a common weed after him.)

Although Linneaus' motto was *Deus creavit, Linnaeus disposuit* ("God created, Linnaeus arranged"), he was not the first naturalist to categorize life; for example, Aristotle classified animals as "blooded" and "bloodless." However, by the tenth edition of *System Naturae* in

1758—which included 7,700 plant species and 4,400 animal species (his first systematic treatment of animal classification)—Linnaeus' system had become the standard. Today, modern taxonomy traces its ancestry directly to this document.

Before Linnaeus, species were generally referred to by either a common name (a problem because the same species may have multiple common names, and the same name might be used for different species in different areas) or by a confusing system of a generic name combined with a lengthy description. Linnaeus created the current system, introduced in *Philosophia Botanica* in 1751, in which each species has a unique combination of generic name (e.g., *Homo*) combined with a specific epithet that identifies a particular species of that genus (e.g., *Homo sapiens*). This system enables scientists worldwide to be certain of the species they are discussing. It has also produced some interesting scientific names, including *Upupa epops* (it really needs to be spoken aloud; a bird), *Parastratiosphecomyia stratiosphecomyioides* (the longest; a fly), *Aa brevis* (the shortest; an orchid), and some with cultural references such as *Masiakasaurus knopfleri* (a dinosaur named after guitarist Mark Knopfler). Linnaeus—who valued nature's diversity for its own sake, not merely for its theological value—is honored in the name of a small woodland plant of the northern hemisphere, *Linnaea borealis*.

Linnaeus was the world's preeminent botanist, and his classification system fit neatly within the Great Chain of Being in which he so ardently believed. During his lifetime, Linnaeus—who once commented that he could not "understand anything that is not systematically ordered"—identified an unbroken chain of organisms that stretched from the simplest to the most complex, with humans just below angels and God. Working in the tradition of natural theology (i.e., that God can be understood by studying nature), Linnaeus, early in his work, believed in the fixity of species, where each type of organism was specially created and immutable. However, over time he accepted interspecific hybridization, and eventually recognized that such events could produce new species. (Biologists now accept that, especially in plants, interspecific hybridization can lead to speciation.) However, Linnaeus believed these new species were part of the Creator's divine plan because the original parent species had been specially created.

For More Information: Blunt, W. 2004. *Linnaeus*. London: Francis Lincoln.

LINNEAN SOCIETY OF LONDON (Est. 1788)

The Linnean Society of London is the world's oldest extant biological society. It was founded in 1788, and takes its name from famed Swedish naturalist Carl Linnaeus, whose biological collections and books have been maintained by the Society. Those collections were bought in 1784 from Linnaeus' widow by Sir James Edward Smith, the Society's first President. The collection includes 14,000 plants, 158 fish, 1,564 shells, 3,198 insects, 1,600 books, and 3,000 letters and documents, most of which can be examined by appointment.

On the evening of July 1, 1858, the Linnean Society hosted the first public presentation of Darwin and Wallace's theory of evolution by natural selection. Darwin did not attend as he had buried his 18-month-old son earlier that day), and Wallace did not know about the presentation. In 2001, the refurbished room that hosted the presentation was adorned with a plaque noting the significance of the reading of the Darwin-Wallace paper.

The Society's role remains as stated in its first charter: "The cultivation of the Science of Natural History in all its branches." The Society maintains historical biological collections, and encourages debate, research, publications, and meetings. The Society also publishes several journals, including *Biological Journal of the Linnean Society, A Journal of Evolution*, which is a direct descendant of the oldest biological journal in the world (*Journal of the Proceedings of the Linnean Society*). This journal published the seminal paper by Darwin and Wallace on August 20, 1858

("On the Tendency of Species to form Varieties; and on the Perpetuation of Varieties and Species by Natural Means of Selection," Volume 3, pp. 45–62).

The Society, which in 2007 commemorated the anniversary of Linneaus' 300th birthday, forms part of Burlington House in Piccadilly, London. The Society's Library includes a portrait of Linnaeus, and its Meeting Room houses portraits of Darwin and Wallace.

JACOB LIPMAN

Jacob G. Lipman was a Rutgers agronomist who was prepared to testify at the Scopes Trial that soil, plants, and animals are linked in the evolution of organic life, and that agriculture cannot effectively serve humankind if legislatures ban the teaching of evolution. Lipman's testimony was read into the court record by Arthur Hays.

R.D. LITCHFIELD

R.D. Litchfield hosted a séance in 1873 at Erasmus Darwin's (Charles' brother) house that was attended by Charles and Emma Darwin and novelist Marian Evans (1819–1880), who wrote under the pseudonym, "George Eliot." Afterward, Darwin commented, "The Lord have mercy on us all if we are to believe such things."

LOTHIAN STREET

Lothian Street in Edinburgh was where Charles Darwin lived from 1825–1827 in Apartment 11. The tablet on the house commemorating Darwin's stay incorrectly numbers the house as Apartment 21.

JOHN LUBBOCK (1834–1913)

Sir John Lubbock was a banker, archaeologist, and Charles and Emma Darwin's neighbor at Downe. At the celebrated Oxford debate in 1860 between Thomas Huxley and Samuel Wilberforce, Lubbock defended Darwinian evolution, and in 1867—at Huxley's urging—Lubbock became the first president of the Royal Anthropological Institute. Lubbock also served as president of the British Association for the Advancement of Science and the Linnean Society, and in 1869 was elected to Parliament (his supporters included Darwin and John Stuart Mill). Lubbock created England's first secular bank holiday (St. Lubbock's Day, which is August 7), was a member of the X-Club, and was fooled by Piltdown Man. In 1865, Lubbock published the influential archaeology book, *Pre-historic Times, as Illustrated by Ancient Remains, and the Matters and Customs of Modern Savages*. Lubbock suggested that Darwin be buried in Westminster Abbey, and was a pallbearer at the funeral. Lubbock was also a pallbearer at Huxley's funeral.

LUCRETIUS (c. 95–55 BC)

Titus Lucretius Carus was a Roman poet and philosopher who wondered whether "there was ever a birth-time of the world and whether likewise there is to be any end." In the austere poem *On the Nature of Things*, Lucretius argued that everything in the universe was not created by deities (as most people believed), but instead consisted of tiny atoms moving in a infinite void. In 55 BC, Lucretius committed suicide.

LUCY

Lucy is a 3.2-million-year-old skeleton of a North African hominin discovered at Hadar in Afar, a triangular region in Ethiopia between the Blue Nile and the Red Sea (Afar is referred to as "Ophir" in the biblical story of Solomon and the Queen of Sheba). At Hadar, three of Earth's plates rub against one another, thereby exposing sedimentary deposits and fossils. On November 24, 1974, American graduate student Donald Johanson found almost one-fourth (47 of 206 bones) of a skeleton that was the most complete fossil of an early hominin ever found. The fossil was filed as A.L. (for "Afar Locality") 288-1, but it became famous as "Lucy"—so-named because some of the fossil hunters were playing the Beatles' song "Lucy in the Sky with Diamonds."

Lucy's discovery attracted worldwide attention to human origins.

Lucy was a diminutive hominin who had a small, ape-sized brain and walked upright. Lucy forced anthropologists to reject earlier claims that erect posture evolved with the enlargement of the brain because, in Lucy's case, standing up did not coincide with a larger brain. Johanson described Lucy as "from the neck up, chimpanzee; from the waist down, human... For all their apeness, Lucy and her kind did share the first human evolutionary marker: They walked. They kept on walking, beautifully adapted to their African environment, for another million years." Although Johanson considered classifying Lucy as a previously unknown species of *Homo*, Johanson's colleague Tim White convinced him that Lucy belonged to a new species of near-humans, which they named *Australopithecus afarensis*, meaning "Southern ape of the afar." The following year, Johanson and White returned to the site and unearthed thirteen more similar skeletons, prompting biologists to refer to the group as "the First Family." Lucy's species appears to be the last common ancestor of several branches of hominins that emerged 2–3 million years ago.

Lucy, who is better known than her discoverer, is a celebrity among fossils; she's appeared in cartoons and on television shows (e.g., *Jeopardy!*), and Ethiopia issued a commemorative stamp with Lucy's Amharic name—"Dinquinesh," which means "wonderful thing." Many hominin fossils are older and more complete than Lucy, but Lucy remains an important reference point to which other hominin discoveries are compared.

Johanson claimed that Lucy is "the oldest, most complete skeleton of any erect walking hominin found anywhere in the world," but South African anatomist Robert Broom's earlier discovery of australopithecine leg bones and pelvises at Sterkfontein showed that early hominins had walked upright. Although Lucy was originally estimated to be 3 million years old, Johanson announced Lucy as the same species as a skull discovered by Mary Leakey in Tanzania (1,000 miles away). Leakey disagreed with Johanson's classification and resented Johanson's "appropriating" her work. This began a long and often bitter feud between Johanson and the Leakeys.

Other famous ancestors of modern humans include Toumaï (*Sahelanthropus tchadensis*, who lived 7–6 million years ago), Taung Baby (*Australopithecus africanus*, who lived 3.5–2.3 million years ago), Nutcracker Man (Zinj; *Australopithecus boisei*, who lived 2.3–1.4 million years ago and was named for its large teeth), Handyman (Lucy's Child; *Homo habilis*, who lived 2.3–1.6 million years ago), Java Man (Peking Man; *Homo erectus*, who lived 1.7–0.4 million years ago), Heidelberg Man (*Homo heidelbergensis*, who lived 0.7–0.2 million years ago), Neanderthal Man (*Homo neanderthalensis*, who lived 0.2–0.03 million years ago), and Hobbit Man (*Homo floresiensis*, a disputed species who lived 0.07–0.012 million years ago). Many of these hominins lived in the Great Rift Valley, a geological split in Earth's crust that extends from eastern Africa to Southwest Asia.

For More Information: Johanson, D., B. Edgar, & D. Brill 2006. *From Lucy to Language: Revised, Updated, and Expanded.* New York: Simon and Schuster.

CHARLES LYELL (1797–1875)

The present is the key to the past.

Charles Lyell was born at his wealthy family's estate near Forfarshire, Scotland, on November 14, 1797, the same year that James Hutton died (Figure 54). Lyell, the oldest of ten children, became interested in geology when at age 15 he read Robert Bakewell's *An Introduction to Geology*. In 1816, Lyell entered Exeter College, Oxford, where he attended lectures by (and went on excursions with) flamboyant geologist William Buckland. Lyell's work would later overturn some of Buckland's claims.

Lyell received a BA from Oxford in 1819, and an MA in 1821. After practicing law for two years, Lyell went to France in 1823, where he met anatomist Georges Cuvier and renowned traveler Alexander von Humboldt. A year later, while in Scotland, Lyell visited Siccar Point with

54. Charles Lyell was a Scottish geologist whose *Principles of Geology* established that Earth has been in a perpetual flux for an inconceivable period of time. No other scientist had as great an impact on Charles Darwin's thinking as Lyell. (*Library of Congress*)

James Hall, who had accompanied Hutton to the famous site thirty-six years earlier.

In 1826, Lyell was elected a Fellow of the Royal Society (from which he would later receive the Royal and Copley Medals), and the following year he abandoned law for geology. Lyell had been introduced to geology through the catastrophism views (and their links with Noah's flood) that prevailed in his day. However, he soon began to question catastrophe-driven geology. In 1828, while visiting Sicily, Lyell found seabeds raised "700 feet and more" above sea level, and abandoned catastrophism for Hutton's uniformitarian view of geology based on the constancy of natural laws over time.

Unlike Hutton, who concluded that Earth went through a grand cycle of creation and destruction, Lyell argued that Earth was in a perpetual, directionless flux for an inconceivable period of time. Lyell was influenced by famed English astronomer Sir John Herschel

(1792–1871), who promoted making every element of a scientific theory a *vera causa* that could be studied directly. Past catastrophes were not observable, and were therefore outside of Lyell's view of geology. Instead, Lyell believed that geology had to be based on Earth's features being produced by "causes now in operation"—for example, rain, volcanic eruptions, and earthquakes. Lyell's uniformitarianism, which substituted vast expanses of time for the violent events claimed by catastrophists, required an immensely old Earth. Lyell was unequivocal: "All theories are rejected which involve the assumption of sudden and violent catastrophes and revolutions of the whole earth."

In July of 1830, Lyell published the first volume of his masterpiece, *Principles of Geology*. *Principles*, a name chosen to echo Newton's *Principia*, used the work of Hutton and Playfair as a starting point to describe an old Earth whose features had been formed by "the slow agency of existing causes" operating for long periods of time. The frontispiece of *Principles* featured a sketch of "The Present State of the Temple of Serapis at Pozzuoli" to show the recent rising and falling of Earth's crust. Lyell acknowledged Hutton on page four of *Principles* as the first scholar to study geology as its own subject, but noted that "although [Hutton's idea of an ancient Earth] was vehemently opposed at first, and although it has gradually gained ground, and will ultimately prevail, it is yet far from being established ... [Hutton] was the first ... to explain the former changes of the Earth's crust, by reference exclusively to natural agents. Hutton labored to give fixed principles to geology, as Newton had succeeded in doing to astronomy." Interestingly, in *Principles*, Lyell misquoted Hutton's most famous claim ("... we find no vestige of a beginning,-no prospect of an end.") Although Lyell admired Hutton's pioneering work, Hutton had dealt only with the physical aspects of geology (e.g., rocks and minerals). Lyell realized that geology would be incomplete if it ignored fossils, which record the changes encountered by past life on Earth.

By the time the second volume of *Principles* appeared in 1832, Lyell had married his wife Mary (who became associated with all of his work) and they honeymooned at the Bavarian limestone quarries at Solenhofen, where the famous *Archaeopteryx* skeleton would later be discovered. In 1831 Lyell became professor of geology at King's College in London, a college founded by members of the Church of England. Lyell remained at King's College for two years.

The full title of Lyell's famous book explained Lyell's thinking: *Principles of Geology: being an Attempt to Explain the Former Changes of the Earth's Surface by Reference to Causes now in Operation*. On the title page of Volume 2 Lyell quoted from Playfair's *Illustrations*: "The inhabitants of the globe, like all the other parts of it, are subject to change. It is not only the individual that perishes, but whole species. A change in the animal kingdom seems to be part of the order of nature, and is visible in instances to which human power cannot have extended."

Monthly Magazine predicted Lyell's book would "form an epoch in the history of science...and marks almost a new era in the progress of our science." More recently, evolutionary biologist Stephen Gould described it as "the most famous geological book ever written." Lyell showed that, if given sufficient time, common events such as erosion and wind could have produced all the changes that are recorded in rocks of the Earth's crust. However, Lyell—who wanted to "free the science from Moses"—admitted how difficult it is to fathom such periods of time ("the imagination was first fatigued and overpowered by endeavoring to conceive the immensity of time required for the annihilation of whole continents by so insensible a process"). Some readers were threatened by Lyell's world, in which the remorseless, ceaseless forces of nature are indifferent to the hopes and beliefs of Earth's inhabitants.

Principles was an immediate success. In 1833, Lyell published the second edition of *Principles*, and by the time of his death in 1872 he had published eleven editions. Like most monumental publications in science (e.g., those of Darwin and Galileo), and unlike those of his predecessor Hutton, Lyell's book was well-written and could be accessed by anyone of average education.

The impact and importance of *Principles* in geology were comparable to those of Adam Smith's *Wealth of Nations*; it finally, and firmly, documented Earth to be ancient. Lyell popularized and refined Hutton's ideas about uniformitarianism, while showing that Earth is something far more than the natural home for humans. Humans have been on a ceaselessly changing Earth for only a tiny portion of its history. *Principles* was the standard reference for geologists for several decades.

While aboard the *Beagle*, Charles Darwin read *Principles*. The book had a dramatic impact on Darwin, who dedicated the second edition of his *Journal of Researches* to Lyell. Asa Gray, Thomas Huxley, and Herbert Spencer modeled their work after that of Lyell, and Darwin modeled his *On the Origin of Species* after Lyell's *Principles*. There was no person whom Charles Darwin admired as greatly as Lyell; as he noted, "I never forget that almost everything which I have done in science I owe to the study of [Lyell's] great works." In a letter inviting Darwin to dinner on January 2, 1837, Lyell urged Darwin "to work as I did, exclusively for yourself and for your science for many years, and do not prematurely incur the honour, or penalty, of official dignities." Three months earlier, Darwin met anatomist Richard Owen at a dinner at Lyell's home.

In 1841, Lyell toured the United States and Canada with John William Dawson (1820–1899), a famous Canadian geologist who served as president of both the British Association and the American Association for the Advancement of Science (the only person to do so). Dawson, a day-age creationist whose opposition to evolution softened in his later years, became Canada's first famous scientist and a protégé of Lyell. While touring with Dawson, Lyell estimated the recession of Niagara Falls.

In 1838 Lyell published *Elements of Geology*, the first modern textbook of geology. Lyell eventually published six editions of *Elements*. In 1848, Lyell was knighted by Queen Victoria for his scientific accomplishments, and in 1864, Lyell became a Baron. In 1851, Lyell helped popularize the phrase *missing link* when he wrote that "newly discovered fossils serve to fill up gaps between . . . types previously familiar to us, supplying often the missing links of the chain, which, if [evolution] is accepted, must once have been continuous." Today, the phrase *missing link* is an outdated term that does not reflect how paleontologists view fossils. Instead of discussing "missing links," biologists discuss intermediate *features*.

In 1858, Darwin asked Lyell for advice when Darwin received Wallace's letter and manuscript about natural selection. On July 1, 1858, letters and papers by Darwin and Wallace were read at a meeting of the Linnean Society, a leading society of professional scientists in England (Wallace did not know of the meeting, and Darwin could not attend because his son had died two days earlier of scarlet fever). The presentation, which was organized by Lyell and Joseph Hooker, generated little interest among those who attended the meeting.

In 1859, Lyell urged his publisher (John Murray) to publish Darwin's "important new work," and Murray agreed. Darwin's *On the Origin of Species*, which to many was an inevitable sequel to Lyell's advocacy of uniformitarianism, troubled Lyell; he did not initially accept the same degree of continuity of life that he claimed for the Earth's surface. However, Lyell finally admitted that Darwin's book was "a splendid case of close reasoning" and that "I have been looking down the wrong road."

Unlike Hutton, Lyell was prominent in life and death. His primary source of income was *Principles*, which he revised continuously. (Lyell also received an annual allowance of $400–500 from his father.) In the early 1860s, Lyell helped document the first clear proof that humans had lived long before the beginning of recorded time. In 1863, Lyell summarized this research in *The Antiquity of Man*, a popular book that later went through three editions in one year, and a fourth edition in 1873. *The Antiquity of Man*, the first scientific book of its kind written in a style accessible to the public, rekindled the public's interest in human history. Darwin was disappointed that *Antiquity* did not endorse evolution as an explanation for human origins, telling Huxley that,

"I am fearfully disappointed at Lyell's excessive caution."

In 1865, Lyell spoke at a dinner at which the Royal Society awarded its Copley Medal to Darwin for Darwin's contributions to geology. Although Darwin did not attend, Lyell noted that his talk "was somewhat a confession of faith as to the 'Origin.' I said I had been forced to give up my old faith without thoroughly seeing my way to a new one." In 1866, Lyell was awarded the Wollaston Medal of the Geological Society.

Like many people who lived in Charles Darwin's time and since, Lyell struggled with Darwin's ideas about life's history. In the earliest editions of *Principles*, Lyell argued that organisms had been created perfectly adapted for local conditions, and although Lyell encouraged Darwin to publish *On the Origin of Species*, he initially rejected Darwin's claims. However, Lyell's own discoveries validated Darwin's ideas, prompting Darwin supporter Thomas Huxley to note that Lyell was "doomed to help the cause he hated." Huxley was right; although the religiously conservative Lyell had argued for special creation in the earliest editions of *Principles*, in the tenth edition, in 1867, Lyell endorsed Darwin's idea. Lyell showed how Darwin's theory explained how introduced species could sometimes out-compete indigenous species, a problem if species arose perfectly adapted to local conditions (as Lyell had suggested in the first edition of *Principles*). Lyell also used Darwin's theory to show why the biota of Australia was so different than that of Asia, and illustrated this with a map showing Wallace's Line. It was no small matter for Lyell to abandon the theory that he had for so long advocated in the book that had established his reputation. By the end of his life, Lyell—who believed that scriptural geologists were "wholly destitute of geological knowledge"—questioned traditional theism.

In June 1874, Lyell received an honorary degree from Cambridge, but soon thereafter he suffered a variety of infirmities. In early February 1875, Lyell finished revising Volume 1 of the twelfth edition of *Principles*, but he did not live to see it published. Thanks to the efforts of Lyell's nephew, Leonard Lyell, it was published in 1876.

Lyell died on February 22, 1875. Nearly forty Fellows of the Royal, Geological, and Linnean Societies immediately requested his burial in Westminster Abbey. The Dean of the famed abbey agreed, and Lyell—"the most philosophical and influential geologist [that] ever lived, and one of the best of men"—was laid to rest near John Woodward, a geologist who had done pioneering work 180 years earlier; Lyell's pallbearers included Joseph Hooker and Thomas Huxley. Darwin, who died seven years later, skipped Lyell's funeral. Lyell's tombstone forms part of the abbey's floor. In 1875 the Geological Society began awarding the Lyell Medal, which features the bust of Lyell on one side and the columns of the Temple of Serapis on the other side.

For More Information: Secord, J.A. 1997. *Introduction to Charles Lyell, Principles of Geology*. London: Penguin.

TROFIM LYSENKO (1898–1976)

Darwin himself, in his day, was unable to fight free of the theoretical errors of which he was guilty. It was the classics of Marxism that revealed those errors and pointed them out.

Trofim Denisovich Lysenko was born September 29, 1898, in the Ukrainian town of Karlovka. Lysenko had to work on the family farm, and did not learn to read or write until he was 13. As a young man, he studied horticulture for two years, and in 1917, entered the Vocational School of Agriculture and Horticulture in Ukraine. However, war prevented classes from being taught regularly, and in 1922, he moved to the Kiev Agricultural Institute, completing most of his remaining studies by mail. Lysenko received a certificate in agronomy in 1925, and immediately accepted a position as a specialist at an agricultural experimental station in Gandzha (Azerbaijan).

At Gandzha, Lysenko studied beans, a crop difficult to cultivate in the area due to the short growing season and long cold winters. Lysenko explored whether "vernalization"—exposure of seeds to an extended period of cold—could help seeds sown in the fall survive the winter. If successful, vernalized seeds would start growing earlier in the spring, providing a longer growing season. The first test appeared promising, although success may have been due to a relatively mild winter. Regardless, media reports heralded the positive results, which encouraged Lysenko to try the same with wheat, results which also appeared promising. He claimed to increase wheat yield by 10 percent, although it is now known that these reports arose from improper statistical manipulation of the data. Lysenko's announcement of his results gained the attention of plant biologists nationally, in Europe, and in the United States.

One scientist intrigued by Lysenko's results was geneticist N.I. Vavilov (1887-1943), a major figure in Soviet agricultural research who became the first director of the Lenin Academy of Agricultural Science in 1929. Although Vavilov was aware of Lysenko's lack of education and scientific expertise, the young man's apparent ability to produce results was attractive, and for several years, Vavilov supported Lysenko's work. Lysenko's success also coincided with Lenin's emphasis on collective farming and transformation of Soviet science from a bourgeois-dominated activity to one controlled by the people. Lysenko's rise from a poor, peasant family combined with his apparent research success made him an ideal candidate to serve in this role.

In 1929, Lysenko moved to the Odessa Institute of Genetics and Plant Breeding. Although Lysenko had a limited understanding of genetics, he became convinced that vernalized seeds could pass on their acquired cold-hardiness to offspring. Lamarckian inheritance had not yet been excised from genetics, and the inheritance of acquired characteristics was accepted by some plant geneticists. Lamarckian inheritance also fit nicely with the Soviet philosophy that the environment determines an organism's traits, which implied that human social progress (as defined by communism) could be a real, biologically based phenomenon. In contrast, the genetic determinism of Darwinian evolution (based on Mendelian genetics) seemed to argue against achieving these social and political goals. These philosophical issues, combined with Lysenko's claims that his research would eliminate famines, allowed Lysenko access to the highest levels of the government, including Joseph Stalin himself.

Not everyone accepted Lysenko's results or his optimistic assessment of future yields. However, by the 1930s, Stalin was purging Soviet science of the "bourgeoisie," so detractors were continually becoming less numerous. Vavilov was arrested in 1940 and sentenced to death; although the death sentence was rescinded, he died in a Soviet prison in 1943. In 1937, Lysenko was appointed deputy to the Supreme Soviet and became director of the Odessa Institute of Genetics and Plant Breeding, and after Vavilov's arrest, he assumed directorship of the Lenin Academy of Agricultural Science. From this post, Lysenko dictated to Soviet science a view of genetics that repudiated Mendel and Darwin.

Lysenko's suppression of genetics in the Soviet Union peaked in 1948 when he dictated that content incompatible with the inheritance of acquired characteristics be expunged from textbooks. In 1949, Lysenko announced that Soviet scientists had succeeded in transforming wheat into rye by planting wheat in areas specifically favorable to rye. However, by the 1950s, facing continued famines despite assurances from Lysenko that the problem had been solved, Soviet scientists began questioning Lysenko's claims and programs. Stalin's death in 1953 removed the final protection upon which Lysenko relied (Nikita Khrushchev did not provide the same level of support for Lysenko). In 1964, a committee established to investigate failures of the Soviet agricultural system determined that Lysenko had lied, and he was exiled. He died in confinement on November 20, 1976.

For More Information: Soyfer, V.N. 1994. *Lysenko and the Tragedy of Soviet Science*. New Brunswick, NJ: Rutgers University.

M

J. GRESHAM MACHEN (1881–1937)

J. Gresham Machen was a fundamentalist who often proclaimed that "I never called myself a fundamentalist." Although Machen urged churches to drive modernists out of their congregations ("Liberalism is not Christianity!"), he questioned many interpretations of Biblical prophecy, and he endorsed the use of alcohol and tobacco (he described tobacco as "a wonderful aid . . . to friendship and Christian patience"). Machen's scholarly approach to theology and science conflicted with that of most other fundamentalists. Machen—a day-age creationist—believed that science and religion all deal with the same thing — "facts"—and that "the church is perishing through a lack of thinking, not through an excess of it." When Machen founded the Westminster Theological Seminar in Philadelphia in 1929, the Presbyterian General Assembly suspended him from the Presbyterian ministry.

Unlike virtually all other fundamentalists, Machen refused to denounce evolution. Some followers of William Jennings Bryan wanted Machen to be the first president of Bryan Memorial University. Machen is buried in Green Mount Cemetery in Baltimore, Maryland.

DUDLEY FIELD MALONE (1882–1950)

Are we to have our children know nothing except what the church says they shall know? . . . Keep your Bible in the world of theology where it belongs and do not try to . . . put [it] into a course of science.

Dudley Field Malone was born in New York City on June 3, 1882 (Figure 55). Malone, the son of William Malone (a Tammany Democratic official) and Rose (McKenny) Malone, became a lawyer and member of the Democratic Party, and in 1913 was appointed Collector of the Port of New York. Four years later, Malone resigned to protest President Woodrow Wilson's failure to advocate a Woman's Suffrage Amendment. In 1920, Malone ran for governor of New York on the Farmer-Labor Party ticket, but his 49,953 votes fell far short of the 1,335,617 votes garnered by the winner, Republican Nathan Miller. When Wilson appointed William Jennings Bryan as Secretary of State, Malone was appointed Third Assistant Secretary under Bryan. Later, Malone developed a thriving international divorce business in New York.

Malone, a witty and well-dressed orator, was in Dayton for the Scopes Trial because of his work with Arthur Hays, his legal partner. Despite his rather accidental presence, Malone

55. Dudley Field Malone was an international divorce attorney who helped defend John Scopes at the Scopes Trial. Malone's "we are not afraid" speech was the most famous speech of the trial. After Scopes' trial, Malone moved to Hollywood and became an actor. (*Library of Congress*)

believed in the cause, noting that: "No more serious invasion of the sacred principle of liberty than the recent act against the teaching of evolution in Tennessee has ever been attempted."

On the fifth day of the trial, Malone delivered a speech that generated the loudest and longest applause. (Reporter H. L. Mencken reported that Malone's twenty-five-minute speech "roared out of the open windows like the sound of artillery.") According to John Scopes and others, Malone's speech was the turning point of the trial; when Malone concluded his speech, Scopes said he could see the "tragedy on [Bryan's] beaten face." The press, breaking its customary silence of neutrality, gave Malone a standing ovation (the court stenographer's transcript noted "profound and continued applause"), and legislator-turned-reporter John Butler described

Malone's speech as "the finest speech of the century." Scopes agreed, noting that Malone's reply to Bryan "was the most dramatic event I have attended in my life." Even Bryan begrudgingly acknowledged that Malone's speech was "the greatest speech I've ever heard," to which Malone responded, "I am sorry it was I who had to make it." Years after the trial, Malone admitted that his famous speech in Dayton was the only extemporaneous speech he ever made.

After Scopes' trial, Malone—the only professing Christian on the defense team—was asked by Will Rogers to join him on stage at the Ziegfeld Follies, where Malone called the trial a "victorious defeat." When the Tennessee Supreme Court set aside Scopes' conviction on a technicality (thereby eliminating the chance of further appeal), Malone condemned the decision as "a typical country lawyer's trick" to prevent the "further exposing of intolerance that prevails among the Tennessee fundamentalists."

When Bryan died in Dayton five days after the trial, Malone—unlike Darrow, Mencken, and others—praised Bryan, noting that "no man in the United States has done more to establish certain standards of conduct in politics." On September 19, 1925, Malone used a speech at the national convention of the Laymen of the Unitarian Church of America in Lenox, Massachusetts, to respond to Bryan's posthumous *Last Message*. Malone noted his contempt for Bryan's views, labeling Bryan "the ablest leader from the most sinister movement in the United States."

Malone, who had been accompanied to Dayton by his wife, returned to his divorce business in New York, and got divorced. His work often took him to Europe, and enroute he often oversaw "ship's pools," in which people would wager on the accuracy of the day's projected mileage. Malone represented some of sports' biggest names, including Jack Dempsey (heavyweight boxing champion) and Gertrude Ederle (the first woman to swim across the English Channel, breaking the men's record by almost two hours). Malone also spoke at the funeral of his friend George Lewis "Tex" Rickard, a promoter who in 1925 built New York's Madison Square Garden. In 1929, Malone's most memorable speeches were published in *Unaccustomed As I Am: Miscellaneous Speeches*. The book included texts of Malone's speeches about Woodrow Wilson's policies, women and suffrage, Russia, prohibition, and his electrifying "There Is Never a Duel with the Truth" speech from the Scopes Trial.

After serving as a delegate from New York to the Democratic National Convention in 1932, Malone used his booming baritone voice to launch a new career in Hollywood as an actor. His most prominent role came in 1943 when he played Winston Churchill in *Mission to Moscow*, a wartime film requested by President Roosevelt to support America's Russian allies.

Malone died on October 5, 1950, in Culver City, California, of a heart attack.

THOMAS MALTHUS (1766–1834)

The power of population is so superior to the power of the Earth to produce subsistence for man, that premature death must in some shape or other visit the human race.

Thomas Robert Malthus was born into a rich family on February 14, 1766, in Dorking, England, just south of London. Malthus went to Cambridge in 1784 where, despite a "marked impediment of speech," he became an ordained minister of the Church of England in 1788, and earned an MA in 1791. Malthus' father was a friend of several intellectuals, including philosopher David Hume. In 1804, Malthus married and began a family that eventually included three children. The following year he became England's first professor of political economy at East India College, where his colleagues and students referred to him as either Robert (as he was called by his parents) or "Pop" (short for "Population") Malthus.

Malthus, who coined the phrase "struggle for existence" to describe the war between tribes in Asia, studied a variety of economic issues, including monetary theories, protectionist laws, rent, and price stability. He was especially concerned about what he perceived as a decline of living conditions in nineteenth-century England. Malthus disputed the utopian views of William Godwin (1756–1836) and Marquis de Condorcet (1743–1794), who suggested that population growth was a blessing because it increased the availability of labor. Malthus blamed England's declining standard of living on overpopulation, the inability of resources to support the growing population, and the irresponsibility of the poor.

In 1798, Malthus anonymously published his epochal 50,000-word *An Essay on the Principle of Population, as it Affects the Future Improvement of Society with Remarks on the Speculation of Mr. Godwin, M. Condorcet, and Other Writers*. Although many believed that Malthus' essay was an analysis of the nature and causes of poverty (twenty-two years earlier Adam Smith had published such an inquiry, *Wealth of Nations*), Malthus' *Essay* was aimed at the optimism of Godwin and Condorcet.

In *Essay*, Malthus made a simple prediction—namely, that the human population would outstrip the supply of food. In later editions of *Essay*, Malthus added that the lower social classes were responsible for many societal ills, and suggested that they practice "moral restraint." "By moral restraint I mean a restraint from marriage, from prudential motives, with a conduct strictly moral.... Delaying the gratification of passion from a sense of duty." Malthus argued that attempts to improve the condition of the lower classes would be pointless, because improvements would be negated by the growing populations. As long as this tendency persisted, Malthus claimed, the romantic notions of societal "perfectibility" would be unattainable.

Before Malthus, high fertility rates were viewed as an economic plus because they increased the number of workers. Malthus' perspective, however, presented high fertility rates in a new way—namely, that although high fertility rates might increase the gross output, they also tend to reduce output per capita. Subsequent editions of *Essay* suggested regulating the size of the human population, but the overall message was unchanged, and became associated with oppressive measures against the poor. The seventh and final edition of *Essay*—now more than five times the length of the first edition—appeared in 1872.

Essay made Malthus famous and controversial. Although Malthus believed that God had established Earth's harsh conditions to improve morality, many people disliked Malthus' pessimistic predictions about the future. Malthus' ideas were condemned by social reformers such as Karl Marx (1818–1883) and Friedrich Engels (1820–1895), who claimed that societal ills could be eradicated with proper social structures. These and other reformers viewed Malthus as a heartless monster and enemy of the working class. Engels, Marx, and others argued that the seemingly innate tendency of the poor to reproduce beyond their means resulted not from a law of nature, but instead from capitalism. Evolutionists Ronald Fisher and John Maynard Smith were also critical of Malthus' theory, but for a different reason; both doubted whether famine had the impact that Malthus claimed.

Malthus' "struggle for existence" was a key part of the theory of evolution by natural selection formulated independently by Charles Darwin and Alfred Wallace. Wallace cited Malthus' essay as "the most important book I read," and Darwin referred to his theory as an application of Malthus' ideas. Unlike Malthus, who focused on humans and used God as a final cause, Darwin and Wallace framed Malthus' idea in purely natural terms, both in outcome and ultimate reason, involving all of life, thereby extending Malthus' theory far beyond what Malthus himself could have imagined. Darwin acknowledged Malthus in Chapter 3 of *On the Origin of Species*.

Many other people were similarly inspired by Malthus' idea. For example, Ernst Mayr referred to Malthus' theory as "the foundation of modern evolutionary theory," and Archdeacon William Paley viewed Malthus' idea as proof of the existence of a deity. Although Malthus believed that famine and poverty were natural events, he also claimed that they were God's way of teaching the virtues of hard work and moral behavior.

Malthus' idea had at least two flaws—namely (1) that human population growth is almost never exponential (it often fluctuates as a function of economic prosperity and is influenced by many factors); and (2) technological advances have dramatically increased the supply of food. Nevertheless, in countries where farming practices and contraception practices have lagged, famine, pestilence, and war continue to tragically vindicate many of Malthus' claims.

Today, Malthus is regarded as the founder of modern demography. In Michael Hart's *The 100: A Ranking of the Most Influential Persons in History* (1978), Malthus was ranked eightieth. Malthus' *Essay* was the first serious study of the welfare of the lower social classes, and many of today's debates about the carrying capacity of the Earth can be traced to Malthus.

Shortly after visiting his in-laws in Bath, Malthus died on December 29, 1834, while Darwin was at sea aboard the *Beagle*. He was buried on January 6, 1835, in the floor of Bath Abbey, and today his grave is covered by church pews. Malthus is memorialized with a tablet on the abbey's northern porch.

For More Information: Chase, A. 1977. *The Legacy of Malthus*. New York: Knopf.

GIDEON MANTELL (1790–1852)

Gideon Algernon Mantell was born on February 3, 1790, in Lewes, Sussex. He apprenticed as a surgeon, and in 1811 became a member of the Royal College of Surgeons. While working at the Royal Artillery Hospital in Lewes, Mantell spent his off-hours studying geology. In 1813, he published his first paper describing the geology of the Lewes area.

By 1819, Mantell began collecting fossils from a nearby quarry. Three years later, just before finishing his first book (*The Fossils of South Downs*), Mantell's wife Mary Ann found several large bones and teeth near Whiteman's Green in Cuckfield. British anatomist Richard Owen dismissed the teeth as being from a mammal, and French anatomist Georges Cuvier claimed that the teeth belonged to a rhinoceros, and the bones to a hippopotamus. However, Mantell believed that the teeth were similar to those of an iguana, but were twenty times larger. Mantel named the creature

Iguanodon ("iguana-tooth"). In 1825, Mantell announced his finding at a meeting of the Royal Society of London. By this time, Cuvier—admitting that "I am quite convinced of my mistake"—agreed that Mantell's fossils were from an unknown monster. Mantell's finding was the first discovery of a dinosaur (although the word *dinosaur* would not be coined by Richard Owen until 1842). Soon thereafter, William Buckland, working near Oxford, also found bones of dinosaurs.

After discovering *Iguanodon*, Mantell became obsessed with dinosaurs. In 1839, his wife left him and his medical practice failed. Facing mounting debt, Mantell sold his fossils to The British Museum (Mantell wanted £5,000, but finally accepted £4,000). In 1841, Mantell—"much broken in health and spirits"—was involved in a carriage accident that damaged his spine and left him in constant pain for the rest of his life. However, he continued to publish books including, *A Pictorial Atlas of Fossil Remains* (1850). Near the end of his life, Queen Victoria gave Mantell a civil gratuity of £100 per year.

On November 10, 1852, Mantell—depressed and morose—mumbled that "I am used up." He then died—in London—of an overdose of opium. Richard Owen removed part of Mantell's spine and stored it at the Royal College of Surgeons. Mantell was buried in West Norwood Cemetery near his daughter Hannah. In accordance with his wishes, no one was invited to his funeral.

Richard Owen's obituary for Mantell, which appeared anonymously in *Literary Gazette*, dismissed Mantell as "in want of exact knowledge." Owen was later denied the presidency of the Geological Society because of his "pointed and repeated antagonism to Gideon Mantell."

The site of Mantell's discovery of *Iguanodon* was marked in 2000 by the unveiling of a monument, and Mantell's house in Lewes is marked by a plaque inscribed "Dr. Gideon A. Mantell F.R.S., Surgeon and Geologist, Born in Lewes 1790, Died in London 1852, Lived Here, He Discovered the Fossil Bones of the Prehistoric Iguanodon in the Sussex Weald." The coat of arms of Maidstone, the county seat of Kent near the quarry where Mantell made his famous discovery, includes an *Iguanodon*. Mantell's *Iguanodon* tooth (Item #MNZGH004839) is displayed in the Museum of New Zealand, Te Papa Tongarewa.

For More Information: Dean, D.R. 1999. *Gideon Mantell and the Discovery of Dinosaurs*. Cambridge: Cambridge University.; Harzog, B. 2001. *Iguanodon and Dr. Gideon Mantell*. Cherry Hill, NJ: Rosen Group.

LYNN MARGULIS (b. 1938)

Disagreements [about evolution] have been misrepresented to the public by creationists as evidence that the theory of evolution is in doubt. On the contrary, they are evidence that what is going on is the pursuit of science and not the shoring up of dogma.

Lynn Alexander was born March 5, 1938, in Chicago, Illinois. When she was 16, Lynn entered the University of Chicago through an early entry program for advanced students. She planned to be a writer, but switched to science after taking a course in genetics. While at Chicago, she also met Carl Sagan, then a graduate student in physics, who became a well-known astronomer and popularizer of science. She and Sagan married in 1957, just before she earned her undergraduate degree in liberal arts.

The couple moved to Madison, Wisconsin, where Lynn began graduate school at the University of Wisconsin and studied inheritance of extranuclear genetic elements. Lynn finished her Master's degree in zoology and genetics in 1960, and moved with her husband to California, where she started PhD work in genetics at the University of California, Berkeley. She found in the scientific literature a decades-old suggestion that a symbiotic relationship among bacterial cells could be the origin of the eukaryotic cell. Here, finally, was an explanation for genetic material outside the nucleus: mitochondria and chloroplasts, which house circular chromosomes like those found in bacteria, were at one time in the evolutionary history of life, free-living cells that had started living within other bacterial cells. Over time, this symbiosis became integrated into a single unit, the eukaryotic cell.

By 1964, the Sagans had decided to end their marriage, with Lynn subsequently noting that: "It's not humanly possible to be a good wife, a good mother, and a first-class scientist ... something has to go." (Her 2007 novel, *Luminous Fish: Tales of Science and Love*, described the personal stresses of a life committed to research.) After finishing her PhD, she tried to publish her ideas about the evolution of the eukaryotic cell. Her bold proposal—the endosymbiotic theory—was rejected by more than a dozen scientific journals before being published in the *Journal of Theoretical Biology* in 1967. Sagan later expanded her discussion of endosymbiosis in the *Origin of Eukaryotic Cells* (1970). The endosymbiotic theory is now a well-established concept. In 1967, Lynn married Thomas Margulis and adopted the last name with which she is most generally recognized. This marriage ended in divorce in 1980. Lynn accepted an adjunct position at Boston University in 1966, and became full professor in 1977.

In *Five Kingdoms* (1982), Margulis and Karlene Schwartz outlined the basis for a classification system that separated fungi from plants. Like the endosymbiotic theory, this proposal gained widespread acceptance. Margulis has also been associated with the proposal that the entire Earth functions as a self-regulating system. This idea—the Gaia hypothesis—remains controversial within the scientific community. Margulis, however, views Earth as merely the largest example of an "autopoietic" or self-regulating entity (bacteria are the smallest).

Synthesizing the entirety of her professional work, from the endosymbiotic theory to Gaia, Margulis proposed that symbiosis is the major evolutionary force on the planet: evolutionary novelties (e.g., species) arise when new symbiotic relationships are created, after which these new associations are "edited" by natural selection. In several books and articles, most notably *Symbiotic Planet: A New Look at Evolution* (1999) and *Acquiring Genomes: The Theory of the Origins of Species* (2002), Margulis further developed the concept of *symbiogenesis* (first introduced by K.S. Mereschkovsky in 1926) as a mechanism for how species arise primarily as a result of newly acquired symbiotic interactions.

Convinced that symbiogenesis is the key to understanding evolution, Margulis predicted that the standard evolutionary model that emphasizes reproductive competition among genetic variants would be viewed as "a minor 20th-century religious sect within the sprawling religious persuasion of Anglo-Saxon biology." Mainstream evolutionary biology, built upon Darwinian evolution, has rejected this proposal. As Ernst Mayr noted, "It's startling to find a reputable scientist arguing such fantasies."

Antievolutionists have used Margulis' work to claim that scientists dispute the importance of natural selection. As an example of how controversy can sometimes lead to unexpected alignments between opposed factions, Young-Earth creationist Henry Morris has claimed common ground with Margulis because both she and the "creation scientists" have been victims of the "evolutionist bigotry" pervasive in science. Margulis, however, is no creationist; she rejects "scientific creationism," "irreducible complexity," and "intelligent design." Margulis describes her religious views by noting that "I'm from a Jewish family, but my personal religion is what I practiced today: I swam nude across a pond."

In 1983, Margulis moved to the Biology Department at the University of Massachusetts, Amherst, where she is now Distinguished University Professor in the Department of Geosciences. In 1983, she was elected to the National Academy of Sciences, and was awarded the Presidential Medal of Science in 1999.

For More Information: Brockman, J. 2004. *Curious Minds: How a Child Becomes a Scientist.* New York: Pantheon Books.

FRANK MARSH (1899–1992)

The main reason why creationism is talked down so generally today is probably the fact that evolutionists do not take the time to read the Bible carefully for themselves.

Frank Lewis Marsh was born on October 18, 1899. He earned a BS in 1929 from Emmanuel Missionary College (today known as Andrews University), and an MS in 1935 from Northwestern University. In 1940, Marsh became the first Seventh Day Adventist to earn a doctoral degree in biology when he earned a PhD in botany from the University of Nebraska.

Marsh, a student of flood geologist George McCready Price and a self-described "fundamentalist scientist," espoused a universal flood and a recent creation. According to Marsh, Satan is a "master geneticist," and the black skin of African Americans is an "abnormality" resulting from Satan's use of hybridization to destroy the original perfection of life. Like many other Seventh Day Adventists, Marsh believed that the world was the site of "a cosmic struggle between the Creator and Satan." In his *Evolution or Special Creation?* (1947), Marsh concluded that "the time is ripe for a return to the fundamentals of true science, the science of creationism."

In 1941, Marsh created a system of "discontinuity systematics" that he called "baraminology," from the Hebrew words *bara* ("created") and *min* ("kind"). Baraminologists, who view the baramin as a taxonomic rank representing the "created kinds" of Genesis, use baraminology to document boundaries between "microevolution" and "macroevolution," as well as to prove the existence of a designer.

Marsh believed that a global flood was responsible for the geological record, and he repeatedly affirmed the scientific validity of the Bible. Geneticist Theodosius Dobzhansky claimed that Marsh's *Evolution, Creation and Science* (1944) was a "sensibly argued defense of special creation," adding that "in rejecting macroevolution, Marsh's book taught the valuable lesson that no evidence is powerful enough to force acceptance of a conclusion that is emotionally distasteful." Marsh later split with Price, and in 1963 was one of the ten founding members of the Creation Research Society. Marsh resigned from the CRS Board of Directors in 1969 because the Board met on Saturdays.

56. Othniel Marsh was a famous fossil-collector in the second half of the nineteenth century. Marsh's discoveries convinced many people of the validity of Darwin's theory, and made Marsh one of the most famous paleontologists of his era. (*Library of Congress*)

Marsh died in 1992. His papers reside at Andrews University.

For More Information: Marsh, F.L. 1976. *Variation and Fixity in Nature*. Mountain View, CA: Pacific Press.

OTHNIEL MARSH (1831–1899)

To doubt evolution today is to doubt science, and science is only another name for truth.

Othniel Charles Marsh was born on October 29, 1831, in Lockport, New York (Figure 56). After graduating from Yale, he studied geology, mineralogy, and other topics in Germany. Marsh returned to the United States in 1866 to be an unpaid professor at Yale, and by the 1870s he had discovered fossils of ancient horses and fossil birds having teeth and other reptilian features. While in school, Marsh met Louis Agassiz, who became interested in Marsh's discoveries.

Marsh, who served as president of the National Academy of Sciences for twelve years, claimed to be "a man of great wisdom" and a friend of Buffalo Bill.

In 1868, Marsh discovered bones of a small horse *Equus parvulus* (now *Protohippus*). Marsh, the first professor of vertebrate paleontology in the United States, teamed with Thomas Huxley to describe the evolution of modern horses from a four-toed ancestor and, in the process, provided crucial evidence for how species evolve over time. Marsh and Huxley predicted that a more ancient, five-toed animal probably existed. Several months later, fossils of *Eohippus* ("the dawn horse")—just such an animal—were discovered. Marsh became a fierce rival of paleontologist Edward Drinker Cope, and their highly publicized rivalry was dubbed "The Bone Wars." Marsh spent $200,000 of his own money (and Cope $70,000 of his) on fossil-hunting expeditions associated with their competition. In one such expedition, Marsh found fifty specimens of *Triceratops*, each weighing an average of a ton (the largest weighed 3.5 tons).

In 1878, Marsh named *Diplodocus*; casts of this dinosaur were distributed throughout the world by Andrew Carnegie. The following year, Marsh named *Brontosaurus*. However, in 1903 (four years after Marsh's death), Elmer Riggs of Chicago's Field Museum determined that Marsh's *Brontosaurus* belonged to the same genus as *Apatosaurus*. According to the rules of scientific nomenclature, the name *Brontosaurus* was retired, although it continues to be used by the public.

As part of a 1868 peace treaty, the Great Sioux Nation was given all of western South Dakota and the sacred Black Hills. However, when gold was discovered in the Black Hills three years later, prospectors flooded into the Black Hills; George Armstrong Custer's (1839–1876) expedition there for gold in 1874 helped produce the lawless town of Deadwood, South Dakota. Although prospectors were violating the peace treaty, the Army protected the prospectors, which infuriated the Sioux. Marsh negotiated a peace treaty with Sioux Chief, Red Cloud, promising that he would take Red Cloud's

grievances to government officials if Red Cloud would allow Marsh to gather fossils. Red Cloud agreed, and Marsh and his crew took away 2 tons of fossils. When Marsh investigated Red Cloud's complaints, he was shocked at the treatment that the Sioux had received. Marsh took his concerns to various government officials and, when he did not get a satisfactory response, took his story to newspapers. The resulting "Red Cloud Affair" rocked Ulysses Grant's presidency, and Secretary of the Interior, Christopher Delano, resigned. Marsh remained friends with Red Cloud, who referred to Marsh as "the best white man I ever saw."

Marsh's discoveries convinced many people of the validity of Darwin's theory, and made Marsh one of the most famous paleontologists of his era. In 1871, Marsh found the first American pterosaur fossils. Charles Darwin wanted to see Marsh's specimens, but he never made the trip. In 1866, Marsh's uncle George Peabody endowed $150,000 for the Peabody Museum; many of the dinosaur fossils in that and many other museums can be traced to Marsh and Cope.

Marsh, who discovered and described hundreds of vertebrate fossils, died on March 18, 1899, and is buried in the Grove Street Cemetery in New Haven, Connecticut. Marsh left his estate and specimens to Yale.

For More Information: Ottaviani, J., Z. Cannon, S. Petosky, K. Cannon, and M. Schultz. 2005. *Bone Sharps, Cowboys, and Thunder Lizards: A Tale of Edward Drinker Cope, Othniel Charles Marsh, and the Gilded Age of Paleontology.* Ann Arbor, MI: G.T. Labs. Plate, R. 1964. *The Dinosaur Hunters: Marsh and Cope.* New York: McKay.

THOMAS MARTIN (1862–1939)

Germans who poisoned the wells and springs of northern France and Belgium and fed little children poisoned candy were angels compared to the teachers, paid by our taxes, who feed our children's minds with the deadly soul-destroying poison of Evolution ... If evolution, which is being taught in our high schools, is true, the Savior was no Deity, but only the bastard, illegitimate son of a fallen woman.

Thomas Theodore "T.T." Martin was born in Smith County, Mississippi, on April 28, 1862. He graduated in 1886 from Mississippi College, where his father preached and taught math. Martin abandoned a career in law for the ministry, and graduated from Southern Baptist Theological Seminary in Louisville, Kentucky, in 1891. After working as a pastor in Cripple Creek, Colorado for three years, he became an itinerant Southern Baptist preacher, teacher, and author.

Martin struck the first blow in the antievolution movement in North Carolina when in 1920 he used a series of articles in the Baptist publication *Western Recorder* to attack William Poteat, the president of Wake Forest College. Martin denounced Poteat's reconciliation of Christianity with Darwin's theory of evolution, but failed to convince the legislature to ban evolution.

At the Scopes Trial, Martin sold William Jennings Bryan's *In His Image* and George Price's *The Phantom of Organic Evolution*, but his biggest seller was his *Hell and the High Schools: Christ or Evolution—Which?* (1923), a book that described many aspects of Bryan's opposition to the teaching of human evolution. Chapter titles of *Hell and the High Schools* told its message—"Exposing Science, Falsely So Called," "Evolution Is Not Science," and "Evolution Repudiated by Great Scientists and Scholars." In Chapter 1, Martin stated his case: "Evolution says that there are ten lies in the first chapter of Genesis." Martin cited Leuba's study showing that most scientists do not believe in a personal god, claimed that the teaching of evolution was "the greatest curse that ever fell upon this Earth," and noted he had "never known a prominent Evolutionist who claimed to be Christian."

Martin defended antievolution laws by claiming that they protected students' religious liberties. After the Scopes Trial, Martin and others from Washburn's Bible Crusaders of America went to Mississippi and helped secure passage of an antievolution law there by likening evolution teachers to German soldiers who poisoned French children during World War I. Martin, who was active in the North Carolina Anti-Evolution League and the Anti-Evolution

League of America, remained active in evangelism until the final weeks of his life.

Martin died on May 23, 1939, in Jackson, Mississippi, and was buried in Glouster, Mississippi.

KARL MARX (1818–1883)

Heinrich Karl Marx was a German communist who corresponded with (but never met) Charles Darwin. Marx saw in Darwinism a scientific basis for his vision of class struggle. In 1880, Marx wanted to dedicate the English translation of *Das Kapital* to Darwin, but Darwin declined the offer. Marx lived most of his life in London and was buried in London's Highgate Cemetery, just a few steps away from rival Herbert Spencer. At Marx's sparsely attended funeral (guests included Thomas Huxley's protégé Edwin Ray Lankester), Friedrich Engels noted that "just as Darwin discovered the law of evolution in inorganic nature, so Marx discovered the law of evolution in human history." In 1873, Marx—who described humans as "apes of a cold God"—inscribed a copy of the second edition of his *Das Kapital* to Darwin "on the part of his sincere admirer," but Darwin did not read Marx's book.

KIRTLEY MATHER (1888–1978)

Kirtley F. Mather chaired the geology department at Harvard when he attended the Scopes Trial as an expert witness for the defense. Mather taught Sunday school for more than thirty years, and believed that God provides only a vague hope of immortality, does not answer prayers without human agents, and does not perform miracles. Mather also believed that evolution does not contradict Genesis, but rather "affirms that story and gives it larger and more profound meaning." Mather's testimony at Scopes' trial concluded with "comparing the body structure of monkeys, apes, and man, it is apparent that they are all constructed upon the same general plan." On the weekend before the end of the trial, Mather helped Darrow prepare for his questioning of Bryan. In 1926, Mather published "The Psychology of the Anti-Evolutionist" in *The*

Harvard Graduates' Magazine, noting the antievolutionists' inconsistencies, such as their opposition to human evolution but not the evolution of plants and other animals. Mather subsequently fought against teachers' oaths, spoke out against governmental abuses during the McCarthy era of the early 1950s, and served as president of the American Association for the Advancement of Science. Mather was also dedicated to helping the public understand science (he wrote some 1,500 book reviews on issues ranging from geology to religion) and at age 87 he incorporated the new idea of plate tectonics into a revision of his popular book, *The Earth Beneath Us.*

PATRICK MATTHEW (1790–1874)

Patrick Matthew was an eccentric naturalist who claimed in *Gardener's Chronicle* (1860) that he had discovered natural selection. Although Matthew had summarized natural selection twenty-six years before Wallace and Darwin in an appendix of *Naval Timber and Arboriculture* (1831), Matthew did not appreciate the importance of natural selection, and provided virtually no evidence for his claims. In later editions of *On the Origin of Species*, Darwin acknowledged Matthew. Although Darwin was not the first to "discover" natural selection, he was the first to grasp its significance.

SHAILER MATHEWS (1863–1941)

Shailer Mathews was Dean of the University of Chicago Divinity School who was ready to testify as an expert witness for the defense at the Scopes Trial. Mathews' *The Faith of Modernism* (1924) was a popular book promoting modernism. Mathews, who believed that "Genesis and evolution are complementary to each other," was frequently attacked by fundamentalist pastors such as William Riley. After the Scopes trial, John D. Rockefeller, Jr., gave $1,000,000 to Mathews' Divinity School.

ERNST MAYR (1904–2005)

The insatiable curiosity of human beings, and the desire for a better understanding of the world they live in, is the primary reason for an interest in science by most scientists. It is based on the conviction that none of the philosophical or purely ideological theories of the world can compete in the long run with the understanding of the world produced by science.

Ernst Walter Mayr was born on July 5, 1904, in Kempten (Bavaria) in southern Germany. Mayr entered medical school at the University of Greifswald in 1923, but did not become a physician. The watershed moment that set Mayr on the path to becoming one of the most influential evolutionary biologists of the twentieth century occurred before he had even started his medical studies. In March 1923, Mayr, a birdwatcher since childhood, and using binoculars he had received as a graduation gift from his mother, spotted a pair of birds he could not identify. After consulting his field guides, Mayr concluded that the birds were red-crested pochards, a species that had not been observed in this area of Germany for many decades. Mayr contacted the ornithologist Erwin Stresemann, at the Zoological Museum in Berlin, and after a lengthy interrogation, Stresemann concluded that Mayr had correctly identified the birds.

Thereafter, Mayr spent much of his time at the Museum and realized (with encouragement from Stresemann) that ornithology was his true calling. Mayr pragmatically completed enough of his medical studies so that he could return to them if his new career choice failed, and then moved to the University of Berlin to study zoology. He completed his PhD on avian biogeography in sixteen months so as to be eligible for a position then vacant at the Museum, a position he secured in 1926.

Stresemann had a close working relationship with Baron Walter Rothschild, who in the 1920s was building one of the world's largest bird collections for his museum in Tring, England. In 1928, Rothschild funded a yearlong collecting expedition to New Guinea and needed a naturalist, due to the untimely death of his most recent collector. Mayr got the position, despite the fact that he had never shot nor skinned a bird. Mayr collected more than 3,400 birds in New Guinea. He ate most of the birds he shot, and claimed to hold the record for eating the most birds-of-paradise.

After the expedition to New Guinea, Mayr was invited to participate in the American Museum of Natural History's (AMNH) Whitney South Sea Expedition to the Solomon Islands. AMNH staff were so impressed with Mayr's work that he was offered a position at the Museum to work on the collections from the Solomons expedition. Mayr moved to New York in 1931 and eventually became curator of the extensive Whitney-Rothschild collections. During his tenure at AMNH, Mayr married his wife Margarete, with whom he had two daughters.

It was while in New York that Mayr began studying avian systematics and published several important books. Mayr identified major problems with existing definitions of species. Mayr ultimately melded Georges Cuvier's older species concept, based on morphology, with newer concepts based on genetics, to define species based on the capacity for interbreeding: members of a species can interbreed to produce fertile, viable offspring. This modern view of species—the *biological species concept*—not only provided a way to identify a species, but also furnished a mechanism for how new species could arise. This new perspective successfully wedded advances in genetics to Darwin's ideas of adaptive evolution, and Mayr, along with Sewall Wright, Theodosius Dobzhansky, George Gaylord Simpson, and Ronald Fisher, became a major architect of the modern synthesis that unified genetics and evolution.

Mayr's *Systematics and the Origin of Species* (1942) used his extensive observations on geographic variation to propose a model for the creation of new species. This model, the *allopatric model of speciation*, required geographic separation of a once cohesive population into discrete units; these now isolated populations no longer represent a single interbreeding unit, and natural selection and chance effects could cause separated populations to diverge into new species. Both the biological species concept and the related allopatric model of speciation have been enormously influential in evolutionary biology by showing that species are real evolutionary

units created by identifiable processes that are open to study. Although there have been problems and complications (e.g., plants, which are known to hybridize quite readily, do not fit this species concept particularly well), Mayr's ideas have profoundly influenced the study of evolution.

In 1953, Mayr moved to the Museum of Comparative Zoology at Harvard University, serving as Director of the Museum from 1961 to 1970, and became Alexander Agassiz Professor of Zoology. At Harvard, Mayr continued to focus on species and speciation, emphasizing how rapid evolution and speciation can occur in isolated populations on the periphery of species' ranges. This peripatric model, discussed at length in *Animal Species and Evolution* (1963), was later used by Stephen Gould and Niles Eldredge in their concept of punctuated equilibrium, in which species exist unchanged for long periods of time followed by bursts of diversification. Mayr rejected reductionism in evolutionary biology, noting that natural selection operates on the entire organism rather than on individual traits, and was especially hostile to "gene level" selection espoused by Richard Dawkins and others. Mayr retired from Harvard in 1975.

Later in his career, Mayr became interested in the history of science, and in 1982 published *The Growth of Biological Thought*. In 1991, Mayr published *One Long Argument*, which discussed the development of evolutionary biology ("modern biologists consider evolution simply a fact"). In 1997, at the age of 93, Mayr authored *This Is Biology: The Science of the Living World*, which traced the development of biology from Aristotle to current times.

Mayr was awarded almost every major scientific award available to an evolutionary biologist: the National Medal of Science (1969), the Balzan Prize (1983), the International Prize for Biology (1994), and the Crafoord Prize (1999). (Mayr was not eligible for a Nobel Prize, a situation Mayr credited to Alfred Nobel being "an engineer and too ignorant about biology.") Mayr died of cancer on February 3, 2005, but this "Darwin of the 20th century" left a lasting imprint on evolutionary biology by demolishing forever the static, typological idea of organisms and providing a way to identify species and understand their formation.

For More Information: Mayr, E. 2001. *What Evolution Is.* New York: Basic Books.

ROBERT McCORMICK (1800–1890)

Robert McCormick was a surgeon and the official naturalist when the *Beagle* sailed from London. When McCormick was "invalided out" at Rio de Janeiro in April 1832, and sent home aboard the H.M.S. *Tyne*, Darwin noted that "he is no loss."

BEN McKENZIE (1866–1938)

People here [in Dayton] know that Genesis tells the full and complete story of creation.

Ben McKenzie was born in 1866, and at the time of the Scopes Trial had practiced law for thirty-two years. During the trial, McKenzie became friends with Darrow, and he described Darrow's two-hour opening speech as "the greatest speech that I have ever heard on any subject in my life." After most days' proceedings, Darrow and McKenzie retired to McKenzie's office to share stories and drink liquor. After the trial, McKenzie—who gave other attorneys military titles (e.g., Captain Malone, Colonel Darrow)—denounced the "interference" of Northerners ("foreigners") and proclaimed that it was better to kill children than to teach them about evolution. He also noted that the "Scopes Trial was a fake, fraud and farce that never convinced anyone that we were men or monkeys." When McKenzie nearly fainted during the trial, defense attorney Dudley Malone offered to pay to install ceiling fans in the hot courtroom. McKenzie, a master of overstatement, claimed that "the modern-day evolutionist is an atheist, and if he is not an atheist he is an agnostic."

Two weeks after the trial, McKenzie praised Darrow's work in Dayton. When McKenzie was arrested for illegally transporting liquor, Darrow

57. Aimee Semple McPherson was a famous preacher who integrated theater, sensationalism, and marketing to attract huge crowds to her church in Los Angeles. Although she was later discredited by an alleged "kidnapping" (that was used to cover up a romance with one of her employees), McPherson's sermons denouncing evolution were among her most famous. Here, McPherson uses her fist and Bible to fight the apelike specter of Darwinian evolution. (*Oregon Historical Society, Image No. 013168*)

offered to represent him, but McKenzie declined the offer, claiming that he'd been framed because his son, Gordon, was campaigning for county judge. McKenzie paid $50 for the offense.

McKenzie visited Darrow in Chicago in 1935, and hosted Darrow's return to Dayton in 1938. McKenzie, a founder of Bryan College, was a popular lawyer and after-dinner speaker throughout his career. McKenzie's son James Gordon McKenzie (1893–1944) was a coprosecutor in the Scopes Trial. Another son who became a lawyer, Oliver Wendell McKenzie, was named for U.S. Supreme Court Justice Oliver Wendell Holmes, before whom Ben had argued a case. When Wendell graduated from law school, Darrow sent him an autographed law book. Wendell later represented the Dayton Coal and Iron Company. Wendell's son James Wendell ("Jimmy") McKenzie served for many years as a family-court judge in Dayton, Tennessee.

McKenzie died on June 27, 1938, and his funeral was one of the largest in Dayton's history. McKenzie was buried in Buttram Cemetery in Dayton, Tennessee, beneath the inscription, "His greatest joy in life was in serving his fellow-men."

AIMEE SEMPLE McPHERSON (1890–1944)

We must either admit God or admit evolution and deny God. If we are the descendants of the ape family, then so was our precious Savior who we love so dearly.

Aimee Kennedy was born on October 9, 1890, in Salford, Ontario, Canada, and by the age of 13 was an agnostic who defended evolution and questioned claims by local preachers. Aimee became a Christian in 1907 during a revival led by Robert Semple, a fiery Pentacostal missionary who died of malaria two years after marrying Aimee. Aimee then married Harold McPherson, an accountant from New York. By 1915, Aimee was an itinerant evangelist. In 1916, McPherson toured the South in her "Gospel Car," a 1912 Packard on which she painted religious slogans such as "Where will you spend eternity?" McPherson, who like many other fundamentalist evangelists accepted donations from the Ku Klux Klan, disliked being compared to fellow evangelist Billy Sunday because of Sunday's use of "slang." However, McPherson, like Sunday, always put on a good show. In the early 1920s, McPherson moved to Los Angeles, where she incorporated theatre and patriotism into her church services and torchlit revivals. During one of her most famous and often-repeated services, McPherson depicted the villains Darwin, Hitler, Mussolini, and Lenin, after which the show climaxed with

McPherson emerging to read the national anthem.

While in California, McPherson became one of the most flamboyant and controversial preachers in the United States. Her theatrical sermons rivaled productions in nearby Hollywood, and her use of spectacle, celebrity status, patriotism, and marketing foretold modern evangelism. Although McPherson seldom delivered "fire and brimstone" sermons like Billy Sunday and Frank Norris, she often spiced her sermons with denunciations of evolution and ritual hangings of "monkey teachers" (Figure 57). McPherson, whose enormous church often hosted William Jennings Bryan, John Straton, and other antievolution crusaders, proudly proclaimed her willingness "to sacrifice science rather than religion." In 1925, McPherson promised Bryan that 10,000 members of her church would be praying for his success at the Scopes Trial. McPherson wanted to abolish all barriers between church and state, and urged Christians to seize control of government by boycotting schools that taught evolution. In 1927, "Sister Aimee" denounced evolution as "the greatest triumph of Satanic intelligence in 5,931 years of devilish warfare against the Hosts of Heaven. It is poisoning the minds of the children of the nation. It is responsible for jazz, bootleg booze, the crime wave, student suicides, Loeb and Leopold, and the peculiar behavior of the younger generation."

McPherson participated in several highly publicized debates with atheist Charles Smith, who also debated fundamentalists William Riley and John Straton. After offering $5,000 to anyone who could find a contradiction in the Bible, McPherson proclaimed that "The Bible can stand any test." Smith, the president of the American Association for the Advancement of Atheism, responded by calling McPherson "the one and only incarnated queen of the backward lookers and heaven seekers." Smith then used Epicurus' (341–270 BC) words to question McPherson's god: "Either God wishes to destroy evil and cannot, or He can but will not, or He both can and will. If he can but will not, he is wicked. If He wishes to destroy evil and cannot, he is impotent. If He both wishes to destroy evil and can destroy it, how come evil exists?" Smith and McPherson repeated their debate to overflowing crowds all along the West Coast. During her career, McPherson spoke directly to more than 2,000,000 people.

McPherson raised large sums of money, often telling her followers "no coins, please." On January 1, 1923, McPherson's award-winning float in Pasadena's Tournament of Roses Parade led worshipers to her newly opened Angelus Temple, which became the home base for McPherson's religious empire. McPherson, who wore make-up, jewelry, and appealing clothes, was flamboyant and attractive, and often preached in a long white gown while holding flowers. Although critics compared her with Cleopatra and complained of her "unashamed use of low-key sex appeal," McPherson attracted giant crowds. Indeed, McPherson held three services per day in Angelus Temple, and most services filled all of the church's 5,300 seats. "Sister Aimee" occasionally spoke in tongues and performed faith healings, and kept a museum of crutches and wheelchairs of people who had been healed in her services (her 1921 revival in Denver included a "Stretcher Day"). In *Elmer Gantry*, Sinclair Lewis modeled Sister Sharon Falconer—an attractive healer—on McPherson.

The often-divorced McPherson, who was the first woman to deliver a sermon on radio and be granted an FCC license (for her "Foursquare Gospel" station KFSG), became a celebrity who socialized with movie stars. When KFSG's broadcasting license was revoked in 1925 for deviating from its assigned frequency, McPherson allegedly sent then-Secretary of Commerce, Herbert Hoover, a telegram demanding that his "minions of Satan" allow her church to continue broadcasting.

Despite her fame as an evangelist, McPherson is best known for her alleged "kidnapping." On May 18, 1926, McPherson disappeared when she and her secretary went swimming at Ocean Park, California. Most people assumed that she had drowned, and two people died while trying to find her body. At about the same time, Kenneth Ormiston—an engineer at KFSG—also

disappeared. About a month later, McPherson's mother (Minnie Kennedy) claimed that she received a ransom note from "The Avengers" demanding $500,000 to avert selling McPherson into "white slavery." On June 23, McPherson reappeared in a Mexico desert just across the border from Douglas, Arizona, claiming to have been kidnapped and tortured. There were many inconsistencies in her story (she disappeared wearing a bathing suit, but reappeared in a gown; she wasn't wearing a watch when she disappeared, but was wearing one when she reappeared), and witnesses later claimed to have seen McPherson and Ormiston at various hotels during the time that she was allegedly kidnapped. When McPherson returned to Los Angeles, she was greeted at the train station by more than 30,000 supporters.

A grand jury investigated McPherson's alleged kidnapping, but adjourned two weeks later without delivering an indictment. The grand jury later reconvened and reviewed hotel documents written in McPherson's handwriting and witnesses' claims that Ormiston had been holed up in a beach bungalow with an unknown, disguised woman (more than 75,000 people later visited McPherson's suspected "love nest"). Although McPherson stuck to her story about the kidnapping, she refused to answer questions about her relationship with Ormiston, and Judge Samuel Blake charged McPherson and her mother with obstruction of justice. However, on January 10, 1927, those charges were dropped, and the $25,000 reward offered by Angelus Temple for anyone who could offer information about McPherson's whereabouts during her kidnapping was never claimed. In August 1930, McPherson—who was more famous than most movie stars—had a nervous breakdown. In 1933, a Broadway play about McPherson's life lasted only a week on Broadway (for which McPherson was paid $17,000).

On September 27, 1944, McPherson died of a drug overdose in a hotel room in Oakland, California. More than 40,000 mourners viewed her body as it lay in state at Angelus Temple for three days amidst $50,000 worth of flowers. McPherson was buried on October 9 (her birthday) in a 1,200-pound bronze casket in an ornate hilltop sarcophagus in Forest Lawn Memorial Park Cemetery in Glendale, California. Rumors circulated that she was buried with a telephone in her casket to ensure her survival if her body was resurrected. In decades following her death, McPherson was often portrayed as a religious hypocrite and sexual vixen.

Today, Angelus Temple—now a federally protected historic landmark—stands opposite Echo Park near downtown Los Angeles. Visitors to the parsonage are greeted by a large photo of McPherson and Bryan. The Foursquare Gospel Church, which McPherson founded, has more than a million members, most of whom live outside the United States. In 1976, the movie *The Disappearance of Aimee* (starring Faye Dunaway as McPherson) described McPherson's alleged kidnapping. McPherson continues to be regarded by her followers as a prophetess.

For More Information: Sutton, M.A. 2007. *Aimee Semple McPherson and the Resurrection of Christian America.* Cambridge, MA: Harvard University.

FRANK McVEY (1869–1953)

Frank LeRond McVey was president of the University of Kentucky who in 1922 opposed a proposed bill that would have banned the teaching of "atheism, agnosticism, or the theory of evolution" in Kentucky's public schools. Thanks to McVey's opposition, the legislation was voted down by the Kentucky legislature on March 9, 1922, by a vote of 42–41. This was the nation's first vote on legislation to ban the teaching of evolution. McVey's courage impressed University of Kentucky undergraduate John Scopes, who in 1925 agreed to be arrested to test the validity of an antievolution law in Tennessee.

HENRY LOUIS MENCKEN (1880–1956)

The evil that men do lives after them...Bryan, in his malice, started something that will not be easy to stop.

Henry Louis "H.L." Mencken was born on September 12, 1880, into a German-American family in Baltimore, Maryland (Figure 58). He was the eldest son of August Mencken, who owned a successful tobacco business. As a child, Mencken read an average of three books per week, and when Mencken was 8 years old he discovered Mark Twain, who would have a strong influence on Mencken (Mencken regarded *Adventures of Huckleberry Finn* to be the finest work of American literature). After his father bought young Henry a camera for Christmas in 1892, Mencken became interested in newspapers and publishing.

In 1896, Mencken graduated as the valedictorian of Baltimore Polytechnic Institute. He wanted to attend college and be a newspaper reporter, but instead began working as a bookkeeper in his father's cigar company. When his father died of a stroke in 1899, Mencken got a job at the *Baltimore Morning Herald*. Less than two years later he was the paper's drama critic, three years later he was the city editor, and four years later he was the paper's managing editor. When the *Herald* closed, Mencken began working for the Baltimore *Sunpapers* (now the Baltimore *Sun*). He continued to work for the *Sun*, on and off, until 1948. Along the way, Mencken edited the influential literary magazine *The Smart Set* (from 1914–1923), which promoted writers such as Eugene O'Neill, James Joyce, Sinclair Lewis, and F. Scott Fitzgerald. *The Smart Set* became known for its pithy epigrams, which were shown in movie theaters to entertain crowds before movies started.

In 1924, the cigar-chomping Mencken founded *The American Mercury*, a "magazine of cleverness" for the "civilized minority." *The American Mercury* was the first magazine edited by white editors to publish the work of African-American writers. Mencken published the worst manuscripts that he received in *Parisienne Monthly Magazine*, *Saucy Stories*, and *Black Mask*, all of which were popular magazines for, in Mencken's words, "the morons." Mencken also wrote books, including *The American Language*, an exhaustive, multivolume study of how English is spoken in the United States. Mencken

58. H.L. Mencken was a famous writer whose coverage of the Scopes Trial is regarded as some of the greatest journalism in American history. Mencken's columns, which enthralled readers and shaped the trial, coined the terms *Bible Belt* and *Monkey Trial*. (*Library of Congress*)

received at least one letter per day about *The American Language* for the rest of his life, and *The American Language* is regularly ranked as one of the 100 most influential books in the United States. Mencken even published a creed: "I believe that religion, generally speaking, has been a curse to mankind—that its modest and greatly overestimated services on the ethical side have been more than overcome by the damage it has done to clear and honest thinking....I believe that the evidence for immortality is no better than the evidence of witches, and deserves no more respect....It is better to know than be ignorant."

On May 14, 1925, Mencken met attorney Clarence Darrow in Richmond, Virginia, and urged him to go to Dayton, Tennessee, to defend John Scopes. When Mencken went to Dayton in July to cover Scopes' trial, he was described as "the most respected, hated, reviled, feared, and loved person" in Tennessee. When members of the prosecution spoke, spectators often looked to Mencken for a response. When a table Mencken

was standing on collapsed, one woman in the audience proclaimed that "It's a judgment. The walls are falling in and Mr. Mencken is the first to go—and he won't go to glory, either." Before Scopes' trial, Mencken suggested that the ACLU "sacrifice" Scopes and "use the case to make Tennessee forever infamous" by luring Bryan onto the witness stand "to make him state his barbaric credo in plain English, and to make a monkey of him before the world."

Mencken's coverage of the Scopes Trial is regarded as some of the greatest journalism in American history. Mencken reported, as well as shaped, the trial; even Scopes admitted that his trial "was Mencken's show," and that "a mention of the Dayton trial more likely invokes Mencken than it does me." Mencken, who coined the enduring phrases *Bible belt* and *Monkey Trial*, wrote thirteen articles about the trial, each of which appeared in the Baltimore *Sun*. More than 200 reporters covered Scopes' "Trial of the Century," but Mencken's articles had the most impact; they were syndicated and quoted throughout the country. He wrote about all aspects of the trial, calling Thomas Stewart (the chief prosecutor) "a convert at a Billy Sunday revival," and Dayton's residents "poor ignoramuses." His articles were alternately satirical, sensible, condemning, and cruel.

Because of his prior agreement with the Baltimore *Sun*, Mencken—who admired philosopher Herbert Spencer—left Dayton after the trial's first week, and before Darrow's questioning of Bryan. Mencken knew that the trial would affect Americans for years to come, and he was right.

Although Bryan had once described Mencken as "the best newspaperman in the country," Mencken despised Bryan; his first comment upon hearing of Bryan's death was: "We killed the son-of-a bitch." Bryan's death did not slow Mencken's attack. Mencken told his readers that if Bryan was sincere, "then so was P.T. Barnum." Mencken's "In Memoriam: W.J.B." became a masterpiece of invective: "[Bryan] seemed deluded by a childish theology, full of an almost pathological hatred of all learning, all human dignity, all beauty, all fine and noble things. He was a peasant come home to the barnyard. Imagine a gentleman, and you have imagined everything that he was not." Jerome Lawrence and Robert Lee took Mencken's unflattering descriptions of Bryan at face value when they wrote *Inherit the Wind*.

In 1930, newspaper headlines across the country announced that the 50-year-old Mencken—"America's Best Known Bachelor"—had married English teacher and Southern writer Sara Haardt, and the couple moved into an apartment on Cathedral Street. However, Sara died of tuberculosis in 1935, after which Mencken moved back to the brick row house at 1524 Hollins Street, where he had lived since he was an infant. The backyard housed a stable and a pet pony, a garden, a brick wall adorned with a plaque for his dog Tessie (who is buried in the garden), and a death mask of Ludwig van Beethoven, one of Mencken's favorite composers. Except for the time he was married to Sara, Mencken lived in the house until his death.

Mencken was an outspoken defender of civil rights and free speech, and an opponent of persecution, pretense, injustice, Puritanism, and self-righteousness. He captured readers with a writing style that was exceedingly confident, satirical, and sometimes hateful, but always entertaining and thought-provoking. Regardless of whether people loved or hated Mencken, they read his columns, and Mencken's words elicited responses. For example, Mencken's claim in 1931 that Arkansas had reached the "apex of moronia" prompted the state legislature to pass a motion to pray for Mencken's soul. Authors such as Sinclair Lewis (*Main Street*) and Ernest Hemingway (*The Sun Also Rises*) mentioned Mencken in their books, and in 1927 Lewis dedicated *Elmer Gantry* (a novel about an itinerant and unscrupulous preacher) to Mencken "with profound admiration."

Mencken, who endorsed Social Darwinism, was the most famous journalist of his era. During his lifetime, Mencken wrote more than 100,000 letters, thirty books, numerous poems, and thousands of articles; he also edited two magazines and played piano in a band, The Saturday Night Club. Mencken's newspaper columns and books made Mencken, in the words of journalist Walter Lippmann, "the most powerful personal influence" in America. Today, Mencken's work continues "to stir up the animals"—he is regularly quoted in newspapers and magazines such as *The New York Times*, has been a character in cartoon strips, and his fight for freedom of speech is often mentioned in Congress and on television shows such as *Law and Order*. At the time of his death, Mencken was one of the leading authorities on the English language, as well as a "violent agnostic"(which he attributed to the writings of Thomas Huxley).

In 1948, Mencken suffered a stroke that left him unable to read or write. As the end drew nearer, Mencken commented simply that "I had a good time while it lasted." Mencken died in his sleep at his home on January 29, 1956, and his ashes were buried in a family plot in the northeastern corner of Baltimore's Loudon Park Cemetery. Mencken's death was announced with a variety of headlines; "We shall not soon see his like again" (*Baltimore Sun*); "As champion of the unfettered mind, he has few equals" (*Boston Globe*); "The arch foe of every sort of sham and hypocrisy. . . . We are impoverished by his death just as we were enriched by his life" (*New York Times*). His epitaph, which Mencken had published thirty-five years earlier in *The Smart Set*, reads, "If, after I depart this vale, you ever remember and have thought to please my ghost, forgive some sinner, and wink your eye at some homely girl." Mencken's epitaph, which was quoted by nearly every obituary writer in America, also appears on a plaque in the lobby of the Baltimore *Sun*.

Although Mencken's house was designated a National Historic Landmark in 1983, it stood empty and shuttered for several years. Recently, The Friends of the H.L. Mencken House have attempted to restore the home and open a Mencken Museum to educate the public about the life and legacy of Mencken. Mencken's papers and manuscripts are preserved at Baltimore's Enoch Pratt Library, Harvard University, Dartmouth College, Princeton University, Yale University, and the New York Public Library.

For More Information: Teachout, T. 2003. *The Skeptic: A Life of H.L. Mencken*. New York: Perennial.

GREGOR MENDEL (1822–1884)

I knew that the results I obtained were not easily compatible with our contemporary scientific knowledge, and that under the circumstances publication of one such isolated experiment was doubly dangerous; dangerous for the experimenter and for the cause he represented.

Johann Mendel was born July 22, 1822, in the village of Hyncice in Moravia of Austria-Hungary (now the Czech Republic). Johann's father was a farmer who hoped his only son would help work the farm as an adult. However, Johann's intellectual abilities and interests in science encouraged the parish priest to suggest that Johann should attend a larger school in a neighboring town. Even though Mendel's parents were of modest means, they agreed. Academic success allowed Johann to move to the Gymnasium in Troppau, from which he graduated in 1840. He was by this time supporting himself financially.

Mendel then entered the University of Olmütz, where he focused on mathematics and physics. After financial problems forced Mendel to leave the university, he entered the Augustinian Order at St. Thomas Monastery in Brünn, the capital of Monrovia. The depth of Mendel's faith has never been questioned, but his decision to become a priest was likely influenced by his interests in science. The city of Brünn was then a center of cultural and scientific activity, and the abbot of the monastery supported scientific inquiry, especially as it related to agriculture. These features undoubtedly made the monastery at Brünn attractive to Mendel; he joined the monastery (and was given the name Gregor) in 1843, and was ordained in 1847.

To his disappointment, Mendel discovered that he was not well-suited to his chosen career, and found visiting the sick and dying—a standard duty for a priest—particularly distressing. Mendel was moved to a teaching position, which required him to pass a certification examination, which he failed. He spent 1851 to 1853 at the University of Vienna in a teacher's training program, taking additional coursework in mathematics and science, including botany. Mendel returned to the monastery and again failed the exam. He was then allowed to serve as a substitute teacher in the local secondary school until he was elected abbot fourteen years later.

In 1854, Mendel was provided space in the monastery's garden to conduct experiments on hybridization in garden peas, *Pisum sativum* (Figure 59). The area surrounding Brünn was agricultural, and because the monastery was a regional center of scientific study, this project to understand the process of hybridization could benefit local farmers. Mendel studied the inheritance of seven different traits in peas, with each trait studied alone or in combination with other traits. Mendel, a meticulous experimentalist and record-keeper, generated large data sets (i.e., he frequently examined hundreds of individuals of each generation) that allowed him to identify the frequencies of the alternate forms of the traits he studied. He was therefore able to

59. Gregor Mendel was a monk, and then an abbot, whose studies of pea plants at this abbey in Brno, Czech Republic, revolutionized genetics. Mendel grew his plants on these grounds; the foundation in the foreground is all that remains of his greenhouse. (*David Fankhauser, University of Cincinnati*)

discern patterns that demonstrated much about the basis of heredity; if he had relied only on qualitative data (i.e., presence/absence), he would have been unable to reach the conclusions he did. Furthermore, Mendel was comfortable applying mathematical models to explain his data, an unusual skill for a natural historian at that time, which reflected his training in mathematics and physics.

In 1865, Mendel reported his findings at two presentations to the Natural Science Society of Brünn, which met at 22 Jánská Street; that building, which is marked with a commemorative plaque, is now a technical college. The following year, a paper ("Experiments in Plant Hybridization") describing Mendel's results was published in the Society's *Proceedings*. (Mendel published only four papers during his brief research career. Besides the 1866 paper, he published one in 1854 on damage to plants by pea beetles, one in 1870 on breeding experiments in hawkweed, and one in 1871 on a tornado that

caused extensive damage in Brünn.) His results revealed several important aspects of the nature of heredity. First, he demonstrated that the "potentially formative elements" (later called *genes* by Wilhelm Johannsen) act like particles by maintaining their integrity across generations, rather than being blended together as predicted by the "blending" model of heredity. Second, Mendel showed that genes exist in different forms (with different forms producing different versions of a trait, e.g., green seeds versus yellow seeds); these alternate forms of a gene are now called *alleles* (i.e., there are two different versions of the gene that determines seed color, one that codes for green, and one that codes for yellow). Third, individuals have two copies of each gene, and these copies segregate into separate cells during the production of sex cells (gametes). This *principle of segregation* predicts that for individuals having two different copies of the same allele (the heterozygous condition), half of the sex cells should contain one allele, and half should contain the other allele. Finally, when examining two traits (e.g., seed color and seed texture) simultaneously, where each trait is determined by a separate gene, Mendel determined that the probability of finding a specific version of one trait (e.g., yellow seeds) is independent of inheriting a specific version of the other trait (e.g., wrinkled seeds). This *principle of independent assortment* applies to genes located on different pairs of chromosomes (i.e., genes that are not linked).

Mendel's results did not immediately create interest. His 1865 presentation was attended by forty people, and apparently little discussion followed his talk. His 1866 paper likewise did not attract attention, as most readers incorrectly interpreted his conclusions as merely confirming that hybridization eventually leads to reversion to the ancestral form. However, after publication of *On the Origin of Species*, the issue of heredity became of greater interest, and in 1900, Mendel's results were "rediscovered" by Hugo de Vries and others. For a while, "Mendelism" was synonymous with de Vries' mutational model of evolution. As Theodosius Dobzhansky has noted, "some log jams had to be cleared before Darwinism and Mendelism could join forces," but by the early 1900s, these obstacles were being removed, principally by Thomas Hunt Morgan's work demonstrating that genes were located on chromosomes. This set the stage for the revolution of the 1920s and 1930s known as the modern synthesis.

Reexamination of Mendel's results has speculated that Mendel's frequency data fit the model of inheritance he identified exceedingly well. Ronald Fisher concluded that Mendel's results had been faked, although he blamed an assistant of Mendel's who may have independently "helped" the monk achieve the anticipated results. Others, however, have suggested that if the sample sizes Mendel used varied from the average values he reported, this could account for the close fit between the expected and observed frequencies. Regardless, there is no evidence that Mendel purposely conspired to either collect his data in a biased manner or that he adjusted his data after collection.

Mendel hoped to replicate his results from peas in another species, and he began studying hybridization in hawkweed (*Hieracium*). Unfortunately, these attempts were frustrated because, unknown to Mendel, hawkweeds reproduce asexually as well as sexually, which meant that the frequencies of traits among offspring were due not only to sexual reproduction between the individuals he crossed. This frustration, combined with his increased administrative duties when he was elected abbot in 1868, ended Mendel's research into hybridization and heredity. However, he did continue studies of meteorology and beekeeping.

The disappointment of his research program on hybridization, combined with ongoing tension between himself and the Austrian government over issues of taxation of the monastery, placed considerable strain on Mendel during the latter part of his life. Mendel, who suffered from kidney problems, died on January 6, 1884, and is buried in the monastery's cemetery in Brünn, near a plaque noting that Mendel "discovered the laws of heredity in plants and animals. His knowledge provides a lasting scientific basis for recent progress in genetics."

Today, Mendel is commemorated with a museum (The Mendelinium) at his Augustinian monastery that includes Mendel's notebooks, frescos of Mendel's plants, and the names of scientists punished during the Lysenko era. Outside the monastery are Mendel's garden (which grows plants in rows labeled P_1, F_1, and F_2), the foundations of his greenhouse, and a large statue—commemorated in 1910—of Mendel and his pea plants (in the 1950s, the statue was hidden from Communists, who wanted to destroy it). Above the entrance to Mendel's museum is the inscription, "My time will come."

For More Information: Henig, R.M. 2002. *The Monk in the Garden*. New York: Houghton Mifflin; Mawer, S. 2006. *Gregor Mendel: Planting the Seeds of Genetics*. New York: Harry N. Abrams.

MAYNARD METCALF (1868–1940)

Maynard Metcalf, a zoologist from Johns Hopkins University, was the first expert witness called by the defense team in the Scopes trial. In 1904, Metcalf had published *An Outline of the Theory of Organic Evolution*, and at the time of the Scopes Trial was chairman of the division of biology and agriculture of the National Research Council. Metcalf, a Christian who taught Bible classes at the Congregational Church in Baltimore, was chosen in part because the defense wanted to demonstrate that an intelligent person could both accept evolution and be a Christian, thereby supporting the defense's claim that the teaching of evolution does not necessarily reject the biblical story of creation. When asked by Clarence Darrow if he knew "any scientific man in the world that is not an evolutionist," Metcalf responded that he was "absolutely convinced from personal knowledge that any one of these men feel and believe, as a matter of course, that evolution is a fact," and that disagreements were only about "the exact method by which evolution had been brought about." Metcalf also noted that: "The fact of evolution is as fully established as the fact that the Earth revolves around the sun," and claimed that ignoring evolution would be "criminal malpractice just as truly as would be a physician's failure to follow established sound methods of treatment because of fear of persecution by ignorant neighbors." Metcalf was the only expert witness allowed to testify at Scopes' trial; summaries of the other expert witnesses were read into the court record by Arthur Hays. Metcalf died on April 19, 1940, in Winter Park, Florida.

STEPHEN MEYER (b. 1958)

Stephen C. Meyer is Vice President of the Discovery Institute and director of the Institute's Center for Science and Culture (CSC). He earned undergraduate degrees in geology and physics from Whitworth College and a PhD in the history and philosophy of science from Cambridge University. While at Cambridge, Meyer met Phillip Johnson, who was then working on a manuscript that eventually became *Darwin on Trial* (1991). When both returned to the United States, they joined with others to coalesce the intelligent design (ID) movement in the 1980s.

Meyer and political scientist John West (now associate director of the CSC) were asked by Discovery Institute founder Bruce Chapman to establish a research center for ID. This was the start of the Center for the Renewal of Science and Culture, later renamed the Center for Science and Culture. The CSC "supports research by scientists and other scholars challenging various aspects of neo-Darwinian theory."

In 2004, Meyer published an article in the *Proceedings of the Biological Society of Washington* that reviewed preexisting information on the Cambrian explosion as evidence for ID. The manuscript did not follow the journal's standard peer-review process, being reviewed and accepted solely by the journal's editor, Richard Sternberg. Sternberg is a proponent of ID and a signatory to the Discovery Institute's "A Scientific Dissent from Darwinism." When they reviewed Meyer's article, the publishers of the journal deemed the article "inappropriate" for its publication. The Discovery Institute cites this response as evidence for "what amounts to a doctrinal statement in an effort to stifle scientific debate" by the scientific community, but also dismisses the importance of peer-review (i.e., "as if such journals represented the only avenue of legitimate scientific publication"). Meyer coedited (with John Angus Campbell) *Darwinism, Design and Public Education* (2004), which "presents a multi-faceted scientific case for the theory of intelligent design." Meyer is also a chief architect of the "teach the controversy" initiative of the Discovery Institute.

MICHAEL POLANYI CENTER (Est. 1999)

The Michael Polanyi Center (MPC) for Complexity, Information, and Design was established at Baylor University in 1999, ostensibly to "foster reflection and conversation between religion and the historical and philosophical nature of science." The MPC originally had two faculty members, Associate Director Bruce Gordon and Director William Dembski. Dembski was invited to join Baylor's Institute for Faith and Learning (IFL) in 1998 by Baylor President Robert Sloan, Jr.

Sloan, appointed president in 1995, had worked to establish the IFL in 1997 to unify religion and science. He was also noted for his doctrinaire management style and efforts to return Baylor ("the largest Baptist university in the world") to its Baptist roots, both of which caused dissension within the university community.

In 1999, the MPC—named by Dembski—separated from the IFL, with Dembski as its director. In April 1999, the MPC hosted a "The Nature of Nature" conference (jointly sponsored by the Discovery Institute) that included well-known academic scientists and ID advocates. Baylor faculty feared that this seemingly official endorsement of creationism would harm Baylor's reputation and would impede Sloan's goal of having Baylor become one of the country's top universities. Outside the university there was also concern that use of the name Michael Polanyi (a Hungarian-born scientist and philosopher [1891–1976] who emphasized the subjective nature of scientific inquiry due to unavoidable biases inherent in what he termed "personal knowledge" of the scientist) in association with a pro-ID organization was inappropriate. Polanyi scholars have argued that he would not have aligned himself with ID, and the Polanyi family did not authorize use of the name.

Sloan convened a committee of outside experts to examine the MPC. In October 2000, the committee advised that the MPC should be folded into the IFL. The report supported continuance of Dembski's ID-related work, although it ironically cited a paper by philosopher Elliot Sober that is critical of ID as an example of how ID was starting to receive academic attention.

Dembski immediately claimed the committee's report vindicated his research and "marks the triumph of intelligent design as a legitimate form of academic inquiry" and that "dogmatic opponents of design who demanded the Center be shut down have met their Waterloo." Sloan asked Dembski to retract his statement, after which Dembski accused Sloan of "intellectual McCarthyism." Dembski was removed as director of the MPC, although he was allowed to remain with the IFL as an associate professor. The MPC was renamed "Program in Science,

Philosophy and Religion," and William Dembski left Baylor in 2000 to become Carl F. H. Henry Professor of Theology and Science at The Southern Baptist Theological Seminary in Louisville, Kentucky. In 2005, Dembski moved to the Southwestern Baptist Theological Seminary in Fort Worth and remains a senior fellow with the Discovery Institute's Center for Science and Culture. Dembski was replaced at The Southern Baptist Theological Seminary by Kurt Wise from Bryan College.

ARTHUR M. MILLER

Arthur M. Miller, who taught geology to John Scopes at the University of Kentucky, published a letter in *Science* supporting Scopes in 1925. While in graduate school, Miller was a student of Henry Osborn at Columbia University.

HUGH MILLER (1802–1856)

Between the Word and the Works of God there can be no actual discrepancies.

Hugh Miller was a Scottish stonemason, writer, and geologist who wrote a variety of popular geology books, including *Footprints of the Creator* (1849), a rebuttal to Robert Chambers' *Vestiges of Natural History*. Miller, like many others, saw parallels between species' embryological development and their entire history, and believed these parallels—along with the beauty and design in nature—were evidence of God. Miller—one of the greatest geological evangelists of his time—attacked the work of Chambers and Lamarck while trying to reconcile geology with biblical accounts of creation. In *Testimony of the Rocks* (published posthumously in 1868), Miller helped popularize day-age creationism when he claimed that biblical "days" might each represent long periods.

Although Miller did not have a college degree, he discovered many important fossils of ancient fish, and he was sought by Adam Sedgwick, Richard Owen, and William Buckland. He published relatively few scientific papers, but became a newspaper editor (*The Witness*) who produced some of Scotland's greatest writing.

Indeed, Miller's autobiographical *My Schools and Schoolmasters* (1854) of life in early nineteenth-century Scotland became a famous example of self-improvement. Charles Darwin cited Miller's work in *On the Origin of Species*, and famed geologist Archibald Geikie (1835-1924) praised Miller for clothing "the dry bones of science with living flesh and blood."

Miller endorsed a position between those of biblical literalists and geologists who saw no role for God. Miller, who believed that Noah's flood was a regional event, often noted that "life itself is a school, and Nature always a fresh study." Many of Miller's specimens—including the type specimen of the fish *Pterichthys milleri*, which Louis Agassiz named in honor of Miller—are now displayed in National Museums Scotland.

On December 23, 1856, after leaving a final letter bidding farewell to his wife Lydia and telling her that "my brain burns," Miller raised his shirt and shot himself in the chest. (Miller's gun later discharged and killed a person investigating Miller's death.) Many of Miller's publications were issued posthumously by Lydia, a children's author. Miller is buried in Edinburgh's Grange Cemetery. In 1975, Miller was commemorated with the opera *Hugh Miller*.

Today, Miller is honored by several memorials, statues, a museum (The Hugh Miller Museum and Birthplace Cottage in Cromarty, Miller's birthplace), and the Hugh Miller Institute, a public library in Cromarty partly funded by Andrew Carnegie. In 2002 (the bicentenary of Miller's birth) when Black Isle Brewery produced Hugh Miller Ale, Miller became the first person to be honored both with a beer and a temperance organization (late in the nineteenth century, Cromarty included the Hugh Miller Lodge of the Good Templars).

For More Information: Taylor, M.A. 2007. *Hugh Miller: Stonemason, Geologist, Writer*. Edinburgh: National Museums Scotland.

KEN MILLER (b. 1948)

Any religious person who is astounded by the cruelty that we see in the world has to find some way to account for the presence of a knowing and loving God alongside that cruelty. I actually think that evolutionary biology helps a Christian to account for that in a remarkable way. Evolutionary biology shows that all life is interrelated and that life, unfortunately, only comes at the expense of death.

Kenneth Raymond Miller was born in Rahway, New Jersey, on July 14, 1948. He was raised a Catholic, a faith he has adhered to for most of his life (as a child, he even considered becoming a priest). During the summer between high school and the start of college, while working as a lifeguard, Miller read several books, including *On the Origin of Species*. Although Darwin's ideas had little immediate effect on Miller, it was clear that "people were afraid of this book," including his father, who cautioned his son that the book was dangerous and to "be careful not to lose your values." The basis for these warnings was not apparent to Miller, who found nothing controversial in a book that discussed seemingly well-established facts.

In the fall of 1966, Miller entered Brown University and began to question his religious faith. As an aspiring poet (he claims his efforts "occasionally rose to the heights of mediocrity"), he read widely, and came across work by poet and Trappist monk Thomas Merton. Merton's autobiography, *The Seven Storey Mountain* (1948), which described his conversion to Roman Catholicism, struck "a resonant chord" with Miller and convinced him that it was possible to be an intelligent person of faith. Miller's faith was reaffirmed and has remained strong ever since.

Miller graduated from Brown with a biology degree in 1970. He then went to the University of Colorado to study cell biology, earning his PhD in 1974. That same year, he became a lecturer at Harvard University and head of the electron microscopy lab, becoming Assistant Professor in 1976. In 1980, he returned to Brown University as Assistant Professor, rising to the rank of Professor in 1986. Miller's research has focused on photosynthesis, particularly the functioning of photosynthetic membranes. He has edited *The Journal of Cell Biology* and *The Journal of Cell Science*.

During Miller's second semester at Brown, a group of students asked him to debate

Young-Earth creationist Henry Morris of the Institute for Creation Research. Miller initially declined, citing the fact that he was not an evolutionary biologist and had not heard of Henry Morris. The students (probably aware of how to goad their professor into agreeing) asked if that meant that Morris was right about evolution. Miller shot back an emphatic "no," and agreed to consider debating Morris after reviewing recordings of Morris' previous debates. In so doing, Miller quickly realized that Morris was a skilled debater who could best eminent scientists through use of "tiny little arguments" that purported to show evolution as a flawed idea. Miller realized that Morris could easily sway a naive audience and he decided that the debate was important enough to require his participation.

Miller spent a month reading, talking with colleagues, and reviewing Morris' earlier debates. Armed with two carousels of slides, Miller confronted Morris in front of nearly 3,000 spectators in April 1981. Unlike in previous debates, Morris faced an opponent that had not taken him lightly, and one that went on the offensive instead of merely dismissing creationists' claims. Afterward, Morris acknowledged Miller's abilities and claimed that he was "deviously sophisticated in argumentation" and "the best evolutionist debater to surface to date." Both creationists and evolutionary biologists noticed Miller's success, and additional debates between Morris and Miller soon followed. Miller also participated in debates with Morris' ICR colleague Duane Gish, as well as intelligent-design advocate Michael Behe.

Miller testified in *Selman v. Cobb County School District*, a case that tested whether the Cobb County, Georgia school board could insert "warning" stickers about evolution in high school biology textbooks. Miller's textbook, coauthored with Joseph Levine, would receive the stickers, and Miller was called to discuss the content and purpose of the book. Miller was also an expert witness for the plaintiffs in the *Kitzmiller v. Dover Area School District* trial over the constitutionality of a statement to be read to ninth-grade biology students that offered intelligent design as an alternative to evolution.

(As in Cobb County, the Dover school district was also using Miller's textbook.) Because Miller was an articulate defender of evolution and devout Christian, he was free of the charge of atheism that antievolutionists frequently level against scientists, and was therefore an ideal authority to speak on behalf of the plaintiffs.

Miller tackled the relationship between science and religion in his 1999 book, *Finding Darwin's God: A Scientist's Search for Common Ground Between God and Evolution*. Miller, an excellent writer, used the first two-thirds of the book to discuss and refute the various forms of antievolutionist thought, ranging from Young-Earth creationism to intelligent design. This section of the book has been consistently reviewed positively, especially by biologists. The remainder of the book melds Miller's perspective as a scientist with his personal religious beliefs; this section has received significantly more criticism, especially from the scientific community. Miller concluded that God created a universe based on contingency rather than one that is deterministic and preordained, and natural processes are the products of that divine decision. Miller claimed that "given evolution's ability to adapt, to innovate, to test, and to experiment, sooner or later it would have given the Creator exactly what He was looking for." These conclusions elicited criticism that Miller was invoking teleology and design, aspects of nature he seemed to have refuted in earlier chapters.

Finding Darwin's God catapulted Miller into the media spotlight. He has discussed God and evolution on a variety of television programs, ranging from William F. Buckley's *Firing Line* to Comedy Central's conservative talk show parody *The Colbert Report*. He also umpires NCAA (National Collegiate Athletic Association) fast-pitch softball games.

For More Information: Miller, K. 1999. *Finding Darwin's God: A Scientist's Search for Common Ground between God and Evolution*. New York: HarperCollins.

STANLEY MILLER (1930–2007)

Stanley Lloyd Miller was born on March 7, 1930, in Oakland, California. He earned a BS from the University of California, Berkeley, after which he enrolled in the doctoral program in chemistry at the University of Chicago. There, he learned about claims from the 1930s by Russian biochemist Aleksandr Oparin and British biologist J.B.S. Haldane that life had originated in the ocean before there was oxygen gas in the Earth's atmosphere.

In 1951, while at the University of Chicago, Miller heard Harold Urey discuss the synthesis of organic compounds in a highly reducing terrestrial atmosphere. Urey suggested that someone study "prebiotic synthesis" using a reducing mixture of gases, and Miller volunteered. Urey, who thought the experiments would probably fail, was reluctant to let Miller do the work, but later helped Miller design an apparatus to simulate the ocean-atmosphere of early Earth. In that apparatus, Miller mixed water vapor (to simulate evaporation from oceans) with methane, hydrogen, and ammonia, which was then subjected to a continuous electrical charge.

Within only two days after starting the experiment, Miller noted that as much as 15 percent of the carbon in the apparatus was in organic compounds, and 2 percent of the carbon was in amino acids (the most abundant being glycine). After sparking the mixture for a week, Miller detected several other amino acids.

Miller wrote a manuscript describing the experiment, but Urey declined to be a coauthor because he believed his name would detract from Miller's involvement. Miller sent the manuscript to *Science* in February 1953, and Urey urged the editors to publish the paper. Miller then noticed an article in the March 8, 1953 issue of *The New York Times* titled, "Looking Back Two Billion Years" that described experiments by W.M. MacNevin and his colleagues at Ohio State University. MacNevin's experiments, like Miller's, involved electrical discharges into mixtures of gases, and reported the formation of "resinous solids too complex for analysis." On March 9, Miller sent Urey a copy of the article, noting that "I am not sure what should be done now, since their work is, in essence, my thesis." Soon thereafter, the editor of *Science* informed Miller that he wanted to publish Miller's paper. Miller's

paper appeared in the May 15, 1953 issue of *Science*, just a few weeks after Watson and Crick announced in *Nature* the double-helix structure of DNA. Miller's results were confirmed by several other groups, after which the phrase "prebiotic soup" became common in movies, books, and cartoons.

Miller's work—which became known as the "Miller-Urey experiments"—began the modern era of studying the origin of life. Today, however, many geoscientists question whether Earth's early atmosphere included all of the highly reducing gases used in Miller's experiments. Instead, some scientists—most notably chemist and patent attorney Günter Wächterschäuser—have argued that life originated in deep-sea vents. Others have claimed that the organic compounds needed for life may have originated from extraterrestrial sources (e.g., meteorites).

Following a series of strokes, Miller died on May 20, 2007, at age 77 in National City, California.

MITOCHONDRIAL EVE

Mitochondrial Eve was a name given by researchers to an alleged "mother" of all humans alive today with respect to matrilineal descent. Mitochondrial Eve lived 175,000 years ago in Ethiopia, Kenya, or Tanzania. The 1986 study by Allan Wilson and his colleagues that proposed Mitochondrial Eve was based on analyses of mitochondrial DNA (mtDNA), which constitutes only about 1/400,000th of the total human genome. Mitochondrial DNA seldom mutates and passes only from a mother to her children; fathers cannot impart their mtDNA. Wilson and his colleagues concluded that "the common ancestor of modern humans lived in Africa" and that when individuals from this population left Africa for Europe and Asia, "they did so with little or no mixing with existing local populations of more primitive humans." Mitochondrial Eve was not the only woman at that time; she merely happened to be the one (among many women) to whom mitochondria trace their ancestry.

GEORGE MIVART (1827–1900)

Useful organs must have developed with a view to the function they would eventually serve.

George Jackson Mivart was a friend of Charles Darwin and Thomas Huxley, but their relationships became strained after Darwin published *On the Origin of Species*. In 1871, Mivart's book *On the Genesis of Species* claimed that Earth was too young for evolution to have produced such a diversity of species; that intermediate steps toward new anatomical structures served no purpose; that natural selection alone could not account for evolutionary history; and that Darwin's theory upset the morals of British society. The sixth edition of *Origin*—the first to include the word *evolution*—included a new chapter that answered Mivart's criticisms (Chapter 7, the only chapter that Darwin added to later editions of *Origin*), but did not end Mivart's attacks. In that edition, Darwin described Mivart as "a distinguished zoologist" and admitted that Mivart had presented objections to natural selection "with admirable art and force."

MOELLER v. SCHRENKO

Moeller v. Schrenko (2001) ended when the Georgia Court of Appeals ruled that using a biology textbook that states creationism is not science does not violate the Establishment or the Free Exercise Clauses of the Constitution.

ASHLEY MONTAGUE (1905–1999)

Ashley Montague was a British anthropologist who in the 1950s questioned the validity of race as a biological concept in *Man's Most Dangerous Myth: The Fallacy of Race*. Montague was born as Israel Ehrenberg, but changed his name after moving to the United States. Montague wrote more than sixty books, and claimed, "Next to the Bible, no work has been quite as influential, in virtually every aspect of human thought, as *The Origin of Species*." Montague was also famous for his critique of creationism: "Science has proof without any certainty. Creationists have certainty without any proof."

DWIGHT MOODY (1837–1899)

Dwight Lyman Moody was a traveling shoe-salesman who became one of the first prominent American theologians to promote biblical inerrancy. Moody, a progenitor of fundamentalism, condemned theatre, disregard for the Sabbath, Sunday newspapers, and atheistic teachings such as evolution. He also believed that understanding Genesis was "the key to the whole [Bible]." Moody's revivals attracted thousands of followers, and prominent politicians such as Abraham Lincoln and Ulysses Grant—who laid the cornerstone of the American Museum of Natural History in 1874—worshipped at his church. In 1864, Moody founded in Chicago what came to be known as Moody Bible Institute, which today includes a museum describing Moody's life. Moody's Illinois Street Church was destroyed in 1871 by the Great Chicago Fire, after which Moody rebuilt the church nearby as the Chicago Avenue Church. It was at Moody's school in Northfield, Massachusetts that "Lady Hope" claimed that Darwin had a deathbed conversion to Christianity. Moody often worked with theistic evolutionist Henry Drummond (1851–1897), whose books helped World Christian Fundamentals Association member James Bole (1875–1956) and others embrace evolution. (Bole's influence faded when he was fired from Wheaton College in 1932 for "showing an improper attitude toward young ladies in his classes and in his home.") Moody influenced a generation of fundamentalists, including Cyrus Scofield, A.C. Dixon, Billy Sunday, and William Riley. (Riley, who idolized Moody, was often referred to as "the second Dwight Moody.") Unlike Riley, however, Moody avoided controversy.

Moody is buried in Northfield, Massachusetts, behind his birthplace, which is now a museum. Ten years after his death, Moody's Chicago Avenue Church was renamed The Moody Church in his honor. In 1922, *Moody Monthly* claimed that the Bible contained scientific and mathematical formulae, and linked the seven days of creation to the seven notes in the octave, the seven sayings of Christ, and the seven parts of Psalm 23. In 1945, the Moody Institute of Science began producing its *Sermons and Science* films, in which founder Irwin Miller asked viewers to accept Jesus; one of those films—the popular *God of Creation* (1946)—was watched by more than 1.5 million people, and was shown (with other Moody films) to servicemen around the world. Today, the nondenominational Moody Church is on North Clark Street in Chicago, about a mile south of Moody Bible Institute.

BARRINGTON MOORE (1883–1966)

Barrington Moore, the first editor of *Ecology* and a past president of the Ecological Society of America, cancelled his subscription to *Ecology* because he had "no use for evolution and do not see how any intelligent person can have." Moore was a pioneer in scientific forestry; the Society of American Foresters honors him with a research award named after him.

JOHN N. MOORE

John N. Moore was a prominent creationist and coeditor (with Harold Slusher) of *Biology: A Search for Order in Complexity* (1971). This creationism-based high school biology textbook, developed by the Creation Research Society (CRS), was adopted by several schools and included a Teachers Guide that listed the "correct" Christian answers. Although most of the textbook is standard biology, it includes strong antievolutionary comments (sections include "Failures of Darwinian Theory" and "Problems for Evolutionists") and concludes that "the most reasonable explanation for the actual facts of biology as they are known scientifically is that of biblical creation." *Order in Complexity* rejects evolution because it "has no observable evidential basis" and endorses flood geology and the dinosaur-human footprints in the Paluxy riverbed. *Order in Complexity*, which references the Scopes Trial, claimed that God put penicillin spores in Alexander Fleming's Petri dishes. The required use of *Order in Complexity* in public schools was declared unconstitutional in

Hendren v. Campbell. CRS planned a less religious edition to circumvent this ruling, but it never materialized because of differences between CRS and the Institute for Creation Research. Moore's *How to Teach Origins (Without ACLU Interference)* (1983) included contributions by creationists Henry Morris and Wendell Bird.

HARCOURT MORGAN (1867–1950)

Harcourt A. Morgan was president of the University of Tennessee during the Scopes Trial. Although he opposed the Butler Law, Morgan was silent about the law, probably because Governor Peay's proposal for the university's budget had not yet been approved. Morgan wrote to Governor Peay that "the subject of evolution so intricately involves religious belief, which the University has no disposition to dictate, that the University declines to engage in the controversy."

HOWARD MORGAN

Howard Morgan was one of John Scopes' students who testified against Scopes at the Scopes Trial. Clarence Darrow and his wife stayed in Morgan's home during the final days of the trial. During the trial, Morgan's mother told Darrow that she wanted her son to learn about evolution.

THOMAS HUNT MORGAN (1866–1945)

The whole subject of human heredity in the past has been so vague and tainted by myths and superstition that a scientific understanding of the subject is an achievement of the first order.

Thomas Hunt Morgan was born September 25, 1866, in Lexington, Kentucky. In 1886, he graduated from the State College of Kentucky (now the University of Kentucky) with a BS in zoology, and then spent a summer in Annisquam, Massachusetts, at the Marine Biology School (the immediate predecessor of the Marine Biological Laboratory at Woods Hole). The summer at Annisquam led to a lifelong interest in marine biology, and Morgan returned to Woods Hole most summers during his career. In the fall of 1886, Morgan entered graduate school at Johns Hopkins University to work with William Keith Brooks (who had trained with Louis Agassiz). Zoology at that time emphasized comparative anatomy, particularly physiology and embryology. Brooks, an avowed Darwinian, approached the study of adaptation through application of Haeckel's now discredited biogenic law ("ontogeny recapitulates phylogeny"). Following in this tradition, Morgan's dissertation research used embryology to reconstruct the evolutionary history of sea spiders.

Morgan became a skilled experimental biologist, initially in embryology and later in genetics. He had demonstrated from a young age skills as a naturalist, and he would retain this focus throughout his life (a student noted that Morgan was never happier than when collecting in the field). However, Morgan contributed significantly to the experimentalist approach in biology, and was skeptical of scientific concepts that could not be directly observed. This perspective explains his lifelong hesitancy of accepting natural selection as the causative agent in adaptive evolution.

After receiving his PhD in 1890, Morgan joined the newly created Bryn Mawr College. His research continued to focus on experimental embryology, particularly tissue differentiation, and in 1901 he produced the book *Regeneration*. Although Mendel's work was being reexamined at this time, the mechanism of heredity was still unknown, which prevented a clear determination of how evolution, especially speciation, operated. Biologists clustered into two groups: those that invoked natural selection, and those that believed large-scale mutations ("saltations") were responsible for species. Morgan, skeptical of the power of natural selection, advocated the saltationist theory, especially after visiting Hugo de Vries in Amsterdam around 1900. This interaction with de Vries also convinced Morgan that the mutational model of evolution could be studied experimentally. It was while at Bryn Mawr that Morgan met Lilian Sampson, who first became his graduate student and then, in 1904, his

wife. The Morgans would have four children, or "F_1s" as Morgan called them, referring to the "first filial generation" designator used by geneticists.

When Morgan moved to Columbia University in 1904, he began experimental research in genetics, using mice and rats, and after 1908, concentrating on *Drosophila*. By 1909, room 613 in Schermerhorn Hall at Columbia had become Morgan's famous "fly room." The room, measuring only $16' \times 23'$, housed at least five researchers who, according to long-time Morgan collaborator A.H. Sturtevant, "practically lived in this room...we slept and ate outside, but that was all."

To study the evolutionary effects of mutations, Morgan exposed flies to radiation and ultraviolet light, but detected few induced mutations. However, in 1910, Morgan discovered a single male fly that had white eyes instead of the typical red. Mating the male with a red-eyed female produced offspring that all had red-eyes; crossing the individuals of this second generation among themselves produced a third generation with a frequency of red to white of 3:1. These results indicated that the white-eyed trait was due to a mutation (it was heritable), but the mutation did not lead to speciation, and these results therefore did not support the saltationist model.

Research by others had suggested that chromosomes were the site of the hereditary factors (genes), and that there were particular chromosomes—sex chromosomes—that determined the sex of individuals. Results of additional crosses with red-eyed and white-eyed flies ultimately allowed Morgan to demonstrate the chromosomal basis of inheritance and sex determination. The discovery of additional traits determined by genes on the X chromosome further enabled Morgan to detect the process of crossing over between homologous chromosomes. Sturtevant, then an undergraduate student of Morgan's, used these results to propose that genes were arrayed along a chromosome and that the rate of crossing-over was a direct measure of the physical distance between linked genes. (The measure of the physical distance between loci based upon recombination frequency came to be measured in units termed *morgans*.) In 1915, Morgan, Sturtevant, Calvin Bridges, and H.J. Muller (winner of the Nobel Prize in 1944), published *The Mechanism of Mendelian Inheritance*. In 1933, Morgan was awarded the Nobel Prize for Medicine for contributions to the understanding of the Mendelian basis of inheritance. Morgan divided the prize money among his children and collaborators.

During Morgan's twenty-four years at Columbia, he and his lab workers spent nearly every summer at Woods Hole. Ongoing research was not interrupted by this seasonal change in location and required shipping fruit fly cultures in barrels to and from Massachusetts; Morgan was frequently conducting other experiments, and he would similarly ship those organisms (e.g., chickens, mice, rats, pigeons). In 1928, Morgan moved to the California Institute of Technology to found a department of biology. There, he returned to his original interest in marine organisms and helped found Caltech's marine laboratory at Corona del Mar. Former students Sturtevant and Theodosius Dobzhansky (both of whom became professors at Caltech) continued research on the genetics of fruit flies.

Thomas Hunt Morgan retired from Caltech in 1941, and died of a ruptured stomach artery in 1945. His influence on the understanding of heredity was monumental, but students point out that he also contributed significantly to a change in how research labs operated. Instead of following the European model then in common usage—where the lab leader determined precisely the direction of research in his lab—Morgan acted as a facilitator who encouraged those working with him to pursue independent lines of inquiry. Former students have commented on his generosity, open-mindedness, and unfailing support.

Honors earned by Morgan, in addition to the Nobel Prize, include the Darwin Medal (1924), the Copley Medal of the Royal Society (1939), and election to the American Association for the Advancement of Science and the National Academy of Sciences. The Genetics Society of America awards the Thomas Hunt

60. Henry Morris was the most influential creationist in the later half of the twentieth century. Morris, a Young-Earth creationist, established the Institute for Creation Research and published numerous books and articles that condemned evolution. *(ICR)*

Morgan Medal in recognition of lifetime contributions to the science of genetics. The building housing the Department of Biology at the University of Kentucky is named in his honor.

For More Information: Allen, G.E. 1978. *Thomas Hunt Morgan: The Man and His Science.* Princeton, NJ: Princeton University.

HENRY MORRIS (1918–2006)

The final and conclusive evidence against evolution is the fact that the Bible denies it . . . Satan himself is the originator of the concept of evolution.

Henry Madison Morris, Jr., was born in Dallas, Texas, on October 6, 1918 (Figure 60). He earned a BS in civil engineering from Rice University in 1939, and a PhD from the

University of Minnesota in 1950. His doctoral degree was in hydraulic engineering because of its importance in understanding Noah's flood. Morris taught engineering at Rice University, the University of Louisiana at Lafayette, Southern Illinois University, and Virginia Polytechnic Institute (VPI), where he served as department chair. Although Morris later condemned Harry Rimmer's views of biblical history, in 1940 Morris was galvanized by Rimmer's *Modern Science and the Genesis Record*. In 1946, Morris published *That You Might Believe*, which began his attempt to establish a scientific basis for creationism (and in which he warned that the teaching of evolution damages students). That book, Morris claimed, was the first book "published since the Scopes trial in which a scientist from a secular university advocated recent special creation and a worldwide flood." The first edition of *That You Might Believe* embraced Gap creationism, but Morris later expunged this material when he became a Young-Earth creationist.

Morris was a productive engineer, and his 1963 book *Applied Hydraulics in Engineering* was a popular and well-respected textbook. Although flood geology had been advocated in the first half of the twentieth century by George McCready Price, in 1961 Morris and Old Testament scholar John C. Whitcomb popularized creation science by publishing *The Genesis Flood: The Biblical Record and Its Scientific Implications* (Figure 61). *The Genesis Flood* emphasized that scientific evidence proves the Bible, and that natural phenomena such as fossils are explained by the Bible. Morris used his knowledge of water movement to explain the impact of Noah's flood.

The opening chapter of *The Genesis Flood* argued that Noah's flood was worldwide; that floodwater covered the tallest mountains; and that the flood destroyed the entire human race, excluding Noah and his family. In the next three chapters, Morris and Whitcomb disputed alternative proposals for the flood (e.g., that it was anthropological rather than universal), and the fifth chapter focused on geological evidence supporting the flood. Morris and Whitcomb also described human footprints alongside dinosaur footprints at Glen Rose, Texas, as a basis for humans and dinosaurs living contemporaneously. After discussing the origin of the universe and a geological timeline of events, Morris and Whitcomb devoted more than 100 pages to problems and unanswered questions about their model. Morris and Whitcomb argued that George Price's flood geology was the only acceptable interpretation of Genesis, and "the fully instructed Christian knows that the evidence for full divine inspiration of Scripture is far weightier than the evidences for any fact of science." Critics have rejected several of the claims in *The Genesis Flood*, including those that Earth would be covered by 182' of meteor dust if it were a billion years old, and that *Archaeopteryx* fossils are fakes, and that scientists have found a missing day (relative to biblical accounts of the sun standing still for Joshua).

The Genesis Flood—the most important contribution to strict creationism since Price's *The New Geology* in 1923—revived flood geology and, in the process, resurrected antievolutionary thought from decades of obscurity. By the mid-1990s, more than 260,000 copies of *The Genesis Flood* had been sold in twenty-nine printings. No event was more important in promoting flood geology than was the publication of *The Genesis Flood*.

By 1969, Morris' research had attracted much attention at VPI, and his relationship with his dean, Willis Worcester (1918–1970), was "more tenuous than ever." After a yearlong sabbatical, Morris retired from mainstream academia in 1970 and moved to California to establish the Creation Science Research Center (CSRC), a creationism-based subsidiary of Tim LaHaye's Christian Heritage College (LaHaye later became famous as the coauthor of the apocalyptic *Left Behind* novels). CSRC promoted "creation science" by legal and political means, but Morris wanted to develop scientific and educational materials supporting "creation science." Morris later left and organized the Institute for Creation Research (ICR). Morris distinguished ICR from CSRC by describing the latter as "a promotional and sales organization," and noted that CSRC cofounder Kelly Segraves' honorary degrees were "false titles."

Morris was president of ICR for twenty-five years, after which his son John became president and Henry became president emeritus until his death in 2006. After retiring, Morris often noted that "[Evolutionary] ideas destroy the foundation for the Gospel and negate the work of Christ on the cross. Evolution and salvation are mutually exclusive concepts." John Morris' views mirrored those of his famous father: "If evolution is right, if the Earth is old, if fossils date from before man's sin, then Christianity is wrong." The younger Morris, who often leads searches for Noah's Ark, admits that he hasn't found it yet, but that "when I get up to heaven I'm going to ask Noah where he parked that thing." John Morris also believes that geology "should be interpreted in light of the Scriptures, rather than distorting Scripture to accommodate current geological philosophy...We don't even try to prove the Bible. We believe it." In 2007, John announced that he would welcome the chance to participate in an evolution-related lawsuit to "mend the damage inflicted at the Scopes Trial." Like many creationists before him, John Morris believes that the teaching of evolution leads to societal ills such as abortion and homosexuality.

In the second half of the twentieth century, Henry Morris published more than sixty books, some of which addressed contemporary issues. For example, in 1972 Morris claimed that the rings of Saturn are associated "with Satan's primeval rebellion," and Morris' *The Bible Has the Answer* addressed abortion, zodiac signs, when a baby receives its soul, and why Christianity is the best religion. However, most of Morris' books and other writings condemned and critiqued evolution. These books, which included *Scientific Creationism* (1974, published in two versions, one for public schools and another for church-related schools), *The Genesis Record* (1976), *Men of Science, Men of God* (1982), *The Troubled Waters of Evolution* (1982), *History of Modern Creationism* (1984), *The Long War against God* (1989), and *Biblical Creationism* (1993), were based on a literal reading of the Bible.

Morris' writings were instrumental in legislative attempts to introduce "creation science"

61. Whitcomb and Morris' monumental *The Genesis Flood* popularized George Price's ideas about flood geology and transformed "creation science" into the most popular type of creationism. The impact of *The Genesis Flood* on modern creationism is exceeded only by that of the Bible. (*ICR*)

into public schools in the 1980s. Morris was unequivocal about evolution, claiming that, evolutionists are "lost to salvation" and "face eternal damnation," evolution is responsible "for our present-day social, political, and moral problems," the phrase " 'Christian evolutionist' . . . is an oxymoron," "evolution is impossible," "evolution could never occur," and "Satan invented evolution at the Tower of Babel." Critics responded that this implicitly equated faith with believing in things without any basis for belief, thereby grouping God with phenomena such as UFOs and Bigfoot.

Courts have ruled repeatedly that creation science is religion, not science (*Edwards v. Aguillard, McLean v. Arkansas Board of Education*). Morris disagreed, claiming that "the Bible is a book of science"; that the Bible "is our textbook on the science of Creation"; that the Bible contains "all the known facts of science"; that "there neither is, nor can be, any proof of evolution"; and that "if the Bible teaches it, that settles it, whatever scientists might say, because it's the word of God."

Morris endorsed Phillip Gosse's claim that God created the universe as if it had an ancient history, noting that "even Adam and Eve [were created] as mature individuals when they were first formed." In 1981, Jerry Falwell urged his followers in the Moral Majority not to read any books except the Bible, but he made an exception for Morris' *The Remarkable Birth of Planet Earth*, which Falwell distributed after Morris appeared on Falwell's *Old-Time Gospel Hour*.

Morris was the most influential creationist of the twentieth century. Antievolutionist Ken Ham, who claims Morris' impact is similar to that of Martin Luther, describes Morris as "the father of the modern creationist movement." However, Morris' scientific claims have often been discredited by mainstream scientists and many other creationists. For example, Kenneth Miller—the author of *Finding Darwin's God*—noted that *The Genesis Flood* contains "contradictions and fallacies and weaknesses of flood geology almost too numerous to mention.... They started with a conclusion—Genesis—and collected facts that appeared to support it, discarding or misinterpreting any that didn't fit." Although Morris' "creation science" has lost much of its popularity to other forms of creationism such as intelligent design, Morris' writings continue to provide the foundation of scientific creationism. According to evolutionary biologist Stephen Gould, *The Genesis Flood* was "the founding document of the creationist movement."

After suffering a series of strokes, Morris died on February 25, 2006, in Santee, California, at age 87.

For More Information: Morris, H. 1982. *Evolution in Turmoil*. San Diego, CA: Creation-Life; Morris, H. 1993. *History of Modern Creationism*. Santee, CA: Institute for Creation Research; Morris, H. 1995. *The New Defender's Study Bible: Understanding the Critical Issues of the Faith from a Literal Creationist Viewpoint*. Nashville, TN: World Publishing.

CAMERON MORRISON (1869–1953)

Cameron Morrison was the governor of North Carolina who in 1924 convinced the North Carolina Board of Education to reject two biology textbooks because they included evolution and were "unsafe." Morrison later noted that "I don't want my daughter or anybody's daughter to have to study a book that prints pictures of a monkey and a man on the same page.... I don't believe in any missing links. If there were any such things as missing links, why don't they keep on making them?" Morrison is buried in Elmwood Cemetery in Charlotte, North Carolina.

WALTER MOXON (1836–1886)

Walter Moxon was a physician called to Down House on April 19, 1882, to help care for Charles Darwin. Darwin was dying when he arrived. Moxon was on the "Personal Friends invited" list for Darwin's funeral.

MOZERT v. HAWKINS COUNTRY BOARD OF EDUCATION

Mozert v. Hawkins County Board of Education was a court case that in 1987 established that students do not have a right to be excused from classroom activities that expose them to competing ideas, even if those ideas contradict their religious beliefs. The plaintiffs—many of whom described themselves as "born again Christians"—were financed by Concerned Women for America, a conservative political organization headed by Beverly LaHaye, the wife of antievolution crusader Tim LaHaye.

JOHN MURRAY PUBLISHERS (Est. 1768)

In 1768, Edinburgh-born John Murray (1745–1793) founded John Murray Publishers of

London, one of the most successful publishing houses in Britain. The company expanded and rose to fame when Murray's son, John Murray II, took over the company. Murray, who knew many of the leading writers of his day, founded *Quarterly Review* in 1809, and published the works of Lord Byron and Jane Austen. His home and office at 50, Albemarle Street, in Mayfair (London) remains the site of the company.

John Murray III (1808–1892) then took over the business, and was responsible for publishing the works of Charles Lyell, David Livingstone, and Charles Darwin. Although Darwin's earliest books were published by Henry Colburn Publishers of London (e.g., *Journal of Researches*) and Smith, Elder Publishers of London (e.g., *Geological Observations of South America*), Darwin's first book to be published by Murray was *On the Origin of Species*, which appeared on November 22, 1859. Darwin had boasted to Murray that "I can say with confidence that the [manuscript] contains many new and very curious facts and conclusions," but added, "I know not in the least, whether the Book will sell." The only proevolution book that Murray published prior to *Origin* was Thomas Hope's *Essay on the Origin and Prospects of Man* (1831).

On the Origin of Species sold well (all of the 1,111 available copies sold on the first day), and today the book remains a steady seller. Darwin was apparently pleased with Murray; with only one exception (the 118-page *The Movement and Habits of Climbing Plants*, published in 1864 by the Linnean Society), all of Darwin's subsequent books were published by Murray. Interestingly, *Insectivorous Plants*, which appeared in 1875, sold faster than *On the Origin of Species*. Murray, who in 1845 paid £150 for the copyright of the second edition of *Journal of Researches*, was on the "Personal Friends invited" list for Darwin's funeral.

Since the eighteenth century, seven generations of the Murray family have guided the company. Today, John Murray is a subdivision of publisher Hodder Headline. The John Murray Archive, which includes more than 150,000 items, is one of the world's most significant literary and cultural collections of the past 300 years. The collection includes manuscripts and letters of novelist Jane Austen, Charles Darwin, Michael Faraday, Charles Lyell, Benjamin Disraeli, Winston Churchill, John Stuart Mill, William Wordsworth, Herman Melville, Arthur Conan Doyle, and J.M. Barrie (the creator in 1904 of *Peter Pan*).

N

NATIONAL ACADEMY OF SCIENCES (Est. 1863)

There is no significant scientific doubt about the close evolutionary relationships among all primates or between apes and humans.

The National Academy of Sciences (NAS), which originated during President Abraham Lincoln's administration, is a self-selected, honorific society of approximately 1,800 members and 350 foreign associates who advise national leaders about scientific issues. In 1916, at the request of President Woodrow Wilson, the Academy established the National Research Council to provide technological recommendations regarding military preparedness. NAS advocates evolution education from kindergarten through college and has published a variety of evolution-related books, including *Teaching about Evolution and the Nature of Science* (1998), *The Future of Evolution* (2002), *Science and Creationism: A View from the National Academy of Sciences* (second edition, 1999), and *Tempo and Mode in Evolution: Genetics and Paleontology 50 Years After Simpson* (1995). Although fewer than 10 percent of NAS members believe in a personal God, NAS notes that many scientists believe in God and accept evolution.

NATIONAL ASSOCIATION OF BIOLOGY TEACHERS (Est. 1938)

The National Association of Biology Teachers (NABT) is the largest national association dedicated to the interests of biology teachers. NABT was conceived by Oscar Riddle of the Carnegie Institute, which provided $10,000 for the new organization. The first edition of NABT's journal, *The American Biology Teacher*, appeared in October 1938. Today, NABT has more than 8,000 members in the United States and abroad. Since its inception, NABT has been active in the evolution-creationism controversy. NABT supported the lawsuits of Susan Epperson and Don Aguillard, and has opposed attempts to legislate creationism and undermine evolution (e.g., *McLean v. Arkansas Board of Education*).

Late in 1970, *The American Biology Teacher* published an article by creationist Duane Gish pleading for teachers to teach creationism. When the journal published another article by Gish three years later, the article was prefaced with a statement noting that biologists reject creationism (this article appeared in the same volume as the famous article by Theodosius Dobzhansky titled "Nothing in Biology Makes Sense Except in the Light of Evolution").

In 1972, NABT helped form state-based "Committees of Correspondence on Evolution" to resist creationists' efforts to undermine science. In 1981, several of these Committees of Correspondence founded the National Center for Science Education. Near the same time, NABT's "Fund for Freedom in Science Teaching" gathered more than $12,000 to oppose the teaching of creationism, but there was also a backlash from NABT's members who were creationists. In response to these complaints, NABT sponsored a creationism panel at its annual meeting the following year that included presentations by creationists such as Duane Gish.

In 1995, NABT adopted a statement about evolution whose preamble described evolution as "an unsupervised, impersonal, unpredictable and natural process." When some scientists complained that the words *unsupervised* and *impersonal* implied theology, the offending words were deleted from the statement. Today, the "NABT Statement on Teaching Evolution" endorses the teaching of evolution and opposes the teaching of creationism.

NATIONAL CENTER FOR SCIENCE EDUCATION (Est. 1981)

The National Center for Science Education (NCSE) is a not-for-profit, membership-based organization that provides information, advice, and resources for schools, parents, educators, and concerned citizens dedicated to keeping evolution in the science curriculum of public schools. NCSE is religiously neutral, and works with local and national religious, educational, and scientific organizations such as the Center for Theology and the Natural Sciences, the National Association of Biology Teachers, and the National Academies of Science.

NCSE's origins can be traced to the late 1970s and early 1980s, when many citizens were appalled to learn that their state legislatures were considering legislation mandating "balanced treatment" for "creation science." In 1980, Iowa high school teacher Stanley Weinberg

(1911–2001) began organizing statewide Committees of Correspondence "committed to the defense of education in evolutionary theory." Like their namesakes in the Colonial era, these Committees shared information about policy-related issues—in this instance, those underlying legislative attempts to undermine the teaching of evolution. In 1981, individuals who had been involved in several of these Committees founded the NCSE, and in 1983 the NCSE was incorporated with Weinberg as its first president. Today, NCSE has more than 4,000 members and an annual budget of $700,000.

NCSE, which is based in Oakland, California, is involved in numerous activities to promote the teaching of evolution in public schools. NCSE also publishes the bimonthly journal *Reports of the National Center for Science Education*.

NATIONAL SCIENCE FOUNDATION (Est. 1950)

The National Science Foundation (NSF) is a federal agency created to promote science for the improvement of life. NSF supports basic research and science education, and has funded many projects that have promoted evolution. NSF's publication in 1970 of the proevolution curriculum, *Man: A Course of Study (MACOS)* produced a national uproar; Congressman John Conlan and Senator Jesse Helms condemned the project, and urged NSF to cut funding for evolution-related projects. In 1973, Congress considered—but did not pass—legislation giving it direct supervision and veto power over every project funded by NSF. In 1980, Ronald Reagan cited *MACOS* as evidence that the government was endorsing subversive values, and urged NSF to develop Christian curriculum materials. In 1989, NSF asked Biological Sciences Curriculum Study director Joe McInerney to remove the word *evolution* from the title of a project because some members of Congress would be unhappy to learn that NSF had funded an evolution-related project. McInerney ignored NSF's request.

NATIONAL SCIENCE SOCIETY OF BRÜNN

The National Science Society of Brünn published Gregor Mendel's manuscript "Experiments in Plant Hybridization" in 1866. Mendel used his own money to purchase and distribute forty copies of his paper. Only seven of those reprints have been found.

NATIONAL SCIENCE TEACHERS ASSOCIATION (Est. 1944)

The National Science Teachers Association (NSTA) originated as a department of the National Education Association, which endorsed the teaching of evolution in 1916. In the 1960s, NSTA supported Susan Epperson's attempt to overturn the Arkansas law banning the teaching of human evolution, noting that evolution "is firmly established even as the rotundity of the Earth is firmly established." In subsequent years, the NSTA continued to support the teaching of evolution, and supported Gary Scott when he was fired for teaching evolution in Tennessee in 1967. NSTA's headquarters are in Arlington, Virginia.

JOHN NEAL, J.R. (1876–1959)

Boy, I'm interested in your case and, whether you want me or not, I'm going to be there...It is not a case of religion against irreligion, not a case of Fundamentalism against Modernism, but a case for the freedom of speech and thought.

John Randolph Neal, Jr., began his law career in 1899 teaching at the University of Denver, but in 1907 moved to Tennessee, where he served two terms in the Tennessee legislature. He joined the faculty of the University of Tennessee Law School in 1917, but he was fired in 1923 after missing classes, failing to assign grades, and ignoring the prescribed curriculum. His personal hygiene was also questionable; he seldom bathed and wore the same shirt for many days, after which he would cover it with a new one. Neal then started the John Randolph Neal

College of Law in Knoxville. Neal's school remained open for more than twenty years, and at its peak enrolled more than seventy students.

Neal, often described as Tennessee's first hippie, was John Scopes' chief counsel of record, but his role was modest. He had the authority to remove Darrow, but he did not. Neal, who headed Scopes' appeal to the Tennessee Supreme Court, was described by Scopes as someone who "rarely impressed others favorably."

Neal was a perennial candidate for governor and United States Senator (in 1946, he ran for both offices simultaneously), but he always lost. He supported the formation of the Tennessee Valley Authority, and represented North Carolina mill workers in strike-related lawsuits.

Neal died on November 22, 1959, and was buried in Brown Cemetery in Postoak, Tennessee.

NEANDER VALLEY

The green, pastoral Neander Valley of Germany was the site in 1856 of the discovery of a skull that looked similar to that of modern humans. These so-called Neanderthals, which were large-brained humans who had protruding jaws and heavy-set brows, challenged the traditional Christian image of humans.

NEBRASKA MAN

Nebraska Man originated in 1917, when rancher and geologist Harold J. Cook found a humanlike molar tooth on his ranch in Nebraska. In 1922, Cook gave the tooth to paleontologist Henry Osborn, who declared that it belonged to *Hesperopithecus haroldcookii* ("Western Ape"). Osborn used the discovery to chide William Jennings Bryan (who had served as a Congressman from Nebraska), noting that "the Earth speaks to Bryan from his own state," and that the discovery should be named *Bryopithecus* to honor "the most distinguished primate which the state of Nebraska has thus far produced." In 1925, however, researchers returned to the site where the tooth had been discovered, and learned that it was a worn tooth of a fossil peccary or javelina. Many creationists have used Nebraska Man as an example of the scientific errors which they claim undermine the credibility of paleontology and evolution.

NELSON, WILBUR A.

Wilbur A. Nelson, Tennessee's state geologist and president of the American Association of State Geologists, attended the Scopes Trial as an expert witness. Nelson claimed in his affidavit that evolution is an important tool for geologists. Nelson's testimony was read into the court record by Arthur Hays.

NEWELL, NORMAN (1909–2005)

Norman D. Newell was professor of paleontology at Columbia University and curator at the American Museum of Natural History (AMNH). He held Bachelor's and Master's degrees from the University of Kansas, and a PhD from the University of Chicago. He joined the AMNH in 1945 after teaching at the University of Wisconsin. Newell was one of the first biologists to understand the importance of mass extinctions, work that was influential on two of his graduate students, Niles Eldredge and Stephen Jay Gould. He was recognized as a "legendary geoscientist" by the American Geological Institute in 2004, and was also a member of the National Academy of Sciences and the American Association for the Advancement of Science. In 1982, he published *Creation and Evolution: Myth or Reality*, which confronted creationists' efforts with particular reference to the geological record.

HORATIO NEWMAN

Horatio Hackett Newman was a zoology professor at the University of Chicago who attended the Scopes Trial as an expert witness for the defense. Newman noted in his affidavit that "once you admit a changing world...you admit the essence of evolution." Newman also challenged creationists to "think what a sensation in the scientific world might be created if someone were to discover even one well-authenticated fact that could not be reconciled with the principle of evolution." Newman's testimony, which was read into the court record by Arthur Hays, concluded with "the evolution principle is thus a great unifying and integrating scientific conception."

NOAH

Noah, the grandson of Methuselah and a tenth-generation descendant of Adam, was a biblical patriarch best known for a worldwide flood that allegedly occurred during his time. Noah's story is told in Chapters 5–9 of Genesis, and in Chapters 6–9 of the Quran.

Ten weeks after Noah reached his 600th birthday, God was upset at Earth's wickedness. God regretted creating humans and decided to kill everything on Earth. However, God decided to spare Noah and seven other humans (i.e., Noah's wife, his three sons—Shem, Ham, and Japhet—and his three sons' wives). God told Noah and his three sons (who Noah fathered when he was 500 years old) to build a three-story boat, "300 cubits long by 50 cubits broad"—a boat that came to be known as Noah's Ark. Noah's Ark had a gross volume of about 1.5 million ft^3, floor space of about 100,000 ft^2, and a total displacement of about 22,000 tons (about half that of *Titanic*). God locked everyone in the Ark, and a week later it started raining. The subsequent flood covered the highest mountains to a depth of more than 20' and killed everything not aboard the Ark. Biblical chronologist James Ussher claimed that the flood occurred in 2349 BC.

The Ark ran aground 110 days after it stopped raining in the mountainous region of Ararat in northeastern Turkey, where Noah built the first altar mentioned in the Bible (the 17,000' Mount Ararat was not mentioned as the Ark's landing-site until the eleventh century). God then promised that he would not again destroy all life with a flood, and told Noah and his descendants to "be fruitful and multiply." Noah got drunk and, when Ham told his brothers of Noah's condition, Noah cursed Ham's son

62. Frank Norris was the most controversial person in the history of the evolution-creationism controversy. Norris (left) is shown here with fellow antievolution crusader John Roach Straton. (*Library of Congress*)

Canaan with eternal slavery. Noah died at the age of 950.

Noah's flood is a certainty for many Young-Earth Creationists, some of whom continue to search for Noah's Ark. Many creationism-based hate-groups such as the Ku Klux Klan justify their racial discrimination by citing Noah's story. In 1964, West Virginia Senator Robert Byrd—a former Ku Klux Klansman who described Genesis as "the greatest scientific thesis that was ever written"—read Noah's story into the Congressional Record as part of a filibuster against the Civil Rights Act of 1964, proclaiming that "Noah saw fit to discriminate against Ham's descendants." In the midst of the Noah's Ark craze in the 1970s, actor Leonard Nimoy—of *Star Trek* fame—narrated a documentary titled *In Search*

of . . . Noah's Flood. Other topics in the *In Search Of* series included UFOs, the Loch Ness Monster, and Bigfoot.

For More Information: Pleins, J.D. 2003. *When the Great Abyss Opened: Classic and Contemporary Readings of Noah's Flood*. New York: Oxford University.

JOHN FRANKLYN 'FRANK" NORRIS (1877–1952)

If you deny the Genesis account of creation, you had just as well throw the Bible in the fire.

John Franklyn "Frank" Norris was born on September 8, 1877, in Dadeville, Alabama, to a devout mother and an alcoholic father who was a sharecropper (Figure 62). After moving to Texas, Norris almost died after being shot by horse thieves. He graduated from Baylor University in 1903, and then from Southern Baptist Theological Seminary in 1905, topping his class. In 1905, Norris became pastor of McKinney Avenue Baptist Church in Dallas; thirteen people attended his first service, but three years later membership exceeded 1,000. In 1909, at age 32, Norris—the owner-editor of the *Baptist Standard* (the leading newspaper of Texas Baptists)—became pastor of First Baptist Church of Fort Worth. At this "Church of the Cattle Kings," Norris hosted numerous visits by antievolution crusaders William Riley, Billy Sunday, and William Jennings Bryan, and increased the membership from 100 to over 12,000 in ten years. Bryan's sermons in 1924 at Norris' church condemning evolution drew crowds of more than 6,000 people, whom Norris told that what Martin Luther was to the Reformation, Bryan was to fundamentalism. Although Norris had little use for many Baptist ideals, he remained pastor of First Baptist Church until his death in 1952. Norris was the highest-paid pastor in the South.

Like fundamentalists such as Sunday and Bryan, Norris campaigned against gambling, prostitution, booze, dancing, Catholics, and Communists. Norris, an unapologetic racist, was sympathetic to the Ku Klux Klan, while calling for Southerners to join his holy war against

63. Norris, who promoted himself as "The Texas Cyclone" and "The Pistol-Packin' Parson," was repeatedly arrested and acquitted for felonies including perjury, arson, and murder. (*Arlington Baptist Church*)

Northern infidelity. Norris brought fundamentalism to the South; he was the only Southern fundamentalist whose stature approached that of Minnesota's Riley, the leader of fundamentalism in the United States. Norris created many of today's unflattering stereotypes of fundamentalists.

Norris, the most controversial figure in the evolution-creationism controversy, had an unquenchable thirst for fame, controversy, and publicity. He was indicted for a variety of felonies, including perjury, several arsons (including the burning of his own church), and murder (Figure 63). After accusing the Catholic mayor of Fort Worth (H.C. Meacham) of corruption and telling his congregation that Meacham "isn't fit to manage a hogpen," Norris was visited on July 17, 1926, at his church office by Dexter Chipps, a wealthy supporter of the mayor. At that meeting, Norris shot the unarmed Chipps. Norris then walked to Chipps body and shot him two more times. Norris claimed that Chipps "was reaching for a gun," but Chipps had no

weapons. On the following Sunday, Norris' followers gave Norris $16,000 to help with his defense. Fellow fundamentalists such as John Straton pledged their support to Norris "no matter what," and the Ku Klux Klan announced that it was "ready to lend whatever assistance" it could.

Despite strong evidence indicating that Norris was guilty of each crime, Norris was always acquitted, after which he used his trials' publicity to boost his fame and attack his detractors. After escaping the death penalty with an acquittal for murder, Norris told his overflowing church that "I offer no apology for what I've done." As a newspaper editor noted after the trial, "In Fort Worth, the 11th Commandment is 'Thou shalt not mess with J. Frank Norris.'"

Norris, who roamed the stage weeping and shouting when he preached, tolerated no dissent, and often told disgruntled members to "go to hell." Anyone who questioned Norris became an enemy who Norris attacked relentlessly. Norris fired entire groups of deacons, slandered and humiliated politicians, and after jailing and publicly denouncing his own son, offered to pay the costs for his son to change his last name. When he was away from the pulpit, Norris would often curse and swear, excusing his coarse language by claiming that "God has no use for sissies."

Norris, a founding member of the World's Christian Fundamentals Association (WCFA), believed that fundamentalism was "the greatest religious revival that time has ever witnessed." Norris, a star of the movement, hired detectives to investigate his enemies, and then published his findings in local newspapers and in his own newspaper, which he called *Searchlight*. (The front-page of each issue showed Norris holding a searchlight in one hand and a Bible in the other, with Satan cowering in the opposite corner.) *Searchlight*, which had editorial offices throughout the United States, published a variety of antievolution articles, as well as articles, advertisements, and occasional announcements about the Ku Klux Klan. In 1927, Norris—in an attempt to make his newspaper appear as an official publication of the WCFA—changed the name of his newspaper to *The Fundamentalist*. This produced an increase in the newspaper's circulation (to beyond 60,000) and a bitter split with Riley, who later said that he regretted ever working with Norris. Other pastors were more blunt; as one noted, "the world would be better if Frank Norris had never lived."

Norris often proclaimed from the pulpit that his enemies would have disastrous accidents, go broke, or die of venereal disease. In a sermon he delivered after being acquitted of arson, Norris showed his followers a broken liquor bottle, adding that it contained the booze and some of the brain tissue of an attorney who had prosecuted him and who had died the previous day in a gruesome automobile accident. Norris, who told his congregation that the accident was an expression of divine judgment, then preached a sermon titled "The Wages of Sin Is Death."

Norris was pastor of a Baptist church, but he denounced other Baptist preachers as "weeds," "worthless bastards," "dirty diapers," "lepers," "old baboons," and, in one instance, "a beer-guzzling bunch of hair." Norris refused to use Baptist literature in his church, and was thrown out of the city, county, state, and denominational associations of Baptists. The Baptist General Convention permanently excluded Norris in 1924; when he showed up at the convention the next year, police escorted him from the meeting. Norris' antics drove away hundreds of church members (600 in 1911 alone), but they were quickly replaced by others wanting to be part of what a local newspaper called "the best show in town."

Late in 1934, Norris—who defended fundamentalism "with both hands loaded with religious TNT"—also became pastor of Temple Baptist Church in Detroit, Michigan. He was pastor of both churches—and their combined membership of over 25,000—for fifteen years. In Detroit, as in Fort Worth, crowds flocked to hear Norris preach. Membership increased from 800 in 1934 to over 6,000 twelve years later. Although Norris was despised by many (e.g., the *Atlanta Constitution* noted that Norris "is one, good, sound, reason why there are 50,000,000 Americans who do not belong to any church at all"), his ultramilitant version of Christianity appealed to legions of fundamentalists who longed for uncompromising certainty. Norris' sermons attracted thousands per service. Novelist Sinclair Lewis attended Norris' church, where Charles "Pretty Boy" Floyd's wife and son were members ("Pretty Boy" himself listened to Norris' sermons on the radio until he was killed in 1934 by FBI agents). Norris' radio station was named KFQB (later renamed KFJZ), which stood for "Keep Folks Quoting the Bible."

Norris aimed many of his attacks at "that hell-born, Bible-destroying, deity-of-Christ-denying, German rationalism known as evolution." He supplemented his sermons with monkeys, denounced colleges for teaching "evolution and infidelity," and demanded that several "modernist" professors be fired. Norris' view of Christianity left no room for interpretation, and Norris urged his followers to "hang the apes and monkeys who teach evolution." Like other antievolutionists, Norris linked evolution with societal ills, claiming that evolution was "made in Germany," that evolution was like the "poison gas of German armies...sweeping through our schools," that "evolution and dance are to each other as cause and effect," and that the "damnable and despicable" biologists who teach evolution "have hands dripping with innocent blood." Norris often reminded his followers that biology teachers were "the finest evidence that [their] ancestors were braying jackasses, screeching monkeys, and yelling hyenas."

Throughout the 1920s, Norris demanded that the Texas legislature ban the teaching of evolution. In 1923, Norris' church hosted the annual meeting of the WCFA, at which Norris arranged a mock trial of evolutionists. The evolutionists were convicted and hanged. Norris was responsible for several resignations of faculty, as well for the University of Texas Board of Regents passing a resolution in 1924 requiring all employees to believe in God. Years later, Norris concluded that he had "given the profs a cleaning" and that Texas was "pretty well cleaned up."

On July 12, 1925, newspapers throughout the United States announced that William Jennings Bryan had invited Norris to Dayton to testify in the Scopes Trial. Norris accepted Bryan's invitation, but he did not attend, instead sending his own stenographer (L.H. Evridge) to Dayton to record the trial. Norris later used Evridge's transcript to produce *The Only Authentic Book on the Dayton Trial*. On the day before he died, Bryan asked Norris if he would "let me correct my part in the trial before you publish it." Victor Matthews, editor of the *Western Recorder*, often told his readers that "Frank Norris saved the Southern Baptists from evolution."

In 1939, Norris organized the Fundamental Baptist Bible Institute to train young preachers. The institute sponsored many missionaries, including John Birch, who had heard Norris preach in Georgia and later enrolled in the institute in Fort Worth. Birch's death prompted Norris to renew his attacks on Communism. Norris' fame attracted invitations from world leaders, including Winston Churchill and Pope Pius XII (Norris was the first Baptist minister to meet with the Pope). The Fundamental Baptist Bible Institute later became the Bible Baptist Seminary, which is today the Arlington Baptist College. The World Baptist Fellowship, founded by Norris, continues to publish *The Fundamentalist*, which lists Norris as the publication's "Founder."

In 1950, after Temple Baptist Church voted 3,000 to 7 to fire Norris, Norris' health began to fail. After being deserted by many of his friends, Norris died on August 20, 1952, in Keystone Heights, Florida. Following a funeral attended by 5,000 people, Norris was buried along the northern edge of Greenwood Cemetery near the Trinity River in Fort Worth, Texas, beneath the inscription "America's Foremost Fundamentalist."

Businessmen of Fort Worth commissioned sculptor Pompeo Coppini to create a life-size statue of Norris, which was displayed downtown. Today that statue welcomes visitors to Arlington Baptist College, a campus located on the former Top O' Hill Casino property. This casino was known as "Vegas Before Vegas" and was frequented by boxer Joe Louis, Bugsy Siegel, Bonnie and Clyde, and other colorful figures of the past. At a tent meeting in the 1930s in Arlington, Texas, Norris vowed one day "to own the place." His prediction came true in 1956 when the institution that he founded purchased the property and converted the casino to a seminary. The McKinney Avenue Baptist Church in Dallas pastored by Norris from 1905 to 1907 housed the Hard Rock Café for many years.

For More Information: Hankins, B. 1996. *God's Rascal*. Lexington, KY: University of Kentucky.

NORTHWESTERN COLLEGE
(Est. 1902)

Northwestern College is a private, nondenominational, Christian college in St. Paul, Minnesota, which was founded as Northwestern Bible and Missionary Training School in 1902 by fundamentalist and antievolution crusader William Riley. The college was housed in First

Baptist Church in Minneapolis during its formative years, after which it moved to St. Paul. Riley was President of Northwestern from 1902 to 1947. On his deathbed, Riley insisted that he be succeeded at Northwestern by the relatively unknown Billy Graham, who reluctantly served as Northwestern's President from 1949 to 1952. Northwestern trained an army of fundamentalist pastors, many of whom continued Riley's campaign against the teaching of evolution.

Today, Northwestern enrolls more than 2,500 students in a variety of academic programs. It also operates fifteen Christian radio stations located throughout the Midwest, including KTIS, on which Billy Graham gave the first on-air prayer. The administrative building in the center of the 107-acre campus is Riley Hall, named in honor of the college's founder.

JACK NOVIK

Jack D. Novik was the ACLU's lead counsel during *McLean v. Arkansas Board of Education*. Novik noted that "our strategy from the beginning of the case through the trial was to avoid challenging the scientific merits of the creationists' claims. It was our position, not that creationism was bad science, but that it was not science at all." Novik admitted that "evidence showing that creationism was not science does not dispositively prove that it is religion, but it does tend toward that conclusion."

O

WILLIE OATES

Willie Oates was an Arkansas state legislator who in 1959 introduced HB 418 ("The Evolution Bill") to repeal the state's 1928 law banning the teaching of human evolution in its public schools. Antievolution activists branded Oates an atheist, and although she soon withdrew her bill, Oates lost her bid for reelection. In the 1960s, Oates supported Susan Epperson's lawsuit to overturn the law.

ADNAN OKTAR (b. 1956)

Adnan Oktar is a leading Muslim creationist who in 1990 founded the Foundation for Scientific Research. Soon after publishing *The Holocaust Lie* (1995), Oktar began a creationism campaign that attacked biology teachers. Oktar, an Old-Earth creationist who served nineteen months in prison, lost a lawsuit initiated by six biology professors, each of whom were awarded $6,000. In 2007, writing under the name Harun Yahya, Oktar began distributing *Atlas of Creation*, a massive (12 pounds, 17″ × 12″, almost 800 glossy pages), lavishly illustrated book that promotes Islam, rejects evolution as "a major hoax," and claims that "those who perpetuate terror all over the world are, in reality, Darwinists." Like many earlier creationists, Oktar believes that intelligent design is a tool of Satan and claims that evolution causes materialism, communism, and Buddhism.

ALEKSANDR OPARIN (1894–1980)

Aleksandr Ivanovich Oparin was a Russian scientist who in 1924 published a small pamphlet describing the "heterotrophic hypothesis," which stated that Earth's first organisms were heterotrophs. Later, in *The Origin of Life* (1936), Oparin argued that life evolved by increasingly complex organization of preliving proto-organisms. Oparin used Darwinian reasoning to propose the synthesis and accumulation of organic compounds and their incorporation into primordial forms of life that used external sources of reduced carbon as sources of energy. Oparin suggested that Earth's early atmosphere was a reducing atmosphere consisting of water vapor, ammonia, and hydrocarbons in which chemical reactions produced organic compounds. When these compounds dissolved in the ocean, they formed a "primordial soup" from which life arose. In 1951, Harold Urey came to a similar conclusion about early Earth's atmosphere, despite the fact that he was unaware of Oparin's work. Oparin's work inspired the work of Stanley Miller, whose research involved sending electrical charges through a re-creation of the primitive atmosphere described by Oparin. Oparin is buried in Novodevichy Cemetery in Moscow.

HENRY OSBORN (1857–1935)

Evolution takes its place with the gravitation law of Newton. It should be taught in our schools simply as Nature speaks to us about it.

Henry Fairfield Osborn was born in Fairfield, Connecticut, on August 8, 1857. He received an AB from Princeton in 1877 and, after studying biology at Cambridge University under Thomas Huxley (and meeting Charles Darwin when Darwin visited Huxley's anatomy class), Osborn received a ScD from Princeton University in 1881. Osborn was a professor at Princeton from 1881 to 1890, after which he worked at the American Museum of Natural History (AMNH).

Although the autocratic Osborn endorsed Lamarckism and denied any substantial power to natural selection, he nevertheless helped popularize evolution. In 1904 he published *From the Greeks to Darwin*, and in 1908 he became president of AMNH. In 1917, Osborn published *The Origin and Evolution of Life*, and the following year he received the Darwin Medal of the Royal Society of London. When British biologist William Bateson told the AAAS in 1921 that scientists had not discovered the process underlying evolution (a speech hailed by creationists as "the swan song of Darwinism"), Osborn snapped back that Bateson was "out of the main current of biological discovery." One of Osborn's doctoral students—Arthur M. Miller—taught geology to John Scopes at the University of Kentucky; a letter from Miller supporting Scopes was published in *Science* in 1925.

When William Jennings Bryan published "God and Evolution" in the *New York Times* in 1922, it caught Osborn's eye. Soon thereafter, Osborn—who visited Charles Darwin shortly before Darwin's death—became Bryan's chief scientific adversary. Osborn used the *New York Times* and a small book titled *The Earth Speaks to Bryan* to denounce Bryan's antievolution views

TO

JOHN THOMAS SCOPES

AND

OTHER COURAGEOUS TEACHERS

OF THE UNITED STATES

WHO ELECT TO FACE SQUARELY THE ISSUE THAT
THE YOUTH OF AMERICA SHOULD BE
FREELY TAUGHT THE TRUTH OF EVOLUTION AND THE FACT
THAT THIS GREAT LAW OF LIVING NATURE IS
CONSISTENT WITH THE HIGHEST IDEALS
OF RELIGION AND CONDUCT

64. Henry Osborn dedicated his book *Evolution and Religion in Education* to John Scopes and "other courageous teachers" who teach evolution. (*Randy Moore*)

and claim that science was the province of professionals, not preachers. Just before the Scopes Trial, Osborn claimed that Bryan was blinded by "religious fanaticism."

Osborn, who believed that evolution was purposeful and that evolution proved the existence of God, claimed that evolution teaches that nothing can be gained without effort. When Scopes visited New York before his famous trial, Osborn posed with Scopes among prehistoric fossils at the AMNH. Although Osborn declined to be an expert witness at Scopes' trial, he greatly admired Scopes' courage (Figure 64).

After the Scopes Trial, Osborn received the Wollaston Medal of the Geological Society of London, and in 1928 he served as president of the AAAS. However, he startled fellow biologists by rejecting the ape ancestry for humans. Osborn used lessons of nature to try to roll back decadence, racial mixing, and effeminacy, noting that "we have taken from our boys and girls the stern element of the struggle with the forces of Nature; we have, as a final step in the emasculation, substituted the woman for the man teacher."

Osborn, who described Thomas Jefferson as "The Father of Paleontology," died at Garrison-on-Hudson, New York, on November 6, 1935. Osborn's son, Henry Fairfield Osborn, Jr. (1887–1969), became a renowned conservationist; his book, *Our Plundered Planet* (1948), was one of his era's most important proenvironment statements.

WILLIAM R. OVERTON (1939–1987)

Creation science has no scientific merit or educational value as science.

William Ray Overton was born on September 19, 1939, in Malvern, Arkansas. After graduating from the University of Arkansas in 1961 and its School of Law in 1964, Overton began a private practice in Little Rock. In 1979, Overton became the U.S. District Judge in the Eastern District of Arkansas. Overton remained in this position until his death in 1987.

Overton is best known for his 1982 ruling in *McLean v. Arkansas Board of Education*, which was the first test of laws giving "balanced treatment" to creationism in science classes of public schools. Overton, a judicial traditionalist who relied heavily on precedent, allowed both sides to present expert witnesses on religion and theology. Despite the publicity associated with the trial, Overton made it clear that his decision would not be affected by public opinion. "I don't read letters that are sent to me about the case.…The application and content of First Amendment principles are not determined by public opinion polls or by a majority vote."

McLean v. Arkansas Board of Education began on December 7, 1981, and ended on December 17, 1981. The creationists had few credible witnesses and one of their witnesses—British astrophysicist Chandra Wickramasinghe—openly scoffed at creation science.

On January 5, 1982, the same day that Mississippi enacted into law its "Balanced Treatment for Creation Science and Evolution Science Act," Overton issued his thirty-eight-page decision. To determine whether Act 590 violated the Establishment Clause of the Constitution, Overton relied on the "Lemon Test" (*Lemon v. Kurtzman*) to rule that "the evidence is overwhelming that both the purpose and effect of the creationism act is the advancement of religion in public schools," and that the law "was simply and purely an effort to introduce the Biblical version of creationism into the public school curricula," and that "if the unifying ideas of supernatural creation by God are removed from [the Act], the

remaining parts of the section explain nothing and are meaningless assertions." Overton's decision established that (1) the Arkansas "balanced treatment" statute is unconstitutional because it violates the First Amendment; (2) "creation science" is not science, but instead is religion masquerading as science; (3) "creation science" has no scientific significance; and (4) "creation science" fails as science because creationists are not scientists and do not function like scientists.

Duane Gish and Henry Morris of the Institute for Creation Research (ICR) condemned Overton as biased, and Wendell Bird—the architect of ICR's "balanced treatment" policy on which the Arkansas statute was based—claimed that Overton's decision was "constitutionally erroneous and factually inaccurate." Nevertheless, *McLean v. Arkansas* doomed future legislative initiatives involving "creation science" because it destroyed the credibility of creation science. Overton, who had read more than sixty creationist texts during his decision-making process, later warned against the teaching of creationism in public schools, noting that creation science "is more like propaganda than educational material."

One month after Overton's decision, Attorney General Steve Clark decided not to appeal Overton's decision. When Overton heard lawyers claim that it was possible to draft a creationist law that would meet constitutional standards, he responded that "it would take a better lawyer than me to do that."

The ACLU asked Overton to award $1.4 million in attorneys' fees, but Overton reduced the award to $357,768. Clark appealed the award, but the award was upheld (along with 8.75 percent interest). The *Arkansas Gazette* estimated that legal costs surrounding Act 590 cost taxpayers almost $1,000,000.

Overton died on July 14, 1987, and is buried in Shadowlawn Cemetery in Malvern, Arkansas.

RICHARD OWEN (1804–1892)

As I do not know the secondary cause by which it pleased the Creator to introduce organised species into this planet, I have never expressed orally or in print an opinion on the subject.

Richard Owen was born July 20, 1804, in Lancaster, England. Owen was a poor student in his early years (at least once being labeled as "lazy and impudent"), and when he became convinced that a university education was not an option available to him, Owen apprenticed with three different surgeons. These experiences highlighted Owen's innate skill in dissection and attention to detail, qualities that would later serve him well as a comparative anatomist. In 1824, he attended the University of Edinburgh, but stayed for only two terms when his anatomy professor, John Barclay, impressed by Owen's skills, advised him to work toward becoming a member of the Royal College of Surgeons in London. Barclay, a staunch antimaterialist who advocated the reality of the soul as well as a "vital principle" of life, influenced Owen greatly, and Owen carried this philosophy and Barclay's favorable letter of recommendation to John Abernathy, a professor at St. Bartholomew's Hospital in London.

Being too young to qualify immediately for the Royal College of Surgeons and still lacking key parts of his education, Owen worked under Abernathy until he turned 22 years old, whereupon he passed the required examination and was admitted to the College. In 1827, he was hired to help the curator at the Royal College catalog the thousands of specimens that comprised the Hunterian Collections of anatomy, and Owen began his long-term interest in comparative anatomy. By 1930, Owen produced his first publication, based on a dissection of an orangutan procured from the zoological gardens. He was also giving the occasional lecture, as well as attending to his small medical practice. What he was not doing was making much money, which was a concern as he had become engaged to the curator's daughter, Caroline Clift. They were married in 1835.

In 1830, renowned paleontologist Georges Cuvier visited the Royal College, and invited Owen to Paris. When in France the following year, Owen attended the debates between Cuvier and Étienne Geoffroy Saint-Hilaire. The contrast between Cuvier's rejection of transmutation and Geoffroy's advocacy of the natural transformation of species was striking, and both perspectives influenced Owen's ideas on evolution. Even more influential was Oxford University geologist William Buckland, who had developed a version of gap creationism that allowed for unspecified lengths of time during the biblical creation week. Buckland, who dubbed Owen "the British Cuvier," excelled at paleontology, and his lectures convinced Owen that fossil organisms could be studied with the same comparative approaches used to study extant organisms.

Owen continued to publish at a furious pace (producing around 800 publications during his career), and described several newly found or little-known species, including the platypus and the pearly nautilus, a "living fossil." This work earned him a fellowship to the Royal Society in 1834. In 1836, Owen was appointed Hunterian Professor of Comparative Anatomy and Physiology at the Royal College of Surgeons, and he ushered in the Hunterian Lectures, a popular series of public lectures. Owen's classification of fossil reptiles introduced the Order Dinosauria to science (and the word *dinosaur* to the rest of the world), combining the Greek words *deinos* ("terrible") and *sauros* ("lizard"). Controversy surrounded this work because Owen was not forthcoming with credit for the original discoverers (e.g., Gideon Mantell) of the fossils he used. During this period, Owen shifted his emphasis to paleontology.

Around this time, Charles Darwin, just back from his voyage on the *Beagle*, asked Owen to help him organize his collection of fossils and specimens. Owen was eager to study Darwin's collections, and he made significant contributions, especially his analyses of the large, extinct ground mammals of South America. Owen and Darwin were collegial, and Darwin relied on Owen's knowledge of anatomy and classification. However, each later held different views on evolution, and their friendly relationship evaporated.

Owen's concept of the "archetype" (developed in his 1848 book, *On the Archetype and Homologies of the Vertebrate Skeleton*) proposed a basic body plan for all vertebrates. Fish were the most basic form, although neither fish nor the archetype itself was considered an ancestral form; instead, the archetype was the work of the Creator. Owen contrasted homology with analogy; homologous structures in two organisms were made of the same tissue or connected to the same body parts, whereas analogous structures only demonstrated similar function. These concepts were, therefore, not based on shared ancestry as they are today. Owen's claim that modifications of the archetype were progressive was influenced by Plato's *eidos* and Aristotle's "Great Chain of Being." Owen claimed that the fossil record showed the gradual appearance of increasingly specific elaborations of the archetype in a branching pattern, but not one that implied common ancestry.

Owen's studies of extinct and living organisms helped him realize that organic change was possible. He did not, therefore, reject evolution, although he did not accept Lamarckian inheritance and would later reject natural selection. Owen also rejected the natural theology of William Paley, which emphasized the perfection of organisms. For Owen, the demonstration of "secondary causes" that reserved a place for the Creator was lacking, and Owen rejected a naturalistic explanation for life itself. Referring specifically to the similarities between humans and apes, however, he eventually admitted that he could no longer "shut my eyes to the significance of that all-pervading similitude of structure," although the specifics of his perspective on evolutionary change would remain vague.

Owen helped both Darwin and Thomas Huxley early in their careers, but by the 1860s, the relationship between Owen and his younger colleagues had deteriorated. Owen often appropriated others' work and tried to sabotage colleagues' careers. Huxley noted in 1851 that "I feel it necessary to be always on my guard" around Owen and remarked that Owen engendered an "intense feeling of hatred." When Owen tried to block funding for Joseph Hooker's research,

Darwin noted that "I used to be ashamed of hating him so much, but now I will carefully cherish my hatred and contempt to the last days of my life." Owen's only child—a son named William—committed suicide in 1886, largely due to Owen's "lamentable coldness of heart."

Darwin, who had used Owen's identification of branching patterns in the fossil record to support his proposal of descent with modification, asked Owen to review *On the Origin of Species* before it was published. Owen praised the work to Darwin, but felt that Darwin had not provided enough evidence for a central role of natural selection. Shortly after *Origin* was released, Owen published a hostile (and anonymous) review of the book, which infuriated Darwin and Huxley. In the review, Owen touted his own research and questioned Darwin's competency in discussing a subject that Owen said was better left to anatomists like himself. Such was the environment just prior to the legendary clash between Huxley and Bishop Samuel Wilberforce at the meeting of the British Association for the Advancement of Science in 1860. Whether Owen actually "coached" Wilberforce is unclear, but he was sympathetic to Wilberforce's goal of showing (as he had written in his own negative review of *Origin*) that for a believer of Darwin's theory "the whole world of nature is laid for such a man under a fantastic law of glamour, and he becomes capable of believing anything."

Owen predicted that Darwin's ideas would soon be forgotten while his own would be upheld. Owen sometimes even suggested that he had proposed what became known as the Darwinian theory of evolution. However, Owen's concept of the archetype was soon replaced by Darwin's evolution by natural selection, and Huxley, commenting after Owen's death, concluded "hardly any of [his] speculations and determinations have stood the test of investigation." This was inaccurate, for some of Owen's contributions are still evident, most notably his concept of homology (although modified from Owen's original formulation) and his identification of dinosaurs. But Owen's most

65. Richard Owen, an ardent opponent of Charles Darwin's theory of evolution by natural selection, was the first director of the British Museum of Natural History. This statue of Owen overlooks the museum's Great Hall (Figure 10). (*Randy Moore*)

enduring legacy is his commitment to the role of natural history museums in research and education. After becoming superintendent of the natural history departments at the British Museum

in 1856, Owen created a new British Museum of Natural History—a "cathedral to nature"—which opened in 1881. On December 18, 1892, Owen accepted his (previously declined) knighthood eight years before his death at Sheen Lodge, a residence Queen Victoria gave him in 1852. In 1897, Owen was memorialized with a statue that today overlooks the museum's Central Hall (Figure 65).

For More Information: Owen, R. 1894. *The Life of Richard Owen*. London: John Murray.

OXFORD UNIVERSITY

Oxford University offered Charles Darwin an honorary degree, but Darwin declined the award, citing poor health. On February 12, 1909, Oxford's celebration of the centenary of Darwin's birth was attended by William, Erasmus, George, Francis, and Leonard Darwin.

P

WILLIAM PALEY (1743–1805)

There cannot be design without a designer.

William Paley was born in Peterborough, England, in July 1743. Paley graduated from Cambridge University first in his class in 1763. Paley became a deacon in 1765, was appointed assistant curate in Greenwich, and taught at Cambridge for ten years. He was ordained in 1767 (after earlier earning an MA), and the remainder of his clerical career included successively more important positions within the Anglican Church. Paley opposed slavery and advocated prison reform, and as a philosopher, he was a utilitarian, believing that humans act morally to increase their overall level of happiness. In 1776, Paley married Jane Hewitt, with whom he had eight children.

Paley was a popular preacher and one of England's most important theologians of his generation. He published his Cambridge lectures in *The Principles of Moral and Political Philosophy* (1785), which outlined his utilitarianism and was used as a textbook at Cambridge for many years. This was followed by *A View of the Evidences of Christianity* (1794) which was a response to David Hume's skepticism of religion and, in particular, Hume's dismissal of miracles. But Paley's best-known book, and the last before his death, was *Natural Theology; or, Evidences of the Existence and Attributes of the Deity, Collected from the Appearances of Nature* (1802).

In *Natural Theology*, Paley—one of the most admired clerics in the English-speaking world—argued that God could be understood by studying the natural world. *Natural Theology* begins with the famous metaphor of "God as watchmaker":

> In crossing a heath, suppose I pitched my foot against a stone, and were asked how the stone came to be there; I might possibly answer, that, for any thing I knew to the contrary, it had lain there for ever: nor would it perhaps be very easy to show the absurdity of this answer. But suppose I had found a watch upon the ground, and it should be inquired how the watch happened to be in that place; I should hardly think of the answer which I had before given, that, for any thing I knew, the watch might have always been there.

Paley argued that the only rational conclusion is that the watch had a designer. Much of *Natural Theology* discusses examples of purported design in "a happy world" that teems "with delighted existence," with many drawn from Paley's own observations, and likely to be familiar—and therefore persuasive—to readers. Paley's designer was his watch-making God.

Charles Darwin read *Natural Theology* while at Cambridge, and was encouraged by his instructors John Henslow and Adam Sedgwick to accept Paley's perspective. Darwin recalled that Paley's work, including *Natural Theology*, "was the only part of the academical course which, as I then felt and as I still believe, was of the least use to me in the education of my mind." When Darwin boarded the *Beagle*, he accepted design in nature. However, after discovering natural selection, he felt differently: "The old argument of design in nature, as given by Paley, which formerly seemed to me so conclusive, fails, now that the law of natural selection has been discovered."

Virtually all biologists have similarly rejected Paley's argument. The most famous of these refutations is Richard Dawkins' *The Blind Watchmaker* (1986), whose title refers to Paley's metaphor. Dawkins agrees that there is a watchmaker, but otherwise concludes that Paley is "gloriously and utterly wrong." The watchmaker for Dawkins (and for contemporary biology) is natural selection. Biologists view the evolution of complexity and apparent design, therefore, simply as the result of the cumulative process of repeated generations of differential reproduction. Dawkins' book motivated Phillip Johnson to write *Darwin On Trial* and to become active in the intelligent design (ID) movement. Although proponents of ID claim that their premises differ from Paley's, and, unlike Paley, do not specify who or what the designer is, most evolutionary biologists see ID as a version of Paley's arguments updated to account for advances in our understanding of biology.

Paley suffered for many years from a serious intestinal ailment. Soon after finishing *Natural Theology*, he suspected that his death was imminent, and he assembled his sermons to be published posthumously and given to anyone "likely to read them." Paley's suspicions of his impending death were borne out, and he died on May 25, 1805. He was buried in the Carlisle Cathedral, next to his wife.

For More Information: LeMahieu, D.L. 1976. *The Mind of William Paley*. Lincoln, NE: University of Nebraska; Paley, W. 1802. *Natural Theology: Or, Evidences of the Existence and Attributes of the Deity, Collected from the Appearances of Nature*. London: R. Fauldner.

PALUXY RIVERBED

In the 1960s and 1970s, a variety of creationists reported human footprints beside tracks of sauropod dinosaurs in the Paluxy Riverbed near Glen Rose, Texas (about 50 miles southwest of Fort Worth). If this were true, it would indicate that humans and dinosaurs lived contemporaneously, and therefore refute the standard evolutionary view of the history of life on Earth.

The Paluxy tracks were first noticed in 1910 by Charlie Moss and his brother Grady, who described the tracks as "giant man-tracks." In the 1930s, Jim Ryals chiseled out and sold some of the tracks to visitors and local residents, and George Adams later began selling rocks in which he carved human tracks. The tracks remained a local legend until fossil-hunter Roland Bird saw some of Adams' carved "man tracks" at a trading post in Gallup, New Mexico, while on an expedition for the American Museum for Natural History. Bird recognized them as fake, but wondered why anyone would carve such prints. Curious, he went to Glen Rose in 1938, where he found footprints of sauropods (probably *Pleurocoelus*, a huge, long-necked, four-footed dinosaur commonly referred to as a "brontosaur"). When Bird asked Ryals to show him more "man tracks," Ryals showed him one that Bird described as "something about 15″ long, with a curious elongated heel." The track was too vague to diagnose. Although Bird never reported human tracks at the Paluxy sites, he did mention the local rumors of giant man-tracks in the area.

Clifford Burdick, a creationist who helped form the Creation Research Society, went to Paluxy, after which he published an article entitled "When Giants Roamed the Earth" in the July 1950 issue of the Seventh-Day Adventist magazine *Signs of the Times*. In that article, Burdick proclaimed the Paluxy tracks are human and that they therefore refute evolution. Burdick's article, which used out-of-context quotes to suggest that Bird had excavated and reported human tracks from the Paluxy River, included photos of tracks carved on loose blocks of limestone by Adams in the 1930s. These tracks became known as the "Burdick tracks" and were featured in *The Genesis Flood* (1961), the book that popularized "creation science." In 1965, A.E. Wilder-Smith's book *Man's Origin, Man's Destiny* repeated the story, and in 1972, Baptist minister Stanley Taylor (of *Films for Christ*) perpetuated the human–dinosaur story with the film *Footprints in Stone*. For many years, the film—produced with the help of Henry Morris—was shown at schools and churches throughout the United States, thereby helping advertise the Paluxy "man tracks." Creationist groups such as the Institute for Creation Research and the Creation Research Society also promoted the dinosaur/human tracks. The tracks featured in *Footprints in Stone* became known as the "Taylor site," which is a few hundred yards west of Dinosaur Valley State Park.

For the past few decades, numerous scientific studies of the alleged "human tracks" that appear alongside the dinosaur tracks have consistently discredited claims that they were made by humans. For example, the Burdick track includes a variety of unnatural features, and tracks at the Taylor site, when cleaned, had impressions not of human feet, but of three-toed feet of dinosaurs. Other tracks in Glen Rose that various creationist groups claim to be human tracks were formed by erosion, and some have been deliberately altered to resemble the tracks of humans. Because of these findings, some creationists—including antievolution organizations such as Answers in Genesis—no longer claim that there are human tracks alongside dinosaur tracks in Glen Rose. However, other creationists—most notably former Baptist minister Carl Baugh—continue to sponsor expeditions and excavations that uncover footprints allegedly left by giant humans who lived with dinosaurs. Baugh's Creation Evidence Museum is just a few hundred yards from the entrance to Dinosaur Valley State Park.

The Glen Rose tracks discovered by Roland Bird are from the lower Cretaceous, and are about 133 million years old. Soon after dinosaurs (including *Iguanodon*) walked in the Paluxy River's moist, slimy mud, their tracks were buried by a different sediment that hardened. The tracks became visible when the overlying layers were removed by the Paluxy's flowing water (or excavators). Today, some of Bird's excavations are displayed at the American Museum of Natural History.

Texas has many sites showing well-preserved dinosaur tracks, including several in nearby Dinosaur Valley State Park just northwest of Glen Rose, Texas. This park, which includes models of a 70′ *Apatosaurus* and a 45′ *Tyrannosaurus rex* commissioned by the Sinclair Oil Company for the New York World's Fair Dinosaur Exhibit in 1964–1965, covers more than 1,500 acres and in 1969 was designated a National Natural Landmark by the National Park Service. You'll also find a track of *Acrocanthosaurus tokensis*, a bipedal, meat-eating dinosaur, in the bandstand at the Glen Rose town square.

PANGLOSS

Pangloss was a character in French Enlightenment author François-Marie Arouet de Voltaire's (1694–1778) novel *Candide* who argued that everything suits a specific purpose. Voltaire modeled Dr. Pangloss on German philosopher Gottfried Leibniz (1646–1716), who claimed that our universe is, in a restricted sense, the best universe God could have made. Evolutionary biologist Stephen Jay Gould used Pangloss to develop the Pangloss Paradigm, which creationists use when claiming that all biological traits are as good and perfect as they can be, just as Pangloss claimed everything on Earth is just as it should be in "the best of all possible worlds" (e.g., "Everything is made for the best purpose... Our nose is formed for spectacles, therefore we wear spectacles"). Gould contrasted the Pangloss Paradigm with the fact that adaptations do not represent perfection, but instead are merely the best that could be achieved with material provided by historical accident. The book *The Privileged Planet* (1994) by Guillermo Gonzalez and Jay Richards (of the Discovery Institute) presents a Pangloss-based argument about Earth.

JOSEPH PARSLOW (1809–1898)

Joseph Parslow was a servant at Darwin's Upper Gower residence and Down House until 1875. Parslow was known by the Darwins as "the venerable Parslow" (after "the aged Parslow" in Dickens' *Great Expectations*), and was described by Hooker as "an integral part of the family, and felt to be such by all visitors to the house." After Charles Darwin died, Parslow received a pension from the Darwin family of £50 per year. Parslow attended Charles Darwins' funeral, and in 1885 he attended Thomas Huxley's unveiling of the statue of Darwin at The British Museum.

JULIA PASTRANA (1834–1860)

Julia Pastrana was a Mexican Digger Indian who many people considered a "throwback" to an ape-like era of humanity. Pastrana was covered by straight black hair and had a large nose and irregular teeth, which gave her a gorilla-like appearance. Wallace, Darwin, and Haeckel described her traits as scientific curiosities, and Pastrana appeared as the "gorilla woman" at circuses and freak shows. When Pastrana died of complications due to childbirth, a showman preserved her body and charged the public to see it. Her mummy was rediscovered in Norway in 1990.

CLAIRE PATTERSON (1922–1995)

Cameron Claire "Pat" Patterson was an American geochemist who, building on the work of Bertram Boltwood (1870–1927), in 1948 refined the uranium–lead method of radioactive dating. When Patterson applied this technique to several meteorites (which he assumed to have originated when the solar system originated), he discovered that all of the meteorites, as well as a modern marine sediment (which was presumably formed from rocks from several different continents), had the same age. In 1956, Patterson ended hundreds of years of speculation and research by scientists and theologians such as James Ussher, Lord Kelvin, and others about the age of the Earth when he reported in a paper titled "Age of Meteorites and the Earth" that Earth has "an age of $4.55 + 0.07 \times 10^9$ years." After making this discovery, Patterson was so excited that he asked his mother to check him into a hospital because he thought he was having a heart attack.

This vast a period of time is extremely difficult to comprehend. For example, most people can appreciate periods up to 100 or so years, because they've seen photos of, and have heard grandparents' stories about, previous decades. However, we know relatively little about life 2,000 years ago (i.e., during the time of Jesus), and we know even less about life 5,000 years ago when Egyptians were building the pyramids. However, if we could go back 5,000 years to watch the Egyptians' work, we would have seen only about 0.0001 percent of Earth's history.

Throughout his career, Patterson acknowledged the pioneering work of geology Arthur Holmes regarding the geological time scale. Later in his life, Patterson campaigned against lead contamination in the environment, and his work led to the Clean Air Act in 1970. By 1986, thanks largely to Patterson's work, lead was out of U.S. gasoline. Patterson died in 1995 of an asthma attack.

AUSTIN PEAY (1876–1927)

Nobody believes that [this law] is going to be an active statute.

Austin Peay IV, a politician who became known as "The Maker of Modern Tennessee," was born on June 1, 1876. A self-described "old-fashioned Baptist," Peay was elected Governor of Tennessee in 1922 and, after he transformed a state budget deficit into a $2-million surplus, he was reelected in 1924. Peay spent liberally on public works such as roads, hospitals, and schools, and was a popular governor.

Peay believed the teaching of evolution converts students to agnosticism. Early in 1924, Peay joined William Jennings Bryan onstage when Bryan urged Tennessee legislators to ban the teaching of evolution. On March 21, 1925, Peay—under pressure by fellow Baptists—signed John

Butler's legislation banning the teaching of human evolution in Tennessee's public schools, noting that "denial of the Bible shall not be taught in our public schools." Peay justified his endorsement of the bill with a statement of faith: "something is shaking the fundamentals of the country, both in religion and morals…an abandonment of the old-fashioned faith and belief in the Bible is our trouble in a large degree… The people have the right and must have the right to regulate what is taught in their schools." Peay was applauded by fundamentalists, but did not attend Scopes' famous trial.

When Bryan died five days after Scopes' trial, Peay proclaimed that Bryan had died "a martyr to the faith of our fathers" and announced a state holiday to commemorate Bryan's funeral. On November 5, 1926, Peay participated in the groundbreaking ceremony for Bryan College.

Peay died in office on October 2, 1927, and was buried in Greenwood Cemetery in Clarksville, Tennessee. To honor Peay, in 1927 Tennessee founded Austin Peay State Normal School for Rural White Teachers in Clarksville, Tennessee. The school's name was later shortened to Austin Peay State University.

PELOZA v. CAPISTRANO UNIFIED SCHOOL DISTRICT

Peloza v. Capistrano Unified School District (1994) ended when the Ninth Circuit Court of Appeals ruled that evolution is not a religion and that a school can require a biology teacher to teach evolution. John Peloza was a high school biology teacher who sued his school for $5,000,000 because it forced him to teach "evolutionism" and that evolution is a religious belief.

LISA PETERS (b. 1951)

Lisa Westberg Peters was an acclaimed author of children's books. In 2004, Peters published *Our Family Tree*, which noted that humans are related to all living things and that "all of us are part of an old, old family. The roots of our family tree reach millions of years to the beginning of life on Earth." Although *Our Family Tree* won several awards, many schools—hoping to avoid exposing students, even incidentally, to evolution—cancelled her appearances.

PFEIFER v. CITY OF WEST ALLIS

Pfeifer v. City of West Allis began in 1999 when Christopher A. Pfeifer of the Genesis Commission (whose purpose is to "educate the public" about creationism) was not allowed to use a room at the West Allis (Wisconsin) public library for a "creation science workshop" that would be "open to the public." In April 2000, Judge J. Adelman of the U.S. District Court, E.D. Wisconsin ruled that the library is a designated public forum and that there was no compelling state interest to exclude Pfeifer from using the room.

JOHN PHILLIPS (1800–1874)

John Phillips was an English geologist who first estimated the age of the Earth using the rate of sediment accumulation as an indicator of time. Phillips, who in 1860 dismissed Charles Darwin's estimate that Earth is several hundred million years old as an "abuse of arithmetic," suggested that Earth is 96 million years old. Phillips warned geologists that trying to investigate the age of the Earth was a waste of time, instead urging them to "be satisfied with the truth, that 'In the Beginning God created the heavens and the Earth.'" Darwin described Phillips' slim *Life on the Earth* as "unreadably dull." In 1859, Phillips—the president of the Geological Society—gave Charles Darwin the Wollaston Medal, which Charles Lyell accepted in Darwin's absence. Phillips coined the terms Mesozoic and Kainozoic (now more often written Cenozoic), and promoted intelligent design, claiming that nature "is the expression of a divine idea."

Phillips, whose uncle was mapmaker William Smith, died in 1874 after falling down a flight of stairs. People who later used sediment accumulation to estimate Earth's age included Thomas Huxley (100 million years), Alfred Wallace (28 million years), Charles Walcott (45–70 million years), Alexander Winchell (3 million years), and John Joly (80–87 million years).

PILTDOWN MAN (1909–1953)

Piltdown Man was a famous fossil forgery that is often traced to Charles Dawson (1864–1916), a country lawyer and amateur geologist who in 1909 claimed to have received a fragment of a human skull excavated from a gravel pit at Piltdown, Sussex, about 30 miles from Charles Darwin's home. When Dawson returned to the pit two years later, he found other pieces of the skull. With the help of paleontologist Arthur Woodward of The British Museum, the enigmatic Dawson continued excavating the site and found more pieces of the skull, animal fossils mixed with human tools, and part of a jawbone with two intact molars. These fragments became known as Piltdown Man.

Dawson and Woodward announced their discovery on December 18, 1912, to the British Geological Society. Their fossil, *Eoanthropus dawsonii* ("Dawson's Dawn-Man"), was named and authenticated by experts at The British Museum, and became front-page news throughout the world. The *New York Times* published a summary titled "Paleolithic Skull Is a Missing Link" of Dawson and Woodward's presentation soon after their announcement, and a few days later published a follow-up titled "Darwin Theory Is Proved True." Although a few scientists questioned Piltdown Man, most accepted the find as an intermediate between humans and apes. A replica of the Piltdown Man skull was brought to the Scopes Trial by the defense, and on July 23, 1938, British paleontologist Arthur Woodward and anatomist Arthur Keith unveiled a memorial at the gravel pit where Piltdown Man was discovered, which in 1952 was designated a Geological Reserve and National Monument by the Nature Conservancy. That monument, now unkept and covered with lichens, still stands.

Charles Dawson died in 1916, and his funeral at St. John Sub Castro Churchyard in Lewes, East Sussex, included many eulogies praising his Piltdown discovery. However, in November 1953, Kenneth Oakley and his colleagues at The British Museum used fluorine testing and microscopy to show that the skull and jaw of Piltdown Man were only about 600 years old, and that Piltdown Man—which had been hailed by Museum director Edwin Ray Lankester as "the first Englishman"—was a fraud consisting of the lower jaw bone of an orangutan combined with the skull of a modern human. Creationists used Piltdown Man to claim that evolutionary scientists are not credible.

Although the perpetrator of the Piltdown Man hoax was never identified, many suspected that it was Dawson; other suspects included French cleric and amateur geologist Teilhard de Chardin and Arthur Conan Doyle, the creator of Sherlock Holmes. In 2003, the Natural History Museum sponsored an exhibit marking the fiftieth anniversary of the exposure of the forgery. Piltdown Man was cited in popular music (on Mike Oldfield's 1973 album *Tubular Bells*), in products produced by Apple Computer, and on episodes of *X-Files*. The comic strip *Peter Piltdown* in U.S. newspapers was a precursor to "The Flintstones." In *Scientology: A History of Man*, Scientology founder L. Ron Hubbard lists Piltdown Man as an ancestor of humans. Today, the waiting room at the offices of the firm of solicitors (Darwon-Hart) where Dawson worked on Lichfield Street in Lewes still displays a replica of the Piltdown skull.

For More Information: Russell, M. 2003. *Piltdown Man: The Secret Life of Charles Dawson and the World's Greatest Archaeological Hoax*. Stroud, Gloucestershire: Tempus; Walsh, J.E. 1996. *Unraveling Piltdown: The Science Fraud of the Century and Its Solution*. New York: Random House.

PLATO (427–347 BC)

Plato, a student of Socrates, was a Greek philosopher who proposed that nature consists of transcendent, ideal forms, and that variations from these forms are illusions. Plato founded The Academy in Athens, the first institution of higher learning in the Western world. Many creationists' insistence on the fixity of species is traceable to Plato.

JOHN PLAYFAIR (1748–1819)

The rivers and the rocks, the seas and the continents, have been changed in all their parts; but the laws which direct those changes, and the rules to which they are subject, have remained invariably the same.

John Playfair was born on March 10, 1748, in Angus, Scotland. After being home-schooled by his father, Playfair enrolled at the University of St. Andrews and began studying for a career in the Scottish church. He received an MA in 1765 and, a year later at age 18, applied to be chair of the math department at Marischal College of Aberdeen. He was turned down. When his father died, Playfair became a minister in the parishes that his father had served. Playfair then studied divinity at the University of St. Andrews, after which he moved to Edinburgh, where he lived the rest of his life. In Edinburgh, Playfair interacted with scholars such as Adam Smith and James Hutton. Playfair, who believed that "the Scriptures are not intended to resolve physical questions," was especially fond of Smith's claim that "science is the great antidote to the poison of enthusiasm and superstition."

In 1779, Playfair presented his first scientific paper ("An essay on the arithmetic of impossible quantities"), and in 1783 was a founding member of the Royal Society of Edinburgh. In 1785, Playfair was appointed a professor of math at the University of Edinburgh, and ten years later he devised an alternative formulation of Euclid's parallel postulate. Although Playfair acknowledged that his postulate was equivalent to the "parallel postulate" derived by Proclus (411–485), it nevertheless became known as Playfair's axiom: "Through any point in space, there is at most one straight line parallel to a given straight line."

When his friend James Hutton died, Playfair wrote *Biographical Account of the Late Dr. James Hutton*. This memoir inspired Playfair to rewrite Hutton's work in a more simple, eloquent, and accessible style. The result in June 1802 was *Illustrations of the Huttonian Theory of Earth*, which addressed Hutton's critics (especially Richard Kirwan), dismissed the theories of Buffon and Burnet, and downplayed Hutton's arguments about God's purpose and design. Playfair's book—which placed Hutton's work "precisely on the same footing with the system of Copernicus"—presented Hutton's ideas to a new generation of scientists and, in the process, popularized Hutton's work. Indeed, it is Playfair's *Illustrations* that Darwin took aboard the *Beagle*, and that Lyell quoted on the cover-page of his *Principles of Geology*. Playfair's *Illustrations*, more than any other book, laid the foundation of modern geology.

In 1805, Playfair became chair of natural philosophy (a position he held until his death), and in 1807 he was elected a Fellow of the Royal Society of London. In 1815, Playfair began a seventeen-month, 4,000-mile geological study of Europe to gather material for the second edition of *Illustrations*. However, Playfair was distracted by other obligations—from 1798–1819 he was the secretary and "the life and soul" of the Royal Society of Edinburgh—and his health began to deteriorate.

Playfair died on July 20, 1819, in Burntisland in Fife, and was buried in an unmarked grave adjacent to the grave of philosopher David Hume. Playfair's grave is overlooked by Calton Hills' Observatory, an observatory that he helped to create. Craters on Mars and the Moon are named in Playfair's honor.

DAVID POOLE (1858–1955)

David Scott Poole was a North Carolina state legislator who, with the backing of evangelist T.T. Martin and governor Cameron Morrison, introduced a resolution on January 8, 1925, declaring that evolution was "injurious to the welfare" of the state's citizens. On February 19, 1925, Poole's bill was defeated by a vote of 67-46. Two years later, Poole introduced legislation drafted by a "capable Christian lawyer" that banned "any doctrine or theory of evolution that contradicts or denies the divine origin of man or of the universe, as taught by the Holy Bible." The House Education Committee rejected Poole's legislation by a vote of 25-11. Poole died in Raeford, North Carolina at age 96.

POPE BENEDICT XVI (b. 1927)

Pope Benedict XVI (born Joseph Alois Ratzinger) succeeded Pope John Paul II as the 264th Pope of the Roman Catholic Church. During Mass on April 24, 2005, in which he was made pope, Benedict noted, "We are not some casual and meaningless product of evolution. Each of us is the result of a thought of God....Each of us is necessary." In 2004, as Cardinal Ratzinger, he endorsed evolution and an ancient Earth, and on July 24, 2007, Pope Benedict proclaimed that evolution must not exclude religious faith, noting that "there are many scientific texts in favor of evolution, which appears as a reality that we must see and enriches our understanding of life and being." In the mid-1980s, while serving as prefect of the Sacred Congregation of the Doctrine of the Faith (the successor of the Inquisition), he defended creationism by insisting that humans are "not the products of chance and error."

POPE JOHN PAUL II (1920–2005)

Pope John Paul II (born Karol Józef Wojtyla), the 263rd Pope of the Roman Catholic Church, reaffirmed to the Pontifical Academy of Science in 1996 that there is no essential conflict between evolution and Catholicism. John Paul noted that "new findings lead us toward the recognition of evolution as more than a hypothesis. In fact it is remarkable that this theory has been progressively accepted by researchers, following a series of discoveries in various fields of knowledge. The convergence in the results of these independent studies—which was neither planned nor sought—constitutes in itself a significant argument in favor of the theory." However, John Paul rejected any theory of evolution that provides a materialistic explanation for the human soul. John Paul insisted that "souls are immediately created by God" and that theories of evolution claiming that the human mind is a product of evolution are incompatible with "the dignity of the person." On October 31, 1992, John Paul acknowledged the Church's wrongdoing to Galileo Galilei, and in 1988 he

beatified Nicolaus Steno—one of the founders of modern geology—on October 23, the same date that Bishop James Ussher had chosen for the creation of the world.

When John Paul died on the evening of April 2, 2005, more than 3 million people came to Rome to pay their respects. On May 9, 2005, John Paul's successor—Pope Benedict XVI—waived the traditional five-year waiting period for a cause for beatification to begin. Soon after John Paul was interred in the Tomb of the Popes beneath St. Peter's Basilica, a Gallup poll reported that John Paul was the eighth most-admired person of the twentieth century. In 2005, Cardinal Christoph Schönborn, the archbishop of Vienna and editor of the 1992 *Catechism of the Catholic Church*, dismissed John Paul's claims about evolution as "rather vague and unimportant" while claiming that "evolution in the neo-Darwinian sense—an unguided, unplanned process of random variation and natural selection—is not...true." Schönborn later claimed that his religion-based argument was "superior to a scientific argument since it was based on more certain and enduring truths and principles." Pope Benedict XVI backed Schönborn, claiming that the universe is the result of an "intelligent project."

POPE PIUS IX (1792–1878)

Pope Pius IX (born Giovanni Maria Mastai-Ferretti), the 250th Pope of the Roman Catholic Church, denounced Darwinism as "a repugnant idea" that "would seem to need no refutation." Pius formally adopted the Immaculate Conception, and his First Vatican Council enshrined papal infallibility. Pope Leo XIII (1810–1903), who succeeded Pius, defended the inerrancy of the Bible and claimed that "there can never, indeed, be any real discrepancy between the theologian and the physicist, as long as each confines himself within his own lines."

POPE PIUS XII (1876–1958)

Pope Pius XII (born Eugenio Maria Giuseppe Giovanni Pacelli), the 260th Pope of the Roman

Catholic Church, wrote *Humani Generis* (1950), an encyclical about human origins. Pius' predecessor—Pope Pius XI (1857–1939)—had tried to reconcile science and religion by claiming that "those who speak of the incompatibility of science and religion either make science say that which it never said or make religion say that which it never taught." However, although Cardinal John Henry Newman wrote in 1868 that "the theory of Darwin, true or not, is not necessarily atheistic," Pius XII's encyclical was the first occasion on which a pope had explicitly addressed evolution at length. In this encyclical, Pius—who urged that evolution "is an open question...[to] be examined and discussed"— acknowledged that evolution might accurately describe the origins of humans, but rejected those who "imprudently and indiscreetly hold that evolution...explains the origin of all things." Pius XII helped Catholics accept human evolution by acknowledging evolution as a serious hypothesis that did not contradict essential Catholic teachings, and that human "souls were created directly by God." Pius' encyclical concluded by noting that the doctrine of evolution is "an open question." When Pius died on October 9, 1958, Riccardo Galeazzi-Lisi—the Vatican's chief physician—tried to sell photos of the dying pope.

WILLIAM L. POTEAT (1856–1938)

William L. Poteat was a self-taught biologist who became president of Wake Forest College (a Southern Baptist school). Poteat was attacked by T.T. Martin, William Riley, Frank Norris, and other fundamentalists for endorsing evolution. Poteat was supported by the Wake Forest Board of Trustees, but the controversy made North Carolina a battleground in the evolution-creationism controversy in the early 1920s.

CHARLES FRANCIS POTTER (1885–1962)

Charles Francis Potter was a Unitarian preacher who faced John Roach Straton in 1924 in a series of highly publicized debates about evolution. The following year, Potter assisted the defense of John Scopes.

BADEN POWELL (1796–1860)

Baden Powell was an English mathematician who wondered why God would have given humans the ability to think if he had not wanted humans to think. Powell used a theological argument to reject miracles: If God made laws of nature, then a miracle would break the laws that God has made; therefore, a belief in miracles would be atheistic. Powell's claim enraged many people, and the notorious *Essays and Reviews* that published Powell's argument sold 22,000 copies in its first two years (far more than Darwin's *On the Origin of Species*). Powell argued that "Mr. Darwin's masterly volume must soon bring about an entire revolution in opinion in favour of the grand principle of the self-evolving powers of nature." Powell was scheduled to be onstage at the British Association's meeting that included the Huxley–Wilberforce debate, but he died of a heart attack two weeks before the meeting. Powell's son Robert Baden-Powell founded the worldwide scouting movement.

PRESBYTERIAN CHURCH GENERAL ASSEMBLY

The Presbyterian Church General Assembly in 1910 affirmed five essential doctrines that became the foundation for fundamentalism: the deity and virgin birth of Christ, the resurrection of Christ, the "second coming" of Christ, the substitutionary atonement of Christ's death, and the inspiration and infallibility of the Scriptures. By 1927, the Assembly had decided that the five fundamentals were no longer binding for ministerial candidates.

JOSEPH PRESTWICH (1812–1896)

Sir Joseph Prestwich was a clergyman-geologist who confirmed the discovery by Boucher de Perthes (1788–1868) of ancient man-made tools with undisturbed bones of mammoths. Because he was a devout Anglican and highly respected geologist, Prestwich's endorsement of de Perthes' claim prompted others to take the discovery seriously.

GEORGE McCREADY PRICE
(1870–1963)

There ought to be no doubt whatever that the popular forms of geology and paleontology should be included as sciences of Satanic origin.

George Edward Price was born on August 26, 1870, in New Brunswick, Canada (Figure 66). As a youth, Price joined the Seventh-Day Adventists, a small fundamentalist sect founded by Ellen White. Price attended the Adventists' Battle Creek College (now Andrews University) from 1891–1893, and then taught English and other subjects while he was a high school principal in the fishing village of Tracadie. In 1902, Price published his first book, *Outlines of Modern Christianity and Modern Science*. On the title page of that manuscript, Price listed his name as George E. McCready Price (McCready was his mother Susan's maiden name). Thereafter, he dropped "Edward" from his name, going by George McCready Price. Price was a self-taught geologist; the degrees often listed after his name were honorary degrees from Adventist schools.

In *Outlines*, Price argued for an uncompromising return to "primitive Christian principles," claiming that biblical literalism was the only explanation of the Bible that a Christian can accept. Price also claimed that "no believer in the Sabbath ... will hesitate to give as the distinct, positive teaching of Genesis that life has been on our globe only some six or seven thousand years; and that the Earth as we know it, with teeming animal and vegetable life, was brought into existence in six literal days." Price worked at the Adventists' Loma Linda Sanitarium in Loma Linda, California, where he met White. White's teachings became a foundation for Price's views of geology.

In 1906, Price published a ninety-three-page book *Illogical Geology, the Weakest Point in the Evolution Theory*, in which Price labeled theistic evolution a "truly pagan ... heathen religion"

66. George McCready Price founded what in the 1970s became "creation science," and his work was the foundation for Whitcomb and Morris' influential *The Genesis Flood*. Price considered Charles Darwin "slow, unimaginative ... and incapable of dealing with the broader aspects of any scientific or philosophic problem." (*Center for Adventist Research, Andrews University*)

and claimed that Darwin's theory is "a most gigantic hoax." Price rejected the fossil evidence for evolution, noting that "the doctrine of any particular fossils being essentially older than others is a pure invention, with nothing in nature to support it." Price paid most of the publication costs of *Illogical Geology* himself (it would take him many years to retire the debt) and sold the book for 25 cents (or ten for $1.75). Despite his financial problems, Price offered a $1,000 reward to anyone who could "prove that one kind of fossil is older than another." Price believed that other geologists "never had the courage to face this problem fairly and squarely" and that they therefore "cannot be trusted." Price considered Charles Darwin "slow, unimaginative ... and incapable of dealing with the broader aspects of any scientific or philosophic problem."

In ensuing years, Price published a variety of books, including *God's Two Books* (1911), *Back* to the Bible, or the New Protestantism (1916), and *Q.E.D., or New Light on the Doctrine of Creation* (1917). In each book, Price continued to promote "primitive Godliness" while claiming that life did not originate from natural causes, that geological evidence supports catastrophism and not uniformitarianism, and true "species" are those described as "kinds' in Genesis, and that fossils in lower strata are not older than those in upper strata. Price also claimed that because "we do not know anything in a scientific way as to how the world was made, or how life or the species of plants and animals came into existence, the conclusion is inevitable that creation was something different, essentially and radically different, from what is now going on." *Q.E.D.* attracted the attention of non-Adventists such as Frank Norris, John Straton, and William Riley, who promoted Price as "one of the real scientists of the day [whose] writings are destined to profoundly influence the thinking of the future." Price claimed that "a belief in the former destruction of the world by water is in accord with a belief in its coming destruction by fire," and that "Christ Jesus, our Lord and Savior, was associated with the Father in all the primary work of Creation." Price denounced evolution as "a fraud, originated and perpetuated by malevolent spiritual powers," and argued that if evolution were true, God is a "tyrant and a fiend."

In 1923, following a yearlong sabbatical, Price published his most influential book, the 726-page *The New Geology*. This textbook described Price's most comprehensive synthesis of "flood geology" and included his "great law of conformable stratigraphic sequences," which claimed that there was no natural order to fossil-bearing rocks. Price then declared his law "the most important law ever formulated with reference to the order in which the strata occur." *The New Geology* contained several arguments that continue to be invoked by creationists (e.g., dating techniques used by geologists to estimate the ages of sediments are unreliable, all fossils are the same age because they were all deposited during the biblical flood). Although Price fared poorly in a highly publicized 1925 debate in London with priest-turned-philosopher Joseph

McCabe (this was Price's only public debate about evolution), in 1926 *Science* declared Price "the principal scientific authority of the Fundamentalists."

In 1925, William Jennings Bryan—at the urging of Norris—asked Price to help him prosecute John Scopes at the famous "Monkey Trial" in Dayton, Tennessee. Although Price could not attend (he was teaching at Stanborough Missionary College, just outside of London), he urged Bryan to avoid scientific arguments and to instead claim that it is un-American to force parents to pay taxes for un-Christian teachings. At the trial, when Bryan cited Price as his chief authority in his testimony, Clarence Darrow snapped back, "You mentioned Price because he is the only human being in the world so far as you know that signs his name as a geologist that believes like you do." After the trial, Price turned against "poor Bryan, with his day-age theory of Genesis," noting that Bryan "really didn't know a thing about the scientific aspects of the case." In the 1940s, Price described the Scopes Trial as a crushing defeat for fundamentalism and "a turning point in the intellectual and religious history of mankind."

Price then published *The Phantom of Organic Evolution*, his most extensive criticism of biological evolution. In this book, Price noted that "the modern theory of evolution is about 95% due to the geology of Lyell and only about 5% to the biology of Darwin.... What is the use of talking about the origin of species if geology cannot prove that there has actually been a succession and general progress in the life upon the globe?" Price, who chafed at the failure of others to endorse his "flood geology," dismissed Darwin, Lyell, and Hutton as "mere children when attempting to handle the larger problems of science."

In 1928, Price returned permanently to the United States and began teaching at Emmanual Missionary College in Michigan. In *Genesis Vindicated*, Price continued to argue for a six-day creation while branding all other claims about creation as distortions of the truth. After discussing "the almost pathetic devotion of a large school of thinkers to the religion founded by Hutton and

Lyell," Price urged that the Sabbath be a divine memorial of creation and demanded that the church "insist that the geologist and the biologist hold steadily to the exact wording" of the Bible.

In the mid-1930s, Price helped found the Religion and Science Association, which was committed to a six-day creation and flood geology, and which repudiated gap theory and day-age theory as "the devil's counterfeit" and "theories of Satanic origin." When the association disbanded in 1938, Price helped organize The Society for the Study of Deluge Geology and Related Sciences, often called the Deluge Society. The society's voting members had to accept a literal, six-day, *ex nihilo* creation and agree that a worldwide flood was responsible for much of the geological record. By the early 1940s, the society included more than 600 members, most of whom were Seventh-Day Adventists. However, theological disagreements led to the collapse of the society in late 1945. By this time, Price equated evolution and similar "laws of nature" with idolatry and "part of the great modern apostasy, predicted a long time ago in the Bible."

Price believed that Satan was "the real instigator of all the mixing and crossing of the races of mankind, and also the mixer of thousands of kinds of plants and animals which God designed should remain separate." Price claimed that racial mixing violated God's intentions and was responsible for a rapid degeneration that produced not only apes, but also "Negroes" and "Mongolians," both of which Price labeled "degenerate or hybridized man." In Price's view, the Tower of Babel triggered reverse evolution—apes evolved from men. Price believed that satanic intervention, not natural selection, explained the origin of many species. In his later years, Price invoked Satan to attack several of his enemies. For example, when Harold Clark—one of Price's former students—substituted a non-Adventist text for Price's *The New Geology* at Pacific Union College, Price denounced Clark as being under the influence of Satan and filed heresy charges against Clark with the church.

Price died on January 24, 1963, in Loma Linda, California. He rests beneath a humble tombstone in Montecito Memorial Park and Mortuary Cemetery near there. In the decades following Price's death, virtually every fundamentalist attack on evolution was based on Price's work. In 1961, Price's claims were resurrected as "creation science" in the book *The Genesis Flood*. Henry Morris, one of the authors of *The Genesis Flood* and the most influential creationist of the twentieth century, first read *The New Geology* in 1943. For Morris, reading Price's book was "a life-changing experience."

For More Information: Price, G.M. 1924. *The Phantom of Organic Evolution*. New York: Fleming H. Revell Co.

PRO-FAMILY FORUM OF FT. WORTH

Pro-Family Forum of Ft. Worth, Texas, distributed thousands of pamphlets titled *Can America Survive the Fruits of Atheistic Evolution?* These pamphlets, which were distributed in the late 1980s and 1990s, claimed that evolution leads to divorce, venereal disease, and communism.

R

MAXWELL RAFFERTY (1917–1982)

Maxwell Lewis Rafferty was a popular Republican politician who, as California's Superintendent of Public Instruction in 1963, ordered that textbooks identify evolution as a theory. Rafferty, who had been convinced by Nell Segraves to exempt her son Kelly from evolution-related instruction, based his order on a vague California Department of Justice ruling that public schools could not prescribe irreligious teaching. Rafferty lost his campaign for the Senate in 1968, and two years later lost reelection for Superintendent of Public Instruction. Rafferty, who proclaimed that he "killed progressive education," drowned on June 13, 1982, when his car crashed into a pond.

BERTRAND RAMM (1916–1992)

Bertrand L. Ramm was an evangelical Baptist whose popular book *The Christian View of Science and Scripture* (1954) described George Price's impact on creationism as "staggering," but rejected the claim that faith in scripture required belief in a recent creation and worldwide flood. Instead, Ramm endorsed progressive creationism, in which God created different types of organisms at widely different times. Ramm, who often claimed that "God cannot contradict His speech in Nature by His speech in Scripture, believed that "creation was *revealed* in six days, not *performed* in six days." Fundamentalists such as Henry Morris condemned Ramm's claims, while Billy Graham and many Christian biologists embraced them.

GEORGE RAPPLEYEA (1894–1966)

Professor J.T. Scopes ... will be arrested and charged with teaching evolution ... Wire me collect if you wish to cooperate and arrest will follow.

George Washington Rappleyea was born on July 4, 1894. After serving in the Corps of Engineers during World War I, he was placed in charge of the six struggling coal and iron mines of the Cumberland Coal and Iron Company (with 400 employees) just outside of Dayton, Tennessee.

In 1925, the 31-year-old Rappleyea noticed an ad in the May Fourth issue of the Chattanooga newspaper placed by the American Civil Liberties Union (ACLU) looking for a teacher to test Tennessee's newly passed Butler Law banning the teaching of human evolution in the state's public schools. Rappleyea went to Robinson's Drug Store, where he showed the ad to F.E. Robinson (a local druggist and head of the county board of education), Walter White (county superintendent of schools), and other local leaders. The group eventually asked coach and substitute science-teacher John Scopes if he would help the community publicize Dayton by being arrested for violating the statute. Scopes agreed, and Rappleyea swore out a warrant against Scopes and wired the ACLU for help. The result was the Scopes Trial, the most famous event in the history of the evolution-creationism controversy. Rappleyea—Scopes' original prosecutor and most vocal and visible local supporter—was represented by attorney Sue Hicks, a friend of Scopes.

Rappleyea gave visitors tours of Dayton, spoke with reporters, and remodeled an abandoned eighteen-room house for the defense lawyers and expert witnesses. That house, located about a mile outside Dayton, was the largest in Rhea County and was known as "The Mansion." Rappleyea also attended Scopes' appeal in 1927.

After the trial, Rappleyea returned to his work at Cumberland Coal and Iron Company. On November 29, 1925, the *New York Times* reported that Rappleyea had been ordained a Bishop in the Liberal Church of Denver, Colorado, the same position that the church had offered to Scopes during his trial. Unlike Scopes, however, Rappleyea accepted the position and announced that his official title was "Doctor of Liberal Religion." Rappleyea did not establish a branch of the church in Dayton, but he liked having the power to perform weddings and other duties ordinarily performed by the clergy.

After declaring he was "as lonely as the ark of truth on Mt. Sinai" and the only "modernist" left in Dayton, Rappleyea and his wife left Dayton for a job in the boating industry in Mobile, Alabama. In January of 1937, Rappleyea went to New York and helped form the American Boat Builders and Repairers, and later that year staged a widely-publicized mock-battle over Long Island Sound between nineteen planes and ten powerboats. In the late 1930s, Rappleyea—then an officer of Higgins Boat Industries in New Orleans—helped charter the United States Power Squadron, and developed equipment to help build landing-strips on beach sand for Marines. Rappleyea then

became a vice president of the American Power Boating Association, and in the following three years wrote several books and pamphlets about boating. In September, 1944, Rappleyea patented an improvement in aerial mapping cameras, after which he became treasurer of Marsalis Construction Company in New Orleans. There, on March 2, 1947, Rappleyea was arrested for conspiring to violate the National Firearms Act. On March 31, 1948, Rappleyea pled guilty in Federal Court in Biloxi, Mississippi, to conspiracy to ship arms and ammunition to British Honduras, and on April 24, 1948, he began serving a 366-day sentence in the Federal Correction Institution at Texarkana, Texas.

After being released from prison, Rappleyea moved to Southport, North Carolina and resumed his work as a chemical engineer. The September 1951 issue of *Popular Mechanics* reported that Rappleyea had developed a way to build houses of molasses (for only $1,000 per house), and in 1955 Rappleyea patented Plasmofalt, an asphalt-molasses stucco-like material used to stabilize adobe.

Rappleyea spent his final years in Miami, Florida, where he directed the Tropical Agricultural Research Lab. Rappleyea died on August 29, 1966, and was buried near the entrance of Arlington National Cemetery.

JOHN RAULSTON (1868–1956)

John Tate Raulston was born on September 22, 1868, in Gizzards Cove, Tennessee, where he spent his entire life. After completing law school, Raulston became a school teacher. In 1918, Raulston—a Republican politician and part-time preacher—was elected circuit judge of seven counties, including Rhea County. Raulston, who viewed the Scopes Trial as a way to garner prestige and votes, urged Dayton's leaders to build a tabernacle that could seat 20,000 people for the trial. Raulston was a religious man who quoted scripture during trial; he felt called by God to preside over Scopes' trial.

On the first day of the trial, after a Christian prayer that Scopes described as "interminable," Raulston read the 31 verses of Genesis I, after which he seated another grand jury that re-indicted Scopes (mindful of the publicity that the trial could produce, Raulston had allowed Scopes to be indicted on May 25 by a grand jury whose term had expired so that another town did not "steal the show"). After the trial, Raulston praised Darrow's courage "to declare a truth or stand for an act that is in controventing to the public sentiment."

Raulston, a conservative part-time lay-minister, used the Scopes Trial as part of his campaign for reelection the following year, but he lost. He then practiced law in South Pittsburg, Tennessee, and gave speeches for several antievolution groups. When speaking at John Straton's church in New York, Raulston told the congregation that "on behalf of my home state, I want to tell you that the charge that we are yokels and ignoramuses is not justified.... But I will say that if more learning will cause us to lose our faith...then I pray to God to leave us in our state of ignorance." Later, Raulston described Bryan as the "most outstanding" man at the trial, and Darrow as "a very pathetic man" because Darrow had "no hope for a future life."

Raulston died on July 11, 1956, and was buried in Cumberland View Cemetery in Marion County, Tennessee.

JOHN RAY (1627–1705)

I know of no occupation which is more worthy or more delightful for a free man than to contemplate the beauteous works of Nature and to honor the infinite wisdom and goodness of God the Creator.

John Wray was born on May 12, 1644, in Black Notley, England. John, who would later change the spelling of his name to Ray so that it could be latinized, attended grammar school in nearby Braintree, and then went to Cambridge. After transferring to St. Catherine's College in 1644, he returned to Cambridge in 1646 and graduated with a BA in 1647 and an MA in 1651.

After graduation, Ray stayed at Cambridge to teach, and worked his way into a low level administrative post. In 1649, he was elected a fellow of Trinity College at Cambridge, which

required that he be ordained. Ray was ordained in 1660, but in 1662 Parliament passed the Act of Conformity, aimed at punishing those dissenting from standard Christian doctrine. At Cambridge, the act required all faculty clergy to either take an oath or forfeit their position. Ray, although a devout Christian, refused to sign a religious oath enforced by the government, and was therefore forced to leave Cambridge.

While still a student, Ray had befriended Francis Willughby, a fellow student from a prosperous family. After forfeiting his Cambridge position, Willughby and Ray, both avid students of natural history, spent years traveling the countryside of Britain and Europe, with Ray studying plants and Willughby studying animals. Both took copious notes, and when their money dwindled, they returned to England. By the mid-1660s, Ray was publishing his botanical observations, and was elected to the newly-formed Royal Society in 1667. Ray saw a great need for a catalog of the plants of Britain, an undertaking he happily pursued. These efforts produced three volumes (1686–1690) of *History of Plants* and *Synopsis of British Plants*.

Unlike previous classification schemes based on single characters, Ray used many morphological traits, including seeds, roots, fruits, and flowers. He also proposed the first definition of a species, and was the first to separate plants into monocots and dicots. The intuitive and empirical nature of Ray's classification system later caused Georges Cuvier to remark that Ray's efforts "are often more accurate and intelligible than those of Linnaeus." Ray also showed that scientific inquiry and piety are not in conflict, a perspective that would have a liberating effect on science in subsequent years.

When Willughby died in 1672, Willughby's estate left Ray an annuity of £60, a sizable sum at that time, but it also tasked him with the education of Willughby's children, which Ray continued until 1677. Ray felt compelled to complete the manuscripts that his friend had either started or had contemplated writing. This would be Ray's primary work for the next twenty years, and he eventually published (with Willughby as author) several books based on Willughby's observations of birds, fish, mammals, and reptiles. The financial independence his friend provided him did, however, allow Ray to marry in 1743 (he would have four daughters) and to build his widowed mother a house, which he inherited upon her death.

By the 1690s, Ray united his two guiding interests, religion and nature. Ray's sermons about the relationship between nature and God were published as *The Wisdom of God Manifested in the Works of Creation* (1691). To Ray, and those that followed, the amazing and intricate adaptations of organisms were a testament to God's power.

Ray, who believed that species do not change over time, was intrigued by fossils. He rejected claims by his contemporaries that fossils were created by some life force within the Earth, or that they were tricks of the Devil. Ray accepted fossils as vestiges of dead organisms, but did not believe they represented extinct species. Instead, he believed that if organisms like those in the fossil record were not yet known, it was simply because they had not yet been discovered.

Ray's last works were two studies of insects that were published posthumously. He died on January 17, 1705.

For More Information: Raven, C.E. 1942. *John Ray, Naturalist, His Life and Works*. London: Cambridge University Press.

RONALD REAGAN (1911–2004)

Ronald Reagan was the fortieth president of the United States who, after claiming that "I don't believe we should ever have expelled God from the classroom," promised to appoint federal judges willing to include religion in public schools. Although several earlier presidents (e.g., Teddy Roosevelt and Woodrow Wilson) accepted evolution, Reagan embraced biblical creationism and school prayer. In August 1982, Reagan announced that evolution is "only a theory [which] is not believed in the scientific community to be as infallible as it was once believed," and that "if evolution is going to be taught in the schools, then I think that also the biblical theory of creation...should also

be taught." During Reagan's presidency, the Department of Education recognized the Transnational Association of Christian Colleges and Schools, which fundamentalists had founded in 1979 to accredit schools that taught "the six literal days of the creation week."

FRANCESCO REDI (1626–1697)

Francesco Redi was an Italian physician who showed in 1668 that fly maggots' appearance on putrefied meat was not due to spontaneous generation, but instead from eggs too tiny for the naked eye to see. Redi was famous for his poem *Bacco in Toscana* (1685).

RED QUEEN

In Chapter 2 of *Through the Looking Glass* (Lewis Carroll's 1872 sequel to *Alice's Adventures in Wonderland*), Alice decides to leave the looking-glass house to see the garden. Alice departs on a path that appears to be straight, but she soon discovers that the path leads her back to the house. When she speeds up along the path, she returns to the house faster. Alice then meets the huffy Red Queen, and they start to run faster and faster. Alice is perplexed, because neither she nor the Red Queen seems to be moving, and when they stop running they are in the same place. The Red Queen then tells Alice, "Now, here, you see, it takes all the running you can do, to keep in the same place."

In 1973, evolutionary biologist Leigh Van Valen proposed the "Red Queen Hypothesis" as an important part of evolutionary theory. The Red Queen Hypothesis states that a constantly changing environment causes continuing evolution, and that species must keep evolving to maintain themselves in constantly changing environments. Accordingly, evolutionary lineages can never finish the process of adapting to their environment because that environment is changing. William Hamilton applied this idea to sexual reproduction, claiming that it persists because it enables many species to rapidly evolve new genetic defenses against parasites. Evolutionary biologist Graham Bell applied the scene of

Alice and the Red Queen to evolutionary thinking about parasites and sexual reproduction: Hosts of parasites that produce different genotypes can stay ahead of the parasites, but they can never "outrun" them and leave them behind.

RALEIGH REECE

Raleigh Reece, a Nashville reporter who covered the Scopes Trial, replaced John Scopes at Rhea County High School after Scopes' trial. When Reece missed the first week of classes in the fall of 1925, his substitute was Darius Darwin.

RELIGION AND SCIENCE ASSOCIATION

The Religion and Science Association (RSA) was founded in 1935 by flood geologist George Price and others hoping to reconcile their Christian faith with science. The short-lived RSA held only one convention—March 27–28, 1936, at Moody Bible Institute in Chicago. By 1938, infighting had destroyed the group. In 1939, several RSA members—including Price—formed The Society for the Study of Deluge Geology and Related Sciences.

EDWARD L. RICE

Mr. Bryan's proposition to delegate to state legislatures or church council the determination of the orthodoxy of scientific theory savors of the Middle Ages rather than of twentieth-century America.

In 1924, Edward L. Rice was a biologist at Ohio Wesleyan University and vice president of the American Association for the Advancement of Science (AAAS). When William Jennings Bryan campaigned to ban the teaching of human evolution, Rice used his vice-presidential address at the annual AAAS meeting in 1924 to confront Bryan's accusations. The address was published the following year in *Science*, and was widely read.

Rice identified himself as not only a scientist who accepted evolution, but also as a

Christian that accepted evolution. Citing the dangers (and fallacy) of Bryan's position that required either acceptance of science or belief in God, Rice argued that both science and religion would suffer as a consequence. Rice exposed Bryan's poor understanding of science and evolution: "To the biologists the evidence seems conclusive for evolution; to Mr. Bryan it has no significance. In large part, doubtless, this difference is due to Mr. Bryan's simple ignorance of the facts." Bryan's campaign against evolution convinced the AAAS to establish a committee to promote the teaching of evolution in public schools. Rice also served as an expert witness in the Scopes Trial. Rice likened the trial to that of Galileo, claiming that the only difference was the severity of the punishment to the individuals involved. Rice understood that public opinion was crucial, and he urged scientists to increase scientific literacy of the public.

THOMAS RICE (1790–1866)

Thomas Spring Rice was the Chancellor of the Exchequer who in 1837 authorized £1,000 for publication of the scientific results gathered during the voyage of the *Beagle*.

ANTHONY RICH (1804–1891)

Anthony Rich, in recognition of Charles Darwin's contribution to science, willed to Darwin nearly all his property, including some in London generating an income of more than £1,000. When Darwin died, Rich's children got the Rich estate except for the house and its contents, which went to Thomas Huxley (who sold it immediately). The final value of the estate was approximately £3,000.

OSCAR RIDDLE (1877–1968)

Oscar Riddle was a biologist at the Carnegie Institution of Washington who reported in 1936 to the AAAS that the antievolution movement was "an educational disgrace" that remained alive because of poor textbooks, poor teaching, and the failure of biologists to educate the public about evolution. Riddle lamented the "flagrant failure of our educational program to comprehend and teach life-science," and closed his essay by asking, "Shall the public that decides the fate of our democracy conceive nature and man as research discloses them, or as uninformed and essentially ignorant masses can variously imagine them?" Riddle helped found the National Association of Biology Teachers, and later noted that "biology is still pursued by the long shadows from the Middle Ages, shadows screening from our people what our science has learned of human origins…a science sabotaged because its central and binding principle displaces a hallowed myth." Riddle, who believed that religion threatens scientific progress, was named Humanist of the Year in 1958 by the American Humanist Association, and on January 9, 1939, was featured on the cover of *Time* magazine. Riddle helped isolate the "pregnancy hormone" prolactin in 1932, and won many awards during his lifetime, including several for his book *Unleashing the Evolutionary Thought* (1955).

WILLIAM BELL RILEY (1861–1947)

As soon as a man accepts the doctrine of evolution he becomes an atheist or an agnostic.

The legend of the Scopes Trial can be traced to a single event: the decision by three-time Democratic presidential candidate William Jennings Bryan to help Tennessee prosecute Scopes for allegedly teaching human evolution. By 1925, Bryan had become the *de facto* leader of the antievolution movement, and the addition of his name to the prosecution team added instant recognition and credibility that no other name could provide. Only after Bryan agreed to participate in the trial did famed attorney Clarence Darrow volunteer to defend Scopes, thereby creating worldwide interest in what Bryan hyped as a "battle to the death" in tiny Dayton, Tennessee.

Bryan's involvement in the Scopes Trial, and therefore the legend of the Scopes Trial, resulted from the work of fundamentalist leader and antievolution crusader William Bell Riley. Unlike other antievolution crusaders such as Billy Sunday and Frank Norris who often focused their work in the South, the austere Riley was based in Minneapolis. In the early 1920s, Riley convinced the Tennessee legislature to pass the Butler Law, which was used to prosecute John Scopes for the crime of teaching human evolution. Concerned that local prosecutors in Dayton could not be trusted to prosecute Scopes and defend the state's antievolution law, Riley sent Bryan a telegram on May 13, 1925, asking him to represent the World's Christian Fundamentals Association (WCFA) at the Scopes Trial "to secure equity and justice, and to conserve the righteous law of the Commonwealth of Tennessee." Although Bryan had twice rejected Riley's plea to lead the WCFA, Bryan agreed to represent the WCFA at the Scopes Trial.

William Bell Riley was born on March 22, 1861, (Figure 67). He was the sixth of eight children, and his tobacco-farming parents were devout Baptists. Riley received a BA from Hanover College in 1885, and an MA three years later, after which he graduated from Southern Baptist Theological Seminary in Louisville, Kentucky, in 1888. Riley pastored churches in Indiana and Chicago before becoming pastor of First Baptist Church in Minneapolis on March 1, 1897. The tall, charismatic Riley remained at First Baptist until his death in 1947, and while there, became the leader of the fundamentalist movement in the United States.

Unlike many other antievolution crusaders, Riley—"The Grand Old Man of Fundamentalism"—was unwaveringly calm and dignified; he preached "like a prosperous banker." Riley espoused a strident, articulate "orthodoxy plus" that attacked the evils of booze, theatre, gambling, dancing, movies, and revealing clothes. Like most other fundamentalists, Riley branded evolutionists as atheists, and rationalized his intolerance as simple honesty. For Riley, truth demanded militancy; this was especially true for evolution, which Riley believed was immoral, destructive, and atheistic. The ambitious Riley gathered a large following; wanting to personally spread his message across

67. While pastor of the First Baptist Church in Minneapolis (Figure 31), William Bell Riley led the American fundamentalist movement in the early 1900s. Riley founded the World's Christian Fundamentalist Association and was responsible for William Jennings Bryan's prosecution of John Scopes at the Scopes Trial. (*First Baptist Church, Minneapolis*)

the country, Riley insisted that First Baptist Church give him four months per year for traveling, evangelism, and writing. Everywhere he went, Riley proudly proclaimed that "I am, by unshaken convictions, a fundamentalist."

One of Riley's attempts to organize fundamentalists was the First World Conference on the Fundamentals of the Faith, which was attended by more than 6,000 people in 1919 in Philadelphia. This meeting established the WCFA, the first organization that united fundamentalists of all denominations. Riley described the Philadelphia meeting as "an event of more historic moment than the nailing up, at Wittenberg, of Martin Luther's Ninety-five Theses. The hour has struck for the rise of a new Protestantism." The Philadelphia conference, which closed with Riley admonishing the faithful that "God forbid that we should fail him in the hour when the battle is heavy," was followed by a massive

publicity campaign that included more than 200 meetings and debates throughout the United States and Canada.

Riley's first antievolution conferences were in 1921 in Kentucky, and in subsequent years the WCFA helped sponsor twenty bills throughout the country to ban the teaching of evolution. All WCFA members had to sign a nine-point creed that Riley helped write. The high-profile Riley was a day-age creationist.

By 1923, Riley had focused almost all of his efforts on driving evolution out of public schools. He knew that the public was interested in his crusade; as he told Bryan that year, "The whole country is seething on the evolution question." Riley was unequivocal: "Let it be understood, fundamentalists propose a war of extermination....We are in the fight to stay until this atheistic religion is forced out of every public school on the American continent." In 1923, Riley formed the Anti-Evolution League, whose goal was "to force the teaching of the evolutionary hypothesis from the public schools." Whereas scientists claimed that nature is organized according to unchanging laws, Riley's fundamentalism presumed that those laws are subject to disruption and miracles.

Like Sunday, Norris, and John Straton—all of whom preached in Riley's church—Riley did not attend the Scopes Trial; instead, he went to battle modernists at the annual convention of the Northern Baptists. When Bryan died in Dayton five days after the trial, Riley denounced Darrow and proclaimed Bryan "the greatest and most godly layman that America has produced." Six weeks after Bryan's death, Riley offered "to travel any reasonable distance to debate evolutionists," and personally challenged Darrow to a debate "anywhere, any time, and under any conditions." Darrow ignored Riley's challenge.

Although less abrasive fundamentalists such as J.C. Massee (1871–1965) urged moderation, Riley became increasingly militant. As the public began to lose interest in the evolution controversy, Riley made a final stand against evolution in his home state of Minnesota, where he addressed both houses of the state legislature. Riley, continuing Bryan's majoritarian message, told his followers that he was fighting "for the God-believing, God-fearing Minnesota majority," and claimed that the teaching of evolution had tripled the population of non-Christian students. Riley had a flair for publicity, and as a vote on the legislation approached, stories about Riley and his crusade covered the front pages of virtually all Minnesota newspapers. Riley's sermons often retold Lady Hope's story of Darwin's deathbed conversion to Christianity. Massee, who believed that Riley's rigidity blocked intellectual honesty and spiritual growth, distanced himself from fundamentalists such as Riley, Norris, and Sunday, noting that "there are fundamentalists and damn fundamentalists."

When the Minnesota Senate voted 55-7 to kill the antievolution bill, Riley claimed that voters would "in the course of time, reverse the decision," but he was wrong. Although Minnesota had come closer than any other northern state to banning the teaching of evolution, no additional bills to ban the teaching of evolution were considered by the state legislature. Membership in Riley's WCFA plunged, and the national antievolution campaign was dead. Despite its strong start, the WCFA had failed to provide an institutional alternative to modernist-influenced denominations. Undeterred, Riley continued to denounce evolution, but soon returned his attention to Northwestern Bible and Missionary Training School (today known as Northwestern College), which he had founded in 1902 to train rural pastors. By the time Riley retired as its president in 1947, Northwestern's graduates ("Riley's Boys") were pastors in almost three-fourths of Minnesota's 125 Baptist churches.

Although Riley began his career as a liberal who denounced class distinctions, corporate wealth, and low wages, in his later years he became a vehement conservative. In 1929, Riley resigned as president of the WCFA and formed the militant Baptist Bible Union with fellow antievolutionist Frank Norris. By the 1930s, Riley was obsessed with an alleged international Jewish conspiracy; as he wrote in *Protocols and Communism* (1934), "[the Jews] believe in and advocate the evolutionary hypothesis; and, they are bitter opponents of the Bible in the public schools." Riley excused Germany's persecution of Jews, linked fundamentalism with anti-Semitism, and claimed that the Jews undermined his earlier antievolution crusade. Only after the start of World War II did Riley stop defending Hitler's policies toward Jews.

In 1947 and near death, Riley split from the Northern Baptist Convention (claiming that it was "too liberal") and put the Northwestern Bible and Missionary Training Schools under the care of a 28-year-old preacher named Billy Graham, who served as the schools' president from 1948–1952. The 86-year-old Riley died at his home in Golden Valley, Minnesota, on the night of December 5, 1947. His funeral at First Baptist Church was officiated by Graham, who noted that "a prince and mighty man has fallen." Today, Riley rests in Lakewood Cemetery in Minneapolis, Minnesota.

For More Information: Trollinger, W.V. 1990. *God's Empire: William Bell Riley and Midwestern Fundamentalism*. Madison, WI: University of Wisconsin.

HARRY RIMMER (1890–1952)

We have not yet discovered one single fact that contradicts or refutes any statement in the Bible.

Harry Rimmer was a flamboyant Presbyterian preacher and self-described "research scientist" who, like many evangelists of the 1920s, often attacked evolution. Rimmer grew up in poverty in California, where he worked as a blacksmith and longshoreman. He quit school after the third grade, and at 19 joined the Army, where he became a champion boxer. After dropping out of a homeopathic medical college one term short of earning a medical degree, Rimmer became a Christian and began trying to use science to prove the literal truth of the Bible. He developed several popular "Bible and Science" lectures, and pledged $1,000 to anyone who could prove evolution to his satisfaction. In 1929, Rimmer's offer was challenged by atheist William Floyd (1871–1943), who argued that God's miraculous feeding of the children of Israel (Numbers 11:31) would have produced 29,613,991,260,171 dead quail (12,266,171 for each Israelite). When Rimmer refused to pay, Floyd sued. During the trial, Rimmer defended Ussher's chronology for creation and denounced fossil evidence for evolution. Rimmer never paid the $1,000.

In 1921, Rimmer founded the Research Science Bureau in Los Angeles "to prove through findings in biology, paleontology, and anthropology that science and the literal Bible are not contradictory." Rimmer, with George Price, was a leading "scientific" authority for fundamentalists. Many people considered Rimmer to be a biblical literalist; for example, he claimed that Jonah could have lived in a whale if he had stayed in the "breathing tanks" in the whale's skull. Rimmer believed that Genesis "is a scientific record of absolute facts," and that there had been two creations separated by a gap in time (the first creation occurring millions of years ago, and the second about 6,000 years ago). Rimmer, whose lecture titles included "The Seamy Side of Evolution" and "The Collapse of Evolution," believed that Earth is very old and that the Noachian flood was a regional event.

In the 1920s, Rimmer—the self-described "noisiest evangelist in America"—served as Field Secretary for the World's Christian Fundamentals Association and held some of the first public debates with evolutionists. Rimmer's popular books included *The Theory of Evolution and the Facts of Science* (1941) and *Modern Science and the Genesis Record* (1940), which influenced famed creationist Henry Morris. After the Scopes Trial, Rimmer defended Bryan's efforts at Dayton, but his support for Bryan waned in later years. Although Rimmer's influence diminished after World War II with the appearance of creationists having academic credentials, Rimmer helped popularize gap creationism, which was first promoted by Scottish naturalist Thomas Chalmers (1780–1847) as an attempt to harmonize science and religion.

In 1934, Rimmer became pastor of the First Presbyterian Church in Duluth, Minnesota, and he twice rejected offers to be president of William Jennings Bryan College. Throughout his life, Rimmer claimed that his ideas about evolution represented "the uniform position of the Christian geologists of the world."

Rimmer, who was awarded an honorary doctorate from Wheaton College, died in March, 1952.

ROADSIDE DINOSAURS

The earliest known appearance of roadside dinosaur replicas was at the Century of Progress Exposition in Chicago in 1933–1934. The popularity of that exhibit prompted the most famous public exhibit of dinosaur replicas, Sinclair Oil's "Dinoland" exhibit at the 1964–1965 World's Fair in New York. That exhibit, which consisted of nine motorized, life-sized fiberglass dinosaur replicas, was built over three years by sculptor Louis Jones and a team of paleontologists, engineers, and robotics experts. After the fair, the Smithsonian Institution declined Sinclair's models, and the replicas were dispersed throughout the country. The most famous of these are the *Apatosaurus* and *Tyrannosaurus rex* that welcome visitors to Dinosaur Valley State Park in Glen Rose, Texas (Figure 68). Throughout the United States, and especially in the Southwest, roadside dinosaurs continue to attract tourists to parks, stores, and Sinclair gas stations.

During the 1990s, creationists began converting roadside dinosaurs into worship areas that promoted creationism and Darwin-free fun. For example, the *Apatosaurus* ("Dinny") and *Tyrannosaurus* in Cabazon, California, promote Young-Earth creationism. The gift shop there, which is in the stomach of the *Apatosaurus*, sells toy dinosaurs with labels that warn, "Don't swallow it! The fossil record does not support evolution!" These dinosaurs are visited by 500,000 people per year, and have been featured in commercials, videos, and movies such as *Pee-Wee's Big Adventure*.

Other antievolution sites that use dinosaurs to promote their message include Dinosaur Adventure Land, Creation Evidence Museum (Glen Rose, Texas), the Museum of Earth History (Eureka Springs, Arkansas), and the Institute of

68. These dinosaurs, which today are displayed at Dinosaur Valley State Park in Glen Rose, Texas, were part of Sinclair Oil's "Dinoland" exhibit at the New York World's Fair in 1964–1965. This exhibit fueled the public's interest in dinosaurs. (*Randy Moore*)

Creation Research (Santee, California), which refers to their dinosaurs as "creation lizards" and "missionary lizards." As Ken Ham—CEO of the antievolution organization Answers in Genesis—has noted, "We're putting evolutionists on notice. We're taking the dinosaurs back. They're used [by scientists] to teach people that there's no God."

For More Information: Rea, T. 2001. *Excavation and Celebrity of Andrew Carnegie's Dinosaur*. Pittsburgh, PA: University of Pittsburgh.

PAT ROBERTSON (b. 1930)

The order of the universe is one of the strongest "proofs" for God's existence.

Marion Gordon "Pat" Robertson was born in Lexington, Virginia, on March 22, 1930. Robertson's father, a Democrat, represented Virginia in the U.S. House and Senate for thirty-four years, and is perhaps best known for being one of nineteen senators who in 1956 signed the "Southern Manifesto" condemning the *Brown vs. Board of Education* decision outlawing the "separate but equal" doctrine in public school education. Pat Robertson claims ancestry with Benjamin Harrison, a signer of the Declaration of Independence, and with William Henry Harrison and Benjamin Harrison, the ninth and twenty-third Presidents of the United States, respectively. His mother's side of the family traces their ancestry through Winston Churchill.

Robertson graduated in 1946 from McCallie School, then a military academy in Chattanooga, Tennessee. He next enrolled at Washington and Lee University, joined the Officer Candidate School of the Marine Corps in 1948, and graduated with a BA in 1950. After a stint in the Marines, Robertson married Adelia ("Dede") Elmer in 1954, with whom he eventually had four children. Robertson graduated from Yale Law School with a JD degree in 1955, but failed the bar exam.

Robertson recounts how over dinner with an evangelist he was converted from "swinger" to "saint" and felt God calling him to the ministry. He enrolled in the New York Theological Seminary, earned a Master of Divinity degree in 1959, and became an ordained Baptist minister in 1961. In 1959, the Robertson family moved from New York to Virginia, where Robertson purchased a bankrupt UHF television station in Portsmouth, Virginia. The following year, Robertson founded the Christian Broadcasting Network (CBN), the first Christian television station in the nation, which began broadcasting in 1961 as WYAH, the call letters referring to Yahweh, the Hebrew name for God.

Commercial advertising was not accepted on the station, which made meeting operating costs a challenge. In response, CBN held a telethon in 1963 to help cover the $7,000 monthly operating expenses. Viewers were urged to support the station by sending in $10 a month: if 700 individuals joined the "club" by pledging such support, the station would remain viable. This was the origin of the *700 Club*, a daily talk show hosted by Robertson, now seen daily in 200 countries and 96 percent of the television markets in the United States. A variety of programs became available through the network, including the talk show *Come On Over* hosted by Jim and Tammy Faye Bakker, who would leave in 1972 to start the *PTL Club* on Trinity Broadcasting. Robertson became one of the world's best-known televangelists (although he prefers "religious broadcaster").

In 1977, Robertson formed the CBN Cable Network, later renaming it the Family Channel. The network was purchased and renamed Fox Family in 1997, and in 2001, it was bought by Disney, which now offers it as ABC Family. Part of the original sales agreement was that the *700 Club* would be shown on the network in perpetuity, regardless of ownership. Robertson continues to host the program, which mixes interviews, news, and discussions framed within Robertson's charismatic Christian theology. Robertson's controversial comments have ranged from endorsing the assassination of Venezuelan President Hugo Chavez to referring to Muslims as "satanic." Robertson also makes predictions that he claims are provided to him by God. Past predictions have included the invasion of Israel by Russia, tsunamis along the coast of North America, and nuclear attacks by terrorists in the United States.

In 1987, Robertson ran for president of the United States. His platform included eliminating the Departments of Education and Energy, and banning pornography. He also stated that only Christians and Jews would be part of his administration. He finished second in the Iowa caucus, but his support faded and he ultimately encouraged his supporters to vote for George Bush.

After his failed presidential bid, Robertson formed the Christian Coalition of America in 1989 and served as its president until 2000, with Christian conservative Ralph Reed acting as Executive Director until 1997. The Christian Coalition sought to "preserve, protect and defend the Judeo-Christian values that made [the United States] the greatest country in history" and is best known for distributing millions of copies of its voter's guide. In 1994, the group was fined for improperly aiding two Republicans running for national office. The Christian Coalition has 2 million members and continues to be a potent political force.

In 1990, Robertson founded the American Center for Law and Justice (ACLJ), which "engages in litigation, provides legal services, renders advice, counsels clients, provides education, and supports attorneys who are involved in defending the religious and civil liberties of Americans." The ACLJ is modeled on and meant to counter the ACLU. Efforts of the ACLJ reflect Robertson's belief that there is no constitutional separation of church and state. The organization claims that "creationism does deserve to be taught along with evolution in the public schools" and has defended public school teachers who have attempted to do so (e.g., Rodney LeVake).

After a federal judge ruled that the Dover School Board's policy about intelligent design violated the First Amendment, Robertson warned Dover that "if there is a disaster in your area, don't turn to God, you just rejected Him from your city." Robertson later suggested that his statements were misinterpreted, but added that "If they have future problems in Dover, I recommend they call on Charles Darwin, maybe he can help them."

FRED ROBINSON (1881–1957)

Dayton has benefited, physically and mentally, by the evolution trial.

At the time of the Scopes Trial, Fred Earle "Doc" Robinson headed the school board and owned Robinson's Drug Store, at which he sold George Hunter's *A Civic Biology* (the textbook used by John Scopes when he allegedly taught evolution in Dayton). When the ACLU placed an advertisement in a Chattanooga newspaper searching for a teacher to test the newly-passed Butler Law banning the teaching of human evolution in Tennessee's public schools, George Rappleyea and other local businessmen met at Robinson's Drug Store to discuss how a test of the law could benefit Dayton's struggling economy. After an editorial in the St. Louis *Post-Dispatch* ridiculed the Scopes Trial and its location, Robinson and W.E. Morgan coauthored a booklet titled *Why Dayton—of All Places?* that used the upcoming Scopes Trial to promote Dayton and the surrounding area. During the trial, Robinson attracted customers with a banner bragging that his store was "where it started."

At Scopes' trial, Robinson testified that he, Scopes, and Rappleyea had discussed the Butler Law in his drugstore, adding that Scopes had said that he could not teach biology without including evolution. Robinson also admitted that he sold the offending textbook at his drugstore. During the trial, Robinson's wife fed many of the visiting journalists, "except Mr. Mencken," who "was not welcome." Robinson arranged for a chimpanzee named Joe Mendi to greet visitors at his store during the Scopes Trial. After the trial, Robinson offered to let Scopes continue to teach at Rhea County High School, provided Scopes would adhere "to the spirit of the evolution law." Scopes declined Robinson's offer, and instead went to Chicago to attend graduate school. Robinson continued to serve as Chairman of the Rhea County School Board and promoted himself as "The Hustling Druggist"

while operating his famous drug store and cultural center in which the initial discussions that produced the Scopes Trial were held. Robinson served as chairman of the Bryan College Board of Trustees during the school's first twenty-six years; in the earliest of those years, Robinson often paid bills and teachers' salaries with personal funds.

Robinson died in 1957 and was buried in Buttram Cemetery in Dayton, Tennessee.

RICHARD ROGERS (1895–1964)

Frederick Richard Rogers worked as a pharmacist at Robinson's Drug Store in Dayton, Tennessee, and rented his home (for $25 per week) to William Jennings Bryan, Bryan's wife, and their entourage during the Scopes Trial. Bryan died in Rogers' home, and Rogers accompanied Bryan's body to Chattanooga.

In 1926, Rogers and Ed Pierce opened the Rogers and Pierce Drug Store at the corner of Market and Main Streets in downtown Dayton. When Rogers later bought out Pierce, he renamed the store Rogers Pharmacy. In addition to serving one term as president of the Tennessee Pharmaceutical Association, Rogers was a founder and trustee of Bryan College until his death.

Rogers died on November 20, 1964, while attending a dinner at Bryan College. Rogers was buried in Dayton's Buttram Cemetery.

THEODORE ROOSEVELT (1858–1919)

Theodore Roosevelt was the twenty-sixth president of the United States who in 1918 recounted that he had "sat at the feet of Darwin and Huxley" while learning natural history. Roosevelt's father was a trustee of the American Museum of Natural History, and his tutor was John G. Bell, who had accompanied John James Audubon on collecting trips to the far West. Roosevelt, who founded America's system of National Parks, became the inspiration for the Teddy Bear. Roosevelt rests in Young's Memorial Cemetery in Oyster Bay, New York.

HERMAN ROSENWASSER

Herman Rosenwasser was a rabbi in San Francisco who was ready to testify at the Scopes Trial about the inaccuracies introduced when the Bible was translated from Hebrew to English.

R.L. ROTTENBERRY

R.L. Rottenberry was an Arkansas legislator who in 1927 introduced legislation that eventually became the basis for that state's antievolution law.

SAMUEL ROWBOTHAM (1816–1884)

Samuel Birley Rowbotham was an eccentric English inventor who founded the modern flat-Earth movement when he published a sixteen-page pamphlet based on his literal interpretation of some verses of the Bible. Rowbotham, using the pseudonym "Parallas," later developed that pamphlet into the 430-page book *Zetetic Astronomy: A Description of Several Experiments which Prove that the Surface of the Sea Is a Perfect Plane and that the Earth is Not a Globe!* Rowbotham's book, which was popular in England around 1860, claimed that Earth is a flat disk centered at the North Pole and bounded by a wall of ice along its southern edge, with the stars, moon, sun, and other planets just a few hundred miles above Earth. Rowbotham and his followers were the most outspoken Bible-scientists of their day; they sponsored several highly-publicized debates with leading scientists, including Alfred Wallace. Other leaders of the Flat Earth Society have included Samuel Shenton and Charles Johnson. Zion, Illinois, was founded by a sect of flat-Earthers led by Wilbur Voliva, who preached on the streets during the Scopes Trial and whose radio show later offered $5,000 to anyone who could prove Earth was spherical. Voliva condemned "the Devil's triplets—Evolution, Higher Criticism and Modern Astronomy" while denouncing "the infidel theory that the Earth is a globe." Other flat-Earthers such as John Hampden denounced the "Satanic device of a round and revolving globe, which sets Scripture, reason, and facts at defiance." The motto of the Flat Earth Society is "Deprogramming the Masses since 1547." Rowbotham rests at Beckenham Cemetery in London, England.

ERNEST RUTHERFORD (1871–1937)

Ernest Rutherford discovered in 1902 the concept of half-life ("half transformation") of radioactive decay, thereby enabling researchers to use radioactivity to determine the ages of fossils and rocks. Rutherford also famously claimed that "in science there is only physics; all the rest is stamp collecting." Rutherford's work was later used by geologists Bertram Boltwood, Arthur Holmes, Claire Patterson, and others to determine the age of the Earth. Rutherford's image appeared on the $100 bill of New Zealand, his homeland. Rutherford, who won the Nobel Prize in Chemistry in 1908, was buried in Westminster Abbey near Isaac Newton and Lord Kelvin.

S

ST. MARY THE VIRGIN CHURCH (Est. c. 1290)

St. Mary the Virgin Church stands along High Street at the main intersection in Downe, Kent, just a short walk from Down House. Charles Darwin was a member of St. Mary Church, attended meetings at the church, and is often mentioned in the Vestry minutes. In 1845, Darwin attended a meeting at which attendees approved money for construction projects at the church, and in 1858 Darwin helped resolve a financial dispute involving the church and a local resident. Darwin's involvement with the church diminished significantly after he published *On the Origin of Species*. The Vestry minutes do not mention Darwin's death.

Surrounding the church are the graves of Emma Darwin, Charles Darwin's brother Erasmus, four of Charles and Emma Darwin's children, and Joseph Parslow, the Darwins' servant for more than thirty-six years. The church memorialized Darwin with a sundial and small sign on one side of its steeple.

RICK SANTORUM (b. 1958)

Intelligent design is a legitimate scientific theory that should be taught in science classes.

Richard John Santorum was born in Winchester, Virginia, on May 10, 1958. Santorum earned a BA in political science from Penn State University in 1980, an MBA from the University of Pittsburgh in 1981, and a JD from Dickinson School of Law in 1986. While attending law school, Santorum worked as administrative assistant to Republican State Senator Doyle Corman and as director of the Pennsylvania Senate's Transportation Committee. After law school, he joined the Pittsburgh law firm of Kirkpatrick and Lockhart and remained there until his election to Congress in 1990.

Santorum married his wife Karen, a nurse (she also has a law degree from the University of Pittsburgh), in 1990, and they have six children. In 1996, the Santorums' son Gabriel was born prematurely and died two hours after birth. The Santorums were determined to have Gabriel be a part of their family and kept the dead child in the hospital bed with them the night after he was born. The following day they took him home to be seen by their other children. Karen Santorum's book about the experience, *Letters to Gabriel* (1998), included "letters" she had written to her deceased child and a foreword by Mother Teresa.

In 1990, Santorum, a Republican, was elected to the U.S. House of Representatives from

Pennsylvania's eighteenth district, besting seven-term Democratic incumbent Doug Walgren. Santorum was reelected to the House in 1992. Santorum was a member of the "Gang of Seven," a group of seven freshman Republican House members elected in 1990 that aggressively pursued corruption scandals involving the House Bank and the Congressional Post Office. In 1995, Santorum was elected to the U.S Senate, and reelected in 2000, ultimately rising to the position of Chairman of the Senate Republican Conference, the third most powerful party leadership position in the Senate. In 2006, he was denied a third term in the Senate after losing to Democrat Bob Casey.

Santorum's relatively brief political career mirrors the political history of the United States during this period: he was elected in 1990 on a rising tide of conservatism and then lost during the 2006 election year that was seen by many as a referendum on the policies of an increasingly unpopular president. When first elected to the House, Santorum advanced his antiabortion, antigay, Christian-based agenda. He allied himself with the "compassionate conservatism" popular during George W. Bush's first presidential campaign. Santorum's religious beliefs (he and his wife are devout Catholics) have played a significant role in his political life; one former aide described him as "a Catholic missionary who happens to be in the Senate." Santorum rejects a strict separation of church and state.

In June 2001, Santorum proposed a Sense of the Senate amendment to the 2001 Elementary and Secondary Education Act Authorization Bill stating "that (1) good science education should prepare students to distinguish the data or testable theories of science from philosophical or religious claims that are made in the name of science; and (2) where biological evolution is taught, the curriculum should help students to understand why this subject generates so much continuing controversy, and should prepare the students to be informed participants in public discussions regarding the subject." This amendment was approved by the Senate in a 91-8 vote. Scientists quickly responded that the

controversy generated by evolution is political and social—not scientific—and that there are much more controversial ideas in science and yet they were not identified. Phillip Johnson, an outspoken proponent of intelligent design (ID) and principal architect of the "wedge strategy" for teaching creationism in public schools, claimed he "helped frame the language" of Santorum's amendment.

Santorum wanted ID in the science curricula because it treats all sides of the issue and encourages "intellectual freedom." He noted the lopsided margin of approval of the amendment and cited Senator Edward Kennedy as someone who "approves of having alternate theories taught in the classroom." Within a week, Kennedy told the *Washington Times* that "unlike evolution, 'intelligent design' is not a genuine scientific theory and, therefore, has no place in the curriculum of our nation's public school science classes." Senators Sam Brownback, Robert Byrd, and others, however, understood and endorsed the antievolution implications of the proposed amendment.

The "Santorum Amendment" was removed by the House–Senate Conference Committee, and the final legislation signed by President Bush—now named the No Child Left Behind Act—did not contain the Senate's amendment (although the language was included in the Joint Explanatory Statement of the Committee of Conference that provides the legislative history of the bill). There is, therefore, no federal requirement to teach ID, although because the amendment appears in congressional documents, ID advocates argued that there is such a federal mandate.

The federal trial involving the Dover School Board's disclaimer about evolution attracted Santorum's attention. Santorum initially supported the school board's decision ("I commend the Dover Area School District for taking a stand and refusing to ignore the controversy"), and even served on the advisory board of the Thomas More Law Center, the Christian-rights legal organization that defended the Dover School Board. Later, however, Santorum distanced himself from the issue stating that "as far as intelligent design is concerned, I really don't believe it has risen to the level of a scientific theory at this point that we would want to teach it alongside of evolution," and accepted Judge Jones' decision that the Dover School Board's actions violated the First Amendment to the U.S. Constitution. Santorum did, however, contribute a foreword to *Darwin's Nemesis* (2006), a volume edited by William Dembski honoring the contributions of Phillip Johnson to the ID movement.

During Santorum's final reelection campaign, his challenger, Bob Casey, cited Santorum's involvement with (and possible backtracking in) the Dover situation and the No Child Left Behind amendment. Prominent Pennsylvania newspapers also cited these activities in their decision to endorse Casey. In 2005, Santorum published *It Takes A Family: Conservatism and the Common Good*. The book, a rejoinder to Hillary Clinton's *It Takes A Village* (1996), details Santorum's view of an appropriate Christian-based conservative social agenda. The release of the book, fairly close to the elections in 2006, and the relatively extreme positions proposed (e.g., the appropriateness of having mothers work is questioned) may have contributed to his eventual loss in the 2006 elections.

In 2007, Rick Santorum joined the Ethics and Public Policy Center, a Washington think tank "dedicated to applying the Judeo-Christian moral tradition to critical issues of public policy." He directs a program called "America's Enemies" that studies "threats posed to America and the West from a growing array of anti-Western forces."

NATHAN SCHOENFELD

Nathan Schoenfeld was an Arkansas state representative who in 1965 introduced legislation to repeal the state's 1928 law banning the teaching of human evolution in its public schools. The *Arkansas Gazette* praised Schoenfeld for trying to bring the state into "the modern era." Although Schoenfeld's bill never came up for a vote, it renewed public interest in the evolution-creationism controversy.

SCIENCE LEAGUE OF AMERICA (Est. 1925)

The Science League of America was a nonsectarian, nonpolitical organization whose founders said was "forced into existence by attacks upon scientists by those who call themselves theologians." The League was later headed by John Maynard Smith and included Luther Burbank, David Jordan, and other prominent scientists on its Advisory Board. The League, which was based in California, opposed bans on the teaching of human evolution, which it claimed "would go far toward plunging us backward into the abyss of the Dark Ages."

CYRUS SCOFIELD (1843–1921)

Scripture gives no data for determining how long ago the universe was created ... Each creative 'day' was a period of time marked off by a beginning and ending.

Cyrus Ingerson Scofield was born on August 19, 1843, in Lenawee County, Michigan. After serving in the seventh Tennessee Infantry during the American Civil War, Scofield—who was awarded the Confederate Cross of Honor—was admitted to the Kansas bar in 1869, and in 1871 was elected to the Kansas legislature as a Republican. The following year, Scofield was appointed U.S. attorney by President Ulysses Grant. However, Scofield accumulated large debts, and he spent six months in a St. Louis jail after being convicted of forgery associated with stealing $1,300 from his mother-in-law. While in jail, Scofield became a Christian.

In 1884, Scofield divorced his first wife (Leontine, the mother of his two daughters), and later married Hettie Hall von Wart, with whom he had a son. In 1883, Scofield—who worked with Dwight Moody—became pastor of the First Congregational Church in Dallas, Texas. Scofield also became a leader in dispensational premillennialism, a forerunner of the Christian fundamentalism that became popular in the twentieth century. Scofield later directed Moody's Northfield Bible Training School, and from 1895 to 1915 headed the Scofield Bible Correspondence School.

In 1909, Oxford University Press published the first edition of the *Scofield Reference Bible*. The *Scofield Reference Bible*, which Scofield wrote because "all of the many excellent and useful editions of the Word of God left much to be desired," was based on Scofield's Bible Correspondence Course. The *Scofield Reference Bible* was immensely popular because it linked related verses of the Bible, thereby allowing readers to follow themes throughout the text. Although Scofield's Bible told readers that Genesis 1:1 referred to "the dateless past," it listed the dates of many events of the Bible (e.g., "The events recorded in Genesis cover a period of 2,315 years"). Scofield later reassured readers that if they would "relegate fossils to the primitive creation, no conflict of science with the Genesis cosmology remains." It was in Scofield's Bible that many Christians first encountered James Ussher's calendar that fixed Creation at 4004 BC. Scofield's notations claimed that the word *day* in Scripture had several meanings, including "a period of time, long or short, during which certain revealed purposes of God are to be accomplished."

The *Scofield Reference Bible* popularized gap creationism, a form of Old-Earth creationism that harmonizes Genesis and an ancient Earth by inserting a gap between Genesis 1 and 2. This gap between the creation of "the heaven and the Earth" and the creation of modern forms of life left time for innumerable geological ages in the Genesis chronology, thereby allowing for an old Earth while simultaneously retaining the literal truth of Genesis. The *Scofield Reference Bible* claimed that "the revealed facts [of Genesis] are [that] Man was created not evolved." Scofield supported this claim by citing Thomas Huxley's claim that there is "an enormous gulf, a divergence practically infinite between the lowest man and the highest beast." Scofield's Bible influenced several generations of evangelists, as well as the prophesies of Hal Lindsey and others about the end of the world.

Scofield died on July 24, 1921, at his home on Long Island, New York, and is buried in Flushing Cemetery in New York City. The Scofield Memorial Church in Dallas is

69. John Scopes (left) was the defendant in the most famous lawsuit associated with the evolution-creationism controversy. Scopes is shown here with George Rappleyea, who instigated the Scopes Trial to improve the economic fortunes of Dayton, Tennessee. (*Smithsonian Institution Archives Image #2005-35069*)

named in Scofield's honor. The 1967 edition of *The Scofield Reference Bible* downplayed gap creationism.

For More Information: Trumball, C.G. 1920. *The Life Story of C. I. Scofield*. New York: Oxford University Press.

JOHN THOMAS SCOPES (1900–1970)

The basic freedoms of speech, religion, academic freedom to teach and to think for oneself defended at Dayton are not so distantly removed; each generation, each person must defend these freedoms or risk losing them forever.

John Thomas Scopes was born in Paducah, Kentucky, on August 3, 1900 (Figure 69). Scopes' socialist father Thomas, who Scopes described as "a very religious man," was especially influential on John. Scopes graduated from high school in 1919 in Salem, Illinois; William Jennings

Bryan, whose hometown was Salem, spoke at Scopes' high school commencement (Scopes remembered Bryan as "one of the most perfect speakers I have heard"). Scopes completed his first year of college at the University of Illinois, but health problems forced him to transfer to the University of Kentucky in 1920. After attending college there sporadically, Scopes received his AB in Arts–Law Degree on June 2, 1924. Just a few days before schools opened in the fall of 1924, the coach and algebra teacher in Dayton, Tennessee, resigned, and Scopes—the first qualified applicant—was offered, and eagerly accepted, the job.

In April 1925, Scopes was asked to substitute-teach for W.F. Ferguson, the high school's biology teacher. Scopes taught students about the various kingdoms of life, and on April 23 he told students to read a chapter in *A Civic Biology* (the course textbook) about evolution. However, Scopes was sick the next day, and never taught the subject.

Two weeks later, Scopes was playing tennis when he was summoned by local engineer George Rappleyea to Robinson's Drug Store to meet with a few civic leaders, including Walter White, Wallace Haggard, and Sue Hicks. Scopes admitted that he included evolution in his biology class, after which he was asked if he would be willing to be arrested to test the newly passed "Butler Law" banning the teaching of human evolution in Tennessee's public schools. When told that a trial could help the Dayton area, Scopes agreed to be charged with the crime, and went back to his tennis game. On May 7, Scopes was charged with violating the Butler Law. When Bryan volunteered to help prosecute Scopes, famed defense attorney Clarence Darrow volunteered to help defend Scopes, and the stage was set for the most famous event in the history of the evolution-creationism controversy.

Scopes had planned to leave Dayton for the summer and sell cars, but was available for his trial because he decided to stay in town to date "a beautiful blond" he met at a church function. Scopes participated in several pretrial publicity stunts (e.g., in New York and Philadelphia), and his trial became a sensational, worldwide event.

During his trial, which pitted science against religion, Scopes took the issues seriously, but he was largely irrelevant; on some days the proceedings began without him. Scopes, who described the trial as "just a drugstore discussion that got past control," did not testify at his trial, and was eventually convicted. He was fined $100, which the Baltimore *Sun* (which had sent reporter H.L. Mencken to cover the trial) agreed to pay. The "Bill of Cost" for the Scopes Trial totaled $343.87, of which $16.10 was charged to Rhea County. Scopes was billed for the rest, but he never paid. In his only statement to the Court, Scopes said, "Your Honor, I feel that I have been convicted of violating an unjust statute. I will continue in the future, as I have in the past, to oppose this law in any way I can. Any other action would be in violation of my idea of academic freedom ... I believe the fine is unjust."

Scopes left Dayton a few days after his famous trial, and was at a train station in Knoxville when he learned of Bryan's death. Scopes returned to Dayton to view the bronze casket containing Bryan's body as it lay in state in the F.R. "Richard" Rogers home.

Scopes received thousands of letters, including proposals for marriage, an offer to become a minister of a new church and "Bishop of Tennessee" with "pontifical powers," lucrative offers to lecture and star in movies, and advice for salvation. Scopes burned the letters (most of them unopened) and refused to cash-in on his accidental fame, noting that he simply wanted "peace and emotional stability." Scopes was content to let his hour of fame pass; as he noted later in his life, "I had only one life in this world and I wanted to enjoy it. I knew I could not live happily in a spotlight ... [I wanted to be] just another man instead of the Monkey Trial defendant."

After the trial, F.E. Robinson offered Scopes a new contract to teach at Rhea County High School (at a salary of $150 per month), provided he would "adhere to the spirit of the evolution law." Scopes declined the offer, and never taught again. He was replaced at Rhea County High School by Raleigh Reece, a reporter from Nashville who asked prosecutor Sue Hicks for some "inside dope" while covering Scopes' trial for the *Nashville Tennessean*. Scopes accepted a scholarship (funded by scientists and newsmen who attended his trial) to study geology at the University of Chicago in September, 1925. While living in Chicago, Scopes became close friends with Clarence Darrow, noting later that except for Scopes' father, "Darrow had a greater influence on my life than any other man I have known."

Scopes hoped to earn a PhD, and in 1927 applied for a fellowship to fund his work (the scholarship money raised by the scientists and reporters from Dayton lasted only two years). However, the president of the University of Chicago responded to Scopes' application with a terse letter: "Your name has been removed from consideration for the fellowship. As far as I am concerned, you can take your atheistic marbles and play elsewhere." Realizing that he had to abandon his goal of teaching and turn to commercial work, Scopes was hired by Gulf Oil and sent to do fieldwork in northwest Venezuela near the city of Maracaibo. There, at a dance, Scopes met Mildred Walker, "a pretty brown-eyed brunette from South Carolina" (and fellow employee of Gulf Oil). Walker, whose aunt "thought that Scopes was something with horns," married Scopes in a Catholic service in February 1930, in Maracaibo. In 1933, Scopes, an agnostic, returned to Texas (and later to Louisiana) to work for Union Producing Company, a subsidiary of United Gas Corporation. Throughout his career, Scopes' modest office included no mementos of his famous trial. John and Mildred had two sons, William and John Jr., both of whom worked in the insurance industry. Their father spoke with them about his famous trial only when asked.

When the movie version of *Inherit the Wind* opened in 1960, Scopes returned to Dayton, where he bought a "Scopes soda" that was "priced now as then in honor of Scopes' return to Dayton." Although a local preacher denounced Scopes as "the devil," other festivities—including a parade, concert, and car show—were more cordial. Scopes listened as Mayor J.J. Rogers—in front of the second-largest crowd in Dayton's history—gave him a key to

the city and proclaimed the day Scopes Trial Day.

Scopes retired in 1963, and four years later published his memoirs, *Center of the Storm: Memoirs of John T. Scopes*. In 1968, Scopes promoted his book with interviews and television appearances (e.g., *The Today Show*, *The Tonight Show with Johnny Carson*), during which Scopes noted that "the Bible simply isn't a textbook of science, and that's all there is to it." When asked about *Inherit the Wind*, Scopes noted that the controversy would "go on, with other actors and other plays."

In 1969, Scopes met in Shreveport, Louisiana, with Susan Epperson, whose challenge to the Arkansas antievolution law (*Epperson v. Arkansas*) had resulted in the U.S. Supreme Court ruling unanimously that laws banning the teaching of human evolution in public schools are unconstitutional. Epperson believed that the Arkansas ban on the teaching of evolution was "a sure path to the perpetuation of ignorance, prejudice, and bigotry," and her principal brief to the Court closed with a dramatic reference to "the famous Scopes case" and to the "darkness in that jurisdiction" that followed Scopes' verdict. When Scopes heard of the Court's decision, he claimed that "this is what I've been working for all along...I'm very happy about the decision. I thought all along—ever since 1925—that the law was unconstitutional."

On April 1, 1970, Scopes returned to a Tennessee classroom for the first time in forty-five years when he accepted an invitation to talk to biology students at Vanderbilt University's Peabody College for Teachers. Scopes, who was greeted "like a returning hero," reiterated his belief that "it is the teacher's business to decide what to teach. It is not the business of the federal courts nor of the state." Forty-five years after his trial, Scopes received an average of one letter per day about his famous trial.

Scopes' talk at Vanderbilt was his last public appearance. In July 1970, Scopes—a heavy smoker—was diagnosed with inoperable cancer. He underwent radiation treatments, but died at his home in Shreveport, Louisiana, on October 21, 1970. Following a service at Shreveport's St.

John's Catholic Church, Scopes' remains were returned to Paducah, Kentucky, and buried beneath the inscription "A Man of Courage," a phrase used to describe Scopes by Clarence Darrow. Scopes rests in a family plot in Oak Grove Cemetery, not far from where he first learned biology in elementary school. He rests beside his parents (Thomas [1860–1945] and Mary Brown Scopes [1865–1957]), wife Mildred (1905–1990), and sister Lela (1896–1989), who lost her job as a math teacher in Paducah when she would not renounce her brother's beliefs. In 1990, Scopes' wife Mildred—also a heavy smoker—died of emphysema in the same room in which her husband had died twenty years earlier.

For More Information: Ginger, R. 1958. *Six Days or Forever? Tennessee v. John Thomas Scopes*. New York: Oxford University; Larson, E.J. 1997. *Summer for the Gods: The Scopes Trial and America's Continuing Debate over Science and Religion*. New York: Basic Books; Metzger, G.O. 1990. *The World's Most Famous Court Trial: Tennessee Evolution Case*. Dayton, TN: Bryan College.; Scopes, J. & J. Presley, 1967. *Center of the Storm*. New York: Holt, Rinehart & Winston.

70. Eugenie Scott is an anthropologist who directs the National Center for Science Education, an organization dedicated to keeping evolution in the science curriculum of public schools. (*Eugenie Scott*)

JOHN THOMAS SCOPES v. STATE OF TENNESSEE

John Thomas Scopes v. State of Tennessee (1927) ended the legal issues associated with the Scopes Trial when the Tennessee Supreme Court upheld the constitutionality of a Tennessee law forbidding the teaching of human evolution, but urged that Scopes' conviction be set aside. Laws banning the teaching of human evolution in Tennessee, Mississippi, and Arkansas remained unchallenged for more than forty years.

EUGENIE SCOTT (b. 1945)

Eugenie Scott was born on October 24, 1945, in Lacrosse, Wisconsin (Figure 70). When she was 10, Scott—a compulsive reader—picked up her older sister's college anthropology textbook. She was fascinated with what she learned, and decided to become an anthropologist.

After earning a BS (1967) and an MS (1968) at the University of Wisconsin-Milwaukee, Scott enrolled as a doctoral student in anthropology at the University of Missouri-Columbia. Scott completed her PhD in 1974, after which she joined the faculty in the Department of Anthropology at the University of Kentucky. While there, Scott took some of her students to attend a debate between Jim Gavan (a former president of the American Association of Physical Anthropology) and antievolutionist Duane Gish of the Institute for Creation Research. That debate, which Scott described as "an eye-opener," showed Scott that creationists were intent on undermining her profession and the scientific point of view.

In 1980, Scott experienced what she called her "true baptism" regarding the social and political importance of the evolution-creationism controversy when a group calling itself "Citizens for Balanced Teaching of Origins" asked the Lexington (Kentucky) Board of Education to include "creation science" in its curriculum. Scott's efforts helped convince the Board to reject the creationists' request (by a 3-2 vote), and taught Scott that creation science is not a problem that can be solved with scientific evidence alone.

In 1986, as the "creation science" movement raged in the United States, Scott became director of the National Center for Science Education, Inc. (NCSE), a small, not-for-profit organization that provides information, advice, and resources for schools, parents, educators, and concerned citizens dedicated to keeping evolution in the science curriculum of public schools. Under Scott's leadership, NCSE soon became the premier organization dedicated to supporting the teaching of evolution. Scott's work has been recognized with awards from organizations such as the National Association of Biology Teachers and the American Association for the Advancement of Science (AAAS). Scott has authored *Evolution v. Creationism: An Introduction* and is coeditor of *Not in Our Classrooms: Why Intelligent Design Is Wrong for Our Schools*. Today, Scott—who views intelligent design as a "soft-core antievolution strategy"—is a self-described "evolution evangelist."

For More Information: Scott, E.C. 2004. *Evolution vs. Creationism: An Introduction*. Berkeley, CA: University of California.

GARY SCOTT

Gary Lindle Scott was a science teacher at Tennessee's Jacksboro High School who was fired on April 14, 1967, for teaching evolution and for allegedly telling students that the Bible is "a bunch of fairy tales." Scott had been hired midway through the 1966–1967 school year to teach physics and general science, and his firing was the first official use of the Butler Law since the Scopes Trial. When Scott's picture appeared on the front page of the *New York Times*, the ACLU offered support if Scott would challenge the law. Scott agreed; as his wife noted, "we need the money." On May 15, 1967, Scott—with the support of the ACLU and the National Education Association—challenged the Butler Law in federal court, where he was represented by William Kunstler. When Scott's attorneys offered to dismiss the lawsuit if Tennessee repealed the Butler Law, the legislature—without debate—repealed the law. Scott then dismissed his lawsuit and was reinstated, with full back-pay, as a teacher.

WOODROW SEALS (1917–1990)

Woodrow Bradley Seals was the district judge who on August 3, 1972, dismissed *Wright v. Houston Independent School District* on the grounds that the free exercise of religion is not accompanied by a right to be shielded from scientific discoveries incompatible with one's beliefs. Seals' decision, which noted that "teachers of science in the public schools should not be expected to avoid the discussion of every scientific issue on which some religions claim expertise," made it clear that scientific findings can be taught, even if some people consider those findings to be offensive. A year after Seals' decision, a federal appeals court issued an unsigned opinion backing Seals' decision. Several subsequent attempts to restrict the teaching of evolution cited the *Wright* decision.

ADAM SEDGWICK (1785–1873)

I have read Darwin's book. It is clever, and calmly written; and therefore, the most mischievous, if its principles be false; and I believe them utterly false.

Adam Sedgwick was born March 22, 1785, in Dent, England. In 1804, Sedgwick enrolled at Cambridge University, and graduated with honors in 1808. In 1810, Sedgwick became a Fellow of Cambridge, a position requiring a substantial amount of teaching. Due to a combination of ill health and dissatisfaction with teaching, Sedgwick left Cambridge in 1813. He returned as a tutor in 1815, and then was elected to the Woodwardian professorship in geology in 1818. (This position required the person to be unmarried. Sedgwick, who retained the position for the rest of his life, was a lifelong bachelor.) Sedgwick admitted that he initially knew little about geology, but his dedication to learning the subject helped him become one of the leading geologists of his time.

Sedgwick soon realized that Cambridge's geology collection was in poor condition. By virtue of his own collections and through judicious purchases, he created an excellent teaching and research collection. Sedgwick also became an excellent field geologist. He enjoyed fieldwork, and most of his major contributions were reports of field observations, rather than large, synthetic works.

One of Sedgwick's early projects was a geological study of the Lakes Region of northern England, which led to his first publications. These studies employed a young Charles Darwin, who had been a student of Sedgwick's. On his later voyage aboard the *Beagle*, Darwin frequently sent specimens to Sedgwick, who then reported on Darwin's geological descriptions of South America and elsewhere to the Geological Society of London.

Sedgwick's most famous collaboration was with Scottish geologist Robert Murchison (1792–1871) on fossil-bearing strata in Devonshire. The two named the time period for the rocks they had been studying the Devonian, after the area in which the rocks were found. Later, Sedgwick and Murchison worked separately in Wales on other strata. Sedgwick named the time period for this strata Cambrian (for "Cambria," meaning Wales), while Murchison used the name Silurian (for the "Silures," a Celtic group of people). Each geologist claimed priority, and each also claimed part of the other's strata. Because they had used different features to describe the rock layers they were studying, there was no clear resolution (except for the two to part ways no longer friends—when Cambridge awarded Murchison an honorary degree in 1861, Sedgwick refused to attend). Eventually, a colleague determined that both time periods were represented, and that there in fact was a third layer—the Ordovician (named for yet another group of Celts)—between the older Cambrian and the newer Silurian. Murchison's 800-page *The Silurian System* was dedicated to Sedgwick.

Sedgwick was a devout Christian (having been ordained in 1816) and he held a prebendary position at Norwich Cathedral from 1834 until his death. Initially, he believed that evidence for the Noachian flood could be found in the geologic column; however, as he learned more about geology, he labeled the flood geology of William Buckland and others "a philosophic heresy." Over time, Sedgwick became a gap creationist. His interpretation of the fossil record most closely approached Georges Cuvier's catastrophism, and he rejected the uniformitarianism of Lyell and Hutton.

Although he accepted geologic change, Sedgwick staunchly opposed the transmutation of species, believing in their immutability and divine creation. Sedgwick condemned Robert Chambers' *Vestiges of the Natural History of Creation* when it was published in 1844. The proposals by Darwin, his former student, were received in the same way, although Sedgwick and Darwin remained on good terms. Sedgwick feared that acceptance of evolution would "sink the human race into a lower grade of degradation than any into which it has fallen since its written records tell us of history," and he told Darwin that "I have read your book with more pain than pleasure." Sedgwick believed that the Bible provides the moral lessons needed for a functioning society, and the materialism he thought inherent to Darwinism undercut such Bible-based morality.

While at Cambridge, Sedgwick wanted to abolish policies that blocked admission of non-Anglicans, and he also urged the university to disseminate scientific information to the public. His public lectures about geology (often demonstrating the compatibility between science and religion) were wildly popular, and he continued giving them well into his eighties. When Prince Albert was appointed chancellor of the university, he selected Sedgwick as his secretary.

Sedgwick was elected to the Royal Society in 1820, and won the Society's Copley medal in 1863. Along with John Stevens Henslow, he helped co-found the Cambridge Philosophical Society, and served as president of the British Association for the Advancement of Science. Adam Sedgwick died at Cambridge on January 27, 1873, and he was buried in Trinity College Chapel, near his friend and sparring-partner

William Whewell. The Sedgwick Museum of Earth Science at Cambridge testifies to his accomplishments, and he is memorialized with a stained glass window in Norwich Cathedral and a fountain along main street in Dent.

Murchison, who rejected most of Lyell's uniformitarianism and could not find "one scintilla of evidence" to support Darwin's theory, died in 1871 following eleven months of progressive paralysis. Murchison is buried in London's Brompton Cemetery.

For More Information: Speakman, C. 1982. *Adam Sedgwick, Geologist and Dalesman, 1785–1873: A Biography in Twelve Themes.* London: Broad Oak Press.

NELL SEGRAVES (b. 1922) AND KELLY SEGRAVES (b. 1942)

If it is unconstitutional ... to teach of God in the public schools, it is equally unconstitutional to teach the absence of God.

During the early 1960s, two Supreme Court decisions significantly altered the role religion could play in public schools. In 1962, *Engel v. Vitale* established that state-prepared prayer in public schools violated the First Amendment. The next year, *Abington School District v. Schempp* upheld *Engel* while adding that statutes (specifically in Pennsylvania and Maryland) requiring the reading of Bible verses in school were similarly unconstitutional. *Abington* further stressed that "there must be a secular legislative purpose and a primary effect that neither advances nor inhibits religion" if religion is to be included in a school's curriculum. This ruling therefore required religious neutrality in public schools, and formed the basis of creationists' immediate challenges to the teaching of evolution.

One of the plaintiffs in *Schempp* was Madalyn Murray (later Madalyn Murray O'Hair), an outspoken atheist who founded the organization American Atheists in the wake of the Court's decisions. The legal rulings, combined with Murray's overt response, prompted Nell Segraves of California (whose son Kelly was then attending public school) to demand that the "scientism" of evolution be balanced with creationist

explanations of origins. She claimed that to do otherwise would "violate the neutrality requirements" and put "the State in the position of having adopted a particular religion of secular humanism." To support their belief that science favors atheism and therefore promotes a particular religious ideology (e.g., "we wanted to identify what was atheistic ... so we chose evolution as an example of atheism"), Segraves and friend Jean Sumrall convinced the U.S. Attorney General that atheism is a religion.

With atheism defined as a religion by the federal government, Seagraves and Sumrall then asked the Curriculum Commission of the California State Board of Education to designate evolution as only a "theory" in textbooks used in the state. The Superintendent of Public Instruction in California, Max Rafferty, was sympathetic to their appeals, and ordered state-approved textbooks to conform to the demands of Segraves and Sumrall. Still needing Board approval for curriculum changes, Segraves and Sumrall petitioned in 1966 for the designation of evolution as only a theory. In 1969, their petition was approved, and creationism became part of the approved curriculum in the state of California—creationism had become "science." However, in 1972, in response to pressure from the scientific community, the Board of Education modified this original stance to one of "anti-dogmatism": science (including evolution) should not be taught as dogma, but rather as statements demonstrating the conditional nature of scientific understanding. The policy also indicated that "hows" (i.e., mechanisms) should be emphasized over "ultimate causes." By 1974, the earlier statements referring to creationism were dropped, leaving only the "anti-dogmatism" policy in place.

The Segraves and Sumrall had been aided by the Creation Research Society led by Henry Morris. After the Board of Education's 1969 decision, the Segraves and Morris established the Creation-Science Research Center (CSRC) at Tim LaHaye's Christian Heritage College in San Diego. The Segraves saw this as an opportunity to publish textbooks and other school materials "founded on creationism and biblical authority" that would replace those developed by the federal-funded Biological Sciences Curriculum Study that emphasized evolution. The Segraves also wanted the CSRC to focus on political and legal action, while Morris was primarily interested in research and education. These differences in goals caused Morris to leave the CSRC in 1972, at which time he formed the Institute for Creation Research.

Under sole leadership by the Segraves, the CSRC continued its political advocacy, including support for "equal time for Genesis" laws across the nation (e.g., Tennessee passed such a law in 1973, but it was overturned in 1975). In 1979, CSRC unsuccessfully tried to block weakening of the "anti-dogmatism" policy by the California Board of Education. Using his mother's earlier actions as a model, Kelly Segraves in 1981 sued the Board of Education, claiming his children's belief in biblical creation was undermined by the school's presentation of evolution. Segraves' suit stressed the supposed hostile attitude toward religion by the school and appealed to the First Amendment protection of religious freedom. In *Segraves v. State of California*, the Court ruled "all that Plaintiffs seek, in the Court's view, presently is contained" in the Board's "anti-dogmatism" policy. The judge, however, noted that the policy was not widely known and required that it be distributed to schools.

In the 1970s and 1980s, Kelly Segraves authored several procreationism books. In 1975, he and chemist Robert Kofahl (CSRC science coordinator who authored the *Handy Dandy Evolution Refuter*) published *The Creation Explanation: A Scientific Alternative to Evolution*, which "presents an impressive array of data from the natural world for anyone who seeks a valid alternative to the theory of evolution." Segraves has also published *Jesus Christ Creator* (1973), *The Great Dinosaur Mistake* (1977), and *The Great Flying Saucer Myth* (1975); the later book discussed how "the concepts of UFOlogy are really Satan's last attempt to falsely fulfill Biblical prophecy in an effort to deceive mankind in the end times." For many years, Nell Segraves advised Ronald Reagan on matters related to evolution.

SELMAN v. COBB COUNTY SCHOOL DISTRICT

Selman v. Cobb County School District (2002) ended when the Cobb County School District agreed to stop pasting stickers into its science books claiming that "this textbook contains material on evolution. Evolution is a theory, not a fact, regarding the origin of living things. This material should be approached with an open mind, studied carefully, and critically considered." In 2005, Judge Clarence Cooper ruled that such stickers convey "a message of endorsement of religion" and "aid the belief of Christian fundamentalists and creationists." The lawsuit is named for Jeffrey Selman, a parent of a Cobb County student.

In 1979, the Cobb County School Board began allowing students to ignore the biology requirement for graduation from high school if they had religious objections to evolution. Two years later, Braswell Dean—Chief Justice of the Georgia Circuit Court of Appeals—claimed that "This monkey mythology of Darwin is the cause of permissiveness, promiscuity, pills, prophylactics, perversions, pregnancies, abortion, pornography, pollution, poisoning, and the proliferation of crimes of all types."

TEMPLE OF SERAPIS

The Serapeum is a structure better known to geologists as the Temple of Serapis (an Egyptian deity worshiped by Romans), which stands along the coast just north of Pozzuoli, Italy. Marble used by Romans to build the temple had originally formed as limestone sediment at the bottom of the sea, where it metamorphosed into marble, and millions of years later was raised as land. Approximately 2,000 years ago, the marble was quarried, carved into pillars, and set into the building, which originally was a marketplace and spa for wealthy Romans. Since that time, sea levels changed several times and, in the process, raised and lowered the Temple. The volcanic features around the Temple of Serapis helped inspire Virgil's account of the entry into the underworld in the *Aeneid*.

At the ruins of the Temple of Serapis, Charles Lyell confirmed the rising and falling of Earth's crust. When Lyell visited the ruins in 1828, the temple's three remaining marble pillars—each some 40 feet high—were still standing (Figure 71; the fourth column lies in pieces on the Temple's marble floor). The lower 12-feet of each column was smooth, but the 10-feet or so above this height was bored by the marine bivalve *Lithodomus* (many of the bored holes still have shells in them). The original temple had been built above sea level, but the presence of the mollusks on the columns meant that the columns had been partially submerged and were standing upright in the ocean. The columns had then been raised to their present level by the volcanic eruption that produced Monte Nuovo just northwest of Pozzuoli. Because the lowest parts of the columns were not bored by bivalves, Lyell suspected that these parts of the columns had been covered by volcanic sediments. He was right; these sediments had been excavated in 1749. While at Pozzuoli, Lyell also noted that two other temples were submerged just offshore northwest of the Temple of Serapis.

John Playfair discussed the Temple of Serapis in his *Illustrations of the Huttonian Theory of Earth* in the chapter titled "Changes in the Apparent Level of the Sea." Lyell made the Temple of Serapis an icon of uniformitarianism when he used a drawing of the Temple's columns as the frontispiece of *Principles of Geology* to emphasize that the geological changes that have been shaping Earth for millennia are observable today. Lyell's logic about the history of the Temple of Serapis prompted Richard Fortney (author of *Earth: An Intimate History*) to describe the ruins as a "holy place for rationalists." Today, the Temple's pillars—which remain standing (Figure 72)—are pictured on the prestigious Lyell Medal, which is awarded annually by the Geological Society of London.

EDWARD SEYMOUR (1775–1855)

Edward Adolphus Seymour, Duke of Somerset, was President of the Linnean Society from

71. The Temple of Serapis near Pozzuoli, Italy, was made famous by Charles Lyell in his *Principles of Geology*. The dark bands on the marble pillars were formed by mollusks that drilled into them after the columns were submerged in the sea. (*Image courtesy History of Science Collections, University of Oklahoma Libraries; copyright the Board of Regents of the University of Oklahoma.*)

1834–1837. Seymour helped obtain money to publish the scientific findings gathered during the voyage of the *Beagle*.

GEORGE BERNARD SHAW (1856–1950)

George Bernard Shaw was a playwright and outspoken supporter of Lamarckian evolution who believed that humans could evolve into a species of "Supermen." Shaw's fictitious Professor Henry Higgins in *Pygmalion* (1912, later the basis for the musical *My Fair Lady*) was based

72. This photo shows the Temple today. The dark bands that Lyell saw are still visible. (*Randy Moore*)

loosely on Thomas Huxley. In 1921, Shaw completed *Back to Methuselah*, a five-play manifesto for Lamarckian evolution that began in the Garden of Eden and ended thousands of years in the future. The preface of *Back to Methuselah* claimed that natural selection "has no moral significance" and that "if it could be proved that the whole universe had been produced by such Selection, only fools and rascals could bear to live." Shaw, who believed that "nothing but a eugenic religion can save our civilization," often condemned the "monstrous nonsense of Fundamentalism," adding that "it is not often that a single state can make a whole continent ridiculous, or a single man set Europe asking whether America has ever really been civilized, but Tennessee and Mr. Bryan have brought off the double event." Shaw was the first person to have been awarded both a Nobel Prize (1925) and an Academy Award (1938).

HARRY SHELTON (1908–1994)

Harry J. "Bud" Shelton was a student who testified at the Scopes Trial that Scopes had reviewed in class the offending pages about evolution in Hunter's *A Civic Biology* textbook. When Clarence Darrow asked Shelton if he had stopped attending church because of Scopes' teaching, Shelton said no. Years later, Shelton claimed that "Evolution should be taught as a theory. Teaching it as a fact, however, is a different matter." Shelton rests atop a hill overlooking Rhea Memory Gardens in Dayton, Tennessee. Shelton was the last surviving witness in the Scopes Trial.

JOHN SHELTON

John A. Shelton was a Tennessee legislator who first proposed banning the teaching of evolution in Tennessee's public schools in 1925.

Shelton's bill was voted down (by one vote by the Senate Judiciary Committee), after which a similar bill proposed by John Butler became law.

MAYNARD SHIPLEY

Maynard Shipley was the founder in the 1920s of the Science League of America, an organization that became the leading scientific opponent to William Jennings Bryan and the antievolution movement. Shipley's *The War on Modern Science: A Short History of the Fundamentalist Attacks on Evolution and Modernism* (1927) claimed that "the armies of ignorance are being organized, literally by the millions, for a combined political assault upon modern science. . . . For the first time in our history, organized knowledge has come into open conflict with organized ignorance."

SHREWSBURY

Charles Darwin was born on February 12, 1809, in picturesque Shrewsbury, Shropshire, a small town in central England near the Welch border. The town and surrounding area include several places important in Darwin's life:

Darwin was born at "The Mount" (now "Darwin House"), a building that now houses government offices.

Darwin was baptized at St. Chad's Church, located just across from The Quarry Park. Growing up, Darwin often collected insects in the park. Not far from the park lived Fanny Owen (1807–1885), who was Darwin's girlfriend.

From spring 1817 to summer 1818, Darwin attended Rev. Case's Grammar School, on Claremont Hill Street.

From September, 1818 to June, 1825, Darwin attended Shrewsbury School, located across the street from Shrewsbury Castle. The school is now the town library. A large statue of Darwin, erected by the Shropshire Horticultural Society in 1897, overlooks the library's courtyard.

Darwin's mother Susannah took Charles to worship at Shrewsbury's Unitarian Church on High Street, just across from the town square. A marble plaque in the church notes that the young Darwin was a member and "a constant worshiper" at the church.

In front of New Shrewsbury School is a statue of a young Charles Darwin at the Galápagos. An iguana is at Darwin's feet.

Darwin's parents Robert and Susannah are buried at St. Chad's Church in Montford, about 7 miles west of Shrewsbury. Susannah is buried inside the church, and Robert is buried outside.

Today, Charles Darwin remains Shrewsbury's most famous son, and the city advertises itself as "The Birthplace of Charles Darwin."

SICCAR POINT

During June 1788, James Hutton—along with mathematician John Playfair and chemist James Hall—visited Siccar Point, a "junction washed bare by the sea" on the rocky Berwickshire coast near Scotland's border with England (Figure 73). Siccar Point is an *unconformity*, a term coined in 1805 by geologist Robert Jameson to describe a surface at which two separate sets of rocks formed at different times come into contact. Sediments at the base of Siccar Point are vertical and, because sediments can only form horizontally, Hutton knew that these sediments had been tilted vertically and raised above land by pressure. Erosion had then worn away the above-ground parts of the vertical sediments, after which they were again submerged and covered by new horizontally deposited sediment. The frontispiece of Charles Lyell's *Manual of Elementary Geology* (1855) featured a sketch of Siccar Point. Siccar Point, which is arguably the most important geological site in the world, provided defining proof for Hutton's revolutionary ideas about the Earth's history.

The vertical sediments at Siccar Point are Silurian greywacke, a gray sedimentary rock formed approximately 425 million years ago when colliding plates created immense pressure that converted the sediment to rock. By about 80 million years later—a period that is more than 10,000 times longer than all of Bishop Ussher's proposed history of Earth—the raised greywacke had eroded, and parts were again submerged in the ocean. Erosion of the nearby

73. Siccar Point, which is arguably the most famous geological site in the world, is an unconformity that was visited in 1788 by James Hutton and John Playfair. This geological formation helped Hutton look into the abyss of time. This view shows Siccar Point from the North, with the North Sea in the background. (*Clifford Ford*)

Caledonian Mountains produced reddish sandstone sediments (of the Devonian) that were deposited horizontally over the vertical greywacke sediments (of the Silurian). When pressure created by moving plates again buckled the sediments, Siccar Point was raised above land for Hutton and others to see. Except for the chiseling by souvenir-seeking geologists who have visited the site, Siccar Point looks today as it did when it was visited by Hutton and his friends in 1788. The geological importance of Siccar Point was summarized by John Playfair: "What clearer evidence could we have had of the different formation of these rocks, and of the long interval which separated their formation, had we actually seen them emerging from the deep. . . . The mind seemed to grow giddy by looking so far into the abyss of time."

GEORGE GAYLORD SIMPSON (1902–1984)

I don't think that evolution is supremely important because it is my specialty; it is my specialty because I think it is supremely important.

George Gaylord Simpson was born on June 16, 1902, in Chicago, Illinois. An eye ailment restricted Simpson's participation in standard childhood activities, but he excelled in school and graduated from high school just before his sixteenth birthday. He entered the University of Colorado in 1918, considering a career as a writer, but a course in historical geology sparked his interest in paleontology. Simpson transferred to Yale University for his senior year, and graduated in 1923.

Simpson stayed at Yale for graduate school and worked with paleontologist Richard Lull, who had earned his PhD under Simpson's future boss, Henry Fairfield Osborn. Among the vast collections of the Peabody Museum, Simpson found himself drawn to the mammalian fossils of the Mesozoic era, a period of evolutionary time when mammals were effectively held in check by the dominant dinosaurs. After finishing his PhD in 1926, Simpson spent a postdoctoral year at the British Museum before becoming an assistant curator at the American Museum of Natural History (AMNH) in 1927.

Simpson married Lydia Pedoja during his final undergraduate year at Yale, and they soon had two daughters, and would have four by the time they divorced in 1938. The marriage was ill-fated, partly because they had to keep it secret from their families and from Yale (who prohibited marriages between students), and also due to Lydia's continuing mental problems. Even while married, the Simpsons lived apart for long periods of time. While in graduate school (and still married), George encountered a friend from Colorado, Anne Roe, who was in graduate school in New York. The two became romantically involved, and several years later were married after Simpson's divorce was final. They remained married until George Simpson's death in 1984.

Simpson began publishing at a furious pace (his lifetime output of journal articles and books surpassed 750), and most of his early research was based on fieldwork and examinations of museum specimens. However, Theodosius Dobzhansky's 1937 book *Genetics and the Origin of Species* forced Simpson to consider how his work fit into the larger context of the still ongoing modern synthesis. In 1938, Simpson began work on a book that became *Tempo and Mode in Evolution* (1944). However, before the book was published, World War II erupted. Simpson, who felt this to be a just (although horrific) war, enlisted in 1942, served in army intelligence in Europe and North Africa, and earned two bronze stars. (General George Patton reportedly once ordered Simpson, who had sported a goatee since the mid-1930s, to shave his "pink whiskers." A

standoff ensued, but Simpson ended up keeping his beard.) During Simpson's absence, the final arrangements for the publication of *Tempo and Mode* were handled by colleagues, and it was published the year he was discharged from the army.

Tempo and Mode was a major contribution to the modern synthesis because of its emphasis on deep evolutionary time. The early architects of the synthesis—Wright, Fisher, and Haldane—described how evolutionary forces cause genetic change in populations. Dobzhansky then integrated experimental and field observations into this theoretical framework during the second phase of the synthesis in the 1930s. Simpson added the crucial paleontological perspective: "[experimental biology] may reveal what happens to a hundred rats in the course of ten years ... but not what happened to a billion rats in the course of ten million years." As the book's title implies, Simpson discussed the causes and rate of evolutionary change ("How fast, as a matter of fact, do animals evolve in nature?"), which were approachable only through paleontology.

After the war, Simpson returned to the AMNH as chairman of the newly formed department of geology and paleontology. He also accepted a faculty position at Columbia University, where he taught graduate courses. Now a giant in the field of evolution, Simpson started writing for general audiences and produced *The Meaning of Evolution* (1949). In the 1950s, he also began studying biogeography. Interpreting the same evidence that Alfred Wegener had used to propose continental drift (e.g., the same fossil organisms being found on opposite sides of the Atlantic Ocean), Simpson deemed the dispersal capabilities of organisms were sufficient to account for these distributions, and therefore rejected the proposition of continental movement. Because Simpson's arguments were so persuasive and his status so great, his dismissal of Wegener was pivotal in defining the resistance to continental drift by American geologists. Although research later demonstrated that continental drift was a reality, Simpson held to his conclusions for some time. But by the 1970s, he admitted that his rejection of continental drift was in error, although he still claimed vindication because his objections had been based on Wegener's original data, which had been shown to be incorrect. (This qualified response by Simpson provides insight into a statement made by Stephen Jay Gould in his memoir of Simpson: "George Simpson was not an easy man to like.")

In 1956, Simpson organized an expedition to the Amazon basin. There, a tree fell on Simpson and he had to be evacuated. Years of surgery and physical therapy ensued, and Simpson was lame for the rest of his life. This ended Simpson's field research. Moreover, Simpson's long recuperation prevented him from maintaining a regular presence at the AMNH, and he was unable to attend to the day-to-day functioning of his department. When the director of the museum suggested that Simpson resign his chairmanship, Simpson took offense, and left the AMNH after thirty-two years for a position as Alexander Agassiz Professor at Harvard University. Simpson continued to work and write, producing *Principles of Animal Taxonomy* (1961), *This View of Life: The World of an Evolutionist* (1964), and a textbook, *Life: An Introduction to Biology* (1957).

By 1967, health problems forced Simpson into semiretirement, and he and Anne (also a faculty member at Harvard) moved to Tucson, Arizona. Simpson formally retired from Harvard in 1970, although he maintained a faculty position at the University of Arizona until 1982. On October 6, 1984, Simpson died of pneumonia. Honors include election to the National Academy of Sciences, election as a foreign member to the Royal Society of London, the Darwin Medal from the Royal Society, and the President's Medal of Science.

For More Information: Laporte, L. 2000. *George Gaylord Simpson: Paleontologist and Evolutionist*. New York: Columbia University Press; Simpson, G.G. 1944. *Tempo and Mode in Evolution*. New York: Columbia University.

HARRY SINCLAIR (1876–1956)

In 1916, Harry Ford Sinclair formed the Sinclair Oil and Refining Company (renamed Atlantic Richfield) from the assets of eleven small petroleum companies. The company made Sinclair exceedingly wealthy, and Sinclair began pursuing hobbies such as racehorses. His horses won one Kentucky Derby and three Belmont Stakes, and two of his horses are in the National Museum of Racing and Hall of Fame.

During the 1920s and 1930s, Sinclair funded several fossil-hunting expeditions in the American West, announcing each year's discoveries with a massive advertising campaign. Sinclair paid Barnum Brown to search for oil while looking for dinosaurs, and one of Brown's discoveries (a previously unknown crocodile) was named *Phobosuchus sinclairi* in Sinclair's honor. Sinclair's life-size models of dinosaurs that were displayed at the World's Fair were donated to parks. Today, the most famous of these models is in Dinosaur Valley State Park, where Brown and his colleagues found tracks of the real thing.

Sinclair incorporated *Apatosaurus*—a giant herbivore named and described by Othniel Marsh—into its corporate logo in 1930, and soon thereafter Sinclair gas stations began giving customers stamps, stickers, and booklets featuring "Dino" the dinosaur (Figure 74). Sinclair's use of a giant dinosaur was meant to portray his company's giantism; he wanted to control all aspects of the production of gasoline. Today, the company logo—which continues to feature Sinclair's name and an outline of a green *Apatosaurus*— is a familiar sight at thousands of gas stations throughout the United States.

In the early 1920s, Sinclair was arrested for allegedly giving U.S. Secretary of the Interior Albert Fall $100,000 for the drilling rights to the U.S. Navy oil reserves in Wyoming that were overlaid by a geological deposit that resembled a teapot. The resulting "Teapot Dome" scandal rocked the administration of Warren Harding, and Fall became the first Presidential Cabinet member to go to prison for his actions in office. During his trial in 1927, Sinclair refused to cooperate with government investigators and was acquitted of bribery charges, but was convicted of contempt of court and of Congress when it was learned that he had paid detectives to investigate jurors. In 1929, Sinclair was fined $100,000 and served six months in prison. A significant

74. Harry Sinclair was an oil tycoon who used his immense wealth to pursue hobbies such as fossil collecting. Today, the logo of Sinclair's famous company includes an outline of an apatosaur. (*Randy Moore*)

outcome of the scandal was *McGrain v. Daugherty*, which explicitly established the right of Congress to compel testimony.

Harry Sinclair died on November 10, 1956, in Pasadena, California. Today, he rests in Calvary Cemetery in East Los Angeles.

HAROLD SLUSHER (1934–)

Harold S. Slusher was the coeditor (with John N. Moore) of *Biology: A Search for Order in Complexity*, an antievolution biology textbook. *A Search for Order*, which was developed by the Creation Research Society, was adopted by a variety of schools, but its required use in science classes of public schools was later declared unconstitutional in *Hendren v. Campbell*. Slusher, who worked for several years at the Institute for Creation Research, rejected most scientific claims about the cosmos and age of the Earth. For example, his *Age of the Earth* (1978) claimed that "the Earth is vastly younger than the 'old' Earth demanded by evolutionists," and he rejected the galactic red shift discovered by Edwin Hubble

by noting that light may get "tired" after traveling long distances. Slusher concluded that the laws of physics prove the universe was created by an "omniscient, omnipotent Creator" and that "the evolutionist lives in a dream world."

ADAM SMITH (1723–1790)

Adam Smith was born on June 5, 1723, in Kirkcaldy, Scotland. When he was 4 years old, Smith was kidnapped by Gypsies, but he was rescued by his uncle and returned to his mother. Smith attended the University of Glasgow, and in 1761 was appointed that university's chair of logic, and later its chair of moral philosophy.

Smith became famous for his second book, *An Inquiry into the Nature and Causes of the Wealth of Nations* (first published in the revolutionary year of 1776), in which he argued that the free market, although seemingly chaotic and unpredictable, is guided by an "invisible hand" that produces the right amount and diversity of products (e.g., if a product is in short supply, its price rises, thereby creating a profit margin and incentive for others to produce the product). Smith claimed that rational self-interest leads to economic prosperity, noting that "it is not from the benevolence of the butcher, the brewer, or the baker that we expect our dinner, but from their regard to their own interest."

Smith, who in Edinburgh lived two blocks from geologist James Hutton, tried to understand "the nature and causes of wealth," just as Thomas Malthus later tried to understand "the nature and causes of poverty." Charles Darwin fashioned Smith's *laissez-faire* economic models into his theory of evolution by natural selection. Just as Smith argued that individuals' struggles for self-interest—undeterred by regulation from on high—would benefit society, Darwin argued that in nature a similar struggle—again with no regulation from on high—would produce adaptation and balance. Darwin recognized nature's order as a consequence of individuals' struggles for personal advantage, not a sign of divine intent and benevolence.

Smith's book, which expounded "the obvious and simple system of natural liberty," became a foundation for free trade and capitalism, and economists such as Thomas Malthus later refined Smith's ideas into classical economics. *Wealth of Nations* remains one of history's most influential books. In Edinburgh, Smith was a close friend of philosopher David Hume, and his literary executors were chemist Joseph Black (1728–1799; the discoverer of CO_2) and geologist James Hutton.

Following a painful illness, Smith died on July 17, 1790, and was buried in Canongate churchyard in Edinburgh. Smith's portrait appears on the Bank of Scotland £50 note and the £20 note in England.

MRS. ARTHUR G. SMITH

In 1926, the Mississippi legislature made it a misdemeanor for a public school teacher to teach human evolution as fact. The law was not challenged until 1969 when, soon after *Epperson v. Arkansas* struck down that state's antievolution law, Mrs. Arthur G. Smith charged in state court that the statute infringed upon her daughter's freedom to learn contemporary science and violated the First Amendment of the U.S. Constitution. Mississippi countered that the law was not enforced and, though still on the books, posed neither an impediment to an adequate education nor a constitutional violation. The state requested dismissal of the case.

Before the court decided on the dismissal motion, the Mississippi House reconsidered the law. By 1970, a bill to repeal the 1926 law had moved through committee with favorable support, and was introduced to the full House for debate. There was considerable opposition to the repeal measure, with some legislators seizing the opportunity to "hold the line as a Christian state." Ultimately, the repeal proposal was defeated. Soon thereafter, the state's motion for dismissal of Smith's suit was granted, and the suit moved to the Mississippi Supreme Court on appeal.

In December, 1970, the Mississippi Supreme Court—citing *Epperson*—ruled that the U.S. Supreme Court has "for all practical purposes already held that our antievolution statutes are unconstitutional." The Mississippi statute was struck down, although the court indicated that "equal time" treatment of creationism and evolution was an appropriate solution.

CHARLES SMITH

Charles Lee Smith was an atheist and advocate of evolution who, while campaigning in 1927 against Arkansas' antievolution law, was jailed in Little Rock for twenty-six days because, as the mayor noted, "no atheist will be permitted to maintain headquarters in Little Rock, Arkansas, if I can prevent it." Throughout the 1920s and 1930s, Smith debated antievolutionists such as William Jennings Bryan and Aimee Semple McPherson. Smith's debates with William Bell Riley in Arkansas played an important role in the public referendum that produced the state's ban on teaching human evolution. In those debates, Riley—who promised a "fight...to a finish"—invoked George Price, Thomas Huxley, Alfred Wallace, Piltdown Man, and others to argue that evolution is "the religion of Atheism" and that "no fact is found in defense of evolution." Smith discussed evidence for evolution and critiques of intelligent design (e.g., "Caesarian operations discredit the design argument.... Observe how well God designed the diphtheria germ for killing babies."). Transcripts of the Riley–Smith debates were sold throughout Arkansas and elsewhere.

When fundamentalist preacher John Straton offered a series of "divine healing services" in New York City, Smith charged Straton with practicing without a license. Smith then sent Straton atheistic literature (that included nude pictures), and Straton took Smith to court. Smith was fined $100.

ELLA SMITH

Ella T. Smith's 1938 textbook *Exploring Biology* was one of the few bright spots for evolutionists in the 1930s. Smith's book was vigorously proevolution, noting that "Evolution is a fact." Virtually all other textbooks, however, ignored evolution, as they had since the Scopes Trial.

WILLIAM SMITH (1769–1839)

The same strata are always found in the same order of superposition and contain the same peculiar fossils.

William Smith was born on March 23, 1769, in Churchill, England. Although Smith received little formal education, he was fascinated by fossils and rocks, and spent much of his time collecting them as a child. He studied mapping and eventually became a surveyor and self-taught engineer. Smith spent six years supervising the digging of the Somerset canal in southwestern England.

When Smith worked at a High Littleton mine called the Mearns Pit, he noted that rock sections on one side of England matched those on the other side, that rock strata were arranged in the same relative positions, that each stratum could be identified by the fossils it contained, and that certain kinds of fossils always seemed to occur—even at distant locations—in a consistent sequence relative to other fossils (i.e., as if they were deposited sequentially). His

conclusion—namely that fossils occur in rocks in a definite and orderly sequence—became known as his "Principle of Faunal Succession." Smith's discovery was important because it showed that (1) rocks formed during specific geologic times could be recognized by their fossils, (2) each layer of rock represents a different stage of geological history, and (3) life had evolved. Smith's discovery enabled biologists to arrange fossils chronologically. As if to emphasize that his conclusion was based neither on evolution nor creationism, Smith noted that "my observations on this and other branches of the subject are entirely original, and unencumbered with theory, for I have none to support." Smith's work led to the development of relative dating techniques and the refinement of geological time. Smith worked at a large, old, three-story manor house at Rugborne Farm in High Littleton, which Smith later said was "the birthplace of geology."

In 1799, Smith published the first geological map—a circular map of the geology around Bath. On that map, Smith colored the various sediments of his maps in colors that were close to the actual colors of the rocks. A similar color-scheme continues to be used on most geological maps to this day.

In 1801, Smith sketched "The Map That Changed the World"—the first geological map of an entire country. That map, which measured $8.5' \times 6'$, was based on fossils as well as the mineral composition of rocks. In 1815, Smith published a more elaborate version of that map that he titled *A Delineation of the Strata of England and Wales with part of Scotland*. Smith's *Strata Identified by Organized Fossils, Containing Prints on Coloured paper of the Most Characteristic Specimens in Each Stratum*, which was published in four parts (of a projected seven) between 1816 and 1819, showed that fossils are not distributed randomly, as in a flood, but instead are arranged in a definite, predictable order.

When George Greenough (1778–1855), the president of the Geological Society, began selling copies of Smith's maps and keeping the profits for himself, Smith ran low on money. Smith was forced to sell his 2,657 fossils to The British Museum to pay his debts; today, Smith's fossils fill more than sixty cabinets in the Museum. In 1819, Smith was sentenced to London's King's Bench Prison II, a debtors' prison (fans of Charles Dickens will recognize this prison as the one that housed Wilkins Micawber of Dickens' *David Copperfield*). When Smith was released on August 31, 1819 (after more than two months in prison), he learned that his home and property had been seized.

Although Smith was never elected a member of the Geological Society of London, in 1831 the Society awarded Smith its highest honor, the first Wollaston Medal. The society's president, Adam Sedgwick, proclaimed Smith to be "the Father of English Geology." Greenough never received the Wollaston Medal.

In 1835, Smith received an honorary doctorate from Dublin's Trinity College, and in 1838 was selected as one of the people to choose the building-stone for the new Palace of Westminster and the House of Commons (they chose Permian limestone, which eroded). Smith's financial problems were remedied by a lifetime pension of £100 per year awarded by King William IV.

Smith founded the science of stratigraphy, and his linking of the clear and consistent appearance of fossils in layered rocks with past geological eras changed fossils from random relics to pages of life's history. Although Smith did not speculate about the age of the Earth, his work established markers in the "abyss of time" first glimpsed by Scottish geologist James Hutton.

Smith, who was called "Strata" by his friends, died on August 28, 1839, in Northampton, England, while en route to a meeting of the British Association for the Advancement of Science. He is buried in Northampton beneath a corroded block of limestone near the west tower of St. Peter's Church. Each year, the Geological Society of London sponsors a lecture in Smith's honor and awards the William Smith Medal (originated in 1977) "for excellence in contributions to applied and economic aspects of science." A sculpted bust of Smith adorns Oxford University's Museum of Natural History, but Smith's house at Rugborne Farm remains unmarked. A portrait of Smith hangs in the offices of the Geological Society; beneath that portrait is a lock of Smith's hair, snipped on the evening that he died.

Smith donated his geological map of Bath to the Geological Society of London in 1831. Only 43 of the original 400 copies of Smith's *A Delineation of The Strata of England and Wales with part of Scotland* remain in existence. Smith's maps are nearly indistinguishable from those issued by the British Geological Survey today.

For More Information: Winchester, S. 2001. *The Map That Changed the World: William Smith and the Birth of Modern Geology.* New York: HarperCollins.

SMITHSONIAN INSTITUTION
(Est. 1826)

The Smithsonian Institution, which began with an 1826 bequest from English scientist James Smithson (who never visited the United States), was sued in 1978 by Dale Crowley, Jr., for using tax money to fund exhibits (*The Emergence of Man* and *Dynamics of Evolution*) that Crowley claimed established the religion of secular humanism and restricted the religious freedom of visitors to its National Museum of Natural History. In December, 1978, District Judge Barrington D. Parker ruled in *Crowley v. Smithsonian Institution* that the Smithsonian's exhibit was "wholly secular" and that the Smithsonian "in no way treats evolution as part of a religion, secular humanism or otherwise." On October 30, 1980, Parker's ruling was upheld in the U.S. Court of Appeals, which noted the exhibit did not establish religion merely because evolution "may coincide or harmonize with a tenet of Secular Humanism or may be repugnant to creationism." The following year, California Republican Congressman William Dannemeyer saw the exhibit and was "offended when they teach evolution as fact. It's a theory." Dannemeyer's subsequent attempt to curtail museum funding until the exhibit was changed failed. Today, the National Museum of Natural History continues to produce evolution-based exhibits that inform its 8 million visitors per year.

SOCIETY FOR THE STUDY OF EVOLUTION (Est. 1946)

The Society for the Study of Evolution was established in March 1946. Its first slate of officers included George Simpson (President), Ernst Mayr (Secretary), Sewall Wright, and Theodosious Dobzhansky (Council Members). Today, the society publishes the journal *Evolution* and promotes the study of organic evolution.

SOLENHOFEN, GERMANY

Solenhofen, Germany, was the site of a quarry of fine-grained Jurassic limestone in which miners found a well-preserved imprint of a feather in 1861. The discovery was newsworthy because, until that time, no birds had been known from the Jurassic period. German paleontologist Hermann von Meyer began monitoring the quarry, and in 1861—two years after the publication of *On the Origin of Species*—a complete skeleton of the feathered creature was discovered. Von Meyer named the fossil *Archaeopteryx lithographica* (limestone from the quarry was often used to make lithographic prints). The fossil had a reptilian skeleton (its bones were heavy, not hollow as in modern flying birds) and a long lizard-like tail, but its feathers and large eyes resembled those of modern birds. The specimen, which lacked a head, was hailed as a "missing link" between birds and reptiles. Darwin's foe Richard Owen dismissed the discovery as an aberrant bird, but Thomas Huxley predicted that *Archaeopteryx*—like ancient reptiles, but unlike modern birds—had a mouth with teeth. Five years later, a second specimen of *Archaeopteryx* was found at Solenhofen, and paleontologist John Evans then noticed what others had overlooked—a set of teeth within the beak. Since then, several more specimens of *Archaeopteryx* have been discovered, some of which are classified by some authorities as a different species, *Archaeoptemx bavarica*. Solenhofen quarries have also produced *Compsognathus lognipas*, a small dinosaur. In 1868, Thomas Huxley claimed that *Compsognathus* and *Archaeopteryx* were two links in an evolutionary chain connecting reptiles and birds. In 1973, antievolutionist Duane Gish dismissed *Archaeopteryx* as "just a bird."

In 2007, there were eight specimens of *Archaeopteryx*—seven reasonably complete and one represented by a single feather.

MARTIN SOUTHERN

Martin Southern was an attorney who, on January 13, 1967, filed suit in Knox County (Tennessee) Chancery Court on behalf of his son Thomas, claiming that Tennessee's Butler Law "limited" his son's education. Southern's lawsuit stimulated efforts to overturn the law.

SOUTHERN BAPTISTS (Est. 1845)

In the 1830s, tensions developed between Baptists in the northern and southern United States over the slavery issue. Baptists in the South embraced slavery, whereas those in the North condemned it. When the Baptist Convention refused to appoint slaveholders as missionaries, Baptists in the South met in Augusta, Georgia, in May, 1845, and seceded from the group, forming the Southern Baptist Convention.

By the early 1900s, Southern Baptists—intent on building a creedal firewall against "modernists"—vehemently opposed the theory of evolution. At their 1922 convention in Jacksonville, Florida, Southern Baptists claimed that the Bible and evolution are irreconcilable, noting that "no man can rightly understand evolution's claim as set forth in the textbooks of today, and at the same time understand the Bible." The following year, Southern Baptists vilified Darwin as a destructive atheist and demanded that scientists concede the authority of the Bible, including the virgin birth, the physical resurrection, and the special creation of man. State Baptist Conventions in the South passed resolutions forbidding Baptist institutions from employing anyone who believed in Darwinism. In 1926, Southern Baptists voted unanimously that "this Convention accepts Genesis as teaching that man was the special creation of God, and rejects every theory, evolution or other, which teaches that man originated in, or came by way of a lower animal ancestry." That same year, Baptists of South Carolina drove Andrew Pickens (1890–1969)—who held degrees in biology and theology—from Furman University. In his letter of resignation, Pickens noted that "it is utterly foolish for a man to talk about teaching biology and not teach evolution."

In more recent years, the teaching of evolution has continued to be opposed by many Southern Baptists, as well as by prominent Baptist leaders such as Jerry Falwell, Tom DeLay, and Pat Robertson. In 1982, the Southern Baptist Convention declared that evolution is unproven, and that "creation science" can be presented without religious concepts.

Today, the Southern Baptist Convention includes more than 16 million members in more than 40,000 churches. The Convention, which is the largest Baptist group in the world and the largest Protestant denomination in the United States, continues to promote creationism and denounce evolution.

For More Information: Rosenberg, E.M. 1989. *The Southern Baptists.* Knoxville, TN: University of Tennessee.

HERBERT SPENCER (1820–1903)

This survival of the fittest, which I have here sought to express in mechanical terms, is that which Mr. Darwin has called "natural selection," or the preservation of favoured races in the struggle for life.

Herbert Spencer was born April 27, 1820, in Derby, England. Although his parents had nine children, Herbert was the only one to survive past infancy. When Spencer was 13, he was sent to live with an uncle who was the rector at a rural parish. Spencer, unhappy with this arrangement, walked the hundred miles back to Derby in three days, but he was immediately returned to his uncle. Spencer's ambivalence toward learning ruled out a university education, and in 1837, he began working in the burgeoning railway industry. By 1841, however, he had quit his position, and he spent the next decade working

sporadically at various jobs. His letters to a local newspaper about social issues were published as *The Proper Sphere of Government* in 1842. After a fitful start, Spencer became a subeditor of *The Economist* in 1848.

In 1851, Spencer's *Social Statics* argued that society is governed by immutable natural laws that produce biological as well as social change. Government's role was to insure equal treatment of all citizens so these natural laws could operate. In summary, Spencer believed that the "liberty of each, limited by the like liberty of all, is the rule in conformity with which society must be organized." His *laissez-faire* approach to society, stressing nonintervention by government, would form the foundation of his later writings. Spencer followed *Social Statics* with two articles demonstrating his acceptance of organic evolutionary change (e.g., "there is nothing absurd in the hypothesis that under certain other influences, a cell may, in the course of millions of years, give origin to the human race"). These writings influenced Darwin's developing ideas of biological evolution.

Spencer inherited a considerable sum of money when his uncle died in 1853. This allowed him to pursue an independent writing career, and he left his position at *The Economist*. His next book, *The Principles of Psychology* (1855), applied a Lamarckian perspective to the evolution of the human mind. A significant backlash to the book ensued, however, because of Spencer's implied atheism. (Spencer was probably not an atheist; he claimed existence of an impersonal "Ultimate Cause" which he considered unknowable. He did, however, reject standard religious dogma.) Soon thereafter, a mysterious but serious mental disorder incapacitated Spencer for nearly two years, and left him permanently unable to work for more than three hours a day. Spencer became increasingly reclusive throughout the remainder of his life, and his illness drained his inheritance. Even though Spencer was forced to live frugally, he continued to write, publishing the first volume (*First Principles*) of his influential *Synthetic Philosophy*. Darwin's *On the Origin of Species* confirmed Spencer's belief in evolution as an all-pervasive process. However, Spencer's perspective on evolution, even after reading Darwin, was primarily Lamarckian. He referred to evolution as "functionally-produced modifications" in his multivolume *The Principles of Biology* (1864–1867), the work in which he described evolution in "mechanical terms" as "survival of the fittest." By this point, Spencer had established evolution as a universal force in biology and psychology; the next step, and his ultimate goal, was to apply it to the social sphere, which he did in *The Principles of Sociology* (1873).

Spencer's financial situation improved when an American admirer had Spencer's books published in the United States. A large market for Spencer's work existed in the United States, and he earned a steady income for the rest of his life. By the 1870s, Spencer was a prolific and popular author, and he published several more books during the next twenty years. Spencer also entered the upper echelon of the scientific and literary communities. He had, in the 1850s, met Thomas Huxley and John Tyndall, who would remain longtime friends (although he grew increasingly estranged from Huxley as "Darwin's bulldog" began to separate ethics from evolution). As his popularity grew, Spencer interacted with members of the exclusive X-Club, a group of eminent scientists that were influential in the Royal Society and British universities. Victorian novelist Marian Evans, who published under the pseudonym, George Eliot, was romantically interested in Spencer, but Spencer did not reciprocate these feelings. Spencer, a lifelong bachelor, died on December 8, 1903, after a long period of increasing isolation, and was buried in an area reserved for religious dissenters in Highgate Cemetery (East), only a few feet away from the towering tombstone of Karl Marx.

Spencer believed that evolutionary change operates at all aspects of the universe, from biology upwards to the social relationships among organisms. He believed that allowing everyone the same degree of autonomy (and responsibility) would produce cooperation, altruism, and personal achievement. However, because such development results from a natural process, it should not be managed by an outside agency such as government. Instead, government should protect individual rights so that the natural laws of progression can operate.

The most onerous aspect of Spencer's proposal is the effect it has on the least able (or least fortunate) in a population. Thomas Malthus' treatise on human population growth influenced Spencer's concept of "survival of the fittest" (just as it had influenced Darwin's "natural selection"). Spencer argued that the purification of a population is enhanced with greater population density, leading to faster societal development. However, there was no escaping that "under the natural order of things society is constantly excreting its unhealthy, imbecile, slow, vacillating, faithless members," a situation many found unacceptable. By the end of the century, governments of Europe and the United States were implementing social programs to aide the unfortunate, and Spencer predictably denounced these efforts (e.g., *The Man Versus the State*, 1884). Faced with widespread suffering, people were unwilling to acquiesce to misery and poverty to "aid the life of a society." However, several notable individuals supported Spencer's philosophy. For example, Andrew Carnegie believed that his rise to power was part of the natural order, and believed that philanthropists should help those most likely to benefit society.

Spencer's proposals have often been viewed as a natural extension of Darwinism. However, Spencer developed his philosophy before the publication of *On the Origin of Species*. Furthermore, Spencer retained a strong Lamarckian perspective in his philosophy, and viewed Darwin's ideas merely as a confirmation of the universal importance of evolution, rather than as a replacement for his mechanism of social change. Most importantly, although Darwin focused on natural selection as an agent of biological evolution, Spencer was interested in a philosophy of social progress.

For More Information: Kennedy, J.G. 1978. *Herbert Spencer*. Boston, MA: Twayne Publishers.

WILLIAM SPOTTISWOODE (1825–1883)

William Spottiswoode was a mathematician and Fellow of the Royal Society who—at the urging of Francis Galton and others—asked the Darwin family if they would allow Charles to be buried at Westminster Abbey. Spottiswoode, who is also buried at Westminster Abbey, was a pallbearer at Darwin's funeral.

STATE OF TENNESSEE v. JOHN THOMAS SCOPES

State of Tennessee v. John Thomas Scopes (1925), the original "Trial of the Century," resulted in coach and substitute science-teacher John Scopes being convicted of the misdemeanor of teaching human evolution in a public school in Tennessee. Scopes' trial, which William Jennings Bryan described as "a duel to the death" between evolution and Christianity, remains the most famous event in the history of the evolution-creationism controversy. The "Scopes Monkey Trial" also provided a framework for the play and movie, *Inherit the Wind*.

For More Information: Larson, E.J. 1997. *Summer for the Gods: The Scopes Trial and America's Continuing Debate over Science and Religion.* New York: Basic Books.

G. LEDYARD STEBBINS (1906–2000)

There is now no need for seeking hidden causes of evolutionary diversification or evolutionary progress.

George Ledyard Stebbins, Jr., was born January 6, 1906, in Lawrence, New York, into a family he later described as "upper middle class white protestant." Stebbins enrolled at Harvard University in 1926 as a reluctant political science major. During the summer between his freshman and sophomore years, however, Stebbins studied the plants around his home in Maine. When classes resumed in the fall, Stebbins decided to become a botanist. He graduated *magna cum laude* in 1928, and immediately entered

graduate school at Harvard, where he earned a PhD in 1931.

Stebbins accepted a position at Colgate University in Hamilton, New York. Although he did not enjoy his time at Colgate, Stebbins did coauthor an introductory biology book and began studies of hybridization in the genus *Paeonia*. He also interacted with giants in the field of genetics (e.g., Thomas Morgan, Barbara McClintock) who were revolutionizing the discipline. These interactions heightened Stebbins' interest in genetics, and in 1935 he accepted a research position at the University of California at Berkeley. Stebbins remained there for the rest of his career, becoming full professor in the Department of Genetics at Berkeley in 1946. Stebbins was an active member of a group of scientists who called themselves the "biosystematists" and met regularly to discuss current topics in evolutionary systematics. He also began a long friendship with Theodosius Dobzhansky, then at the California Institute of Technology.

Dobzhansky, who moved to Columbia University in 1940, was instrumental in inviting Stebbins to give the prestigious Jesup Lectures in 1946. This series of lectures, named after Morris Ketchum Jesup (former director of the American Museum of Natural History), is "an honorary public lectureship to focus on timely topics in biology and to be given biannually by a learned scientist of high stature." Stebbins' lectures were published in book form as *Variation and Evolution in Plants* (1950). This book joined Dobzhansky's *Genetics and the Origin of Species* (1937), Ernst Mayr's *Systematics and the Origin of Species* (1942), and George Gaylord Simpson's *Tempo and Mode of Evolution* (1944) as the major treatises for the second phase of the modern synthesis, and built upon the pioneering work in the 1920s and 1930s of Fisher, Wright, and Haldane. Stebbins affirmed that the basic tenets of evolutionary biology applied to plants (up until this time, there was lingering acceptance of Lamarckism in botany) and discussed how speciation can operate differently in plants as compared to animals, due to the frequency of polyploidy.

In 1950, Stebbins moved to the Davis campus of the University of California system to found a new Department of Genetics. He chaired the department until 1963, and was instrumental in having Dobzhansky join the department in 1971. Before retiring, Stebbins helped the Biological Sciences and Curriculum Study develop textbooks that returned evolution to the nation's public schools during the 1960s. Stebbins advocated teaching evolution, and publicly opposed efforts to include creationism in public school science curricula. Stebbins alone and with colleagues (including Dobzhansky) wrote several textbooks about evolution in the 1970s.

Stebbins died in Davis, California, on January 19, 2000. His legacy is a synthetic approach to biology that combined basic understanding of plants with genetics and evolution. He was elected to the National Academy of Sciences in 1952, and was awarded the National Medal of Science in 1980.

NICOLAS STENO (1638–1686)

The present state of anything discloses the past state of the same thing.

Niels (Nicolas) Steno was born into a wealthy family in Copenhagen, Denmark, on January 20, 1638. He left Copenhagen in 1660 to study medicine at the University of Leiden in the Netherlands, and in 1665 went to Florence, where he changed his name to Nicolas Steno. In Florence, Steno discovered the parotid salivary duct and used geometry to show that a contracting muscle changes shape but not volume.

In 1667, Steno dissected a shark and noted the similarity of its teeth to so-called "tongue stones" found in rocks. Although Roman writer Pliny the Elder (23–79) had claimed that tongue stones came from the sky or moon, Steno believed the tongue stones looked like shark teeth because they *were* shark teeth. Steno also argued that seashells in rocks atop mountains—which had been noted by ancient Greeks and others— came from organisms that had been buried in sediment that was later transformed into rock.

Before Steno, scholars believed that the Earth had appeared as a fiat of divine intervention. As a result, they based their claims about

Earth's history on human artifacts and written documents. Natural historians studied objects to classify them, and others studied them to determine their function. But in 1669, Steno changed everything with his seventy-eight-page *De solido intra solidum naturaliter contento*. In this book, Steno studied objects to determine their history and, in the process, stated a foundation of historical science: "Given a substance endowed with a certain shape, and produced according to the laws of nature, to find in the substance itself clues disclosing the place and manner of its production." Unlike Redi's controlled experiments and Descartes' abstract deductions, Steno's work showed that Earth's history was written in the layers of Earth's crust.

In the Apennine Mountains near Florence, Steno noted that the lower layers of sediment had no fossils, but the upper layers were rich in fossils. Steno explained this finding by claiming that the upper layer had formed in the Flood after life had been created, and that the lower layers formed before life existed on Earth. This was the first use of geology to distinguish different epochs in Earth's history. Steno's work, which produced "The Stenonian Revolution" by opening geology to the dimension of time, included three principles of geology:

Steno's Principle of Original Horizontality stated that because water is the source of sediments, and because the upper surface of water is always parallel to the horizon, sediments always form in horizontal layers (any deviations are due to later disturbances of the rocks). Similarly, the largest and heaviest grains of the sediment settle first and are at bottom, thereby marking each layer of sediment "this side up."

Steno's Principle of Lateral Continuity elaborated his principle of original horizontality, and stated that water deposits sediments as laterally continuous sheets that end at the edges of basins.

Steno's Principle of Superposition stated that when one finds sediments deposited atop each other, the layer on the bottom was deposited first and is the oldest layer. That is, strata form a narrative sequence that records Earth's history. This principle established stratigraphy, documented that rocks are a testimony to the past, and became Steno's most famous contribution to geology.

De solido, which was published while Steno explored the Austrian Alps, received a lukewarm reception, but two years later was noticed by the Royal Society of London. *De solido* was intended to be an abstract of a longer, more detailed book, but that book never materialized.

De solido was Steno's last published geological work, and in 1673 he made his last public appearance as a scientist when he dedicated a revamped anatomical theatre in Denmark. During this dedication, Steno stated his famous aphorism:

> Beautiful is what we see,
> More beautiful is what we understand,
> Most beautiful is what we do not comprehend.

Steno believed that science was a step to spirituality, and he gave up science for religion as a sacrifice to God. In April, 1675, the Church waived the customary theological exam and ordained Steno a priest. Steno said his first mass a few days later at Easter. In 1677, Steno was made a Bishop.

Steno died the morning of November 25, 1686, largely because of the rigors of his self-imposed life of poverty, fasting, and self-denial. Steno's only worldly possessions were books, religious materials, a black garment, tunic, cloak, two shirts, some handkerchiefs, and a nightcap. He had sold all of his possessions and given the money to the poor. Steno's funeral was delayed two weeks while officials searched for proper burial clothing. Steno's friends called for his canonization, but the Church did not respond. Six months later, Steno's remains were sent from Germany to Italy aboard the ship *Saint Bernhard*. Because superstitious captains often refused to allow corpses on their boats, Steno's body was transported in a box disguised as a crate of books. In Florence, Steno was buried in a bishop's robe atop a small limestone sarcophagus in the basilica of San Lorenzo, near Donatello's pulpits, Michaelangelo's staircase, and Brunelleschi's sacristy. Less than a mile away, in the Church of Santa Rosa, rests Steno's hero, Galileo.

In 1938, on the tricentennial of Steno's birth, a group of Danish pilgrims appealed to Pope Pius XI to make Steno a saint. Fifteen years later, Steno's coffin was found in a crypt of the Basilica of San Lorenzo and relocated to a chapel off the nave, which was renamed "Capella Stenoniana." There, Steno's body was interred in a ninth-century sarcophagus donated by the Italian government.

In 1988, before 20,000 worshipers at St. Peter's Basilica, Pope John Paul II said a mass of beatification praising the life and works of Steno. Steno was beatified on October 23, the same date that Irish Bishop James Ussher had chosen as the first full day of the creation of the world.

For More Information: Cutler, A. 2003. *The Seashell on the Mountaintop: How Nicolaus Steno Solved an Ancient Mystery and Created a Science of the Earth*. New York: Dutton Adult; Lutz, F. 1995. *Nicholas Steno*. New York: Dover.

ARTHUR STEWART (1892–1972)

[The Butler Law] was formed and passed by the legislature, because they thought they saw a need for it. And who, forsooth, may interfere? ...Bar the door, and do not allow science to enter.

Arthur Thomas Stewart was born on January 11, 1892. After finishing college and law school, Stewart became Attorney General of Tennessee's eighteenth Judicial District (fellow prosecutor Ben McKenzie was the first Attorney General of the district). During the Scopes Trial, Stewart—who endorsed William Jennings Bryan's majoritarianism—led the prosecution. During the trial, Tennessee Governor Austin Peay wired Stewart that "you are handling the case like a veteran and I am proud of you." When Stewart was asked if the Butler Law favored Christianity over other religions, Stewart snapped back, "We are not living in a heathen country.... If science could not be taught without teaching evolution, it should not be taught."

After the trial, the no-nonsense Stewart—who had described Darrow as "the greatest menace present-day civilization has to deal with"— hosted a visit by Bryan to Winchester, Tennessee, and continued to serve as Attorney General of Tennessee's eighteenth Judicial District. In 1942, Stewart was elected to represent Tennessee in the U.S. Senate, after which he returned to Nashville to practice law. Stewart was also a delegate to the Democratic National Convention in 1940.

Stewart died on October 10, 1972, and was buried in Memorial Park Cemetery in Winchester, Tennessee. Today, Stewart's grandson Jeffrey services the Rhea County Courthouse as Chancellor.

LYMAN STEWART (1840–1923)

Lyman Stewart cofounded (with his brother Milton [1838–1923]) the Union Oil Company. After attending Chicago's Moody Church, Lyman worked with A.C. Dixon to produce *The Fundamentals* between 1910 and 1915. These twelve booklets, funded with $300,000 from the Milton Stewart Evangelistic Fund, listed a variety of topics inimical to Faith—inerrancy of Scripture, the virgin birth and divinity of Christ, substitutionary atonement, the bodily resurrection of Christ, and the authenticity of Christ's miracles. *The Fundamentals*, which included articles from sixty-four British and American theologians and educators, were sent to every pastor, missionary, theology professor, theology student, YMCA-YWCA worker, college professor, Sunday school superintendent, and religion editor in the "English-speaking world, so far as the address of these [could] be obtained." *The Fundamentals* did not single out evolution as a particular threat, and some articles—most notably the contributions by Scotish divine James Orr (1845–1913) and George Frederick Wright (1838–1921)—endorsed theistic evolution. Indeed, all of the four science-related articles in *The Fundamentals* endorsed an ancient Earth. However, the booklets listed many enemies, including atheism, modernism, Mormonism, and spiritualism. Nearly 3,000,000 copies of *The Fundamentals* were distributed throughout the United States and elsewhere. *The Fundamentals* were ignored by academic journals and denominational presses, but became a symbolic reference point in the fundamentalist movement in the United States.

Stewart also provided funds to help publish *The Scofield Reference Bible*.

JOHN ROACH STRATON (1875–1929)

Evolution is a colossal error [that is] utterly incompatible with the Christian religion ... The terms "Christianity" and "evolution" are mutually exclusive and self-contradictory.

John Roach Straton was born on April 6, 1875, in Evansville, Indiana. He attended Mercer University, where he taught speech for two years until 1900. Straton won several regional and national speaking contests, and he wore two of his first-place medals every day for the rest of his life. He was then ordained a minister, and preached at a suburban Baptist church just outside Louisville, Kentucky. Straton attended Southern Baptist Theological Seminary from 1901–1903, after which he taught at Baylor University until 1905. While at Baylor, Straton tired of the academic life, and decided to return to evangelism as pastor of churches in Chicago (1905–1908), Baltimore (1908–1913), and Norfolk (1914–1917). However, it was at his final church—New York City's Calvary Baptist, located across the street from Carnegie Hall on West 57th Street—that Straton achieved fame.

At Calvary Baptist, Straton took his militant fundamentalism "to the people" in highly publicized crusades against a variety of societal ills, including dancing, evolution, theater, booze, and gambling (Figure 75). Soon after arriving at Calvary in May, 1918, Straton aggressively confronted sinners in bars, theaters, and brothels, and helped obtain the state's first conviction under the prohibition laws that had been passed in 1920. In 1922, Straton advocated a legislative ban on the teaching of evolution, and on March 4, 1923, Straton began broadcasting his sermons—many of them denouncing evolution—on WQAO, the first church-owned and operated radio station in the United States. Although Straton was a charter member of the World's Christian Fundamentals Association, he opposed the Ku Klux Klan, and he often praised

75. John Roach Stratton was a militant fundamentalist who often denounced evolution while shouting from his car in the streets of New York. (*Courtesy of the American Baptist Historical Society, American Baptist–Samuel Colgate Historical Library, Rochester, NY.*)

the work of Catholics. Like Billy Sunday, Straton advocated "old-time Americanism."

In the early 1920s, Straton began attacking evolution, and in 1923 he debated the validity of evolution with Harvard biologist Kirtley Mather, Unitarian minister Charles Potter, and Maynard Shipley in Carnegie Hall. The debates were attended by thousands, heard by tens of thousands on Calvary's radio station, and reported in newspapers across the country. In the debates,

Straton rejected evolution as "unsafe" while equating creationism with "monogamy, the sacredness of the marriage vow...and the sanctity of the home." Some of the debates were judged by New York Supreme Court Justices, who usually declared Straton the winner. Straton, who was often mistaken for Woodrow Wilson, claimed to be "a prophet of the Lord," and was soon labeled "The Fundamentalist Pope." Straton had no tolerance for uncertainty, proclaiming that "People will not get any doubts or question marks from my pulpit."

In early 1925, Straton joined fellow fundamentalists William Riley and Frank Norris as they toured the country denouncing evolution. Straton's speeches, which drew standing-room-only crowds, condemned evolution while declaring George Price's *The New Geology* as a "great and monumental" contribution from a "thoroughly up-to-date scientist." Straton's apocalyptic speeches often concluded with his claim that "the great battle of the age is now on between Christianity and evolution." As Straton noted, "The United States is shooting like a rocket . . . down the greased ways to hell."

In 1925, as the Scopes Trial approached, William Jennings Bryan asked Straton to be a witness in the trial. Straton immediately agreed, "I am ready to come [to Dayton] at the drop of a hat." Although Straton did not go to Dayton, he often cited the famous trial in his sermons, debates, and radio broadcasts that condemned evolution. When Bryan died in Dayton five days after the trial, Straton declared himself to be Bryan's God-appointed replacement. Straton challenged Clarence Darrow and Dudley Malone to debates, but was rebuffed.

In 1926, Straton accepted $30,000 to give sixty lectures about the evils of evolution for the Supreme Kingdom, an antievolution group committed to combating evolution and putting the Bible in every public school. However, the Supreme Kingdom's director—Edward Young Clarke—was soon implicated in a financial scandal, and Straton was embarrassed by his involvement with the group. As Straton's prominence waned, he focused his efforts on constructing a new building for Calvary Baptist Church.

In September 1929 Straton had a nervous breakdown. On "Black Tuesday"—October 29, 1929, the day on which the stock market crashed—the athletic Straton was killed by a stroke while at a sanitarium in Clifton Springs, New York. He is buried in a family plot in a small, rural cemetery in Warwick, New York. His epitaph, like those of several of his fundamentalist friends, invokes Straton's warrior-like approach to evangelism: "A good soldier

for Jesus Christ" (II Timothy 2:3). Today, Straton's Calvary Baptist Church occupies the lower five floors in a sixteen-story building in the heart of midtown Manhattan; the upper floors of the building comprise the 320-room Hotel Salisbury, which is owned and managed by the church.

For More Information: De Plata, W.R. 1997. *Tell It From Calvary*. New York: Calvary Baptist Church.

WILLIAM A. "BILLY" SUNDAY (1862–1935)

I don't believe the old bastard theory of evolution . . . If you believe your great, great grand-daddy was a monkey, then you can take your daddy and go to hell with him.

William A. "Billy" Sunday was born on November 19, 1862, near Ames, Iowa. In 1883, after spending his formative years at the Iowa Soldier's Orphan Home, Sunday began playing professional baseball with the Chicago White Stockings (later named the Chicago Cubs; Figure 76). Sunday was never a good hitter (he struck out his first thirteen times at bat, and his lifetime batting average was .248), but he was a good fielder and an exceptional base runner. Sunday could circle the bases in 13 seconds, and in 1891, he set a record by stealing 90 bases in 116 games. In 1888, Sunday was traded to the Pittsburgh Alleghenies (later renamed the Pittsburgh Pirates). After quitting baseball to become "Secretary of the Religious Department" at the Chicago YMCA (for $83 per month), Sunday became a traveling revival preacher.

Sunday's sermons were acrobatic, theatrical, and often violent—he leaped from the piano, ran up and down the aisles, slid on the stage as if he were sliding into home, smashed chairs, and screamed from atop the pulpit. Although Sunday's critics considered him "the worst thing that ever happened to America," branded him a hypocrite, and labeled his services "circus salvation," Sunday's impassioned and unwavering world-saving message of civic cleanup, no-nonsense fire-and-brimstone, patriotism, and simplified "old-time religion" made him the most popular religious figure of his time.

76. Billy Sunday was a former professional baseball player who became the most famous preacher of his era. (*Library of Congress*)

Sunday promised the greatest show around, and he delivered. During a 1907 revival, he even hired a former Barnum & Bailey "giant" to be an usher.

During his prime, Sunday's crusades were conducted in enormous, specially constructed tabernacles (Figure 77). For example, "The Glory Barn" for Sunday's 1917 revival in New York City was 344' long, 247' wide, and seated almost 20,000 people; its construction required 400,000 feet of lumber and 250 barrels of nails. All of Sunday's expenses in New York were paid by his friend John D. Rockefeller, Jr., and Sunday collected a "love offering" of over $120,000 for his ten weeks of work. Sunday donated it all to the

YMCA and the Red Cross to help with the war effort.

During a typical crusade, Sunday preached two or three times per day, six days per week for three to eight weeks, and drew up to 40,000 people per day. For his 1917 crusade in New York, conservative estimates projected that Sunday spoke to nearly a quarter of the city's 5,000,000 residents. Sunday's production costs were high, as were his needs for volunteers—for example, his Boston revival in 1916 required 35,000 local people for ushers, choir members, security staff, and outreach workers. Sinclair Lewis' *Elmer Gantry*—in which a fraudulent preacher takes money from the masses in exchange for promises of heaven—was loosely based on Sunday, and a poll conducted by *American Magazine* in 1914 ranked Sunday the eighth-greatest man in America. Sunday was sought by celebrities such as William "Buffalo Bill" Cody, H.J. Heinz, Woodrow Wilson, William Taft, Warren Harding, Herbert Hoover, and Cecil B. DeMille, who described Sunday as "the only man who works harder than I do." Although in baseball Sunday had been a minor star, in evangelism he was king.

Sunday, who idolized Dwight Moody, popularized revivalism; famous preachers such as Robert Schuller, Rex Humbard, Jim Bakker, Oral Roberts, Billy Graham, and Jimmy Swaggart emerged from Sunday's revivalist tradition. Sunday was a trustee of Bob Jones College, which gave him an honorary degree in 1935. Soon thereafter, Billy Graham—who had heard Sunday preach in North Carolina—enrolled at the conservative school.

Sunday crusaded against a variety of social ills, and especially against the evils of theater, dancing, gambling, and liquor. In 1915, Sunday teamed with William Jennings Bryan to lead a national campaign for temperance; in Philadelphia, Bryan and Sunday told a crowd of more than 25,000 that they were forming a 10,000-man "Abstinance Army" and declared the first Sunday in November "World Temperance Day." Sunday's famous "Booze Sermon" was printed in books and newspapers across the country, and resulted in Sunday receiving hundreds of

77. Sunday's theatrical services, such as the one shown here in Decatur, Illinois, attracted thousands of followers and often focused on the evils of evolution and booze. (*Image courtesy of the William & Helen Sunday Archives, Grace College, Winona Lake, Indiana*)

death threats from antiprohibition activists. Sunday saw the destructive powers of alcohol firsthand in baseball as well in his own family. His son George died in 1933 after falling from a highrise apartment; son Billy Jr. died in 1938 when he crashed his car into a telephone pole after an all-night party; and Paul died in 1944 in a plane crash. Alcohol was involved in each accident. A line in Frank Sinatra's famous song "Chicago" (written in 1922 by Fred Fisher) declared that Chicago is "the town that Billy Sunday couldn't shut down."

Sunday was proudly anti-intellectual, often proclaiming that when research and scholarship say one thing and the Bible says another, "scholarship can go to hell." Sunday, who never graduated from high school, flaunted his lack of theological education; "I know no more about theology than a jackrabbit does about ping pong."

Sunday saved many of his most venomous attacks for evolution, linking it with prostitution, eugenics, and crime in the early 1900s. Sunday rejected evolution, claiming that it was for "godless bastards and godless losers." Like most other fundamentalists, Sunday believed that the teaching of evolution poisoned minds, destroyed faith, perverted education, and destroyed society. Always a proponent of a masculine, vengeful God ("I have no interest in a God who does not smite"), Sunday's calls for a moral purge often included prayers for the slaughter of atheist evolutionists. As he did with many of his enemies who Sunday claimed were not "pure 100 percent American," Sunday promised his followers that Charles Darwin was spending eternity in hell's flames.

In June 1925, Bryan asked Sunday to come to Dayton, Tennessee to testify in the Scopes Trial. Sunday was also invited to Dayton by Walter White, Dayton's superintendent of schools. Sunday declined, but urged Bryan to equate evolution with atheism, and remind people that a person can't simultaneously be an evolutionist and a Christian. Sunday closed his response to Bryan by noting that "all the believing world is back of you in your defense of God and the Bible."

Sunday also generated scandal. For example, he was accused of plagiarizing several writers (including agnostic Robert Ingersoll), accepting money from businesses to help subvert labor strikes, and raking in enormous amounts of money from his followers. Indeed, between 1907 and 1918, Sunday earned $1,139,315 from his crusades' "love offerings," and even more from the sales of Bibles, photographs, postcards, books, sermons, and other materials he sold in his tabernacles. Sunday and his wife Helen ("Ma," who was also his business manager) wore the finest clothes, sported fur coats, and traveled in private Pullman cars. Sunday may not have lived up to the demands he made of others; when he died, archivists found jazz records, cordial glasses, and brandy snifters in his home. Just as Lewis' Elmer Gantry plagiarized Ingersoll and drank booze while preaching the virtues of abstinence, Sunday may have also enjoyed an occasional drink.

During his remarkable career, Sunday conducted more than 300 revivals and preached to more than 100 million people (without the aid of radio or microphones); no person in history has spoken directly to so many people. However, in his later years, Sunday was increasingly viewed as a relic and his appeal began to fade. Like many other fundamentalist leaders who opposed evolution, Sunday did not distance himself from groups such as the Ku Klux Klan, and critics denounced his doctrine as materialistic, perverted, and plagiarized. America had changed, but Sunday had not. Sunday campaigned for the reelection of his friend Herbert Hoover in 1932, but Hoover lost. Sunday also became more extreme, claiming at one point that he would "stand [his enemies] up before a firing squad." Crowds dwindled, forcing Sunday to make increasingly desperate pleas for money: "Don't let me hear any coins fall into those buckets; I want to hear the rustle of paper."

Sunday preached his last sermon on October 27, 1935, at First Methodist Church in Mishawaka, Indiana. After suffering a heart attack, Sunday died in Chicago at the home of his brother-in-law on November 5, 1935, just two weeks shy of his seventy-third birthday. Sunday's death was marked by memorial services across the country; his wife Helen even got a telegram of condolences from President Franklin Roosevelt, whose policies Billy had denounced. More than 3,500 mourners attended Sunday's funeral at Chicago's Moody Church, and his pallbearers included former U.S. District Judge Kenesaw Mountain Landis (1866–1944), who was professional baseball's first commissioner. The most prominent floral arrangement at Sunday's funeral was from the Chicago Cubs.

Billy Sunday is buried beside his wife along the eastern edge of Forest Home Cemetery in Forest Park, Illinois. Next to Billy and "Ma" are their three sons; only their daughter Helen—the closest of the Sunday children to their parents—is buried elsewhere (in Oaklawn Cemetery in Sturgis, Michigan). Billy Sunday rests beneath an inscription similar to the one that adorns the tombstone of fellow antievolution crusader William Jennings Bryan, "I have fought a good fight; I have finished my course. I have kept the faith."

For More Information: Bruns, R.A. 1992. *Preacher: Billy Sunday amd Big-Time American Evangelism.* New York: W.W. Norton; Firstenberger, W.A. 2005. *In Rare Form: A Pictorial History of Baseball Evangelist Billy Sunday.* Iowa City, IA: University of Iowa; Martin, R.F. 2002. *Hero of the Heartland: Billy Sunday and the Transformation of American Society.* Bloomington, IN:: Indiana Unievrsity.

JIMMY SWAGGART (b. 1935)

Evolution degrades God's image to nothing more than a mere beast.

Jimmy Lee Swaggart was born on March 15, 1935, in Ferriday, Louisiana. When he was 8 years old, Swaggart began speaking in tongues. By the time he was a teenager, Swaggart was an evangelist, and in the 1980s he established within his Baton Rouge church a broadcasting network and publishing house. At the peak of his popularity in the late 1980s,

Swaggart employed 1,500 people, had a church that seated 7,500, enrolled 1,451 students at his Bible college, and generated more than $150 million per year.

Swaggart often condemned evolution, which he described as "a bankrupt speculative philosophy, not a scientific fact. Only a spiritually bankrupt society could ever believe it." Swaggart also denounced anyone who accepts evolution as morally impure and corrupt, noting that "only atheists could accept this Satanic theory." In books such as *The Pre-Adamic Creation and Evolution* (1984), Swaggart promoted Gap creationism, denounced scientists, and ridiculed evolution with several of the arguments used by William Jennings Bryan. Swaggart claimed that biologists "teach that man came from monkeys. They teach that the vast universe came from a few molecules. They actually teach that nothing working on nothing by nothing through nothing for nothing begat everything!" Swaggart urged his followers to combat secular humanism by seeking help from the Institute for Creation Research.

In 1988, Swaggart accused fellow televangelist Marvin Gorman of "immoral dalliances." Soon thereafter, Gorman's private detective photographed Swaggart with a prostitute at a seedy hotel. While tearfully apologizing to his followers, Swaggart compared himself to King David. The scandal, which generated worldwide interest, placed Swaggart on the cover of *Time* magazine on April 6, 1987. When the Assemblies of God insisted that Swaggart take a yearlong disciplinary hiatus, Swaggart refused, and made his church an independent business. Swaggart later blamed his behavior on demons, and claimed that evangelist Oral Roberts had used a telephone call to "cast out the demons." In 2004, Swaggart threatened to kill homosexuals, noting that "if one ever looks at me like that, I'm gonna kill him and tell God he died."

Jimmy Swaggart Ministries includes television programs, radio programs, and the World Evangelism Bible College & Seminary. Swaggart—an accomplished piano-player who condemns rock music—is the cousin of country musician Mickey Gilley and rock legend Jerry Lee Lewis. Swaggart has recorded several dozen gospel albums and has written several books, including an account of his 1988 scandal. Swaggart's television audience of 500 million viewers in 140 nations was second only to that of Pat Robertson.

T

WILLIAM TAYLOR

William Taylor was a Tennessee Congressman who was asked before the Scopes Trial by Rhea County School Superintendent Walter White to introduce an antievolution bill in the U.S. Congress. After discussing the idea with Georgia Congressman W.D. Upshaw, Taylor decided that such legislation would do "more harm to the church and Christianity than it would do good," adding that such restrictions were best left to states.

PIERRE TEILHARD DE CHARDIN (1881–1955)

Evolution is a light illuminating all facts, a curve that all lines must follow.

Marie-Joseph Pierre Teilhard was born in Sarcenat, France on May 1, 1881. He later adopted "de Chardin" from his maternal grandmother. Teilhard graduated from a Jesuit boarding school in 1888, after which he joined the Jesuit order and began his religious studies. In 1905, the seminary sent Teilhard to Cairo to teach science in the Jesuit college of St. Francis, and while there, he began paleontological research that led to his first scientific publication. From Egypt, Teilhard moved to Hastings, England, for several years of theological study. There,

Teilhard read Henri Bergson's book *Creative Evolution* (1907), which discussed the possibility of evolution being guided by an *élan vital*, a perspective that subsequently influenced Teilhard as he developed a philosophy based on evolution. Teilhard also worked with amateur geologist Charles Dawson at Piltdown, close to Hastings. Teilhard helped Dawson with excavations, and reportedly discovered a canine tooth at the site. In 1953, Piltdown Man was exposed as an elaborate hoax, and speculation was rampant about the identity of the perpetrators. Although evidence supports at least some role for Dawson (and even author Arthur Conan Doyle arose as a suspect), Teilhard was also implicated (particularly by Louis Leakey and Stephen Gould), although the actual parties involved and their motives remain unclear.

After being ordained in 1911, Teilhard went to the University of Paris to study geology. In the midst of his education, World War I began, and Teilhard worked in North Africa from 1915–1918. Declining the opportunity to be an officer by serving as a chaplain, Teilhard was a stretcher bearer during the war, and was decorated for bravery. Teilhard's thoughts on the war were recorded in his letters that were collected into *The Making of a Mind* (1965). After his military service, Teilhard finished his doctorate in paleontology. In 1919, Teilhard's "The Spiritual

Power of Matter" outlined his emerging belief that evolution occurs at a number of levels (geological, biological, spiritual) and that Earth remains in the midst of evolutionary change that is leading to a spiritual unity between man and God. Teilhard presented these ideas in public lectures, and further developed them in *The Divine Mileau* (1957) and in his best-known book, *The Phenomenon of Man* (1955), both of which appeared after Teilhard's death (the Jesuits had blocked their earlier publication).

Teilhard argued that evolution was consistent with the Bible and brought humans closer to God. To Teilhard, God is the central point toward which all creation is evolving; everything in the universe is in constant flux, evolving as decreed by God. To account for this evolution, Teilhard proposed that matter contains differing amounts of two kinds of energy, "tangential" and "radial." Tangential energy is detectable by science and operates through the natural and observable laws of the universe. Radial energy, in contrast, is the energy of "within," and therefore of a spiritual nature. Inanimate objects are dominated by tangential energy, and nonhuman life has some form of radial energy, but humans have a special form known as consciousness. Hence, science can explain the functioning of inanimate matter and even nonhuman life within a purely mechanistic framework; humans, however, can only be understood by invoking the spiritual aspects.

Teilhard believed that everything in the universe would eventually converge to form the body of Christ, a state that Teilhard called the Omega Point. The evolution of the geosphere and the bioshere have already occurred, and we are in the midst of the development of the last layer, the *noosphere*, where human thought will be united. Teilhard cited the rise of global communication that facilitates communal thought as evidence for the rise of the noosphere (from the Greek word *noos*, meaning *spirit*). Evolution was elevated to "a general condition to which all theories, all hypotheses, all systems must bow," but it was not sufficient to account for all aspects of progress toward the Omega Point, and divine intervention must have occurred at particular points.

Teilhard criticized the church for "verbal theologizing" and urged people to "save Christ from the clerics, in order to save the world." Not surprisingly, Teilhard's proposals—including his support for evolution and his questioning of original sin—were controversial, and he was censured by the Church. The scientific community rejected his redefinition of evolution as "the continual growth of radial energy," and dismissed his ideas as yet another attempt at finding a role for a creator within the ever-narrowing gaps left by advances in scientific understanding. (Julian Huxley, intrigued by Teilhard's proposals, wrote an introduction to *Phenomenon of Man*, although he disagreed with the metaphysical interpretation of evolution.) In *Mankind Evolving*, evolutionary biologist Theodisius Dobzhansky described Teilhard's ideas "as a ray of hope" and quoted from *The Phenomenon of Man*: "Man is not the centre of the universe as was naively believed in the past, but something much more beautiful—Man the ascending arrow of the great biological synthesis. Man is the last-born, the keenest, the most complex, the most subtle of the successive layers of life. This is nothing less than a fundamental vision. And I shall leave it at that."

In 1920, Teilhard began teaching geology at the Institute Catholique. A move toward conservatism within the Vatican, including a less favorable view of evolution, forced Teilhard to repudiate his views in order to remain in the Church. In 1923, demoralized and unsure of his fate, Teilhard left to do field work in China. Except for brief periods when he traveled to Europe, he remained in Asia for the next twenty years. Despite his exile, Teilhard continued to write and conduct anthropological research.

Teilhard was part of the December 2, 1929 discovery of "Peking Man," now classified as *Homo erectus*, in the village of Zhoukoudian, China. In 1939, during Japan's occupation of China, the Peking Man fossils were moved to the Peking Union Medical College, after which they were packed in two crates for shipment to the American Museum of Natural History. The famous fossils never made it to New York; today, no one knows their whereabouts.

By the mid-1940s, Teilhard was frustrated by the Vatican's obstructionist efforts—including its decision to block publication of his works—and in 1951 he began collaborating with the Wenner-Gren Foundation for Anthropological Research in New York. Teilhard suffered declining health in the 1950s, including a heart attack in 1953. On April 10, 1955 (Easter Sunday), he died of a stroke, and was later buried in the Jesuit cemetery at Saint Andrews-on-Hudson in New York. No one attended Teilhard's burial. Soon after his death, manuscripts that the Vatican had censored began to be published. In 1962, the Vatican warned of serious errors in Teilhard's theology regarding the relationship between humans and nature. However, in 1996, Pope John Paul II celebrated the fiftieth anniversary of Teilhard's priesthood with *Gift and Mystery*, which included Teilhard's "Mass of the World" from 1923.

While he was in China, Teilhard was pursued romantically by American artist Lucile Swan (1890-1965). Teilhard resisted Swan's advances. Today, Swan's sculpture of Peking Man welcomes visitors to the Zhoudoudian site where Peking Man was discovered.

For More Information: Aczel, A.D. 2007. *The Jesuit and the Skull: Teilhard de Chardin, Evolution, and the Search for Peking Man*. New York: Riverhead Hardcover; Birx, J. 1993. *Interpreting Evolution: Darwin and Teilhard de Chardin*. Amherst, NY: Prometheus Books.

ALFRED TENNYSON (1809–1892)

Lord Alfred Tennyson was the Victorian poet who succeeded William Wordsworth as poet laureate of England. Tennyson was a student of William Whewell. Tennyson's masterpiece, the 131-stanza *In Memoriam A.H.H.*, was published on June 1, 1850, and quickly became one of Britain's most popular poems. Tennyson's poem, which was partly inspired by Robert Chambers' *Vestiges of Creation*, was an elegy mourning the death of Tennyson's friend Arthur Hallam. The poem, which took seventeen years for Tennyson to write, describes a world in which natural laws, rather than God, govern life, and

decries the relentless struggle that scientists such as Darwin were seeing in nature. Although Tennyson decried the lack of direction of Lyell's geology, he regained hope at the end of the poem through evolution's progress. Tennyson's line in the fifty-sixth stanza of *In Memorium*, "Nature, red in tooth and claw," became a famous description of the struggle for existence. Today, many people attribute the poem's most famous lines to Shakespeare: "'Tis better to have loved and lost/Than never to have loved at all." Tennyson, an agnostic, died on October 6, 1892, and is buried in Poets' Corner of Westminster Abbey.

CHARLES THAXTON (b. 1939)

One can empirically detect the products of an intelligent agent without specifying who that agent is.

Charles Thaxton is a physical chemist credited with being one of the main originators of the modern intelligent design (ID) movement. Born in Dallas, Texas, Thaxton earned BS (1959) and MS (1962) degrees in chemistry from Texas Tech University, and a PhD in physical chemistry from Iowa State University in 1964. He completed postdoctoral work in the history of science at Harvard University and molecular biology at Brandeis University. He has taught at several universities in the United States, was a visiting professor at Charles University (Prague) from 1992 to 1997, and has published over two dozen papers in the primary scientific literature.

In 1981, Thaxton read an article by physicist Hubert Yockey that concluded "self organization must yield only genetic message ensembles of information content much too low to constitute a genome." Thaxton interpreted this as meaning the complex information contained in DNA is evidence for the activity of an intelligent agent (Yockey was more equivocal: "This means that *at present* there is no valid scientific scenario for the origin of life"), and in 1984 Thaxton, along with Walter Bradley and Roger Olson, published *The Mystery of Life's Origin*, which proposed that DNA has the same complex structure as language. Because DNA is the molecule of heredity used by all known life, the "specified complexity" of DNA indicates that life was designed by an intelligent agent. The book, self-described as "an advanced college-level work on chemical evolution," was published by the Foundation for Thought and Ethics, where William Dembski was Academic Editor.

While developing his ideas, Thaxton initially used the word "create" to describe the directed process he imagined, but concluded that this description was simultaneously "too broad and too specific." Additionally, "create" was already closely associated with biblical literalists. Another term would be needed. After searching the scientific literature, Thaxton finally decided that the phrase "intelligent design" was the best description, thereby creating a name for the fledgling movement.

Thaxton—who claims that ID does not specify the designing agent—edited a new ID-based textbook written by biologists Percival Davis and Dean Kenyon. The book, *Of Pandas and People* (1989), contained a section entitled "A word to the teacher" written by Thaxton. Because the Dover Area School District proposed to make *Pandas* available in the district's ninth-grade biology courses, Thaxton was deposed during that trial (*Kitzmiller v. Dover Area School District*). He also testified in 2005 during proceedings that arose after the Kansas State Board of Education changed that state's science standards.

Thaxton has argued that "Christians . . . must insist that whenever origins are discussed, public schools allow the teaching of the evidence for creation alongside instruction in the naturalistic concept of evolution." Thaxton's concern that "naturalistic evolution and moral relativism are like two rooms in the same house" is in accord with Phillip Johnson's efforts to topple materialism, and Thaxton and Johnson conferred frequently when Johnson was writing *Darwin On Trial* (1991). Charles and Carole Thaxton operate Konos Connections, a resource for homeschoolers that works "to connect people with their Creator, Jesus Christ."

THEODOSIUS (347–395)

Theodosius was a Roman emperor who made Christianity the only approved religion. As a result, several western ideas critical to what would become biology can be traced to Genesis, including the idea that all species are the direct, intentional, and sudden creation of God, and that life has not changed much since the Creation.

D'ARCY THOMPSON (1860–1948)

D'Arcy Wentworth Thompson was a Scottish biologist who wrote *On Growth and Form* (1917), one of the most celebrated literary works ever produced by a scientist. In *On Growth and Form*, Thompson described the similarities that one encounters in biological form, and how these similarities obey the basic principles of physics and mathematics. Thompson, who became known for claiming that "form is a diagram of forces," presented a prototype for adaptation resulting from physical forces governed by invariant laws of nature, not by functionalist mechanisms such as natural selection. *On Growth and Form* influenced biologists as well as architects, engineers, and artists. Thompson died at St. Andrews, Scotland, on June 21, 1948.

W.O. THOMPSON (1781–1861)

W.O. Thompson was Clarence Darrow's law partner and part of the team of lawyers that defended John Scopes in Dayton, Tennessee, in 1925. Thompson joined the defense when Bainbridge Colby resigned before the trial started.

WILLIAM THOMSON (LORD KELVIN) (1824–1907)

The argument of design has been greatly too much lost sight of in recent zoological speculation . . . Overwhelming strong proofs of intelligent and benevolent design lie around us.

William Thomson was born on June 28, 1824, in Belfast, Ireland. He began studying at

Glasgow University when he was 10, and later became a physicist and engineer who helped establish the science of thermodynamics. As a result of his work in laying the first transoceanic cable, Kelvin became rich and famous, and in 1892 was made Lord Kelvin of Largs (so-named for the River Kelvin that flowed past Glasgow University). Kelvin is best known for developing in 1848 the Kelvin temperature scale. Kelvin's *Treatise on Natural Philosophy* (1867) helped define modern physics.

While an undergraduate at Cambridge, Kelvin began wondering how long it would take an Earth-size "red-hot globe" (7,000° F) to cool to today's temperature. Kelvin concluded—in a now-famous one-paragraph paper—that Earth was probably about 98 million years old. Several other investigators (including Irish geologist John Joly) reached similar conclusions. Even Cambridge astronomer George Darwin—Charles Darwin's son—supported Kelvin; as the young Darwin noted to Kelvin in 1878 (four years before Charles Darwin's death), "he cannot quite bring himself down to the period assigned by you, but does not pretend to say how long may be required." By 1899, Kelvin had lowered his estimate of Earth's age to 20–40 million years, and "probably much nearer 20 than 40." Although Kelvin was a Christian and not concerned with defending biblical literalism, he rejected Lyell's gradualism and Darwin's "transmutation of species," all while claiming to be "on the side of the angels."

Charles Darwin—who referred to Kelvin as an "odius spectre"—realized that life's timeline ultimately relied on geology. Consequently, Kelvin's conclusion was one of Charles Darwin's "sorest troubles," for it did not provide sufficient time for natural selection to produce life's diversity. Darwin suspected that Earth was much older than Kelvin claimed, but Kelvin's reputation and mathematical calculations gave his claim much credibility (even Mark Twain noted that "we must yield to [Kelvin] and accept his view"). To accommodate Kelvin's timeline, Darwin proposed pangenesis as a solution to inheritance (every sperm and egg contained "gemmules thrown off from each different unit throughout the body"). Darwin's Lamarckian explanation sped the process while avoiding Lamarck's quasi-spiritual sources of acquired traits. In the third edition of *On the Origin of Species*, Darwin deleted references to an absolute age of the Earth, and later changed the phrase "incomprehensibly vast" to "how vast." Thomas Huxley attacked Kelvin's conclusions, but abandoned gradualism for saltation to fit the allotted time for life on Earth.

When he made his calculations, Kelvin did not know that Earth's internal radiation was a source of Earth's heat (Joly and George Darwin were among the first to note that radioactivity could produce some of Earth's heat). Although radiation was discovered near the end of Kelvin's life, Kelvin continued to defend his claims of Earth's chronology. Although Kelvin won virtually every award given to scientists, he also made some remarkably inaccurate claims, including "heavier-than-air flying machines are impossible," "x-rays will prove to be a hoax," and "radio has no future."

Kelvin taught at Glasgow University for fifty-three years, and his house in Glasgow was the first house in Scotland to be lit by electric light. Soon after retiring (Queen Victoria attended his three-day retirement celebration), Kelvin enrolled as a research student at the university, thereby becoming one of its youngest and oldest students. Kelvin died after a month-long illness (the physician described it as "a severe chill to the liver") at his home in Glasgow on December 17, 1907. He was buried in Westminster Abbey, next to Isaac Newton. Kelvin is memorialized in Glasgow by several buildings and statues.

For More Information: Lindley, D. 2004. *Degrees Kelvin: A Tale of Genius, Invention, and Tragedy*. New York: Joseph Henry.

FRIEDRICH TIEDEMANN

Friedrich Tiedemann was a German anatomist who in 1816 claimed that the embryonic human brain passes through stages that resemble those of simpler vertebrates. Later, Ernst Haeckel expanded Tiedemann's conclusion by claiming that humans go through fish, amphibian, and reptile stages during embryonic development, summarizing his idea with the famous (and now discredited) claim that "ontogeny recapitulates phylogeny."

TORCASO v. WATKINS

Torcaso v. Watkins was a lawsuit in which the U.S. Supreme Court in 1961 ruled that plaintiff Torcaso, an atheist, had been unduly denied the office of notary public by Maryland because he would not affirm a belief in God. In a minor footnote in the decision, Justice Hugo Black popularized "secular humanism" when he noted that "among the religions in this country which do not teach what would generally be considered a belief in the existence of God, are Buddhism, Taoism, Ethical Culture, Secular Humanism, and others." Black did not define secular humanism or give it special attention. *Torcaso* was later cited by antievolutionists such as Jack Chick, Max Rafferty, William Willoughby, John Conlan, and Tim LaHaye as a decision that made secular humanism an "official" religion in the United States. These creationists then argued that because Darwinism was the same as secular humanism, and because secular humanism was a religion, then schools should not be allowed to teach Darwinism. In 1984, Orrin Hatch—a Republican Senator from Utah—sponsored a provision that denied funding for the teaching of secular humanism. However, Hatch's amendment did not define secular humanism, and it expired before affecting public education.

DAVID TREEN (b. 1928)

David T. Treen was the Republican governor of Louisiana who signed into law the statute requiring "equal time" for the teaching of evolution and creationism in the state's public schools. Although Treen had concerns about the law, he argued that the bill "simply 'permits' competing theories to be covered, rather than mandating that one be covered if the other is included. Academic freedom can scarcely be harmed by

inclusion. It can be harmed by exclusion." The law was later overturned by the U.S. Supreme Court in *Edwards v. Aguillard*.

TULSA ZOO (Est. 1927)

The Tulsa Zoo is a publicly funded attraction owned by the City of Tulsa, Oklahoma. The zoo houses nearly 1,500 animals and includes exhibits featuring elephants, polar bears, primates, penguins, and a tropical rain forest. Nearly 600,000 people visit the zoo annually, and in 2004, it was voted "America's Favorite Zoo" in a contest sponsored by Microsoft to advertise the video game *Zoo Tycoon 2*.

The Tulsa Zoo strives to educate as well as entertain, and includes information on ecology, conservation, and evolution in its exhibits. While visiting the zoo in 1995, Tulsa resident Dan Hicks noted that the chimpanzee exhibit described that chimps and humans had descended from a common ancestor millions of years ago. Hicks complained to the zoo that this information offended his religious beliefs. He also predicted that his beliefs were held by most Tulsa residents, who would oppose a publicly supported zoo promoting an anti-Christian, naturalistic philosophy. Under pressure from Hicks, zoo administrators removed the reference about the shared ancestry of humans and chimps and added a disclaimer at the zoo's entrance stating, "There are many views on the origin of biological species and their behaviors. The information that accompanies our displays is based on compelling evidence of the natural sciences. Because scientific knowledge is subject to change, these displays may be revised as new information becomes available."

Emboldened by this victory, Hicks asked the zoo to remove all references to evolution because evolution conflicted with a literal interpretation of the Bible. Through the local Southern Plains Creation Society (for which Hicks eventually served as vice president), Hicks offered free seminars about creationism, including creation-based tours of the Tulsa Zoo. By 2005, Hicks complained to the Tulsa Park and Recreation Board (which administers the zoo) that the Tulsa

Zoo included references to several religions, but none to Christianity.

Hicks argued that including Christianity in the zoo's exhibits is an "issue of fairness" because the "majority of Oklahomans are creationist." He proposed that a display illustrating the biblical story of creation be erected at the zoo, and he offered $3,000 to support such an effort. After heated public debate about the proposal, the five-member Park and Recreation Board (composed of the mayor plus mayoral appointees) voted 3-1 to accept the proposed 3' x 5' exhibit detailing the Genesis account of creation. The mayor of Tulsa, Bill LaFortune, who supported the proposed display, directed the Tulsa city attorney to determine the legality of such a display. The city attorney responded that the display was acceptable if it was part of a broader exhibit of creation stories.

Tulsa's decision made national news. Ken Ham, president of Answers In Genesis, claimed that "we need more people like Dan Hicks who are willing to boldly lead the battle to tell people the truth concerning the creation of the universe." Americans United for Separation of Church and State condemned the decision while urging the Park Board to reverse its decision, and the newly organized Friends of Religion and Science collected 2,000 signatures on a petition to have the Board reconsider its vote.

Within a month, donations to the zoo decreased and the proposed display was reconsidered by the Board. Acknowledging significant public resistance to the display (despite a poll showing that nearly 60 percent of those polled approved of the proposed display), the Board reversed itself; only LaFortune continued to support the display. After the vote, LaFortune suggested that a committee consider removing all displays that contain religious themes. Hicks grudgingly accepted defeat, remarking that it was time to "have a zoo that's just about animals."

TURKANA BOY

Turkana Boy was a 5' 3", 12-year-old boy (*Homo erectus*) who lived 1.5 million years ago near Lake Turkana in Kenya. He is the most complete skeleton of a prehistoric human ever found. When Turkana Boy was displayed publicly for the first time in Kenya in 2007, Bishop Boniface Adoyo—head of Kenya's thirty-five evangelical denominations—organized a protest, claiming that he "did not evolve from Turkana Boy or anything like it." Famed fossil-hunter Richard Leakey, whose team discovered Turkana Boy in 1984, responded that "the bishop is descended from the apes and these fossils tell how he evolved. Turkana Boy is a distant relative of his, whether the bishop likes it or not."

JOHN TYNDALL (1820–1893)

John Tyndall was a physicist and president of the British Association for the Advancement of Science who was an outspoken advocate of evolution and opponent of religious orthodoxy. During the association's meeting in Belfast, Ireland, in 1874, Tyndall created controversy when he claimed that the church's long monopoly on issues related to human origins would soon be replaced by those based on evolution.

U

SAMUEL UNTERMEYER (1858–1940)

Samuel Untermeyer, Vice-President of the American Jewish Congress, was nominated by William Jennings Bryan to serve on the prosecution team at the Scopes Trial. Bryan believed that "being a Jew, he ought to be interested in defending Moses from the Darwinites."

HAROLD UREY (1893–1981)

Harold Clayton Urey was an American chemist who, in 1950, began studying the origin of life in reduced terrestrial atmospheres. In 1951, Urey's seminar at the University of Chicago was attended by Stanley Miller, who later used electrical charges to synthesize organic compounds (including amino acids) in what was then believed to be a simulation of Earth's early atmosphere. Urey, who won the Nobel Prize in Chemistry in 1934 for his discovery of deuterium, contributed to the Manhattan Project to develop an atomic bomb. One of Urey's students was famed science-fiction writer Isaac Asimov. Urey died on January 5, 1981, and was buried in Fairfield Cemetery in DeKalb County, Indiana.

JAMES USSHER (1581–1656)

James Ussher was born into a rich family on January 4, 1581, in Dublin, Ireland. When he was 13, Ussher entered Ireland's Trinity College, from which he earned a BA in 1598, and an MA in 1601. At age 20, Ussher was ordained a priest in the Anglican Church. Ussher, an expert in Semitic languages, received his doctor of divinity degree at age 26, and the following year became a professor at Trinity. In 1615, Ussher drafted doctrinal statements for the Irish Protestant church and was appointed Vice-Chancellor of Trinity College; the next year, he became Vice-Provost. In 1625, Ussher was named archbishop of Armagh. Ussher enjoyed travel, and often went to England to meet with scholars and statesmen. In 1641, he was in England when the Irish rebellion broke out. Ussher never returned to Ireland.

In Ussher's day, determining the exact date of creation was an important topic because many people believed that Earth's potential was 6,000 years (4,000 before Christ, 2,000 after). This potential corresponded to the six days of creation, with the logic being that "one day is with the Lord as a thousand years, and a thousand years as one day" (II Peter 3:8). This belief was abandoned in 1997, 6,001 years after 4004 BC.

Many people had calculated a date of creation; for example, the Benedictine monk Venerable Bede (672–735) claimed that creation occurred in 3952 BC, famed German astronomer Johannes Kepler (1571–1630) claimed that creation occurred in 3993 BC, Ussher's near contemporary Joseph Scaliger (1540–1609) claimed that creation occurred in 3949 BC, and Martin Luther (1483–1546) claimed that creation occurred in 3961 BC. By the mid-1700s, there had been more than 200 computations of the date of creation, all based on various interpretations of the Bible.

In 1650, after more than a decade of research, Ussher announced his calculations in his 2,000-page *Annals of the Old Testament, deduced from the first origins of the world* (*Annales veretis testamenti, a prima mundi origine deducti*). Ussher relied on the Hebrew Bible for his calculations and interpretations. Although that Bible provides a direct, unbroken lineage from Creation to Solomon, Ussher used additional sources for dates between Solomon and the destruction of the Temple (e.g., the death of Chaldean King Nebuchadnezzar). Ussher also made several assumptions, including that creation must have occurred near the autumnal equinox because that corresponded to the harvest time of fruits present in the Garden of Eden. Ussher also accepted the ages of biblical patriarchs listed in the Bible—for example, that Adam lived 930 years, that Seth lived 912 years, and that Methuselah—Adam's great-great-great-great-great-grandson—lived 969 years. Although many people in Ussher's era questioned these ages, others claimed that "God afforded [these people] a longer lifespan on account of their virtue" and that "the Earth then produced mightier men."

Ussher calculated the dates of numerous events, including the Great Flood (2348 BC), the Exodus from Egypt (1491 BC), the founding of the Temple in Jerusalem (1012 BC), the destruction of Israel by Babylon (586 BC), the birth of Jesus (4 BC), Adam and Eve being driven from Paradise (Monday, November 10, 4004 BC), and Noah's ark landing at Ararat (Wednesday, May 5, 2348 BC). However, Ussher remains most famous for his calculation of the days of creation. Apparently not a believer in building suspense, Ussher stated his famous conclusion in *Annals'* first paragraph (Figure 78):

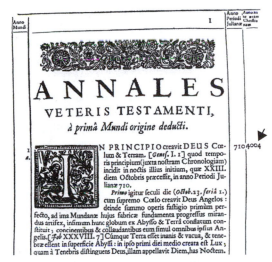

78. Although several people (including Martin Luther) had stated a date for creation, James Ussher's claim that God created heaven and earth in 4004 BC became the most famous date associated with Young-Earth creationism. Ussher announced his claim on the first page of his *Annals of the World*, which included three columns. The column on the left was the year of the world; these dates began at year 1. The inside right column listed dates according to the Julian calendar, which starts at 710. The outside column, which listed "the year before Christ," listed Ussher's famous date (arrowhead).

> In the beginning God created Heaven and Earth, Gen. 1, v. 1. Which beginning of time, according to our chronologie, fell upon the entrance of the night preceding the twenty third day of Octob[er], in the year of the Julian [Period] 710. The year before Christ 4004. The Julian Period 710.

Ussher's calculation that Creation occurred during the night of Saturday, October 22, 4004 BC became the most famous "creation date" when London bookseller Thomas Guy included it in annotated versions of the King James Version of the Bible in 1675 (the popularity of these Bibles may have also been aided by their engravings of topless women). In 1701, the Church of England—at the urging of Oxford cleric John Feel and William Lloyd (Bishop of Winchester)—began including Ussher's dates adjacent to the beginning of Genesis in its official Bibles (Ussher's date remained in these Bibles until 1900). Later, some of Ussher's dates were also included in the influential *Scofield Reference Bible*, as well as in Bibles placed by Gideons in hotel rooms.

William Henry Green (1825–1900) became a prominent critic of Ussher's calculations. Green, who endorsed Arnold Guyot's belief in day-age creationism, claimed that biblical chronology is uncertain—and therefore undeterminable—before Abraham. Green believed that modern science and religion are harmonious, and his 1890 article titled "Primeval Chronology" attacked John Draper's claim that religion and science are in perpetual conflict.

In 1958, Ussher's book was translated into English with the title *The Annals of the World. Deduced from the Origin of Time, and continued to the beginning of the Emperour Vespasians Reign, and the total Destruction and Abolition of the Temple and Common-wealth of the Jews. Containing the Histoire of the Old and New Testament, with that of the macchabees. Also all the most Memorable Affairs of Asia and Egypt, and the Rise of the Empire of the Roman Caesars, under C. Julius, and Octavianus. Collected from all History, as well as Sacred, as prophane, and Methodically digested.* Understandably, most people referred to the book as *Annals of the World*. Ussher ended his chronology at 70 AD because he believed that the history of the Jews ended then.

For nearly three centuries, most King James versions of the Bible listed Ussher's historical date in the margin. Ussher's date later appeared elsewhere, including in Shakespeare's *As You Like It*, in which Rosalind laments, "The poor world is almost six thousand years old." The Bible introduced into evidence at the Scopes Trial included Ussher's chronology, which was cited during Clarence Darrow's questioning of William Jennings Bryan. In that testimony, Bryan responded that "it would be just as easy for the kind of God we believe in to make the Earth in six days as in six years or in 6,000,000 years or in 600,000,000 years. I do not think it important whether we believe one or the other." This testimony shocked many fundamentalists, and is why Jerry Falwell noted that Bryan "lost the respect of Fundamentalists when he subscribed to the idea of periods of time for creation rather than twenty-four hour days."

Today, Ussher's chronology is not included in most Bibles, and is often ridiculed. However, Ussher's *Annals of the World* was one of the most influential books of the seventeenth century, and it represented some of the best scholarship of its time; Ussher had a personal library of over 10,000 books, and *Annals* included more than 10,000 footnotes. Although Ussher's chronology has been abandoned by most Judeo-Christian religions, his creation date of 4004 BC remains the citation that most Young-Earth Creationists use to date God's creation of the universe. *The MacArthur Study Bible* continues to claim that "scripture does not support a creation date earlier than about 10,000 years ago," and the *King James Study Bible* tells readers that Genesis refutes evolution, that there is no evidence for an ancient earth or gap creationism, and that "God created the world in six literal days."

Ussher, who never predicted when the world would end, died on March 21, 1656, at Lady Peterborough's house in Reigate. Oliver Cromwell provided £200 for Ussher's funeral and burial in St. Paul's Chapel in Westminster Abbey. In Dublin, the Ussher family name adorns several streets and lanes. Thousands of Ussher's books are at Ussher Library at Trinity College, Dublin.

For More Information: Ussher, J. 1650. *Annals of the World: James Ussher's Classic Survey of World History*. (Modern English republication, ed. Larry & Marion Pierce). Green Forest, AR: Master Books, 2003.

V

JO VIENS

Jo Viens was a 51-year-old, 4'-tall man who promoted himself as "The Missing Link" during the Scopes Trial. He was joined by Lewis Levi Johnson Marshall ("Absolute Ruler of the Entire World"), Elmer Chubb (who claimed he could "withstand the bite of any serpent"), and flat-Earth advocate Wilbur Voliva.

FRIEDRICH VIEWEG

Friedrich Vieweg was a publisher in Germany who in 1844 published *Journal of Researches*, the first translation or printing abroad of any of Charles Darwin's books. When Darwin's *Origin* was translated into German in 1860, Darwin's sentence about human evolution ("Light will be thrown on the origin of Man and his history") was omitted by the first translator.

RUDOLF VIRCHOW (1821–1902)

Rudolf Ludwig Carl Virchow was a pathologist and politician who linked evolution with socialism. In 1878, Virchow seconded Charles Darwin's election to Berlin's Koeniglich-Preussische Akademie der Wissenschaften. Virchow, who coined the term *embolism*, was the first to recognize leukemia, and in 1858 claimed that every cell originates from another cell (*Omnis cellula e cellula*). In 1892, Virchow—who helped found cellular pathology and comparative pathology—was awarded the Copley Medal. Virchow urged physicians to act as "attorneys for the poor."

KARL VON BAER (1792–1876)

Karl Ernst von Baer was a German embryologist who in the second volume of *Entwickelungsgeschichte der Thiere* (*Animal Development*; 1828) argued against recapitulation. Although von Baer rejected evolution, his work influenced Darwin. Von Baer was described by Huxley as "a man of the same stamp" as Darwin, and was awarded the Copley Medal of the Royal Society in 1867. Today, von Baer—who also was the first to describe the notocord and the mammalian egg—is considered a founding father of modern embryology.

ALBRECHT VON HALLER (1708–1777)

Albrecht von Haller was a Swiss biologist and poet who in 1744 coined the word *evolution* (from the Latin *evolvere*, meaning "to unroll") to refer to embryos that grow from preformed homunculi enclosed in egg and sperm. Von Haller believed that all generations had been created (and stored like Russian dolls) in the ovaries of Eve or the testes of Adam. Interestingly, Darwin, Haeckel, and Lamarck—the most prominent nineteenth-century evolutionary biologists of England, Germany, and France, respectively—did not include the word *evolution* in the first editions of their famous books; Darwin wrote about *descent with modification*, Haeckel of *Transmutations-Theorie*, and Lamarck of *transformisme*. The use of the word *evolution* to describe Darwin's descent with modification was borrowed not from an earlier technical use of the word, but instead from the vernacular. Darwin first used the word *evolution* in 1871 in *Descent of Man*. *Evolved* is the last word in all editions of *On the Origin of Species*.

ALEXANDER VON HUMBOLDT (1769–1859)

Friedrich Heinrich Alexander von Humboldt, with botanist Aime Bonpland, traveled the New World from 1799–1804. Humboldt and Bonpland collected more than 60,000 specimens, and noted that similar climates support similar types of plants. Humboldt returned to Paris in 1804 and spent the next twenty-five years preparing a twenty-three-volume description of his work. Edgar Allan Poe was inspired by Humboldt's attempt in *Kosmos* to unify the sciences, and dedicated *Eyreka: A Prose Poem*—his last great work—to Humboldt. With the exception of Napoleon Bonaparte, Humboldt was the most famous man in Europe in the early 1800s. Darwin was fascinated with Humboldt's books, and he often referred to Humboldt's work in *Voyage of the Beagle*. Although Humboldt did not address evolution, his findings contributed to biogeography and evolutionary understanding. East Germany honored Humboldt in 1964 by placing his image on its five-mark banknote. Humboldt is buried at Tegel Castle in Berlin, Germany.

W

ALFRED RUSSEL WALLACE (1823–1913)

Evolution, even if it is essentially a true and complete theory of the universe, can only explain the existing conditions of nature by showing that it has been derived from some preexisting condition through the action of known forces and laws. It may also show the high probability of a similar derivation from a still earlier condition; but the farther back we go the more uncertain must be our conclusions, while we can never make any real approach to the absolute beginnings of things.

Alfred Russel Wallace (the unusual spelling of his middle name arose from an error made during the recording of his birth) was born on January 8, 1823, in the small town of Usk in southeast Wales (Figure 79). In 1836, Wallace's parents were forced to end his formal education, after which he was sent to London to live with his older brother, John.

While staying with John, Alfred studied the philosophy of social reformer Robert Owen that combined agnosticism with the belief that a person's character is determined by their social environment. Owen condemned the established caste system as an obstacle that prevented members of the lower classes from realizing their potential. These ideas, plus the worker-centered ideals discussed at the Mechanic's

Institute in London where Alfred learned a variety of trade skills, influenced Wallace as he trended towards socialism throughout his life. By 1837, Alfred had moved to Bedfordshire to live with his brother William, who was a self-employed surveyor. Alfred learned much about geology through surveying, and started collecting and preserving plants. In 1841, the brothers moved to Neath, a town in the south of Wales; there, Alfred joined the local scientific societies, and later gave public lectures about various aspects of science.

William released Alfred in 1843, due to a lack of work. The death of their father the same year obligated the young men to support their mother and younger brother. Alfred became a schoolmaster at the Collegiate School in Leicester in 1844, which provided him access to the school's library, where he read Thomas Malthus' *An Essay on the Principle of Population* (1798), Charles Lyell's *Principles of Geology* (1830), and Robert Chambers' recently and anonymously published *Vestiges of the Natural History of Creation* (1844). Malthus affected Wallace's thinking, but the sensational *Vestiges* convinced Wallace of the reality of biological evolution.

While at Leicester, Wallace also met Henry Bates, a young entomologist who shared Wallace's agnosticism and interest in natural history, and the two quickly became close friends.

ALFRED RUSSEL WALLACE
O.M., LL.D., D.C.L., F.R.S., F.L.S.

Naturalist - Scientist - Explorer - Writer
Social Campaigner - Humanitarian

•CO-DISCOVERER OF EVOLUTION BY NATURAL SELECTION•
•FOUNDER OF THE SCIENCE OF ZOOGEOGRAPHY•

THIS MONUMENT WAS RESTORED IN THE YEAR 2000 BY THE A. R. WALLACE
MEMORIAL FUND; IT IS CARED FOR BY THE LINNEAN SOCIETY OF LONDON.

79. Alfred Wallace was the codiscoverer, with Charles Darwin, of evolution by natural selection. This plaque at Wallace's grave also notes Wallace's important contributions to biogeography. When considering humans, Wallace advocated what is now known as "intelligent design," and later in his life became a strong supporter of spiritualism. (*George Beccaloni*)

Spiritualism also attracted Wallace's attention during this period. Through his brother William, Wallace had been exposed to (and intrigued by) phrenology. After seeing a demonstration of mesmerism, Wallace concluded that not only was it a real phenomenon, but that he had mesmeric abilities. Although aware of the likely reaction from the scientific community to such conclusions, Wallace resisted the temptation to "accept the disbelief of great men, or their accusations of imposture or imbecility." When William died in 1845, Wallace resumed surveying, joined by his brother John. But Wallace disliked running a business, and during a visit by Bates, the friends decided to go to the tropics. Wallace was also still intrigued by Chambers' *Vestiges* and eager to study evolution; the biological diversity of the tropics seemed the ideal setting for such work.

After studying the insect collections at the British Museum and tropical plants at the Royal Botanic Gardens, Wallace and Bates left for the mouth of the Amazon River on April 26, 1848, and Wallace planned to "take some one family to study thoroughly, principally with a view to the theory of the origin of species." They arrived in Brazil on May 26, and began to explore. Initially slowed by their inexperience, they eventually accumulated collections, and soon

sent their first shipment of hundreds of specimens (mostly insects) to England. An argument between Bates and Wallace after only a few months of work ultimately caused them to part company, whereupon Wallace was joined by his younger brother Herbert. Wallace and Bates met again briefly in 1850, and later that same year, Herbert left for home. Although Herbert died of yellow fever before leaving Brazil, Alfred remained in South America until 1852, suffering from frequent bouts of dysentery and malaria, but collecting thousands of specimens and interacting with the indigenous people.

In July 1852, as Wallace was planning his passage home, he discovered that most of the specimens he thought had been forwarded to England had not been shipped. Finally arranging passage, he and his collections sailed for England on July 12, 1852. Wallace was exhilarated and exhausted by his four-year expedition (as well as still suffering from periodic bouts of malaria), and looked forward to sorting through his collections. However, three weeks into the voyage, the ship caught fire, and all on board evacuated into lifeboats 700 miles from shore. Wallace only had time to retrieve a few drawings and personal items, and all of his collections and notes burned with the ship. After ten days, Wallace and the others were rescued by a passing merchant vessel, and he finally returned to England on October 1, 1852.

Deprived of his biological collections, Wallace was forced to rely solely on the observations he had made while in South America. In 1853, he self-published two books, *Palm Trees of the Amazon* and *Narrative Travels on the Amazon and Rio Negro*, and also contributed several scientific papers. This work attracted the interest of the British scientific community, but Wallace remained unemployed and needed money; although he received some insurance money for his lost collections, it did not equal what he would have made by selling the specimens. Moreover, Wallace was already missing the adventure of an expedition, and he had made no progress in the study of evolution. Believing that "the very finest field for an exploring and collecting naturalist was to be found in the great Malayan Archipelago," Wallace decided to go to Asia, an area even more foreign to Europeans than South America. Having secured support from the Royal Geographic Society and passage from the Admiralty, Wallace, along with his 16-year-old assistant Charles Allen, sailed on March 4, 1854, for Singapore.

After arriving on April 20, Wallace collected on the island of Borneo, and soon thereafter sent specimens, especially insects (he had been told to "not neglect the small things") back to his broker in England. After recognizing patterns in the distribution of species and higher taxonomic units, Wallace concluded that "the natural sequence of species by affinity is also geographical," and that "every species has come into existence coincident both in space and time with a preexisting closely-allied species." Eventually, he interpreted these patterns as reflecting the evolutionary process, and he composed a paper (informally known as the "Sarawak law" paper) entitled "On the Law Which has Regulated the Introduction of New Species," published in 1855, that reported these findings. Wallace used biogeographical evidence to infer what Charles Darwin referred to as "descent with modification."

Wallace knew that he had made a major discovery, and he was eager for feedback, including from Darwin, whose observations from the Galápagos Islands Wallace had referenced in his paper. To his disappointment, Wallace heard back that, upon reading the paper, Darwin had merely written "Nothing very new here" in the margin. However, Wallace's paper prompted Darwin to inform his friend Charles Lyell about his own theory of evolution by natural selection. Lyell encouraged Darwin to publish quickly to avoid being scooped by Wallace, who was clearly thinking along the same lines, but Darwin continued to hesitate.

Wallace left Borneo in May 1856 for Bali and the neighboring island of Lombok, leaving behind his assistant in the care of missionaries. While exploring the islands—separated by less than twenty miles—Wallace noted the striking distinctness of the birds and mammals of each island. Wallace eventually marked the channel between Lombok and Bali as the divide between two great zoogeographic regions, the Asian and Australian. This line—which divides the Indonesian archipelago, starting in the South between Bali and Lombok, continuing northward between Sulawesi and Borneo, and curving towards the Pacific just south of the Philippines Islands and north of the Hawaiian Islands—is now referred to as "Wallace's Line" (Figure 80). A century later geologists confirmed that the line is at the edge of the Indo-Australian plate.

In January 1858, Wallace was on the small island of Ternate. While slowed by malaria, he pondered the puzzling biogeographical observations he had made on Borneo, Bali and Lombok. Recalling the book by Malthus he had read years before, he considered the factors that determine the fates of individuals in a population:

> Why do some live and some die? And the answer was clearly, that on the whole the best fitted live.... Then it suddenly flashed upon me that this self-acting process would necessarily improve the race, because in every generation the inferior would inevitably be killed off and the superior would remain—that is, the fittest would survive.

80. While in the Malay Archipelago, Wallace noted the striking differences in the zoogeographic regions of Asia and Australia. Today, these differences are marked by Wallace's Line. (*Jeff Dixon*)

By February 1858, Wallace had written "On the Tendency of Species to form Varieties; and on the Perpetuation of Varieties and Species by Natural Means of Selection," which described the same basic evolutionary process that Darwin had identified two decades earlier. Wallace sent the paper to Darwin asking him to show it to Lyell, who had responded favorably to Wallace's Sarawak paper. Upon receiving the paper, Darwin realized that Wallace had, as Darwin feared he would, independently discovered the process of natural selection, and in desperation he wrote to Lyell that "all of my originality, whatever it may amount to, will be smashed." Lyell and botanist Joseph Hooker conferred, and decided that papers by Darwin and Wallace should be presented together publicly.

On July 1, 1858, at a meeting of the Linnean Society attended by only a few dozen people (Darwin did not attend), George Busk presented a letter from Hooker and Lyell that credited the discovery of evolution by natural selection to both Darwin and Wallace, but gave priority to Darwin. The following items were read, in order: an unpublished manuscript about natural selection by Darwin, an 1857 letter by Darwin to Asa Gray that outlined his concept of natural selection (establishing priority), and Wallace's paper. Darwin immediately began writing his treatise on evolution. On November 24, 1859,

John Murray published *On the Origin of Species by Means of Natural Selection, or the Preservation of Favoured Races in the Struggle for Life*. Wallace was unaware of the presentation of his ideas, and he had not planned on publishing his paper until hearing from Lyell. When Wallace learned what had happened, he was disappointed that his conclusions were not unique and put aside his plans for a book-length treatment of the subject. But he was gracious in accepting his reduced role; as he wrote to Hooker in 1858, "I cannot but consider myself a favoured party in this matter."

Wallace collected in the Malay Archipelago until 1862, traveling more than 14,000 miles and collecting 125,000 specimens. Once back in England, he championed what he now referred to as "Mr. Darwin's principle," and Wallace and Darwin corresponded frequently. (Wallace so relinquished intellectual ownership of the concept of natural selection that when he published a book about evolution in 1889, he titled it *Darwinism*.) While returning to England, Wallace was elected a fellow of the Zoological Society. Wallace had kept tens of thousands of specimens (including a thousand species of birds) for his own use, and he began studying his collections and publishing his findings. In 1866, Wallace married Annie Mitten, with whom he had three children (although one died in infancy). Wallace's 1869 book, *The Malay Archipelago* (dedicated to Charles Darwin), discussed the major scientific findings of his travels in Asia. This was followed by *The Geographical Distribution of Animals* (1876) and *Island Life* (1880), which became standard authorities in biogeography.

Sales of Wallace's books generated income, but because he never secured stable employment after his return (he frequently worked short-term jobs as an instructor and as editor of others' work), and because his investments did not perform well, Wallace's financial situation deteriorated over time. Needing money (he was living on only about £60 per year), Wallace even answered an advertisement by a biblical literalist offering £500 to anyone who could prove Earth is spherical. Wallace won the bet, but he was publicly harassed by the loser for years, forcing Wallace to sue for libel. By the 1880s, thanks to friends and colleagues (including Darwin and Huxley), Wallace was granted a civil pension of £200 per year. Wallace continued to publish scientific articles and books for many more years. He delivered the prestigious Lowell Institute lectures in Boston in 1886, after which he toured the United States, speaking on a variety of subjects.

By the 1860s, Wallace embraced spiritualism, harkening back to his dabblings in phrenology and mesmerism. In 1865, he started attending séances, convinced that genuine phenomena (including a dancing Native American he witnessed at a séance in New York) occurred at such events. Wallace wrote about spiritual matters, several of his essays were collected in *On Miracles and Modern Spiritualism* (1875), and in 1882 he joined The Central Association of Spiritualists. Wallace even served as an expert witness in the trial of Henry Slade, a purported psychic who claimed to communicate with spirits via a writing slate. At one of Slade's séances, a student of Thomas Huxley (Huxley had also investigated séances incognito) grabbed the slate from Slade's hands before questions had been asked, only to find words already on the slate. Slade was charged with being a "common rogue," but Wallace testified at the trial that Slade was "as sincere as any investigator in a university department." (Darwin, meanwhile, contributed money to the prosecution.) Slade was convicted, but was released on a technicality and fled the country.

Wallace and Darwin also discussed the applicability of natural selection to humans. In the late 1860s, Wallace claimed that natural selection alone could not explain the origin of humans, especially because our large brains seemed too powerful to have been produced by selection. Believing that the intellectual abilities and moral sense of *Homo sapiens* are fundamentally apart from nature, Wallace urged others to accept that an "Overruling Intelligence" had created humans. Darwin responded by writing to Wallace that "I differ grievously from you," and in 1871

Darwin published the *Descent of Man*, in which he applied a strict evolutionary perspective to human origins and traits. (Lyell, interestingly, sided with Wallace.)

Wallace's interactions with indigenous peoples during his expeditions had convinced him of the inherent equality of all people, leading him to fully embrace socialism. In the 1880s, Wallace advocated making land "a common property," and became the first president of the Land Nationalization Society, a position he held until his death. Wallace also resisted vaccination campaigns, seeing them as a violation of individual rights. Late in life, he published social commentaries including *Social Environment and Moral Progress* and *The Revolt of Democracy* (both 1913).

On November 7, 1913, Wallace died in his sleep. He was buried three days later in Broadstone, Dorset. Wallace's neglected grave, which was restored in 2000, consists of a 7'-tall, 146-million-year-old fossilized treetrunk (probably *Protocupressinoxylon*) mounted on a block of Purbeck limestone (now surrounded by black granite). In 2000, the lease for Wallace's grave was transferred from the Wallace family to the Linnean Society. Today, the Linnean Society maintains Wallace's grave.

In 1915, a medallion bearing Wallace's name was dedicated and placed in Westminster Abbey. Awards given to Wallace include the Royal Medal from the Royal Society (1868), the Linnean Society's Gold Medal (1892), the Copley Medal (1908), and the Order of Merit (1908). In 2006, Wallace's admirers erected a monument to him outside the church where Wallace was baptized in 1823 and near the cottage in which he was born. Although Wallace was the codiscoverer of evolution by natural selection, it was Darwin who became the focal point of the subsequent evolution-creationism controversy.

For More Information: Oosterzee, P.V. 1997. *Where Worlds Collide: The Wallace Line*. Ithaca, NY: Cornell University; Raby, P. 2001. *Alfred Russel Wallace*. Princeton, NJ: Princeton University; Shermer, M. 2002. *In Darwin's Shadow: The Life and Science of Alfred Russel Wallace*. Oxford: Oxford University.

JOHN WALTON (1881–1949)

John C. Walton was a progressive Democrat governor of Oklahoma who, on March 24, 1923, signed America's first antievolution law. That law, an amendment to House Bill No. 197, was championed by the Oklahoma Baptist Convention and provided that "no copyright shall be purchased, nor textbook adopted that teaches the 'Materialist Conception of History (i.e.) the Darwin Theory of Creation versus the Bible Account of Creation.'" Although William Jennings Bryan had not seen the legislation and did not go to Oklahoma to campaign for the legislation, he urged other states to pass similar laws, adding that "it is very important that there should be such a simple declaration, declaring [the teaching of evolution] unlawful." Walton was impeached in 1925, soon after which the state's antievolution law was repealed.

GEORGE F. WASHBURN

George F. Washburn was a Boston capitalist and friend of William Jennings Bryan who founded the Bible Crusaders of America five months after Bryan's death. The Crusaders published *Crusaders' Champion*, which later merged with other fundamentalist publications, including T.T. Martin's *Conflict*. Washburn, like many others, claimed that God had appointed him to succeed Bryan in the antievolution crusade. Like Bryan, Washburn believed that "the great battle of the age is now on between Christianity and evolution."

GEORGE WATERHOUSE (1810–1888)

George Waterhouse collected insects with Charles Darwin at Cambridge, and in 1833 founded the Entomological Society. Waterhouse catalogued many of the animals that Darwin brought back from the *Beagle* voyage. Darwin helped Waterhouse get a job at The British Museum. Waterhouse, who traveled to Germany to buy the *Archaeopteryx* for The British Museum, named one of his sons after Darwin.

JAMES WATSON (b. 1928)

The book of the DNA sequence would in time be regarded as more relevant to human life than the Bible. It tells us who we are. I've never read the Bible, so I'm not sure I've missed much.

James Dewey Watson was born on April 6, 1928, in Chicago, Illinois. Watson attributed his self-confidence, libertarian spirit, and trademark frankness—what some would call impertinence—to his parents' supportive influence. Above all, inquiry and knowledge were valued in his home; as Watson recalled, "I was sort of trained to get pleasure from understanding the world around me, not from material things." Watson's father was an avid birdwatcher, and father and son took regular walks to look for birds. When 15 years old, Watson entered the University of Chicago, intent on a career in ornithology. However, he became interested in genetics, and after graduating in 1947 with a BS in zoology, decided to study genetics in graduate school.

In 1947, Watson earned a fellowship to Indiana University and started working in microbiologist Salvador Luria's lab. Luria was part of the "Phage Group" that included Max Delbrück and Alfred Hershey (both of whom, along with Luria, would share a Nobel Prize in 1969). These biologists used bacteriophage (viruses that infect bacteria) to study inheritance. The group met regularly at Cold Spring Harbor Laboratory, and Luria included Watson in these meetings. The Phage Group was convinced that DNA was the molecule of heredity (rather than proteins), and Watson later credited the early opportunity to work closely with such pioneers as being crucial to his ultimate success. Watson finished his PhD in 1950, keen on studying genetics. This meant analyzing the structure of DNA, and he accepted a postdoctoral position in Copenhagen.

Although his postdoc was a disappointment (his new advisor was not studying DNA's structure), Watson heard a presentation by Maurice Wilkins of King's College, London, which described how X-ray diffraction could be used to study the structure of DNA. This was exactly the research Watson wanted to do, and he transferred to the Cavendish Laboratory at Cambridge, the leading center for x-ray diffraction studies. Soon after arriving in Cambridge in 1951, Watson befriended graduate student Francis Crick, who was interested in the same questions.

Watson and Crick were assigned a project on the structure of proteins but they remained obsessed with DNA. Watson in particular was driven to solve the mystery of DNA as quickly as possible because he knew that Linus Pauling in the United States was working on the same question. Following Pauling's lead, Watson and Crick not only approached the task through theory and experiment, they also built elaborate models out of wire, metal, and cardboard. But it was not until Wilkins let the two glimpse results produced by Wilkins' estranged colleague Rosalind Franklin that they identified the simple yet elegant structure of the DNA double helix. Watson and Crick published their findings in a short paper in *Nature* in 1953, followed by three others over the next year that detailed their work. This research earned Watson, Crick, and Wilkins the Nobel Prize for Physiology or Medicine in 1962.

In 1953, Watson moved to the California Institute of Technology to perform similar studies of RNA. He returned to the Cavendish Laboratory briefly in 1955 before settling in at Harvard in 1956, attaining the rank of professor five years later. Watson continued to study RNA and ribosomes, forming a long-term collaboration with Walter Gilbert (winner of the Nobel Prize in 1980). Watson's bluntness also earned him a dubious reputation among his colleagues, especially from organismal biologists, whom Watson labeled "stamp collectors" left behind by the new era of molecular biology. Noted Harvard ecologist Edward O. Wilson in turn later declared Watson "the most unpleasant human being I had ever met."

While at Harvard, Watson published *The Double Helix: A Personal Account of the Discovery of the Structure of DNA* (1968), a frank account of the race to determine the structure of DNA. The book was considered so inflammatory (including by Crick) that Harvard University Press rescinded its agreement to publish it. (The book was eventually released by Atheneum.) Watson was also criticized for his harsh treatment of Rosalind Franklin, who had done the x-ray diffraction work critical to Watson's and Crick's conclusions. (Franklin died of ovarian cancer in 1958.) Regardless, the book, which became a bestseller, is included in the Modern Library's list of 100 best nonfiction books of the twentieth century, and was made into a 1987 movie starring Jeff Goldblum as Watson. Watson also helped write the textbook *The Molecular Biology of the Gene* (1965), which provided a contemporary discussion of molecular biology and set a new standard for textbook design.

In 1968, Watson became director of the Cold Spring Harbor Laboratory (CSHL) and also married his wife, Elizabeth, with whom he eventually had two sons. Watson left Harvard in 1976 for permanent residence at CSHL, became president of CSHL in 1994, and then chancellor in 2004 when CSHL became a graduate-degree-granting institution with the opening of the Watson School of Biological Sciences. Watson is credited with rescuing CSHL, primarily by emphasizing cancer research that attracted large grants, when the Carnegie Institution removed its financial support in the 1960s. From 1988 to 1992, Watson also served as the first director of the Human Genome Project (HGP). At a news conference early in his tenure with the HGP, he made an on-the-fly decision to devote 3 percent (later increased to 5 percent) of the program's budget to study ethical and social implications of the project. This was hailed as a major endorsement of the need for bioethical analysis in science, especially genetics. Watson stepped down (he says he was "fired") as director of the HGP over disagreements with the director of the National Institutes of Health about granting patents for DNA sequences of unknown function.

In *DNA: The Secret of Life* (2003), Watson supported the development of genetically-modified foods, discussed the need for confidentiality of personal genetic information, and rejected the current system of patenting genetic information. Watson also endorsed "human germline engineering," noting that "we have really got to worry that a genetic underclass exists." Later, he

was even more brash when he commented, "If you are really stupid, I would call that a disease...So I'd like to get rid of that, to help the lower 10 percent." Watson also hopes that genetic engineering will make people more compassionate; "We'll understand why people can't do certain things. Instead of asking a child to shape up, we'll stop having unrealistic expectations."

Scientists, especially cognitive scientists, have objected to Watson's implied genetic determinism of a complex trait like intelligence, and religious leaders have warned of the dangers of "playing God." Watson's militant atheism—"The biggest advantage to believing in God is you don't have to understand anything"—brought condemnations by believers. Bioethicists highlight both the ethical problems of designing children and the sinister nature of past eugenic movements. Undaunted, Watson urged a group of 1,000 German scientists to "put Hitler behind us" and use genetics to improve humans. Ironically, one of Watson's sons suffers from a serious cognitive disorder, and Watson was chancellor of an institution—Cold Spring Harbor Laboratory (CSHL)—that, through the Eugenics Record Office, led twentieth century eugenics efforts in the United States. In 2005, Watson agreed to have his genome sequenced and made public; Watson's sequence is now accessible through a National Institutes of Health database. Watson resigned his position at CSHL in late October, 2007 after he provoked widespread outrage by claiming that Africans are intellectually inferior to Westerners.

For More Information: Watson, J.D. 1968. *The Double Helix: A Personal Account of the Discovery of the Structure of DNA.* New York: Athenuem.

WEBSTER v. NEW LENOX SCHOOL DISTRICT #122

Webster v. New Lenox School District #122 established in 1990 that (1) a teacher does not have a First Amendment right to teach creationism in a public school, and (2) a school district can ban a teacher from teaching creationism. Ray Webster was a social studies teacher in New Lenox School District who claimed that the school's prohibition against teaching creationism violated his constitutional rights.

JOSIAH WEDGWOOD (1730–1795)

Josiah Wedgwood was born on July 12, 1730, in Burslem, England. When he was 14 years old, Wedgwood apprenticed for five years with his brother to learn the pottery trade. Soon thereafter, he contracted smallpox, which permanently damaged his left knee. Despite this handicap, Wedgwood became a skilled potter. He focused on managing the business after his left leg was amputated in 1768.

By 1759, Wedgwood was working independently and developed a reputation for innovation and quality. He created designs for Queen Charlotte, which were so successful that he was appointed potter to the Queen. Wedgwood's "Queen's Ware" was popular throughout Europe and represented a major advance in English pottery. Later, Wedgwood invented jasperware, a popular medium for creating vases and decorative pieces.

In 1764, Josiah Wedgwood married Sarah Wedgwood (a third cousin). They would have five children that survived childhood. Their eldest child, Susannah, married Robert Darwin (son of Erasmus Darwin) in 1796, and Robert and Susannah had six children, including Charles Darwin. In 1839, Charles Darwin married into the Wedgwood family when he married Emma Wedgwood (his cousin), the youngest child of Josiah Wedgwood II. The younger Josiah worked in his father's pottery business, and was a Member of Parliament in the 1830s. The great-grandson of Josiah II, Josiah "Jos" Wedgwood IV, was also a Member of Parliament, and an opponent of the eugenics espoused by Francis Galton, a half-cousin of Charles Darwin.

In 1794, Wedgwood fell ill with severe jaw pain, originally diagnosed by family physician Erasmus Darwin as an abscessed tooth. Removal of the tooth revealed untreatable problems, probably cancer, and Wedgwood died on January 3, 1795.

ALFRED WEGENER (1880–1930)

Scientists still do not appear to understand sufficiently that all Earth sciences must contribute evidence toward unveiling the state of our planet in earlier times.

Alfred Lothar Wegener was born in Berlin on November 1, 1880. Although he earned a PhD from the University of Berlin in astronomy in 1904, he was most interested in meteorology. By 1905, Wegener was working at the Royal Prussian Aeronautical Observatory, where he pioneered the use of kites and balloons to collect meteorological data. Wegener also used balloons to take himself aloft; he and his brother set a record in 1906 when they remained airborne for fifty-two hours. In 1906, Wegener was the meteorologist on a Danish expedition to the northeast coast of Greenland. (This was the first of four trips he would make to Greenland.) In 1909, after returning to Germany, Wegener taught astronomy and meteorology at the University of Marburg. In 1911, he published the results of his research as *The Thermodynamics of the Atmosphere*.

During World War I, Wegener was drafted into the army and was injured in combat. While recuperating, he remembered an article that reported identical fossils being found on opposite sides of the Atlantic Ocean. This discovery, combined with the observation that the continents were shaped like jigsaw puzzle pieces that could be fit together into a single unit, prompted Wegener to wonder if the continents have moved across the face of the Earth. (This was not an entirely new proposal. Austrian geologist Edward Suess, using paleobotanical evidence, had suggested that the continents of the southern hemisphere were at one time joined to form Gondwanaland.)

At the 1912 meeting of the Geological Association in Frankfurt, Wegener tentatively outlined his theory of continental drift. His proposal was met by reactions ranging from polite skepticism to outright hostility. Undeterred, he scoured the literature and discovered supporting evidence, including strata of rock shared between eastern North America and Scotland, fossils indicating the presence of different climates than what is currently found in that locale (e.g., fossils of tropical plants in the Arctic), and coal bands that would align if the continents were brought together. In *The Origin of Continents and Oceans* (1915), Wegener concluded that "it is just as if we were to refit the torn pieces of a newspaper by matching their edges and then check whether the lines of print ran smoothly across." Wegener proposed that his supercontinent, Pangaea (Greek for "all Earth"), existed approximately 300 million years ago.

By the end of World War I, *The Origin of Continents and Oceans* had been translated into several languages and was available to the larger scientific community. Wegener's concept of continental drift did not receive a warm reception, partly because Wegener could not provide a convincing explanation for continental movement. Wegener won some converts to his view, but during his relatively short life, he did not see widespread acceptance of his proposal. However, based on the strength of his other contributions, he secured a permanent faculty position at the University of Graz (Austria) in 1924.

In 1930, Wegener participated in another expedition to Greenland to search for proof of continental drift. This trek suffered numerous setbacks, and by the time the research party left base camp, they were five weeks behind schedule. A colleague, anxious to establish an outpost to study the jet stream during the winter, left with a small group that was later resupplied by Wegener. The conditions Wegener's party experienced on the way to the camp were so severe (i.e., temperatures commonly below $-50°$C, with a constant oncoming wind) that of the fourteen people who started the journey, all but Wegener and two others gave up and returned to base camp. After five weeks, the three men, along with dogsleds and thousands of pounds of supplies, finally reached the remote post. Wegener, anxious to return, set out with another researcher a couple of days later (just after celebrating his fiftieth birthday), but neither man survived the return trip. Wegener's body was discovered in his sleeping bag six months later; he may have suffered a heart attack brought on

by the extreme exertion of the trip. His remains, which were later marked by a 20'-tall iron cross, were left where they were found. All have since disappeared.

By the 1950s, growing evidence indicated that Wegener was correct about continental movement. Analyses of the sea floor documented the existence of mid-ocean ridges where volcanoes form crust that spreads outward from the ridge. These findings were integrated in the 1960s to form the now well-supported theory of plate tectonics, in which plates are moved by convection in the underlying athenosphere within which the plates are embedded or "float." Wegener's original proposal, although incorrect in its details, remains a remarkable example of how seemingly unrelated sets of observations make sense when interpreted relative to the great age and changing nature of the planet.

For More Information: McCoy, R. 2006. *Ending in Ice: The Revolutionary Idea and Tragic Expedition of Alfred Wegener*. New York: Oxford University.

AUGUST WEISMANN (1834–1914)

I read [On the Origin of Species] first in 1861 at a single sitting and with ever-growing enthusiasm. When I finished I stood firm on the basis of evolution theory, and I have never seen reason to forsake it.

August Weismann was born on January 17, 1834, in Frankfort, Germany. He studied medicine at the University of Göttingen, and graduated with an MD degree in 1856. After working for several years as a research assistant, Weismann established a medical practice in Frankfort, and, from 1861 to 1862, he was the private physician to Archduke Stephan of Austria. In 1863, Weismann joined the medical school faculty and became director of the zoological institute at the University of Freiberg, where he would remain until his retirement in 1912.

Weismann read *On the Origin of Species* soon after it was published, and immediately accepted evolution and natural selection. Weismann became interested in heredity, a central concept for evolution, because he disagreed with Darwin's proposed hereditary mechanism, pangenesis, due to its Lamarckian nature. Weismann believed that the hereditary essence—the "germ plasm"—occurs in every cell, but is completely separate from it: the cell uses the information within the germ plasm, but the cell has no effect on it (i.e., there is no possibility of acquired characteristics). This meant that in the sex cells he had studied, "each reproductive cell potentially contains two kinds of substances." Weismann hypothesized that the germ plasm is stable, incurring no changes from one generation to the next, and represents a connection with an organism's ancestors.

Weismann also concluded that the germ plasm was associated with chromosomes that were known to be housed in the nucleus, a level of cellular detail that had recently become apparent with improvements in microscopy. He further proposed a hierarchical structure for the genetic material, ranging from the chromosomes ("idiants") that were composed of "determinants" (the actual units that determine individual morphological characters), which in turn were composed of "biophores." The variation upon which natural selection acts was created by shuffling the germ plasm each generation during sexual reproduction. Weismann—an ardent proponent of selection—even suggested that some types of biophores may be better able to use resources, which resulted in these biophores becoming more frequent within a cell. (He suggested that such a system could account for vestigial organs.) Weismann also predicted the process of meiosis, based on the need to maintain a constant amount of germ plasm per cell across generations. His proposals on the nature of heredity were introduced in *The Germ-Plasm, A Theory of Heredity* (1892).

Weismann was so intent on exorcizing Lamarckian inheritance from evolution that he removed the tails from several generations of mice. Not surprisingly, each mouse, even those in later generations, grew well-developed tails. Critics argued that this was an unfair test because the loss of the tail was not benefiting the mouse, so there is no reason to expect the lack of a tail to be inherited. (In *Back to Methuselah*, George Bernard Shaw—a Lamarckian—referred to these experimental tests as demonstrating Weismann was "reduced to idiocy by Neo-Darwinism.") Regardless, Weismann concluded that however tails are specified in mice, having a tail wasn't necessary to pass the trait to offspring.

Weismann's emphasis on selection was poorly received by a scientific community struggling with understanding the process of evolution without much understanding of heredity. Thomas Hunt Morgan, who later substantiated another of Weismann's claims—that the hereditary units, now called genes, were arrayed linearly along chromosomes—dismissed Weismann's proposals due to lack of supporting evidence. However, Weismann identified several important aspects of heredity and evolution, and Ernst Mayr considered him second only to Darwin for furthering evolutionary thought in the nineteenth century.

Weismann died on November 6, 1914.

H.G. WELLS (1866–1946)

Herbert George Wells was George Rappleyea's first choice to present the defense's case for evolution at John Scopes' trial. Wells, a former student of Thomas Huxley, declined the invitation to participate in the trial. Wells was an outspoken socialist, and was one of the first writers to describe time travel and lost worlds. Antievolution crusaders such as William Riley often condemned Wells' books. Wells' *The Shape of Things to Come* (1933), in which a world government destroys Christianity, inspired Christian apologist C.S. Lewis (1898–1963) to include Jules—a caricature of Wells—in *That Hideous Strength* (1945). Today, Wells—an advocate of eugenics—is best known for his earlier work (e.g., *The War of the Worlds, The Invisible Man*, and *The Island of Doctor Moreau*).

JONATHAN WELLS (b. 1942)

Jonathan Wells is a Senior Fellow at the Discovery Institute. In 1976, Wells entered Sun Myung Moon's Unification Theological Seminary. Moon, who opposed evolution, selected a handful of seminary students to pursue PhD degrees to provide them with credentials to "destroy Darwinism." Wells was one of those selected, and in 1978 he entered Yale University "to prepare myself for battle." Wells earned a PhD from Yale in 1986 in religious studies, but decided that a background in religion alone would be insufficient because "Darwinism was clearly winning the ideological battle . . . largely because it claimed to be supported by scientific evidence." He decided to earn a PhD in biology, which he did in 1994, from the University of California at Berkeley. Since then he has worked to fulfill his hope "that the Darwinist establishment will come apart at the seams, just as the Soviet Empire did in 1990."

In 2002, Wells' *Icons of Evolution: Science or Myth? Why Much of What We Teach About Evolution is Wrong* discussed ten evolutionary exemplars (e.g., "Haeckel's embryos," peppered moths) that are frequently found in biology textbooks. Wells claimed that these "icons" are unsupported by evidence and concluded that "science now knows that many of the pillars of Darwinian theory are either false or misleading. Yet biology texts continue to present them as factual evidence of evolution." The scientific community excoriated Wells for misleading readers by purposely discussing information that had long been rejected by science, and by doubting evolution simply because textbooks sometimes contain unintended errors. Undeterred, Wells has continued his attack on evolution with *The Politically Incorrect Guide to Darwinism and Intelligent Design* (2006).

For More Information: Wells, J. 2002. *Icons of Evolution.* Washington, DC: Regnery.

ABRAHAM WERNER (1749–1817)

Abraham Gottlob Werner was an exceedingly confident German geologist who rejected uniformitarianism in favor of Neptunism, the belief that sedimentary rocks originated in water via precipitation. Although Werner was wrong about the formation of rocks, he was nevertheless one of the most influential geologists at the beginning of the Industrial Revolution.

Werner—who rejected the claims of Nicolaus Steno—never discussed the actual age of Earth, but in 1786 he noted the "enormously great time spans, which perhaps far exceed our imagination." His students included famed traveler Alexander von Humboldt.

WESTMINSTER ABBEY (Est. 616)

Westminster Abbey, officially known as The Collegiate Church of St. Peter at Westminster, is a gothic church just west of the Palace of Westminster in London, England (Figure 81). Although it was first created at its current site in 616, the stone abbey was built in 1045–1050 by King Edward the Confessor, who had chosen the site for his burial (he was buried there at dawn on January 6, 1066, the day after he died). The cross-shaped abbey is the tallest Gothic building in the British Isles. (The abbey is more than 500' long, and its west towers are more than 225' high.) In the Nave, the ceiling reaches more than 100' high.

When Henry VIII assumed control of the abbey in 1539, he made the abbey a cathedral, thereby saving it from destruction. The phrase "robbing Peter to pay Paul" likely referred to money meant for the abbey (which was dedicated to St. Peter) being diverted to St. Paul's Cathedral. Queen Elizabeth I made the abbey The Collegiate Church of St. Peter at Westminster.

The abbey has hosted innumerable historic events. The first third of the King James Bible's Old Testament, and the first half of the New Testament, were translated at the abbey, and the New English Bible was later compiled at the abbey. With few exceptions (Lady Jane Grey, Edward V, and Edward VIII), all English and British monarchs have been crowned at the abbey.

More than 3,000 people are buried or memorialized in the abbey. For example, the abbey's Nave includes the graves of William Thomson (Lord Kelvin), Ernest Rutherford, and James Clerk Maxwell, and Poet's Corner includes the graves of Robert Browning, Geoffrey Chaucer, Charles Dickens, Alfred Tennyson, George Frederick Handel, and Lawrence Olivier. The abbey

81. Westminster Abbey, one of the greatest churches in Christendom, enshrines the history of England. The famed abbey houses the graves of several people associated with the evolution-creationism controversy, including James Ussher, Charles Lyell, William Thomson (Lord Kelvin), Alfred Tennyson, and Charles Darwin. (*Randy Moore*)

also includes commemorations to (but not the graves of) Winston Churchill, William Shakespeare, Oscar Wilde, Henry Wadsworth Longfellow, and Lewis Carroll (Charles Lutwidge Dodgson, 1832–1898), a famed mathematician whose stories told to Alice Liddell and her sisters on a boating trip became *Alice's Adventures in Wonderland*. Oliver Cromwell was given an elaborate funeral at the abbey in 1658, but three years later his remains were disinterred and hanged.

Irish prelate and biblical chronographer James Ussher, who lived at Westminster during some of Ireland's political unrest, is buried in the abbey's Chapel of St. Paul.

When Charles Darwin died on April 19, 1882, newspapers announced that he would be buried beside his infant children in St. Mary's churchyard at Downe, just a short walk from his home. However, his friends—including Thomas Huxley—arranged for him to be buried in Westminster Abbey, beside astronomer Sir William Herschel. Near Darwin's grave are monuments to Alfred Wallace and botanist Joseph Hooker, and the graves of Isaac Newton and Darwin's friend, geologist Charles Lyell. Lyell's bust adorns a window bay near the grave of physicist John Woodward in the North Aisle.

Westminster Abbey, which is under the jurisdiction of the Crown rather than a diocese, hosts church services daily, and is one of the greatest churches in Christendom. It is visited by people from around the world, and enshrines the history of England.

For More Information: Dean and Chapter of Westminster. 2002. *Westminster Abbey: Official Guide*. London: Westminster Abbey.

WGN RADIO (Est. 1924)

On June 1, 1924, radio station WGN (formerly WDAP) began broadcasting in Chicago. Within a few weeks, WGN had reported live from the Republican National Convention in Cleveland, from the Democratic National Convention in New York, and from the sentencing of thrill-killers Leopold and Loeb, who had been defended by Chicago attorney Clarence Darrow. In spring of the following year, WGN broadcast a live "Prohibition Debate" that also featured Darrow. The call letters of WGN, a subsidiary of the *Chicago Tribune*, stood for "world's greatest newspaper."

In July 1925, at a cost of more than $1,000 per day, WGN aired a live broadcast of the Scopes "Monkey Trial" from Dayton, Tennessee. This was the first trial in history to be broadcast on radio. To produce the broadcast, WGN rented cables stretching from Chicago to Dayton and

was allowed to rearrange the courtroom as it saw fit. Quinn Ryan, a *Tribune* editor famous for creating broadcasts that were "almost as good as being there," and engineer Paul Neal covered the trial in Dayton with four microphones in the courtroom. WGN's coverage gave listeners a front-row seat to the trial's proceedings.

In 1925, radio was still relatively new, especially in tiny Dayton. Ryan and Neal were treated as celebrities; as Ryan later noted, "We were like moon men there." The mayor of Dayton asked Ryan and Neal to stay at his house. For most of the trial, Ryan sat in a windowsill and let the trial's dialogue speak for itself, only occasionally adding commentary.

In 1925, there was no technology available to record broadcasts. As a result, there are no recordings of Ryan's famous coverage of the Scopes Trial.

WILLIAM WHEWELL (1794–1866)

The arm and hand of man are made for taking and holding, the wing of the sparrow is made for flying; and each is adapted to its end with subtle and manifest contrivance. There is plainly Design.

William Whewell was born May 24, 1794, in Lancaster, England. In 1811, he entered Cambridge University with classmates Charles Babbage and John Herschel. Adam Sedgwick, already teaching at Cambridge by 1811, became a lifelong friend, although there would be some minor altercations between Whewell and the somewhat nonconformist Sedgwick when Whewell became an administrator. In one well-known incident, Whewell chastised Sedgwick for his "frequent appearance in the [Trinity] College courts accompanied by a dog."

Whewell graduated in 1816, second in his class, and remained at Cambridge for the rest of his life. In 1817, he was elected fellow (a teaching position awarded for life), and then assistant tutor in mathematics in 1818, and tutor in 1823. He was later appointed professor of mineralogy (1828), Knightbridge professor of moral philosophy (1838), and finally, Master of Trinity College (from 1841 onwards). He was also ordained an Anglican priest in 1826, as required of

Trinity fellows. He married Cordelia Marshall in 1841; they had no children.

Whewell's contributions are distinguished by their breadth and synthetic nature. Whewell's work spanned an amazing range of subjects, including mathematics, mechanics, astronomy, educational reform, philosophy and history of science, mineralogy, and architecture. In the sciences, his most influential research involved understanding tides. Although this work earned him a medal from the Royal Society in 1837, Whewell did not consider himself a major scientist. Ironically, it was Whewell himself that coined the term *scientist*, at a meeting of the British Association for the Advancement of Science in 1833, and he used it in print the following year. Other familiar terms Whewell introduced include *cathode* and *anode* (after Michael Faraday asked for help devising descriptive terms), and *uniformitarianism* and *catastrophism*, for the dominant but opposing perspectives in geology at the time.

Whewell advocated natural theology, not as a source of spiritual guidance, but as a way of understanding how the universe works. In the third *Bridgewater Treatise*, *Astronomy and General Physics Considered with Reference to Natural Theology* (1833), he argued that the universe exhibits incontrovertible evidence for design. (The Earl of Bridgewater had commissioned *Bridgewater Treatises* to unite natural theology with current scientific understanding.) The astronomical and physical structure of the universe proclaims an "Idea of Organization" which Whewell equated with "a Final Cause." Therefore, for Whewell, study of nature does not simply allow us to glimpse the mind of the Creator; rather, the only way to understand nature is to accept that it has been specially created. Design is not an end product, it is a causal force. Not surprisingly, Whewell rejected evolution, and he refused to allow the Cambridge library to include Darwin's *On the Origin of Species*.

One of Whewell's last works, *A Dialogue on the Plurality of Worlds* (1854), argued against the existence of intelligent life elsewhere in the universe, a proposal he had earlier considered possible. In 1855, his wife died. Whewell

remarried in 1858, but his second wife also died a few years later. In February of 1866, Whewell fell from his horse, and was seriously injured. He died on March 6, 1866, and is buried at Trinity College.

For More Information: Fisch, M., & S. Schaffer (eds.). 1991. *William Whewell: A Composite Portrait*. Oxford: Clarendon Press.

JOHN C. WHITCOMB, JR. (b. 1924)

John C. Whitcomb, Jr., was an Old Testament scholar at Grace Theological Seminary. In 1961, Whitcomb and Henry Morris developed Whitcomb's 450-page doctoral thesis into *The Genesis Flood*, a popular book that denounced mainstream geology and evolution as badly flawed. *The Genesis Flood* generated little attention in the secular world, but it galvanized evangelicals, led to the formation of the Creation Research Society (CRS), and began the modern "creation science" movement in America. Whitcomb's *The World That Perished* (1973)—a sequel to *The Genesis Flood*—included advertisements for a "documentary" by *Films for Christ* about the Flood.

Whitcomb advocated biblical creationism, and therefore resisted Morris' scientific creationism. When Morris helped found the CRS, Whitcomb claimed its religious message was compromised by attempting to use scientific data to "prove" creation; as Whitcomb noted in 1992, "one might just as well be a Jewish or even a Muslim creation scientist as far as [the CRS] model is concerned." In 2007, Whitcomb—who believes that intelligent design undermines the Bible and wants to "purge schools of evolutionary perversion"—was a speaker for Answers in Genesis and president of Whitcomb Ministries, Inc.

For More Information: Whitcomb, J.C., & H. Morris. 1961. *The Genesis Flood*. Philadelphia, PA: Presbyterian and Reformed.

ELLEN WHITE (1827–1915)

Ellen G. White was a cofounder of the Seventh-Day Adventists in 1864 who claimed that science could not contradict the Bible. White claimed that God had carried her back to creation and revealed to her that "creation week was only seven literal days, and . . . the world is now only about six thousand years old." White also claimed that fossils are the remains of creatures killed in the Noachian flood, a claim that remains popular among many Young-Earth creationists. White condemned evolution as "satanic" and claimed that "the Bible is not to be tested by men's ideas of science, but science is to be brought to the test of this unerring standard." She also attacked "infidel geologists [who] claim that the world is very much older than the Bible record makes it." White influenced flood geologist George McCready Price, who in turn influenced Henry Morris and John Whitcomb's *The Genesis Flood*, the foundation of modern "creation science."

During her lifetime, White claimed that God gave her approximately 2,000 visions and dreams, and her followers believed White was a modern-day prophetess who helped prepare people for Christ's return. Critics accused White of plagiarism and of being mentally ill. White is buried alongside her husband in Oak Hill Cemetery in Battle Creek, Michigan. Today, the Seventh-Day Adventist Church sponsors the Geoscience Research Institute and its journal *Origins*, which were established in 1958 to reconcile evolution with biblical creation. In 1978, creationist Duane Gish claimed that the institute was "doing more in the way of active research than any other group of creationists in the world."

FRANK WHITE (1933–2003)

If we're going to teach evolution in the public school system, why not teach scientific creationism? Both of them are theories.

Frank Durward White was born on June 4, 1933, in Texarkana, Texas. He graduated from the U.S. Naval Academy and served in the U.S. Air Force, during which time he flew a division of the 101st Airborne from Kentucky to Arkansas to help calm the racial unrest caused by

integration of Little Rock's Central High School (where Susan Epperson would later teach and test the state's law banning the teaching of human evolution). White later became a banker, and from 1975–1977 directed the Arkansas Industrial Development Commission. White then became president of the Capital Savings and Loan in Little Rock.

In 1980, White—a Republican who claimed to be "born again"—defeated incumbent Bill Clinton to become the forty-first governor of Arkansas. A few days after the election, White raised eyebrows by claiming his political triumph was a "victory for the Lord." In 1982, Clinton beat White in a rematch. After White's political career ended, he returned to banking, and from 1998 until his death he directed the Arkansas State Bank Department.

On March 19, 1981, White repaid Jerry Falwell's Moral Majority for its support by signing into law Act 590, which required Arkansas' public schools to give "balanced treatment" to creationism. White signed the legislation, despite the fact that he had not read it. The Arkansas statute was later overturned by *McLean v. Arkansas Board of Education*.

White died of a heart attack at his home on May 21, 2003, and is buried in Little Rock's Mount Holly Cemetery.

WALTER WHITE (1880–1951)

Something has happened that's going to put Dayton on the map.

Walter White was born on December 24, 1880. After college, White became involved in public service and politics in Dayton, Tennessee. White, who replaced George Rappleyea as the complainant when Rappleyea resigned the position, was the first witness in the Scopes Trial. White testified that Scopes had said that he had reviewed Hunter's *A Civic Biology* in class and that he could not teach biology without teaching evolution. When Darrow cross-examined White, White admitted that Hunter's book had been adopted by the state textbook commission in 1919, and readopted in 1924, and that he had not warned Scopes or any other teacher against using Hunter's book. Nor had anyone objected to the book's use until Rappleyea signed the complaint against Scopes.

After the trial, White required all of the teachers he hired at Rhea County school to be fundamentalists, noting that they had to "declare themselves believers in the Bible as it is written and must teach nothing that is contrary to that book." White suggested that Dayton establish its own university and name it after Bryan; that university became Bryan College, which opened in 1930. White, whose business cards described himself as "Prosecutor of John T. Scopes," sold insurance and served as Superintendent of Schools of Rhea County until 1944, when he began practicing law. White described Bryan as "the greatest man in the world and its leading citizen."

In the 1930s and early 1940s, White served five terms in the Tennessee House of Representatives. In 1926, White was the Republican nominee for governor, but lost to Austin Peay. In 1940, White was a delegate to the Republican National Convention.

White died on February 14, 1951, and was buried in Dayton's Buttram Cemetery.

SAMUEL WILBERFORCE (1805–1873)

That Mr. Darwin should have wandered from this broad highway of nature's works into the jungle of fanciful assumption is no small evil.

Samuel Wilberforce was born September 7, 1805, in Clapham, England. In 1826, Wilberforce graduated with a strong background in mathematics and the classics, and earned an MA in 1829, the same year he was ordained. He married Emily Sargent in 1828, and the couple had five children that survived infancy.

Immediately after being ordained, Wilberforce accepted a curacy at Checkendon in Oxfordshire. Wilberforce distinguished himself as a gifted cleric, and attracted the attention of his diocese. Throughout his career, Wilberforce opposed slavery and supported a number of socially-relevant measures, including laws to limit the length of the work day for factory workers, and increased support for the poor. Yet as a cleric, he was conservative (he opposed relaxing divorce laws and allowing public places to be open on Sundays), and was especially concerned about preventing scientific advances from changing Church orthodoxy.

In 1841, Emily Wilberforce died, an event that forever haunted her husband. However, Wilberforce kept working, and became known as an eloquent public speaker. This attracted the attention of Prince Albert, who made Wilberforce a court chaplain in 1841. Wilberforce's oratorical skills, combined with his penchant to engage in controversial topics (especially related to church matters), followed by attempts to extricate himself from these situations, earned him the nickname "Soapy Sam."

Wilberforce became Bishop of Oxford in 1845 at a time of conflict within the Church. In 1847, he opposed elevating a colleague to bishop, based on what Wilberforce considered heterodox statements written by the individual. Unknown to Wilberforce, the nominee had already retracted those statements, meaning there was no reason to level such public charges against him. To his credit, Wilberforce apologized, but his handling of the affair was interpreted as self-interested promotion. He blamed this event for his failure to ascend to the position of Archbishop of Canterbury. Controversy occurred in his personal life as well, as members of his family became Roman Catholics (the Anglican Church was resisting an increasing role for both mainstream Protestantism and Roman Catholicism).

Science had always interested Wilberforce; he considered himself an amateur natural historian, and he had served as vice-president of the British Association for the Advancement of Science (BAAS). Wilberforce's review of *On the Origin of Species*, published in the *Quarterly Review* in 1860, attracted widespread attention, including from Darwin, who called it "uncommonly clever." Wilberforce rejected Darwin's theory because of its theological implications.

Wilberforce also condemned evolution at the annual meeting of the BAAS in July 1860. One of the last events of the week-long meeting was a discussion of Darwin's influence on social progress by American John Draper, whose *History of the Conflict Between Religion and Science* (1874) was among the first books to use a conflict metaphor to discuss religion and science. After consulting Darwin's foe Richard Owen (who had written a scathing and personal critique of *Origin*), Wilberforce made arrangements to speak after Draper's presentation. As word spread about Wilberforce's upcoming talk, hundreds of spectators gathered to see him "smash Darwin." Thomas Huxley had not planned to attend the event, even though he knew Wilberforce would be speaking, and only decided to go after being goaded by Robert Chambers (author of the *Vestiges of the Natural History of Creation*.) Years before, Wilberforce had led the attack on *Vestiges*.

The "Huxley–Wilberforce debate" has become legendary, aided by the lack of a transcript of the proceedings. Following Draper's long presentation, Wilberforce argued that humans are the product of divine creation. He allegedly then asked Huxley if he was related to apes on his grandmother's or grandfather's side. Huxley rose, while whispering, according to legend, "The Lord hath delivered him into mine hands," and responded that he was not ashamed of being related to apes, but was ashamed of being associated with people who use "aimless rhetoric" and "skilled appeals to religious prejudice." Whether these statements were actually said, and who "won" the debate is unclear. We do know that in letters written by Huxley and Wilberforce soon after the confrontation, both claimed victory. However, Joseph Hooker, a friend and supporter of Darwin, claimed that he actually bested Wilberforce (a statement corroborated by others in attendance), and felt that Huxley had failed to "put the matter in a form or way that carried the audience."

Although the "debate" accomplished little for either science or religion, the event was significant because it was a public refusal by the scientific community to allow the Church to dictate matters of science. As this trend continued during the later part of the 1800s, the "debate" was appropriated and used to further this separation. Most published discussions of the event did not appear until twenty or more years later.

Wilberforce's reputation was unharmed by his confrontation with Huxley. In 1869, he was appointed Bishop of Winchester, and remained busy with Church matters, including a revised translation of the New Testament. When Lord Palmerston suggested that Charles Darwin be knighted, Wilberforce opposed the idea, and the proposal was withdrawn.

In the early 1870s, Wilberforce suffered a series of heart attacks, which only partially slowed his work schedule. In 1873, Wilberforce—one of the most prominent theologians of the nineteenth century—was injured in a riding accident, and died a few days later. He is buried in Lavington, next to his wife Emily.

For More Information: Meacham, S. 1970. *Lord Bishop: The Life of Samuel Wilberforce*. Cambridge, MA: Harvard University Press.

GEORGE WILLIAMS (b. 1926)

I am convinced that [evolution] is the light and the way.

George Christopher Williams was born on May 12, 1926. After receiving his PhD from UCLA in 1955, he began a distinguished career as an ichthyologist and evolutionary biologist. Williams taught biology for more than twenty-five years at the State University of New York at Stony Brook.

Williams is best known for his 1966 book *Adaptation and Natural Selection*, in which he argued that almost every aspect of biology resulted from natural selection working on individuals. In doing so, Williams criticized V.C. Wynne-Edwards' ideas about "group selection," which claimed that selection might act for the good of the group instead of the good of an individual. Stephen Gould described Williams' *Adaptation and Natural Selection* as "the founding document for Darwinian fundamentalism."

Williams, who contributed significantly to the "gene's eye view" of evolution that was later developed by W.D. Hamilton and others, was one of the first biologists to deplore the societal implications of the cruelty of natural selection. Like Thomas Huxley, Williams claimed that societal progress depends not on imitating or denying natural selection, but instead on combating natural selection.

Williams also realized that evolutionary theory has important practical implications for society. Williams' and Randolph Nesse's *Why We Get Sick: The New Science of Darwinian Medicine* (1995) discussed the importance of understanding the adaptive significance of the body's response to disease. Williams argued that because medicine compensates for the failings of our adaptations, physicians need a thorough understanding of evolutionary biology, and that this understanding of evolution could lead to better treatments.

Williams, who shared the Crafoord Prize in 1999 with Ernst Mayr and John Maynard Smith, was a member of the National Academy of Sciences, and was awarded its Elliot Medal. In 1998, he served as president of the Society for the Study of Evolution and was named an Eminent Ecologist by the Ecological Society of America.

TOM WILLIS

Thompson "Tom" Willis is President of the Creation Science Association for Mid-America, an antievolution group that urged members of the Kansas Board of Education to delete evolution and cosmology from its science standards in 1999. Willis claimed that "Christians or anybody who teaches evolution as science is likely to be causing harm. Some of [the teachers] are evil." Willis led yearly "creation safaris" to see blasphemous exhibits at The University of Kansas' Natural History Museum. Willis' books include *Real Scientists Just Say NO! (to Evolution)* and *Ape Men: Science or Myth* (co-authored with David Brown).

WILLOUGHBY v. STEVER

Willoughby v. Stever (1973) established that government agencies such as the National Science Foundation can use tax money to disseminate scientific findings, including evolution.

Willoughby began in 1972 when William Willoughby, an evangelist and religious writer for the *Washington Evening Star*, sued H. Guyton Stever (the director of the National Science Foundation) and others for funding the proevolution textbooks produced by the Biological Sciences Curriculum Study (BSCS). Willoughby, who claimed to be acting "in the interest of forty million evangelical Christians," argued that the government was establishing "secular humanism" as "the official religion of the United States" and claimed that creationists should receive the same amount of money as BSCS to promote creationism. Willoughby received much public support, but the U.S. District Court in Washington, DC ruled that (1) books supported by taxes allocated to the National Science Foundation disseminate scientific findings, not religion, and (2) the First Amendment does not allow the state to demand that teaching be tailored to particular religious beliefs. Willoughby's case ended in February of 1975, when the U.S. Supreme Court refused to hear his appeal. Willoughby later edited *The Crusader: The Voice of Religious Freedom*, the official publication of the Religious Freedom Crusade.

ALLAN WILSON

Allan Wilson, along with Vincent Sarich, used analyses of DNA to show in 1967 that humans are more closely related to African apes than those apes are to orangutans. Four years later, Wilson and Mary-Claire King estimated that chimps and humans share more than 98 percent of their genes. A difference of 1.2 percent (as indicated by a more recent study) translates into 36 million base pairs.

EDWARD O. WILSON (b. 1929)

The final decisive edge enjoyed by scientific naturalism will come from its capacity to explain traditional religion, its chief competition, as a wholly material phenomenon. Theology is not likely to survive as an independent intellectual discipline.

Edward O. Wilson

Edward Osborn (E.O.) Wilson was born on June 10, 1929, in Birmingham, Alabama, and decided as a boy that he wanted to be a biologist. Wilson lost sight in his right eye in a boyhood fishing accident and, as a teen, suffered some hearing loss due to a hereditary condition. These challenges prompted him to decide that, if he were to become a biologist, it would make sense to study organisms that he could see close up. By the time Wilson was in high school, he knew that he wanted to study ants.

Wilson earned a BA in biology in 1949 and a master's degree in 1950, both from the University of Alabama. He published his first scientific paper (about ants) while still an undergraduate, and in 1955 earned a PhD in biology from Harvard University. Wilson joined the Harvard faculty in 1956, eventually becoming curator of entomology at Harvard's Museum of Comparative Zoology in 1973, and the Frank B. Baird, Jr. Professor of Science in 1976. Wilson is now Pellegrino University Professor Emeritus, having retired from Harvard in 1996. He married Irene Kelley in 1955 and has one daughter.

Wilson is the author or coauthor of hundreds of scientific papers, and his work has influenced several areas of biology, particularly evolutionary biology and ecology. But Wilson is most well-known, both inside and outside of the scientific community, for his examinations of social behavior (including that of humans); his efforts to protect biodiversity; and his analysis of the relationship between religion and science. His many awards include the National Medal of Science (1977) and two Pulitzers for non-fiction (*On Human Nature*, 1979; *The Ants*, 1990, with Bert Hölldobler).

Wilson, one of the world's foremost myrmecologists, has been interested in all aspects of ant biology; his years of studying ant colonies eventually produced the book *The Insect Societies*, in which Wilson argues that the abilities and accomplishments of these "little things that run the world" are on a par with those of humans. He also proposed that much of what is learned from other species can be applied to understanding ourselves.

Wilson applied the work by evolutionary biologist W.D. Hamilton and others to animal behavior, ultimately producing the book *Sociobiology: The New Synthesis* (1975). *Sociobiology* primarily deals with nonhuman animals, but his claim that the question "is no longer whether human social behavior is genetically determined; it is to what extent" produced controversy. Several prominent scientists and teachers—including Stephen Jay Gould and Richard Lewontin, both colleagues of Wilson at Harvard—equated Wilson's ideas with the those that led to forced sterilization and restricted immigration laws, racism, and even the Holocaust. Wilson understood that "dislike of human sociobiology" is a powerful force, and one that he believes is at the core of antievolution efforts. In 1978, Wilson's examination of human biology was extended in his book *On Human Nature*, which blended social and biological understandings of human behavior. The book was less controversial than *Sociobiology*, and some people even grumbled that Wilson had now taken an overly conciliatory position.

Wilson's claim that human behavior is at least partly a product of evolution led him to identify adaptive traits that could be used to improve society. One of these characteristics he labeled *biophilia*: the need for humans to be surrounded by and interact with the natural world. Wilson linked this trait (most notably in his 1984 book *Biophilia*) with preserving biodiversity. Wilson has also claimed that religion is an adaptive trait. According to Wilson, groups of individuals that share a religion were more successful, thereby perpetuating that religious belief. But as science has increasingly taken over that explanatory power, the tensions between religion and science have increased. Unlike some scientists (e.g., Richard Dawkins), he has called for religion and science to cooperate to achieve common goals such as preserving life on the planet. In *Consilience* (1998), Wilson argued that there is only one class of explanation (science) and that "the central idea of the consilience world view is that all tangible phenomena...are based on...the laws of physics."

For More Information: Wilson, E.O. 2000. *Sociobiology: The New Synthesis—25ᵗʰ Anniversary Edition*. Cambridge, MA: Harvard University Press.

ALEXANDER WINCHELL (1824–1891)

Alexander Winchell was an eminent American scientist who helped found the American Geological Society (he was its first president) and the journal *American Geologist*. Winchell praised the work of James Dwight Dana (1813–1895) and Edward Hitchcock, and was a theistic evolutionist. Winchell's 1878 book *Adamites and Pre-Adamites* argued for the existence of humans before Adam, and prompted Vanderbilt University to fire him "for holding questionable views on Genesis." Winchell defined and introduced the Mississippian subseries of the Carboniferous Period of the geological column.

GERALD WINROD

Gerald Burton Winrod was a Kansan who formed the Defenders of the Christian Faith in 1926 to continue William Jennings Bryan's antievolution crusade. The Defenders, who advocated ultraconservative views of race and religion, included more than 3,000 members and recruited speakers who toured the Midwest as the "Flying Fundamentalists." Winrod, who ran unsuccessfully for a seat in the U.S. Senate, often worked with Frank Norris, and in 1930 Billy Sunday headlined the Defenders' World-Wide Congress. Winrod's magazine *Defender* often published articles by antievolutionists such as William Riley and Frank Norris. Winrod denounced evolution, and was one of the most vocal anti-communist, anti-Semetic, and pro-Nazi fundamentalists. By 1930, Winrod's organization had shifted its focus from evolution to the "Negro menace."

KURT WISE (b. 1959)

To accept the entire evolutionary model would mean one would have to reject Scripture. And because I came to know Christ through Scripture I couldn't reject it ... Scripture trumps interpretations of physical data.

Kurt Patrick Wise received a BA in geology from the University of Chicago, and an MA and PhD in geology in 1989 from Harvard University. Wise's doctoral degree (about molluscan systematics) was directed by Stephen Jay Gould. After graduating, Wise taught for many years at Bryan College in Dayton, Tennessee, where he directed the Center for Origins Research and Education.

Wise, who Richard Dawkins has called the "most honest" creationist, is a Young-Earth creationist "because that is my understanding of the Scripture." According to Wise, "the fact that God created the universe is not a theory—it's true. Some issues—such as creation, a global Flood, and a young age for the Earth—are determined by Scripture, so they are not theories." Wise's views are based on his religious faith, and are not affected by evidence. As Wise has noted, "If all the evidence in the universe turns against creationism ... I would still be a creationist because that is what the Word of God seems to indicate." At the Answers in Genesis Creation Museum, Wise claims that Behemoth and Leviathan from Job were probably dinosaurs.

Wise, who claims that "evidence from Scripture is by far the best evidence for creation," hopes "to replace the evolutionary tree with the creationist orchard, separately created, separately planted by God." In *Faith, Form, and Time* (2002), Wise notes that nature "must be reinterpreted ... from a Christian perspective so all these things can be taken captive under the mind of Christ." Wise acknowledges that mainstream scientists consider his claims to be "absolute bunk." As he has noted, "If humans really date back [as far as geologists claim], and Adam lived far enough in the past to be their ancestor, then the genealogical record of Genesis 5 is wrong, and thus the Bible and its author, God, are wrong."

Unlike most other creationists, Wise—who has taught at the Institute for Creation Research and has assisted their explorations of the

Grand Canyon—admits that Darwin's theory is "a very good theory." However, Wise also claims that "if you don't believe in a young Earth, you really cannot believe in the truth of much of Genesis 1–11." Wise understands that most Christians accept that Earth is much older than 6,000 years, yet he maintains that many biblical doctrines—including marriage and the end of times—depend on belief in a young Earth. As Wise has noted, "The most important thing is that you ought to be able to trust your God and the claims the Bible makes...To trust the scientists...[is] trusting in man's reason rather than God."

In 2006, Wise moved from Bryan College to the Southern Baptist Theological Seminary, where he leads the Center for Theology and Science. Wise, who replaced intelligent design activist William Dembski, also served as a consultant for the Answers in Genesis Creation Museum.

CARL WOESE (b. 1928)

If one's representation of reality takes evolution to be irrelevant to understanding biology, then it is one's representation, not evolution, whose relevance should be questioned!

Carl Woese was born in Syracuse, New York, on July 15, 1928. He graduated from Amherst College with a degree in physics and mathematics in 1950, and earned a PhD in biophysics from Yale University in 1953. After working at the General Electric Research Laboratory (1960–1963), Woese secured a faculty position at the University of Illinois at Urbana in 1964 and remained there throughout his career. He now holds the Stanley O. Ikenberry Endowed Chair in the Department of Microbiology.

Woese describes himself as a "molecular biologist turned evolutionist" and has frequently disparaged the reductionism in biology that began in the 1960s with advances molecular biology. As of the 1960s, microbes—which comprise most of the biomass on Earth and which represented life for most of Earth's history—had not been integrated into the tree of life because little

evolutionary information can be gleaned from comparing the morphology of bacteria. Starting in 1966, Woese used ribosomal RNA from several bacterial species to create an evolutionary tree for these species.

In 1976, Woese examined a methanogenic ("methane-producing") bacterial species. When he discovered that the DNA sequences from this organism were unlike those from the other sixty species he had studied, Woese concluded that "these things aren't even bacteria." Woese had discovered a third major form of life, unlike either bacteria or eukaryotes. In 1977, his revised classification of life proposed three major groups (domains), the eubacteria, the archaebacteria ("ancient bacteria," later changed to archaea), and the eukaryotes. Woese's findings were corroborated by other labs, and his conclusions were eventually accepted. Woese's work is now considered among the most significant of the twentieth century, and colleagues have nominated him for a Nobel Prize. He was awarded a MacArthur "genius" award in 1984, elected to the National Academy of Sciences in 1998, and awarded the National Medal of Science in 2000. In 2003, he also won the Crafoord Prize, which is viewed as an unofficial Nobel Prize for areas not considered by the Nobel committee.

Not everyone, however, has embraced Woese's new tree of life. Harvard biologist Ernst Mayr, who developed the biological species concept, declared that Woese had "greatly overemphasized the importance of the archaea" and that "there was no justification...for classifying them as a new domain." Mayr argued that morphological data clearly indicate that the eubacteria and archaea should be classified together as prokaryotes, and that Woese was making too much of molecular differences that were minor compared to these shared morphological similarities. Woese responded by suggesting that Mayr, a zoologist, failed to recognize that the Earth is a "microbial planet" where multicellular organisms are a recent and perhaps insignificant addition.

Woese later discovered that the three domains diverged long ago. He also proposed that before this split, life did not begin as a single, genetically discrete cell, but went through a period where "precells" exchanged genetic material through horizontal gene transfer and functioned as a larger semi-integrated gene pool. Evolution at this stage would therefore be "communal, not individual." Once precells exhibited the process of translation (conversion of genetic information into proteins), a tipping point was reached which favored vertical over horizontal gene transfer.

Woese claims that life on Earth has three common ancestors and that the early evolutionary unit was a supergenomic entity. Creationists have frequently used Woese's conclusions to claim that science has rejected the idea of common descent and that Darwinian evolution has been disproven.

WILLIAM HYDE WOLLASTON (1659–1724)

William Hyde Wollaston, the discoverer of palladium and rhodium, in 1831 became the namesake for the Wollaston Medal, the highest award given by the Geological Society of London. The first Wollaston Medal was given to William Smith (1831), and subsequent winners included several people in the evolution-creationism controversy, including Agassiz (1836), Owen (1838), Buckland, (1848), Sedgwick (1851), Darwin (1859), Lyell (1866), Huxley (1876), Osborn (1926), Broom (1949), and Holmes (1956).

JAMES WOODROW

James Woodrow was a former student of Louis Agassiz (and uncle of future president Woodrow Wilson) who, while a professor of Natural Science at the Columbia (South Carolina) Theological Seminary, argued in the July 1884 issue of *Southern Presbyterian Review* that Adam's soul had been imparted by God but that his body had probably been created by evolution. Woodrow claimed evolution is compatible with Christianity, that science does not contradict the literal truth of the Bible, and that the study

of nature helped him appreciate God's plan, but the seminary asked him to resign. In 1896, when Woodrow refused to resign, he was fired. Four years later, the Presbyterian General Assembly voted 139-31 to uphold Woodrow's firing, declaring that Adam had been formed from dirt "without any natural animal parentage of any kind."

WORLD'S CHRISTIAN FUNDAMENTALS ASSOCIATION (Est. 1919)

The World's Christian Fundamentals Association (WCFA) was founded by William Bell Riley, and its inaugural conference in Philadelphia in May 1919, attracted more than 6,000 conservative Christians. The WCFA had ambitious goals, including the accreditation of Bible schools, the establishment of a foundation to rival the Rockefeller Foundation, the formation of fundamentalist theological seminaries and conservative colleges in every state and Canadian province, and the eradication "not by regulation, but by strangulation" of the teaching of evolution. Although many modernist Protestants tried to reconcile science with their Christian beliefs, the WCFA campaigned against the science that they believed threatened their faith.

By 1920, the WCFA—which was headquartered in Riley's First Baptist Church in Minneapolis—had organized more than 100 "Conferences on Christian Fundamentals" throughout the United States and Canada. In 1922, the WCFA declared, "As taxpayers we have a perfect right... to demand the removal of any teacher who attempts to undermine... the Christian faith of pupils." The following year, while meeting at Frank Norris' church in Texas, the WCFA staged mock trials (and convictions) of biology teachers who taught evolution (Figure 82). By the time the WCFA met in Minneapolis in June 1924, Riley's goal was to enroll 100,000 new members. That same year, Riley—who had twice "personally pled [with Bryan] to accept the presidency" of the WCFA—met with Bryan, Norris, and other fundamentalists at Memphis' Hotel Claridge, at which time Riley "secured

82. The World's Christian Fundamentals Association (WCFA), which was organized in 1919 by William Bell Riley (Figure 67), tried to eradicate the teaching of evolution "not by regulation, but by strangulation." In this photograph from 1923, Frank Norris (left) welcomes William Jennings Bryan to the annual conference of the WCFA, which was being held at Norris' church in Ft. Worth, Texas. When Bryan died, Norris sold copies of this photo for 25 cents. The WCFA was responsible for Bryan's participation in the Scopes Trial. (*Arlington Baptist College*)

[Bryan's] partial pledge to [become president of the WCFA at the next WCFA] convention." At that meeting, Bryan also endorsed the WCFA's doctrinal statement and donated $500 to the organization.

On the same day that John Scopes agreed to be tried for teaching evolution in Dayton, Tennessee, the WCFA—meeting in Memphis with the Southern Baptist Convention—thanked Governor Peay and the state legislature for "prohibiting the teaching of the unscientific, anti-Christian, atheistic, anarchistic, pagan rationalistic evolutionary theory." On May 13, 1925, Riley asked Bryan to represent the WCFA at the upcoming Scopes Trial. When Bryan agreed

to serve the WCFA "without compensation" in the "battle royal between the Christian people of Tennessee and the so-called scientists," the Dayton trial became a national story. Norris considered the trial to be "the greatest opportunity ever present to educate the public," and claimed that it would "accomplish more than ten years' campaigning."

At its annual meeting in Atlanta in 1927, the WCFA began formulating an antievolution bill that was to be presented to legislatures in every state, as well as in Europe, China, and South America. This meeting was the high-water mark of the WCFA, for its membership soon began to decline. In 1930, Riley resigned from the WCFA, after which the WCFA conference included no scheduled speeches about evolution. Within a few years, the WCFA disappeared.

GEORGE FREDERICK WRIGHT (1838–1921)

George Frederick Wright was an American cleric, geologist, and coauthor (with Asa Gray) of *Darwiniana* (1876). By the 1870s, Wright was the leader of Christian Darwinists, and claimed that the "Creator first breathed life into one, or more probably, four or five, distinct forms." Early in his career, Wright—like Gray—was a theistic evolutionist who reconciled religion and evolution, and he believed that God's purposes were understandable. In *Studies in Science and Religion* (1882), Wright urged scientists to accept evolution because it agrees with observed facts, as does Christianity. Wright embraced Arnold Guyot's belief in day-age creationism.

Wright, who believed that Darwinism was "the Calvinistic interpretation of nature," was asked by A.C. Dixon to contribute a chapter about evolution to *The Fundamentals*; that chapter, titled "The Passing of Evolution," appeared in Volume 7 and was the most detailed analysis of evolution in *The Fundamentals*. Although Wright and Gray helped calm some people's fears about evolution, Wright later condemned Darwin, Lyell, and geologists while promoting day-age creationism and a recent, supernatural origin of humans in Asia.

SEWALL WRIGHT (1889–1988)

I am especially interested in the question as to how far there is subdivision of species into smaller local strains differentiated in the random fashion expected of inbreeding, instead of in adaptive ways by natural selection ... such a condition is the most favorable for progressive evolution of the species as a single group.

Sewall Green Wright was born in Melrose, Massachusetts, on December 21, 1889. In 1906, Wright entered Lombard College. His initial, although lukewarm, interest was chemistry, but in his senior year he chose biology when he took his first course in the subject. The mathematics he learned from his father (an instructor at Lombard) represented the extent of his formal training in the subject, an astounding situation given the mathematical sophistication Wright would later demonstrate. Wright's father also taught him surveying, which led to a job helping to build a railroad through South Dakota.

After graduating from Lombard in 1911, Wright was awarded a $250 scholarship to the University of Illinois for graduate training. He earned an MS from Illinois in 1912, submitted his thesis for publication, and went on to Harvard University to pursue a PhD with geneticist William Castle. Castle, who worked primarily with mammals, encouraged all of his students to work with different species, and it was here that Wright began studying the genetics of guinea pigs. Wright believed that science was concerned only with the "external side" of the universe, while the mind was completely internal and thus not open to scientific inquiry. He viewed "mind" as a fundamental aspect of nature (on a par with matter) that was not simply an emergent property of biology. Most colleagues reacted coolly to this proposal, but because Wright was never dogmatic, it did not interfere with his work or his relationships with others.

In 1915, Wright earned his PhD and became Senior Animal Husbandman at the USDA Animal Husbandry Division in Washington, DC. Three main areas of Wright's work—statistics, animal breeding, and inheritance—were developed during his ten years with the USDA. (The fourth, integration of his earlier work into evolutionary biology, would wait until he moved to the University of Chicago.) Wright's studies of inbreeding introduced his inbreeding coefficient, which became a central concept in population genetics and animal breeding. In 1921, Wright met Louise Williams, an instructor at Smith College, while both she and Wright were at Cold Spring Harbor Laboratory. The two were married in 1921 and had three children. Louise Wright died in 1975.

In 1926, Wright moved to the University of Chicago. There, Wright became convinced that adaptive evolution depends on particular, favorable combinations of genes. If true, however, this creates a problem: if a certain favored combination of genes is present within a population, but another combination actually has higher fitness, how can the population evolve towards the fitter set if the only way to get there is by breaking up the current set, which would lead to an intermediate state of lower fitness?

Wright proposed that populations often exist as smaller, semi-isolated units (demes) that are more likely to experience random changes in genetic composition via genetic drift. These random changes can move the genetic makeup of a deme through the intermediate fitness valleys and up to peaks of higher fitness; individuals in these now better-adapted demes would reproduce faster and pass on the new combination of genes to the larger population. Wright's proposal of adaptive evolution being mediated through periodic random genetic changes was dubbed the "shifting balance" model, calling the totality of possible fitness peaks and valleys a population's "adaptive landscape."

Wright's model was controversial when introduced and remains so. One of its chief opponents was Ronald Fisher, who in *The Genetical Theory of Natural Selection* (1930) espoused a view of evolution based on selection operating effectively within large populations in which random genetic changes and gene interactions were minor and unimportant complications. Fisher and Wright feuded until Fisher's death in 1962. Hence, the main architects of the early phase of the modern synthesis—Wright, Fisher, and Haldane—did not work in concert to produce a unified theory. Wright, undeterred by Fisher's criticisms, developed the concept of effective population size, which is the size of an ideal population that would experience the same degree of random change as experienced by a population under study. Wright also developed his *F*-statistics, based on his earlier work on inbreeding that quantified the hierarchical nature of a population's genetic structure. These perspectives have become central to contemporary population genetics.

In 1955, Wright moved to the University of Wisconsin at Madison, from where he retired in 1960. At Madison, he continued writing, but finally abandoned research because of his worsening eyesight. At the age of 70, he started summarizing and compiling his life's work, and from 1968 to 1978, he produced the four-volume *Evolution and the Genetics of Populations*. Stephen Jay Gould pointed out that because Wright's publications were filled with mathematical treatments of the topic under consideration, Wright's original papers were seldom read. (Theodosius Dobzhansky, who collaborated with Wright for several years, put it more succinctly: "He is a remarkably difficult writer.") Consequently, Wright's ideas have often been discussed in simplified form, as if nonadaptive change via genetic drift was the only evolutionary mechanism he believed was important, leading some to dismiss the importance of his contributions. Wright responded that "I have never attributed any evolutionary significance to random drift except as a trigger that may release selection toward a higher selective peak."

On March 3, 1988, Wright died of a pulmonary embolism that resulted from a broken pelvis he received a few days earlier after falling on an icy sidewalk. Wright earned numerous awards including election to the National Academy of Sciences (1934), the Elliot Award (1947) and the Kimber Award (1956) from the National Academy of Sciences, the National Medal of Science (1966), the Darwin Medal from the Royal Society of London (1980), and the Balzan Prize (1984). Just three days before his death at the age of 98, he received the reprints of what would be his final paper.

For More Information: Provine, W.B. 1986. *Sewall Wright and Evolutionary Biology*. Chicago, IL: University of Chicago Press; Wright, S. 1968–1978. *Evolution and the Genetics of Populations*. Chicago, IL: University of Chicago.

WRIGHT v. HOUSTON INDEPENDENT SCHOOL DISTRICT

Wright v. Houston Independent School District (1972), the first lawsuit initiated by creationists, established that (1) the teaching of evolution does not establish religion, (2) the state has no legitimate interest in protecting particular religions from scientific information "distasteful to them," and (3) the free exercise of religion is not accompanied by a right to be shielded from scientific findings incompatible with one's beliefs. *Wright* is named for Rita Wright, a student in the Houston Independent School District. After filing the lawsuit on Wright's behalf in November, 1970, Wright's mother (Leona Wilson) had trouble retaining an attorney for her case.

X-CLUB

The X-Club was founded in 1864 by George Busk, Joseph Hooker, Thomas Huxley, and others as a small dinner-club for scientists who wanted to do research without the interference of religious dogma. When a curious journalist asked what the club did, a professor replied: "They run British science and, on the whole, they don't do it badly." The last surviving member was John Lubbock, who died in 1913. Charles Darwin was never a member of the club.

XENOPHANES (560-478 BC)

Xenophanes was a Greek philosopher who believed that fossils showed that land had once been submerged. He rejected the claim that gods resembled humans, noting that if oxen had gods, then those gods would resemble oxen.

Y

EVELLE J. YOUNGER

Evelle J. Younger was California's Attorney General who in 1975 ruled that the state "could not repair its entrance into the sectarian arena by balancing its 'religious' treatment of evolution with a 'religious' treatment of creation." This ruling was one of the many legal decisions associated with challenges to California's *Science Framework*, which gave equal recognition to creationism and evolution, and required textbooks to do the same.

Appendix: A Guide to the Sites of the Scopes Trial

Thanks largely to the efforts of Bryan College professor emeritus Richard Cornelius (b. 1934), many of the sites related to the Scopes Trial have been preserved and marked with "Scopes Trial Trail" plaques (designated with an asterisk in the legend below).

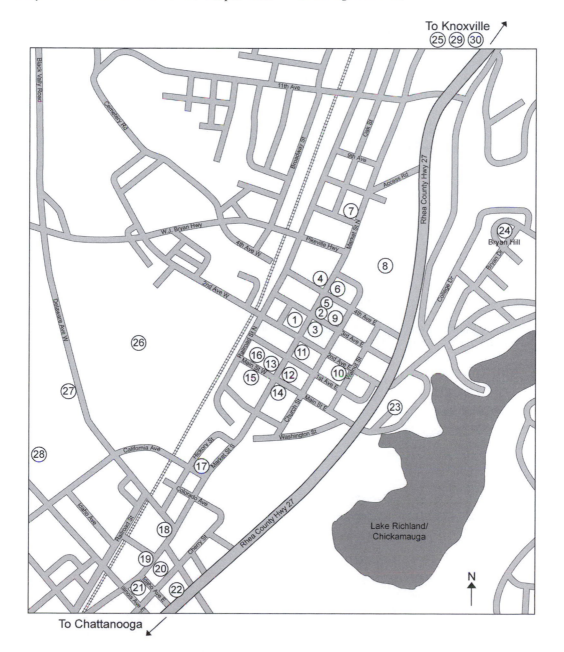

① The Rhea County Courthouse and Scopes Trial Museum (Figure 25) is in the center of Dayton. Scopes' trial was held in the second-floor courtroom, which still contains several items from the famous trial (e.g., the judge's desk, dais rail).*

② F.E. Robinson's home on the corner of 3rd and Market Streets was home of "The Hustling Druggist," who helped initiate the Scopes Trial.*

③ Former home of F.E. Robinson was occupied by photographers during the Scopes Trial.

④ Darwin-Cunnyngham home on Market Street housed journalists during the Scopes Trial.

⑤ McKenzie Law Office, which is adjacent to the F.E. Robinson home, was formerly used by Jim McKenzie, the nephew of J.G. McKenzie and grandson of Ben McKenzie. In 2007, Jim McKenzie was a judge in the Rhea Family Court.*

⑥ W.C. Bailey's boardinghouse on the northeastern corner of 4th and Market Streets was where John Scopes lived when he worked in Dayton. Scopes' father, journalist Bugs Baer, and briefly the chimpanzee Joe Mendi also stayed at the house during Scopes' trial.*

⑦ A.M. Morgan home at the southwest corner of 7th Street and the adjacent alley was where journalist H.L. Mencken lived during the Scopes Trial. After Scopes' trial, Morgan was a founder of Bryan College.

⑧ Rhea County High School was where John Scopes taught and coached in 1924–1925. Bryan College used the building from 1930–1935.*

⑨ Ballard-Bailey house at the northwest corner of 3rd and Church Streets was where chimpanzee Joe Mendi stayed during the Scopes Trial after being evicted from Bailey's boardinghouse.

⑩ Luke Morgan home, located at the southwest corner of 2nd and Walnut Streets, is where Clarence Darrow and his wife Ruby stayed during the Scopes Trial. Luke Morgan, a former student of John Scopes, testified during the trial.

⑪ Morgan Furniture Company on Market Street housed reporters during the Scopes Trial. The business has been open since 1909.*

⑫ Bailey Hardware housed more than 100 reporters during the Scopes Trial. Until recently the building—on Market Street between 1st and Main Streets—housed an antique store.*

⑬ Thomison Hospital, Wilkey Barbershop, and Rogers Pharmacy were all in this area. Rogers worked at Robinson's Drug Store during the Scopes Trial, and later opened a pharmacy here. West of Rogers Pharmacy was the Wilkey Barbershop. On May 19, 1925, barbers Virgil Wilkey and Thurlow Reed staged a fake fight at the courthouse with George Rappleyea to promote the upcoming Scopes Trial. Above Rogers Pharmacy was a hospital operated by Walter Agnew Thomison, whose father, Walter F. Thomison, was the attending physician at William Jennings Bryan's death. A sign for Thomison's office remains on the wall of the building near the intersection of Main and Market Streets.

⑭ Hicks Law Office, located in the second lot from the southeast corner of Main and Market Streets, was used by Scopes prosecutors Herbert and Sue Hicks.

⑮ Robinson's Drug Store was where several of Dayton's businessmen devised the Scopes Trial. Adjacent to the drug store was the three-story Aqua Hotel, where John Neal, John Raulston, Arthur Hays, Dudley Malone, and Clarence Darrow stayed, met, or ate during the trial.*

⑯ Cumberland Presbyterian Church was built two years after the Scopes Trial; F.E. Robinson was a member of that church. When Clarence Darrow returned to Dayton after the Scopes Trial and saw this church, he commented that "I guess I didn't do much good here after all." The church is no longer a Cumberland Presbyterian church.

⑰ First United Methodist Church was where William Jennings Bryan made his last public appearance. During Scopes' trial, the church at this site—the northwest corner of California and Market Streets—was a Southern Methodist church.*

⑱ Smith's Crossroads historical marker designates where, around 1820, Cherokee being moved to the southwest came through the area on "The Trail of Tears." This marker is at the southwest corner of Delaware and Market Streets.*

(19) Richard Rogers home was where William Jennings Bryan and his entourage stayed during and after the Scopes Trial. Bryan died in his sleep in the Rogers' home on July 26, 1925. Only the retaining wall of the property is as it was in 1925.*

(20) A.P. Haggard home was built across the street from the Richard Rogers home by A.P. Haggard, the father of Scopes prosecutor Wallace Haggard.*

(21) Walter F. Thomison home was built by Walter F. Thomison for Ella Darwin, his 16-year-old wife. Thomison's house is now called Magnolia House.*

(22) Broyles-Darwin home is on the National Historic Register, and during the Scopes Trial housed reporters. S.D. Broyles, who built the house in 1861, was the first resident of Smith's Crossroads.*

(23) Cedar Hill, the first hospital in Dayton, was built in 1929 by Walter Agnew Thomison. The building was used by Bryan College from 1932–1938 and 1967–1984.*

(24) Bryan College opened in 1930 as a memorial to Scopes prosecutor William Jennings Bryan. The campus includes several exhibits related to the Scopes Trial, and several of the college's founders were involved in the trial.*

(25) Dayton Coal & Iron Company is a former mining operation that was managed by George Rappleyea, an instigator of the Scopes Trial. The land is now a recreational area, but coke ovens remain visible.

(26) Blast furnaces of the Dayton Coal and Iron Company were at this site, which now is covered by sports fields.

(27) St. Genevieve's Academy at 449 Delaware Road was where some children of Dayton Coal and Iron Company employees were educated before the Scopes Trial. Today the school building—which was built in 1891—houses Fehns 1891 Restaurant.

(28) The Mansion was an eighteen-room house renovated by George Rappleyea to house several members of the defense team during the Scopes Trial. The house, which was atop a knoll over-looking the furnaces and company store of the Dayton Coal and Iron Company, had been vacant for more than a decade. Before he moved into the Morgan's home, Clarence Darrow stayed at the Mansion, and it was at the Mansion that Darrow and Kirtley Mather prepared for Darrow's questioning of William Jennings Bryan. In *Inherit the Wind*, several of the participants stayed at a hotel named "The Mansion."

(29) Buttram Cemetery just outside Dayton is where many of the participants in the Scopes Trial are buried.

(30) Dayton Drive-In Theater, which was 2.5 miles north of Dayton, was the site of the U.S. premiere of *Inherit the Wind*.

Bibliography

Alters, Brian, and Sandra Alters. *Defending Evolution in the Classroom: A Guide to the Creation/Evolution Controversy*. Sudbury, MA: Jones & Bartlett, 2001.

Ayala, Francisco J. *Genetics and the Origin of Species: From Darwin to Molecular Biology, 60 Years After Dobzhansky*. Washington, DC: National Academy of Sciences, 1997.

Badash, L. (Ed.). *Rutherford and Boltwood: Letters on Radioactivity*. New Haven, CT: Yale University Press, 1999.

Baker, Catherine, and James Miller (Eds.). *The Evolution Dialogues: Science, Christianity, and the Quest for Understanding*. Washington, DC: American Association for the Advancement of Science, 2006.

Baugh, Carl E. *Why Do Men Believe Evolution against All Odds?* Bethany, OK: Bible Belt Publishing, 1999.

Behe, Michael J. *Darwin's Black Box: The Biochemical Challenge to Evolution*. New York: Free Press, 1996.

———. *The Edge of Evolution*. New York: Free Press, 2007.

Berra, Tim M. *Evolution and the Myth of Creationism: A Basic Guide to the Facts in the Evolution Debate*. Stanford, CA: Stanford University Press, 1990.

Bird, Wendell. *The Origin of Species Revisited*. Nashville, TN: Thomas Nelson, 1991.

Bjornerud, Marcia. *Reading the Rocks: The Autobiography of the Earth*. Cambridge, MA: Westview Press, 2005.

Bowler, Peter J. *The Eclipse of Darwinism: Anti-Darwinian Evolution Theories in the Decades around 1900*. Baltimore, MD: Johns Hopkins University Press, 1983.

———. *The Non-Darwinian Revolution. Reinterpreting a Historical Myth*. Baltimore, MD: Johns Hopkins University Press, 1988.

———. *Charles Darwin: The Man and His Influence*. Cambridge, UK: Blackwell Scientific, 1990.

———. *Evolution: The History of an Idea*, 3rd ed. Berkeley, CA: University of California Press, 2003.

Browne, Janet. *Charles Darwin: A Biography*, 2 vols. New York: Knopf, 1995–2002.

Burkhardt, Richard W., Jr. *The Spirit of System: Lamarck and Evolutionary Biology*. Cambridge, MA: Harvard University Press, 1977.

Cadbury, D. *Terrible Lizard: The First Dinosaur Hunters and the Birth of a New Science*. New York: Henry Holt, 1995.

Campbell, John Angus, and Stephen C. Meyer. *Darwinism, Design and Public Education*. East Lansing, MI: Michigan State University Press, 2004.

Carpenter, Joel A. *Revive Us Again: The Reawakening of American Fundamentalism*. New York: Oxford University Press, 1997.

Carroll, Sean B. *Endless Forms Most Beautiful*. New York: Norton, 2006.

Chapman, Matthew. *40 Days and 40 Nights: Darwin, Intelligent Design, God, OxyContin®, and Other Oddities on Trial in Pennsylvania*. New York: HarperCollins, 2007.

Coleman, Simon, and Leslie Carlin (Eds.). *The Cultures of Creationism: Anti-Evolution in English-Speaking Countries*. Burlington, VT: Ashgate, 2004.

Conkin, P.K. *When All the Gods Trembled: Darwinism, Scopes, and American Intellectuals.* New York: Rowman and Littlefield, 1998.

Conway Morris, Simon. *Life's Solution: Inevitable Humans in a Lonely Universe.* New York: Cambridge University Press, 2003.

Corey, Michael A. *The God Hypothesis: Discovering Design in Our "Just Right" Goldilocks Universe.* Lanham, MD: Rowman & Littlefield, 2001.

Craig, G.Y., D.B. McIntyre, and C.D. Waterston. *James Hutton's Theory of the Earth: The Lost Drawings.* Edinburgh, Scottish Academic Press, 1978.

Dalrymple, G. Brent. *The Age of the Earth.* Stanford, CA: Stanford University Press, 1991.

Darwin, Charles. *On the Origin of the Species by Means of Natural Selection: Or, The Preservation of Favoured Races in the Struggle for Life.* London: J. Murray, 1859.

Davis, P., and D.H. Kenyon. *Of Pandas and People,* 2nd ed. Dallas, TX: Haughton Publishing, 1993.

Dawkins, Richard. *The Selfish Gene.* Oxford, UK: Oxford University Press, 1976.

———. *The Blind Watchmaker: Why the Evidence of Evolution Reveals a Universe Without Design.* New York: Norton, 1986.

———. *River Out of Eden: A Darwinian View of Life.* New York: Basic Books, 1995.

———. *Climbing Mount Improbable.* New York: Norton, 1996.

Dembski, W.A. *The Design Inference.* Cambridge, UK: Cambridge University Press, 1998.

———. *Intelligent Design: The Bridge Between Science and Theology.* Downers Grove, IL: InterVarsity, 1999.

———. *No Free Lunch: Why Specified Complexity Cannot be Purchased without Intelligence.* Lanham, MD: Rowman Littlefield, 2002.

———. *The Design Revolution: Answering the Toughest Questions About Intelligent Design.* Downers Grove, IL: InterVarsity, 2004.

——— (Ed.). *Uncommon Dissent: Intellectuals Who Find Darwinism Unconvincing.* Wilmington, DE: ISI Books, 2004.

Dembski, William, and Michael Ruse (Eds.). *Debating Design: From Darwin to DNA.* New York: Cambridge University Press, 2004.

Dennett, Daniel C. *Darwin's Dangerous Idea: Evolution and the Meaning of Life.* New York: Simon & Schuster, 1995.

Diamond, Jared. *The Third Chimpanzee: The Evolution and Future of the Human Animal.* New York: HarperCollins, 1992.

Dobzhansky, Theodosius G. *Genetics and the Origin of Species.* New York: Columbia University Press, 1937.

Dugatkin, L.A. *The Altruistic Equation: Seven Scientists Search for the Origins of Goodness.* Princeton, NJ: Princeton University Press, 2006.

Eldredge, N. *The Monkey Business: A Scientist Looks at Creationism.* New York, Washington Square, 1983.

———. *The Triumph of Evolution and the Failure of Creationism.* New York: Freeman, 2000.

———. *Darwin: Discovering the Tree of Life.* New York, W.W. Norton, 2005.

Falwell, Jerry. *The Fundamentalist Phenomenon: The Resurgence of Conservative Christianity.* Garden City, NY: Doubleday, 1981.

Fastovsky, David E. *The Evolution and Extinction of the Dinosaurs.* New York: Cambridge University Press, 1996.

Fenton, C.L., and M.A. Fenton. *Giants of Geology.* Garden City, NY: Doubleday, 1952.

Fischer, Robert B. *God Did It, But How?* 2nd ed. Ipswich, MA: American Scientific Affiliation, 1997.

Forrest, Barbara, and Paul R. Gross. *Creationism's Trojan Horse: The Wedge of Intelligent Design.* New York: Oxford University Press, 2004.

Fowler, Thomas B., and Daniel Kuebler. *The Evolution Controversy: A Survey of Competing Theories.* Grand Rapids, MI: Baker Academic, 2007.

Futuyma, Douglas J. *Evolution.* Sunderland, MA: Sinauer, 2005.

———. *Science on Trial: The Case for Evolution,* rev. ed. New York: Pantheon Books, 1995.

Geisler, N. *Creation and the Courts: Eighty Years of Conflict in the Classroom and the Courtroom.* Wheaton, IL: Crossway Books, 2007.

Gish, Duane T. *Evolution: The Fossils Say No!* San Diego, CA: Creation-Life, 1972.

———. *Evolution: The Fossils Still Say No!* San Diego, CA: Creation-Life Publishers, 1985.

Gould, Stephen Jay. *Ever Since Darwin: Reflections in Natural History.* New York: Norton, 1977.

———. *The Panda's Thumb: More Reflections in Natural History.* New York: Norton, 1980.

Gunn, Angus. *Evolution and Creationism in the Public Schools: A Handbook for Educators, Parents and Community Leaders.* Jefferson, NC: McFarland & Company, 2004.

Ham, Kenneth. *The Lie: Evolution.* El Cajon, CA: Master Books, 1987.

Hamilton, W.D. *Narrow Roads of Gene Land.* New York: Oxford University Press, 1997.

Haught, John F. *Responses to 101 Questions on God and Evolution.* New York: Paulist Press, 2001.

Hayward, James L. *The Creation/Evolution Controversy: An Annotated Bibliography*. Metuchen, NJ: Scarecrow, 1998.

Herbert, S. *Charles Darwin: Geologist*. Ithaca, NY: Cornell University Press, 2005.

Hunter, Cornelius G. *Darwin's Proof: The Triumph of Religion Over Science*. Grand Rapids, MI: Brazos Press, 2003.

Huxley, Julian. *Evolution: The Modern Synthesis*. London: Allen and Unwin, 1963.

Isaak, Mark. *The Counter-Creationism Handbook*. Westport, CT: Greenwood Press, 2005.

Jackson, Patrick W. *The Chronologers' Quest: The Search for the Age of the Earth*. Cambridge, UK: Cambridge University Press, 2006.

Jennings, W.H. *Storms Over Genesis: Biblical Battleground in America's Wars of Religion*. Minneapolis, MN: Fortress Press, 2007.

Johanson, D., B. Edgar, and D. Brill. *From Lucy to Language: Revised, Updated, and Expanded*. New York: Simon and Schuster, 2006.

Johnson, Phillip E. *Darwin on Trial*. Downers Grove, IL: InterVarsity, 1991.

———. *Defeating Darwinism By Opening Minds*. Downers Grove, IL: InterVarsity, 1997.

———. *The Wedge of Truth: Splitting the Foundations of Naturalism*. Downers Grove, IL: InterVarsity, 2000.

Kitcher, Philip. *Abusing Science: The Case against Creationism*. Cambridge, MA: MIT Press, 1982.

———. *Living with Darwin: Evolution, Design, and the Future of Faith*. New York: Oxford University Press, 2007.

Kofahl, Robert E. *The Handy Dandy Evolution Refuter*. San Diego, CA: Beta, 1977.

Larson, Edward J. *Summer for the Gods: The Scopes Trial and America's Continuing Debate Over Science and Religion*. New York: Basic Books, 1997.

———. *Evolution's Workshop: God and Science on the Galápagos Islands*. New York: Basic Books, 2001.

———. *Evolution: The Remarkable History of a Scientific Theory*. New York: Modern Library, 2004.

Leeming, David, and Margaret Leeming. *A Dictionary of Creation Myths*. New York: Oxford University Press, 1994.

LeGrand, H. *Drifting Continents and Shifting Theories*. New York: Cambridge University Press, 1998.

Levine, G. 2006. *Darwin Loves You: Natural Selection and Re-enchantment of the World*. Princeton, NJ: Princeton University Press, 2006.

Lewin, R. 1988. *In the Age of Mankind: A Smithsonian Book on Human Evolution*. Washington, DC: Smithsonian Institution Press, 1988.

Lindberg, David C., and Ronald L. Numbers (Eds.). *When Science and Christianity Meet*. Chicago, IL: University of Chicago Press, 2003.

Livingstone, David N. *Darwin's Forgotten Defenders: The Encounter between Evangelical Theology and Evolutionary Thought*. Grand Rapids, MI: Eerdmans, 1987.

Lubenow, Marvin. *Bones of Contention*. Grand Rapids, MI: Baker, 1992.

Lurquin, Paul F., and Linda Stone. *Evolution and Religious Creation Myths: How Scientists Respond*. New York: Oxford University Press, 2007.

Margulis, Lynn. *Symbiotic Planet: A New Look at Evolution*. New York: Basic Books, 2000.

Marks, J. *What It Means to be 98% Chimpanzee*. Berkeley, CA: University of California Press, 2002.

Marsden, George M. *Fundamentalism and American Culture*. New York: Oxford University Press, 1980.

Martin, R.A. *Missing Links: Evolutionary Concepts and Transitions Through Time*. Sudbury, MA: Jones and Bartlett, 2004.

Mayr, Ernst. *One Long Argument: Charles Darwin and the Genesis of Modern Evolutionary Thought*. Cambridge, MA: Harvard University Press, 1991.

———. *What Evolution Is*. New York: Basic Books, 2002.

McCalla, A. *The Creationist Debate: The Encounter Between the Bible and the Historical Mind*. London: T & T Clark International, 2006.

McGowen, Chris. *In the Beginning: A Scientist Shows Why the Creationists Are Wrong*. Buffalo, NY: Prometheus, 1984.

McIver, Tom. *Anti-evolution: An Annotated Bibliography*. Jefferson, NC: McFarland, 1988 (paperback reprint: Johns Hopkins University Press, 1992).

Miller, Kenneth. *Finding Darwin's God: A Scientist's Search for Common Ground*. New York: HarperCollins, 1999.

Mindell, D.P. *The Evolving World: Evolution in Everyday Life*. Cambridge, MA: Harvard University Press, 2006.

Moore, James R. *The Post-Darwinian Controversies: A Study of the Protestant Struggle to Come to Terms with Darwin in Great Britain and America, 1870–1900*. Cambridge, UK: Cambridge University Press, 1979.

———. *The Darwin Legend*. Grand Rapids, MI: Baker Books, 1994.

Moore, John A. *From Genesis to Genetics: The Case of Evolution and Creationism*. Berkeley, CA: University of California Press, 2002.

Moore, Randy. *Evolution in the Courtroom*. Santa Barbara, CA: ABC-CLIO, 2002.

Moore, Randy, and Janice Moore. *Evolution 101*. Westport, CT: Greenwood, 2006.

Morris, Henry M. *Scientific Creationism*. El Cajon, CA: Master Books, 1974.

———. *The Trouble Waters of Evolution*. San Diego, CA: Creation Life, 1975.

———. *A History of Modern Creationism*. San Diego, CA: Master Books. 1984.

———. *Creation and the Modern Christian*. El Cajon, CA: Master Books, 1985.

———. *The Long War Against God: The History and Impact of the Creation/Evolution Conflict*. Grand Rapids, MI: Baker, 1989.

Morris, Henry M., and John D. Morris. *The Modern Creation Trilogy: Scripture and Creation, Science and Creation, Society and Creation*. Green Forest, AR: Master Books, 1996.

National Academy of Sciences. *Science and Creationism: A View from the National Academy of Sciences*. Washington, DC: National Academy Press, 1984.

Nelkin, Dorothy. *The Creation Controversy: Science or Scripture in the Schools?* New York: Norton, 1982.

Newell, Norman. *Creation and Evolution: Myth or Reality*. New York: Columbia University Press, 1982.

Numbers, Ronald L. *Darwinism Comes to America*. Cambridge, MA: Harvard University Press, 1998.

———. *The Creationists: From Scientific Creationism to Intelligent Design*. Cambridge, MA: Harvard University Press, 2006.

———. *Science and Christianity in Pulpit and Pew*. New York: Oxford University Press, 2007.

Pennock, Robert T. *Tower of Babel: The Evidence against the New Creationism*. Boston, MA: MIT Press, 1999.

——— (Ed.). *Intelligent Design Creationism and Its Critics: Philosophical, Theological, and Scientific Perspectives*. Cambridge, MA: MIT Press, 2001.

Perakh, Mark. *Unintelligent Design*. Amherst, NY: Prometheus, 2003.

Petto, Andrew, and Laurie Godfrey. *Scientists Confront Intelligent Design and Creationism*. New York: W.W. Norton, 2007.

Pigliucci, Massimo. *Denying Evolution: Creationism, Scientism, and the Nature of Science*. Sunderland, MA: Sinauer, 2002.

Pleins, J. David. *When the Great Abyss Opened: Classic and Contemporary Readings of Noah's Flood*. New York: Oxford University Press, 2003.

Preston, D.J. *Dinosaurs in the Attic: An Excursion into the American Museum of Natural History*. New York: St. Martin's Press, 1986.

Prothero, D.R. *Evolution: What the Fossils Say and Why It Matters*. New York: Columbia, 2007.

Quinn, S. *Marie Curie: A Life*. New York: Simon & Schuster, 1996.

Richards, Robert J. *The Meaning of Evolution: The Morphological Construction and Ideological Reconstruction of Darwin's Theory*. Chicago, IL: University of Chicago Press, 1992.

Ridley, Mark. *Evolution*. Cambridge, MA: Blackwell Science, 1996.

Ridley, Matt. *The Red Queen: Sex and the Evolution of Human Nature*. New York : Macmillan, 1994.

———. *The Origins of Virtue: Human Instincts and the Evolution of Cooperation*. New York: Viking, 1997.

Rose, Michael. *Darwin's Spectre: Evolutionary Biology in the Modern World*. Princeton, NJ: Princeton University Press, 2000.

Ross, Hugh. *Creation and Time: A Biblical and Scientific Perspective on the Creation-Date Controversy*. Colorado Springs, CO: NavPress, 1994.

———. *The Genesis Question: Scientific Advances and the Accuracy of Genesis*. Colorado Springs, CO: NavPress, 1998.

———. *Creation As Science: A Testable Model Approach to End the Creation/evolution Wars*. Colorado Springs, CO: NavPress, 2006.

Roughgarden, J. *Evolution and Christian Faith: Reflections of an Evolutionary Biologist*. Washington, DC: Island Press, 2006.

Ruse, Michael. *But Is It Science?: The Philosophical Question in the Creation/Evolution Controversy*. Amherst, NY: Prometheus Books, 1996.

———. *The Darwinian Revolution: Science Red in Tooth and Claw*, 2nd ed. Chicago, IL: University of Chicago Press, 1999.

———. *The Evolution Wars*. Santa Barbara, CA: ABC-CLIO, 2000.

———. *Can a Darwinian Be a Christian? The Relationship between Science and Religion*. New York: Cambridge University Press, 2001.

———. *Darwin and Design: Does Evolution Have a Purpose?* Boston, MA: Harvard University Press, 2003.

———. *The Evolution – Creation Struggle*. Cambridge, MA: Harvard University Press, 2005.

Russell, C. Allyn. *Voices of American Fundamentalism*. Philadelphia, PA: The Westminster Press. 1976.

Sapp, Jan. *Genesis: The Evolution of Biology*. New York: Oxford University Press, 2003.

Schopf, J.W. *Life's Origin: The Beginnings of Biological Evolution*. Berkeley, CA: University of California Press, 2002.

Scott, Eugenie C. *Evolution vs. Creationism: An Introduction*. Berkeley, CA: University of California Press, 2004.

Scott, Eugenie, and Glenn Branch (Eds.). *Not in Our Classrooms: Why Intelligent Design Is Wrong for Our Schools*. Boston, MA: Beacon Press, 2006.

Secord, James A. *Victorian Sensation: The Extraordinary Publication, Reception, and Secret Authorship of Vestiges of the Natural History of Creation*. Chicago, IL: University of Chicago Press, 2000.

Shanks, Niall. *God, the Devil, and Darwin: A Critique of Intelligent Design Theory*. New York: Oxford University Press, 2007.

Shipman, Pat. *Taking Wing: Archaeopteryx and the Evolution of Bird Flight*. New York: Simon & Schuster, 1998.

Skybreak, Ardea. *The Science of Evolution and the Myth of Creationism: Knowing What's Real and Why It Matters*. Chicago, IL: Insight Press, 2006.

Slack, Gordy. *The Battle Over the Meaning of Everything: Evolution, Intelligent Design, and a School Board in Dover, PA*. San Francisco, CA: Jossey-Bass, 2007.

Slifkin, N. *The Challenge of Creation: Judaism's Encounter with Science, Cosmology, and Evolution*. Brooklyn, NY: Zoo Torah/Yashar Books, 2006.

Smith, Cameron M., and Charles Sullivan. *The Top Ten Myths About Evolution*. Amherst, NY: Prometheus, 2007.

Spetner, L.M. *Not By Chance! Shattering the Modern Theory of Evolution*. New York: Judaica Press, 1998.

Stoner, Don. *A New Look at an Old Earth: Resolving the Conflict between the Bible and Science*. Eugene, OR: Harvest House, 1997.

Strahler, Arthur. *Science and Earth History: The Evolution/Creation Controversy*. Buffalo, NY: Prometheus Press, 1999.

Stringer, C., and C. Gamble. *In Search of Neanderthals: Solving the Puzzle of Human Origins*. London: Thames & Hudson, 1993.

Tattersall, I., and J. Schwartz. 2000. *Extinct Humans*. New York; Westview, 2000.

Thomson, Keith. *Before Darwin*. New Haven, CT: Yale University Press, 2005.

Toumey, Christopher. *God's Own Scientists*. New Brunswick, NJ: Rutgers University Press, 1994.

Van Oosterzee, Penny. *Where Worlds Collide: The Wallace Line*. Ithaca, NY: Cornell University Press, 1997.

Webb, George F. *The Evolution Controversy in America*. Lexington, KY: University of Kentucky Press, 1994.

Weiner, J. *The Beak of the Finch: A Story of Evolution in Our Time*. New York: Alfred Knopf, 1994.

Wells, Jonathan. *Icons of Evolution: Science or Myth?* Washington, DC: Regnery, 2000.

Wilson, David B., and Warren D. Dolphin. *Did The Devil Make Darwin Do It?: Modern Perspectives On the Creation-Evolution Controversy*. Ames, IA: Iowa State University Press, 1996.

Wilson, David S. *Darwin's Cathedral: Evolution, Religion, and the Nature of Society*. Chicago, IL: University of Chicago Press, 2002.

———. *Evolution for Everyone*. New York: Bantam Dell, 2007.

Witham, Larry A. *Where Darwin Meets the Bible: Creationists and Evolutionists in America*. New York: Oxford University Press, 2002.

Woodmorappe, John. *Noah's Ark*. Santee, CA: Institute for Creation Research, 1996.

Woodward, Thomas. *Doubts about Darwin: A History of Intelligent Design*. Grand Rapids, MI: Baker Books, 2003.

Young, Christian, and Mark Largent. *Evolution and Creationism: A Documentary and Reference Guide*. Westport, CT: Greenwood Press, 2007.

Young, Matt, and Taner Edis. *Why Intelligent Design Fails: A Scientific Critique of the New Creationism*. New Brunswick, NJ: Rutgers University Press, 2004.

Zimmer, Carl. *Evolution: The Triumph of an Idea*. New York: HarperCollins, 2001.

Index

About the Authors

RANDY MOORE (b. 1954) earned a PhD in biology from UCLA, after which he worked as a biology professor at several large universities. He edited *The American Biology Teacher* for twenty years, teaches courses about evolution and creationism, and has written several books about the evolution-creationism controversy (e.g., *Evolution in the Courtroom, Evolution 101*). Today, Moore is the H.T. Morse-Alumni Distinguished Teaching Professor of Biology at the University of Minnesota. He enjoys music, running marathons, and visiting sites associated with the evolution-creationism controversy. Randy is shown here with a disinterested pair of Nazca Boobies (*Sula granti*) at Isla Española (Hood Island) in the Galápagos Archipelago. (*Randy Moore*)

MARK DECKER (b. 1960) has a PhD in conservation biology from the University of Minnesota, where he is now Associate Director for Scholarship and Teaching in the Biology Program. He is interested in all aspects of science teaching, particularly science literacy among non-science college majors. Away from the office, Mark plays guitar, skydives, and enjoys outdoor activities with his family. He is shown here with one of the sixty dinosaurs the Science Museum of Minnesota placed around St. Paul in 2007 as part of its 100th anniversary celebration. (*Patricia Fettes*)